Chemical Kinetics
Beyond the Textbook

Other World Scientific Titles by the Author

First-Passage Phenomena and Their Applications
edited by Ralf Metzler, Gleb Oshanin and Sidney Redner
ISBN: 978-981-4590-28-0

Fractional Dynamics: Recent Advances
edited by Joseph Klafter, S C Lim and Ralf Metzler
ISBN: 978-981-4340-58-8

Chemical Kinetics
Beyond the Textbook

editors

Katja Lindenberg
University of California at San Diego, USA

Ralf Metzler
University of Potsdam, Germany

Gleb Oshanin
Sorbonne Université & Centre National de la Recherche Scientifique (CNRS), France

 **World Scientific**

NEW JERSEY · LONDON · SINGAPORE · BEIJING · SHANGHAI · HONG KONG · TAIPEI · CHENNAI · TOKYO

Published by

World Scientific Publishing Europe Ltd.

57 Shelton Street, Covent Garden, London WC2H 9HE

Head office: 5 Toh Tuck Link, Singapore 596224

USA office: 27 Warren Street, Suite 401-402, Hackensack, NJ 07601

Library of Congress Cataloging-in-Publication Data

Names: Lindenberg, Katja, editor. | Metzler, Ralf, editor. | Oshanin, G. (Gleb), editor.

Title: Chemical kinetics : beyond the textbook / edited by Katja Lindenberg (University of California at San Diego, USA), Ralf Metzler (University of Potsdam, Germany), Gleb Oshanin (Sorbonne Université & Centre National de la Recherche Scientifique (CNRS), France).

Description: New Jersey : World scientific, 2019. | Includes bibliographical references.

Identifiers: LCCN 2019015573 | ISBN 9781786347008 (hc)

Subjects: LCSH: Chemical kinetics. | Reactivity (Chemistry)

Classification: LCC QD502.2 .C44 2019 | DDC 541/.394--dc23

LC record available at https://lccn.loc.gov/2019015573

British Library Cataloguing-in-Publication Data

A catalogue record for this book is available from the British Library.

For any available supplementary material, please visit
https://www.worldscientific.com/worldscibooks/10.1142/Q0209#t=suppl

Desk Editors: Herbert Moses/Jennifer Brough/Shi Ying Koe

Typeset by Stallion Press
Email: enquiries@stallionpress.com

Preface

Chemical reactions involving concentrations (or finite numbers) of particles which move randomly and react either upon encounter or via distance-dependent reaction rates among reactants are ubiquitous in nature. A few stray examples include recombination of ions, quenching of excitations, reactions involved in various catalytic processes, trapping of particles by other species, coagulation, polymerization, growth of dendrites or nuclei of a new phase, binding of proteins to DNA, and other (quite numerous) processes in physics, chemistry, and molecular and cellular biology. In some cases, the reactions are perfect and do not necessitate penetration through or motion over a barrier against reaction; in other instances, such a barrier is present and an elementary reaction event can take place only when the barrier is crossed or surmounted. In some systems, the particles involved in a reaction process perform conventional Markovian diffusion; in others, diffusion is anomalous and non-Markovian, as, for instance, when reactions involve chemically active particles attached to polymers. In turn, the reaction bath can be rather large and homogeneous, or very small and heterogeneous, for example, an interior part of a cell. Moreover, it may be crowded by other mobile or immobile species that are inert with respect to the reaction.

The kinetics of such processes have been studied for a long time, and some examples can be found in virtually every textbook on chemical physics. The conventional textbook picture is, however, somewhat outdated because the systems of current interest, such as biological cells, include molecules that may be present at nanomolar "concentrations". The traditional approaches are based on mean-field theories that discard fluctuations and many-particle effects as well as effects of anomalous diffusion, crowding, and internal geometry. In recent years, along with a growing interest in chemical processes occurring in biological systems or cellular environments, numerous advances have been made and considerable knowledge has been acquired. These seminal contributions are, however, scattered among many journals, and no attempt has been made so far to present a unified picture.

In this special volume on *Chemical Kinetics: Beyond the Textbook* we present a general overview of different contemporary facets of chemical kinetics in a variety of different environments. This volume comprises 23 seminal works and reviews on different aspects of reaction processes in chemical, physical, and biophysical systems, both theoretical and experimental.

We believe that this special volume will become a source of information and inspiration for many researchers. We thank all the contributing authors for their effort in support of this project.

Katja Lindenberg, *University of California San Diego, USA*
Ralf Metzler, *University of Potsdam, Germany*
Gleb Oshanin, *Sorbonne University, Paris, France*

About the Editors

**Katja Lindenberg** was born and brought up in Ecuador. She came to the USA for her postsecondary school education, got her BA degree in Mathematics from Alfred University, her PhD in Theoretical Physics from Cornell University (the only woman among 64 graduate students in her class), and completed her postdoc at the University of Rochester. There she had the opportunity to immerse herself in a new area for her, one she has pursued for her entire academic life — Statistical Mechanics — under the tutelage of a giant in the field, Elliott Montroll. She then came to the University of California San Diego, Department of Chemistry and Biochemistry, where she was the first woman and where she will celebrate her 50th anniversary in the summer of 2019. She has been celebrated at UCSD for her research (she holds the title of Distinguished Professor and holds a Chancellor's Associates Endowed Chair), her teaching (she received a Distinguished Teaching Award), and her extensive service, and she was granted the University of California System-wide (10 campuses) Oliver Johnson Award for her service to the University. She has published over 350 papers and greatly enjoys and values her scientific collaborations around the world.

Ralf Metzler obtained his PhD at the University of Ulm, Germany, in 1996. He completed his postdoc at Tel Aviv University and MIT. Ralf was appointed as the Assistant Professor at the Nordic Institute for Theoretical Physics (NORDITA), then in Copenhagen. He moved to the University of Ottawa as an Associate Professor and Canada Research Chair in Biological Physics before his appointment as W2 Professor ("extraordinariate"), at the Technical University of Munich. Since 2011, Ralf has been a Chair Professor for Theoretical Physics at the University of Potsdam. Ralf was a Finland Distinguished Professor of the Academy of Finland at Tampere University of Technology (2010–2015) and is currently an Alexander von Humboldt Polish Honorary Scholar of the Foundation of Polish Science.

Gleb Oshanin graduated from the Physics Department of the Moscow State University in 1986, and obtained his PhD Degree in Theoretical and Mathematical Physics from the Institute of Chemical Physics of Russian Academy of Sciences in 1989. He made postdoctoral studies in France (University Pierre and Marie Curie, Paris) and at the University of Freiburg, Germany, and also had a Visiting Professor position at the University of Mons, Belgium. In 1997, Gleb Oshanin has been admitted to CNRS first as a Researcher, and eventually promoted in 2014 to the position of the first-class CNRS Research Director, affiliated to Sorbonne University, Paris, France. He published nearly 200 scientific papers and reviews, co-organized more than 15 conferences and workshops worldwide, and enjoys numerous fruitful collaborations with his colleagues in France, Germany, Spain, Italy, USA, Russia, Poland, etc.

Contents

Chapter 1

Fluctuations and Correlations in Chemical Reaction Kinetics and Population Dynamics

Uwe C. Täuber

Department of Physics and
Center for Soft Matter and Biological Physics,
Virginia Tech, 850 West Campus Drive,
Blacksburg, VA 24061, USA
tauber@vt.edu

This chapter provides a pedagogical introduction and overview of spatial and temporal correlation and fluctuation effects resulting from the fundamentally stochastic kinetics underlying chemical reactions and the dynamics of populations or epidemics. After reviewing the assumptions and mean-field-type approximations involved in the construction of chemical rate equations for uniform reactant densities, we first discuss spatial clustering in birth–death systems, where nonlinearities are introduced through either density-limiting pair reactions, or equivalently via local imposition of finite carrying capacities. The competition of offspring production, death, and nonlinear inhibition induces a population extinction threshold, which represents a non-equilibrium phase that separates active from absorbing states. This continuous transition is characterized by the universal scaling exponents of critical directed-percolation clusters. Next we focus on the emergence of depletion zones in single-species annihilation processes and spatial population segregation with the associated reaction fronts in two-species pair annihilation. These strong (anti-)correlation effects are dynamically generated by the underlying stochastic kinetics. Finally, we address noise-induced and fluctuation-stabilized spatio-temporal patterns in basic predator–prey systems, exemplified by spreading activity fronts in the two-species Lotka–Volterra model as well as spiral structures in the May–Leonard variant of cyclically competing three-species systems akin to rock–paper–scissors games.

1. Introduction

The kinetics of chemical reactions, wherein the identity or number of reactant particles changes either spontaneously or upon encounter, constitutes a highly active research field in non-equilibrium statistical physics of stochastically interacting particle systems, owing both to the fundamental questions it addresses as well as its broad range of applications. Of specific interest are reaction-diffusion models that for example capture chemical reactions on catalytic solid surfaces or in gels where convective transport is inhibited. This scenario of course naturally pertains to genuine reactions in chemistry or biochemistry, and in nuclear, astro-, and particle physics. Yet reaction-diffusion models are also widely utilized for the quantitative description of a rich variety of phenomena in quite distinct disciplines that range from population dynamics in ecology, growth and competition of bacterial colonies in microbiology, the dynamics of topological defects in the early universe in cosmology, equity and financial markets in economics, opinion exchange and the formation of segregated society factions in sociology, and many more. Reactive "particles" also emerge as relevant effective degrees of freedom in other physical applications; prominent examples include exciton kinetics in organic semiconductors, domain wall interactions in magnets, and interface dynamics in stochastic growth models.

The traditional textbook literature, e.g., in physical chemistry and mathematical biology, almost exclusively focuses on a description of reacting particle systems in terms of coupled deterministic nonlinear rate equations. While these certainly represent an indispensable tool to characterize such complex dynamical systems, they are ultimately based on certain mean-field approximations, usually involving the factorization of higher moments of stochastic variables in terms of products of their means, as in the classical Guldberg–Waage law of mass action. Consequently, both temporal and spatial correlations are neglected in such treatments, and are in fact also only rudimentarily captured in standard spatial extensions of the mean-field rate equations to reaction-diffusion models. Under non-equilibrium conditions, however, spatio-temporal fluctuations often play a quite significant role and may even qualitatively modify the dynamics, as has been clearly established over the past decades through extensive Monte Carlo computer simulations, a remarkable series of exact mathematical treatments (albeit mostly in one dimension), and a variety of insightful approximative analytical schemes that extend beyond mean-field theory. These include mappings of the stochastic dynamics to effective field theories and subsequent analysis by means of renormalization group methods.

In the following, we shall discuss the eminent influence of spatio-temporal fluctuations and self-generated correlations in five rather simple particle reactions and, in the same general language, population ecology models that however display intriguing non-trivial dynamical features:

(1) As in thermal equilibrium, strong fluctuations and long-range correlations emerge in the vicinity of continuous phase transitions between distinct non-equilibrium steady states, as we shall exemplify for the extinction threshold that separates an active from an absorbing state. In this paradigmatic situation, the dynamical critical properties are described by the scaling exponents of critical directed-percolation clusters which assume non-mean-field values below the critical dimension $d_c = 4$.

(2) Crucial spatio-temporal correlations may also be generated by the chemical kinetics itself; this is indeed the case for simple single-species pair annihilation reactions that produce long-lived depletion zones in dimensions $d \leq 2$, which in turn slow down the resulting algebraic density decay.

(3) For two-species binary annihilation, the physics becomes even richer: For $d \leq 4$, particle anti-correlations induce species segregation into chemically inert, growing domains, with the reactions confined to their interfaces.

(4) Spatially extended stochastic variants of the classical Lotka–Volterra model for predator–prey competition and coexistence display remarkably rich noise-generated and -stabilized dynamical structures, namely spreading activity fronts that lead to erratic but persistent population oscillations.

(5) Cyclic competition models akin to the rock–paper–scissors game too produce intriguing spatio-temporal structures, whose shape is determined in a subtle manner by the presence or absence of conservation laws in the stochastic dynamics: Three species subject to cyclic Lotka–Volterra competition with conserved total particle number organize into fluctuating clusters, whereas characteristic spiral patterns form in the May–Leonard model with distinct predation and birth reactions.

2. Chemical Master and Rate Equations

2.1. *Stochastic reaction processes*

To begin, we consider simple *death–birth* reactions,[1–5] with reactants of a single species A either spontaneously decaying away or irreversibly reaching a chemically inert state $\emptyset$: $A \rightarrow \emptyset$ with rate μ; or producing identical offspring particles, e.g., $A \rightarrow A + A$ with rate σ. Note that we may also view species A as indicating individuals afflicted with a contagious disease from which they may recover with rate μ or that they can spread among others with rate σ. We shall consider these reactions as continuous-time Markovian stochastic processes that are fully determined by prescribing the transition rates from any given system configuration at instant t to an infinitesimally later time $t + dt$. Assuming mere local processes, i.e., for now ignoring any spatial degrees of freedom, our reaction model is fully characterized by specifying the number n of particles or individuals of species A at time t. The death–birth reactions are then encoded in the transition rates $w(n \to n-1) = \mu n$ and $w(n \to n+1) = \sigma n$ that linearly depend on the instantaneous particle number n. Accounting for both gain and loss terms for the configurational probability $P(n,t)$ then immediately yields the chemical *master equation*[1,4-7]

$$\frac{\partial P(n,t)}{\partial t}\bigg|_{db} = \mu\,(n+1)\,P(n+1,t)$$
$$- (\mu+\sigma)\,n\,P(n,t) + \sigma\,(n-1)\,P(n-1,t). \quad (1)$$

This temporal evolution of the probability $P(n,t)$ directly transfers to its moments $\langle n(t)^k \rangle = \sum_{n=0}^{\infty} n^k\,P(n,t)$. For example, for the *mean particle number* $a(t) = \langle n(t) \rangle$, a straightforward summation index shift results in the exact linear differential equation

$$\frac{\partial a(t)}{\partial t}\bigg|_{db} = \sum_{n=1}^{\infty} n\,\frac{\partial P(n,t)}{\partial t}\bigg|_{db} = \sum_{n=1}^{\infty} [\mu\,n(n-1)$$
$$- (\mu+\sigma)\,n^2 + \sigma\,n(n+1)]\,P(n,t) = (\sigma-\mu)\,a(t), \quad (2)$$

as the terms $\sim n^2$ in the bracket all cancel. Its solution $a(t) = a(0)\,e^{(\sigma-\mu)t}$ naturally indicates that if the particle decay rate is faster than the birth rate, $\mu > \sigma$, the population will go extinct and reach the inactive, *absorbing* empty state $a = 0$, whereupon all reactions irretrievably cease. In stark contrast, Malthusian exponential population explosion ensues for $\sigma > \mu$.

In order to prevent an unrealistic population divergence, one may impose a nonlinear process that effectively limits the reactant number; for example, we could add coagulation $A + A \to A$ with reaction rate λ that can be viewed as mimicking constraints imposed by locally restricted resources. In the context of disease spreading, this scenario is often referred to as *simple epidemic process*. The frequency of such binary processes annihilating one of the reactants is proportional to the number of particle pairs in the system, whence the associated transition rate becomes $w(n \to n-1) = \lambda\,n(n-1)$, and the master equation reads[4,5,8,9]

$$\frac{\partial P(n,t)}{\partial t}\bigg|_{an} = \lambda[n(n+1)\,P(n+1,t) - n(n-1)\,P(n,t)]. \quad (3)$$

Proceeding as before, one now finds for the mean particle number decay

$$\frac{\partial a(t)}{\partial t}\bigg|_{\text{an}} = \lambda \sum_{n=1}^{\infty} \left[n(n-1)^2 - n^2(n-1) \right] P(n,t)$$

$$= -\lambda \langle [n(n-1)](t) \rangle. \tag{4}$$

As expected, it is governed by the instantaneous number of particle pairs, that quantity involves the second moment $\langle n(t)^2 \rangle$, whose time evolution in turn is determined by the third moment, etc. Consequently, one faces an infinite hierarchy of moment differential equations that is much more difficult to analyze than the simple closed Eq. (2).

2.2. *Mean-field rate equation approximation*

A commonly applied scheme to close the moment hierarchy for nonlinear stochastic processes is to impose a mean-field-type factorization for higher moments. The simplest such approximation entails neglecting any fluctuations, setting the connected two-point correlation function to zero, $C(t,t') = \langle n(t)\,n(t') \rangle - \langle n(t) \rangle \langle n(t') \rangle \approx 0$. This assumption should hold best for large populations $n \gg 1$, when relative mean-square fluctuations $(\Delta n)^2 / \langle n \rangle^2 = C(t,t)/a(t)^2$ should be small; Eq. (4) then simplifies to the kinetic *rate equation*

$$\frac{\partial a(t)}{\partial t}\bigg|_{\text{an}} \approx -\lambda\, a(t)^2, \tag{5}$$

or $\partial a(t)^{-1}/\partial t \approx \lambda$. It is readily integrated to $a(t)^{-1} = a(0)^{-1} + \lambda t$, i.e.,

$$a(t) = \frac{a(0)}{1 + \lambda\, a(0)\, t}, \tag{6}$$

which becomes independent of the initial particle number $a(0)$ and decays to zero algebraically $\sim 1/\lambda t$ for large times $t \gg 1/\lambda a(0)$. Note that the right-hand side of the rate equations resulting from mean-field factorizations of nonlinear reaction terms encode the corresponding stochiometric numbers as powers of the reactant numbers, precisely as in the Guldberg–Waage law of mass action describing reaction concentration products in chemical equilibrium.[5,7] Already in non-spatial systems, these factorizations disregard any temporal fluctuations; in spatially extended systems, they moreover assume well-mixed and thus homogeneously distributed reactants.

3. Population Dynamics with Finite Carrying Capacity

3.1. *Mean-field rate equation analysis*

Next we combine the three reaction processes of the preceding section to arrive at the simplest possible population dynamics model for a single species that incorporates death with rate μ, (asexual) reproduction with rate σ, and (nonlinear) competition with rate λ to constrain the active-state particle number.[2,3] Adding the right-hand sides of Eqs. (2) and (5) yields the associated mean-field rate equation for the particle number or population

$$\frac{\partial a(t)}{\partial t} \approx (\sigma - \mu)\, a(t) - \lambda\, a(t)^2 = \lambda\, a(t)[r - a(t)]. \qquad (7)$$

In the last step, we have cast the rate equation in the form of a *logistic model* with *carrying capacity* $r = (\sigma - \mu)/\lambda$. Its stationary solutions are the absorbing state $a = 0$ and a population number equal to the carrying capacity $a = r$. Straightforward linear stability analysis of Eq. (7) establishes that the latter stationary state is approached for $\sigma > \mu$ or $r > 0$, whereas of course $a(t) \to 0$ for $\sigma < \mu$ ($r < 0$).

For $\sigma \neq \mu$, the ordinary first-order differential equation (7) is readily integrated after variable separation:

$$\lambda\, t = \frac{1}{r} \int_{a(0)}^{a(t)} \left(\frac{1}{a} + \frac{1}{r-a} \right) da = \frac{1}{r} \ln \left[\frac{a(t)}{a(0)} \frac{r - a(0)}{r - a(t)} \right].$$

Solving for the particle number at time t gives

$$a(t) = \frac{a(0)}{e^{(\mu - \sigma)t}\left[1 - a(0)/r\right] + a(0)/r}. \qquad (8)$$

As anticipated, this results in exponential decay for $\mu > \sigma$ with characteristic time $\tau = 1/|\mu - \sigma| = 1/\lambda|r|$; in the active state, the particle number also approaches the carrying capacity r exponentially with rate $1/\tau$, and monotonically from above or below for $a(0) > r$ and $a(0) < r$, respectively. Right at the extinction threshold $\sigma = \mu$ ($r = 0$) separating the active and inactive and absorbing states, Eq. (7) reduces to Eq. (5) for pair annihilation, and hence the exponential kinetics of Eq. (8) is replaced by the power-law decay (6), as follows also from taking the limit $\mu - \sigma \to 0$ in Eq. (8).

3.2. *Diffusive spreading and spatial clustering*

At least in a phenomenological manner, the above analysis can be readily generalized to spatially extended systems by adding (e.g., unbiased nearest-neighbor) particle hopping or exchange in lattice models, or diffusive spreading in a continuum setting. Still within a mean-field framework that entails mass action factorization of nonlinear correlations, in the continuous representation this leads to a *reaction-diffusion equation*

$$\frac{\partial a(\vec{x}, t)}{\partial t} \approx \left(\sigma - \mu + D\nabla^2 \right) a(\vec{x}, t) - \bar{\lambda} \, a(\vec{x}, t)^2, \qquad (9)$$

for the *density field* $a(\vec{x}, t)$ with diffusion constant $D > 0$. In the population dynamics context, this nonlinear partial differential equation is referred to as the Fisher–Kolmogorov(–Petrovskii–Piskunov) equation.[3] In one spatial dimension, Eq. (9) admits solitary traveling wave solutions of the form $a(x, t) = u(x - ct)$ that interpolate between the active and inactive states, i.e., $u(z \to -\infty) = 0$, while $u(z \to \infty) = \bar{r} = (\sigma - \mu)/\bar{\lambda}$, if $\bar{r} > 0$; their detailed shape depends on the wave velocity c.

Away from the extinction threshold at $\sigma = \mu$, we may linearize this equation by considering the deviation $\delta a(\vec{x}, t) = a(\vec{x}, t) - a(\infty)$ from the asymptotic density $a(\infty)$. Upon neglecting quadratic terms in the fluctuations δa, one obtains near both the active (where $a(\infty) = \bar{r}$) and inactive states (with $a(\infty) = 0$)

$$\frac{\partial \delta a(\vec{x}, t)}{\partial t} \approx \left(D\nabla^2 - |\sigma - \mu| \right) a(\vec{x}, t), \qquad (10)$$

with characteristic length scale $\xi = \sqrt{D/|\sigma - \mu|}$ and corresponding time scale $\tau = \xi^2/D$ as to be expected for diffusive processes. The correlation length ξ describes the extent of spatially correlated regions, i.e., density clusters, in the system, whereas the rate $1/\tau$ governs their temporal decay.

Since $a(t) = \int a(\vec{x}, t)\, d^d x$, the mean-field logistic equation (7) follows from Eq. (9) only under the assumption of extremely short-range spatial correlations $\sim\delta(\vec{x} - \vec{x}')$, i.e., in the limit $\xi \to 0$, which is definitely violated near the extinction threshold. This Dirac delta function also indicates that the nonlinear reaction rate in the continuum description and the corresponding dimensionless mean-field rate are related to each other via the volume b^d of the unit cell in an ultimately underlying discrete lattice model $\lambda \sim b^d\, \bar{\lambda}$. It is important to realize that the connection between microscopic reaction rates and their continuum counterparts is not usually direct and simple, but depends on the details of the involved coarse-graining process. With this caveat stated explicitly, which also applies to the relationship between lattice hopping rates and continuum diffusivities, we shall for notational simplicity henceforth drop the overbars from the continuum rates.

3.3. *Extinction threshold: Directed-percolation criticality*

On a lattice with N sites, where locally spontaneous particle death, reproduction, and binary coagulation (or annihilation) can take place, and which are coupled through particle hopping processes, the single-site extinction bifurcation discussed above translates into a genuine continuous non-equilibrium phase transition in the thermodynamic limit $N \to \infty$, or in the corresponding continuum model with infinitely many degrees of freedom. Note that in any finite stochastic system that incorporates an absorbing state, the latter is inevitably reached at sufficiently long times. A true absorbing-to-active phase transition hence requires taking the thermodynamic limit first in order to permit the existence of a stable active phase as $t \to \infty$. In addition, this asymptotic long-time limit must be

considered prior to tuning any control parameters. Yet characteristic
extinction times tend to grow exponentially with N, and active
states may survive in a quasi-stationary configuration as long as
$\log t \ll \mathcal{O}(N)$. Hence, absorbing phase transitions are in fact easily accessible numerically in computer simulations with sufficiently
many lattice sites. As in the vicinity of critical points or second-order
phase transitions in thermal equilibrium, nonlinearities and fluctuations become crucial near the extinction threshold and cannot be
neglected in a proper mathematical treatment. Consequently, mean-
field approximations that neglect both intrinsic reaction noise and
spatial correlations become at least questionable.

The phenomenological description of active-to-absorbing (and
other non-equilibrium) phase transitions closely follows that of
near-equilibrium critical dynamics.[5,10–14] As the phase transition
is approached upon tuning a relevant control parameter $r \to 0$,
spatial correlations become drastically enhanced. For the correlation function of an appropriately chosen order parameter field that
characterizes the phase transition (e.g., the particle density in our
population dynamics model), one expects a typically algebraic diver-
gence of the associated correlation length: $\xi(r) \sim |r|^{-\nu}$ with a criti-
cal exponent ν. Consequently, microscopic length (and time as well
as energy) scales are rendered irrelevant, and the system asymptotically becomes scale-invariant. This emergent critical-point symmetry
is reflected in power-law behavior for various physical quantities that
are captured through additional critical indices. The *dynamic critical
exponent* z links the divergence of the characteristic relaxation time
to that of the correlation length: $\tau(r) \sim \xi(r)^z \sim |r|^{-z\nu}$, describing
critical slowing-down. The stationary (long-time) order parameter
sets in algebraically: $a(t \to \infty, r) \sim r^\beta$ for $r > 0$, while it decays to
zero as $a(t, r = 0) \sim t^{-\alpha}$ precisely at the critical point. These two
power laws are limiting cases of a more general *dynamical scaling*
ansatz for the time-dependent order parameter,

$$a(t, r) = |r|^\beta \, \hat{a}\big(t/\tau(r)\big), \tag{11}$$

where the scaling function on the right-hand side satisfies $\hat{a}(0) =$
const. For large arguments $y = t/\tau(r) \sim t\,|r|^{z\nu}$, one must require

$\hat{a}(y) \sim y^{-\beta/z\nu}$ in order for the r dependence to cancel as $r \to 0$. Hence, we obtain the critical decay exponent $\alpha = \beta/z\nu$. This *scaling relation* is of course fulfilled by the mean-field critical exponents $\alpha = \beta = 1$, $\nu = 1/2$, and $z = 2$ (indicating diffusive spreading) found in our previous population model analysis.

As for equilibrium critical phenomena, there exists an (upper) *critical dimension d_c* below which fluctuations are strong enough to modify not just amplitudes and scaling functions, but alter the critical scaling exponents. We invoke a simple scaling argument to determine d_c for competing birth–death–coagulation processes: Let us set an inverse length (wave vector) scale κ, i.e., $[x] = \kappa^{-1}$, and corresponding time scale $[t] = [x]^2 = \kappa^{-2}$ (implying that we choose $[D] = \kappa^0$), where the square bracket indicates the scaling dimension. Equation (9) then enforces $[\sigma] = [\mu] = [r] = \kappa^2$; moreover, $[a(\vec{x},t)] = \kappa^d$ since a represents a density field in d spatial dimensions, whence we find $[\lambda] = \kappa^{2-d}$ for the coagulation (or annihilation) rate. Nonlinear stochastic fluctuations that will affect particle propagation and (linear) extinction incorporate subsequent branching and coagulation processes, and hence scale like the rate product $[\sigma\lambda] = \kappa^{4-d}$. The critical dimension $d_c = 4$ indicates when this effective nonlinear coupling becomes scale-invariant. For $d < d_c$, it attains a positive scaling dimension and is considered *relevant* (in the renormalization group sense), along with the "mass"-like parameter r. In dimensions beyond d_c, the nonlinearity becomes irrelevant and does not alter the fundamental scaling properties of the model, yet of course fluctuations still contribute numerically to various observables. At $d = d_c$, one usually finds *logarithmic corrections* to the mean-field power laws.[5,14]

In order to heuristically include fluctuations, one may add (for simplicity) Gaussian white noise with vanishing average $\langle \zeta(\vec{x},t) \rangle = 0$ to the reaction-diffusion equation (9), turning it into a stochastic partial differential equation[5,12]:

$$\frac{\partial a(\vec{x},t)}{\partial t} = \left(\lambda r + D\nabla^2\right) a(\vec{x},t) - \lambda a(\vec{x},t)^2 + \zeta(\vec{x},t). \quad (12)$$

Note that the deterministic part of the right-hand side can be interpreted as an expansion of a very general reaction functional in terms of the small fluctuating local particle density near the extinction threshold, and even the diffusive spreading term may be viewed as the leading contribution in a long-wavelength expansion for spatially varying fluctuations (in systems with spatial inversion symmetry). Equation (12) is thus quite generic, provided there are no additional special symmetries that would enforce the coefficients r or λ to vanish. For $r > 0$ and if $\lambda > 0$, the system resides in an active phase, whereas it reaches the empty, absorbing state for $r < 0$. For negative λ, one would have to amend Eq. (12) with a cubic term in the density field. The absence of a constant particle source (or sink) term on its right-hand side is mandated by presence of the absorbing state $a = 0$.

This constraint must be similarly reflected in the noise correlations: as the mean particle number $a(t) \to 0$, all stochastic fluctuations must cease. To lowest non-vanishing order in a, one would therefore posit

$$\langle \zeta(\vec{x}, t)\, \zeta(\vec{x}', t') \rangle = v\, a(\vec{x}, t)\, \delta(\vec{x} - \vec{x}')\delta(t - t'). \qquad (13)$$

Stochastic dynamics with such a multiplicative noise correlator is properly defined through a corresponding functional integral representation[5,12]; intriguingly, the ensuing path integral action turns out to be equivalent to a well-studied problem in nuclear and particle physics, namely Reggeon field theory,[15] which has in turn been shown to capture the universal scaling properties of critical *directed-percolation* clusters.[16,17] This sequence of mathematical mappings lends strong support to the Janssen–Grassberger conjecture,[17,18] which states that the asymptotic critical scaling features of continuous non-equilibrium transitions from active to inactive, absorbing states for a single scalar order parameter governed by Markovian stochastic dynamics, and in the absence of any quenched disorder and coupling to other conserved fields, should be described by the directed-percolation universality class. In fact, this statement applies generically even for multi-component systems.[19] Indeed, if we represent the basic stochastic death, birth, nearest-neighbor hopping, and coagulation processes on a lattice through their "world lines" in a

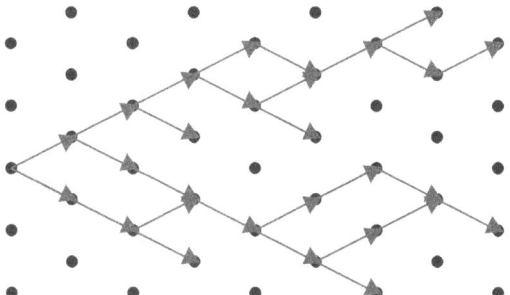

Fig. 1.   Elementary death, birth, hopping, and coagulation processes initiated by a single particle seed generate a directed-percolation cluster; bonds connecting lattice sites are formed only by advancing along the "forward" (time-like) direction to the right. Figures reproduced with permission from Ref. 12, copyright (2005) by Elsevier Inc.

space-time plot as depicted in Fig. 1, it becomes apparent how they generate a directed-percolation cluster.[12,14]

   More rigorously, the Doi–Peliti formalism allows a representation of "microscopic" stochastic reaction kinetics as defined through the associated master equation in terms of a coherent-state path integral.[5,13] For the combined reactions $A \to \emptyset$, $A \to A + A$, and $A + A \to A$ in the continuum limit, augmented with diffusive spreading, the resulting action assumes precisely the form of Reggeon field theory, and hence the "mesoscopic" Langevin description (12, 13), in the vicinity of the absorbing-state transition.[5,12,13] As a consequence of a special internal "rapidity reversal" symmetry of Reggeon field theory, the directed-percolation universality class is fully characterized by a set of only three independent critical exponents (say ν, z, and β); all other critical indices are related to them via exact scaling relations.

   The critical exponents for directed percolation are known to high precision from dedicated Monte Carlo computer simulations[10,11,14]; recent literature values for α, β, ν, and z in two and three dimensions are listed in Table 1. For $d = 1$, the most accurate exponent values were actually obtained semi-analytically via ingenious series expansions.[20] A dynamical renormalization group analysis based on the (Reggeon) field theory representation allows a systematic perturbation expansion for the fluctuation corrections to the mean-field

Table 1. Critical exponents for directed percolation obtained from series expansions in one dimension[20]; Monte Carlo simulations in two and three dimensions[14]; first-order perturbative renormalization group analysis $(d = 4 - \epsilon)$[12]; and experiments on turbulent nematic liquid crystals $(d = 2)$[21] (numbers in brackets indicate last-digit uncertainties).

| Critical exponent | Dimension | | | | Liquid crystal experiment |
	$d = 1$	$d = 2$	$d = 3$	$d = 4 - \epsilon$	
α	0.159464(6)	0.4505(10)	0.732(4)	$1 - \frac{\epsilon}{4} + \mathcal{O}(\epsilon^2)$	0.48(5)
β	0.276486(8)	0.5834(30)	0.813(9)	$1 - \frac{\epsilon}{6} + \mathcal{O}(\epsilon^2)$	0.59(4)
ν	1.096854(4)	0.7333(75)	0.584(5)	$\frac{1}{2} + \frac{\epsilon}{16} + \mathcal{O}(\epsilon^2)$	0.75(6)
z	1.580745(10)	1.7660(16)	1.901(5)	$2 - \frac{\epsilon}{12} + \mathcal{O}(\epsilon^2)$	1.72(11)

exponents in terms of the deviation $\epsilon = 4 - d$ from the critical dimension d_c.[5,12,13,15] The first-order (one-loop) results are also tabulated below; as compared to mean-field theory, critical fluctuations effectively reduce the values of $\alpha, \beta, 1/\nu$, and z in accord with the numerical data, and increasingly so for lower dimensions. Note that $z < 2$ implies subdiffusive propagation at the extinction transition for $d < 4$. Several experiments have at least partially detected directed-percolation scaling near active-to-absorbing phase transitions in various systems.[14] Perhaps the most impressive confirmation originates from detailed and very careful studies of the transition between two different turbulent states of electrohydrodynamic convection in (quasi-)2D turbulent nematic liquid crystals carried out by Takeuchi et al., at the University of Tokyo,[21] who managed to extract 12 different critical exponents (four of them listed in Table 1) along with 5 scaling functions from their experimental data.

4. Dynamic Correlations in Pair Annihilation Processes

4.1. Depletion zones and reaction rate renormalization

We next return to simple pair annihilation $(A + A \rightarrow \emptyset$, rate $\lambda')$ or fusion $(A + A \rightarrow A$, rate $\lambda)$ processes, but in spatially extended systems. The on-site master equation for the former looks like Eq. (3), with the gain term replaced with $\lambda' (n + 1)(n + 2) P(n + 2)$. This merely results in a rescaled reaction rate $\lambda \to 2\lambda'$ in both the exact Eq. (4) and the mean-field equations (5), (6). However, as the particle density drops toward substantial dilution, the rate of further pair annihilation processes will ultimately not be determined by the original microscopic reactivity λ (λ'), and instead be limited by the time it takes for two reactants to meet.[4,8,9,13] If we assume diffusive spreading with diffusion constant D, and hence relative diffusivity $2D$, the typical time for two particles at distance l to find each other is $t \sim l^2/4D$. In the diffusion-limited regime, l and t set the relevant length and time scales, whence the particle density should scale according to $a(t) \sim l(t)^{-d} \sim (Dt)^{-d/2}$ in d spatial dimensions. This suggests a slower decay than the mean-field prediction $a(t) \sim (\lambda t)^{-1}$ in dimensions $d < 2$. Indeed, the previously established scaling dimension $[\lambda] = \kappa^{2-d}$ too indicates that $d_c = 2$ sets the critical dimension for diffusion-controlled pair annilation. In one dimension, the certain return of a random walker to its origin ensures that the ensuing annihilations carve out *depletion zones*, generating spatial *anti-correlations* that impede subsequent reactions. In contrast, for $d > 2$ diffusive spreading sustains a well-mixed system with largely homogeneous density, and mean-field theory remains valid.

This simple scaling argument is quantitatively borne out by Smoluchowski's classical self-consistent approach.[4,13] In a continuum representation, we need to impose a finite reaction sphere with radius b: two particles react, once their distance becomes smaller than b. In a quasi-stationary limit, we then need to solve the stationary diffusion equation

$$0 = \nabla^2 a(r) = \frac{\partial^2 a(r)}{\partial r^2} + \frac{d-1}{r} \frac{\partial a(r)}{\partial r}, \tag{14}$$

where we have invoked spherical symmetry and written down the Laplacian differential operator in d-dimensional spherical coordinates. General solutions are then linear combinations of a constant term and the power r^{2-d}. For $d > 2$, we may impose the straightforward boundary conditions $a(r \leq b) = 0$, whereas the particle density approaches a finite asymptotic value $a(\infty)$ far away from the reaction center located at the origin, which yields $a(r) = a(\infty)\left[1 - (b/r)^{d-2}\right]$. The effective reactivity $\widetilde{\lambda}$ in the diffusion-limited regime is then given by the incoming particle flux at reaction sphere boundary: $\widetilde{\lambda} \sim Db^{d-1}a(\infty)^{-1}\left[\partial a(r)/\partial r\right]|_{r=b} \sim D(d - 2)b^{d-2}$, which replaces the annihilation rate λ in the rate equation (5). Consequently, one obtains the large-time density decay $a(t) \sim (Dt)^{-1}$ with the same power law (6) as in the reaction-controlled region at large densities.

However, at $d_c = 2$ the effective reaction rate $\widetilde{\lambda}$ vanishes; indeed, in low dimensions $d < 2$ one needs to impose a different boundary condition $a(R) \approx a(\infty)$, where R denotes the mean particle separation: since the density $a(R) \sim R^{-d}$, one has $R(a) \sim a^{-1/d}$. The boundary condition hence depends on the actual density in the quasi-stationary limit, resulting in the profile $a(r) = a(\infty)\left[(r/b)^{2-d} - 1\right] / \left[(R/b)^{2-d} - 1\right]$. From the diffusive flux at the reaction sphere, one obtains the effective reactivity $\widetilde{\lambda}(a) \sim D(2 - d)b^{d-2}/[(R/b)^{2-d} - 1] \to \lambda_R a^{-1+2/d}$ as $a \to 0$ and $R \to \infty$, with a constant $\lambda_R \sim D$ (that again tends to zero as $d \to 2$). Upon self-consistently replacing λ with the density-dependent effective rate $\widetilde{\lambda}(a)$ in Eq. (5), we arrive at

$$\frac{\partial a(t)}{\partial t} \sim -\lambda_R\, a(t)^{1+2/d}. \tag{15}$$

Its solution through variable separation yields the anticipated slower density decay and in turn the time dependence of the effective reaction rate:

$$a(t) \sim (Dt)^{-d/2}, \quad \widetilde{\lambda}(t) \sim (Dt)^{-1+d/2}. \tag{16}$$

At the critical dimension $d_c = 2$, one finds logarithmic slowing down relative to the rate equation power law: $a(t) \sim (Dt)^{-1}\ln(8Dt/b^2)$.

Dynamical renormalization group calculations based on the Doi–Peliti field theory representation of the associated master equation confirm these findings.[5,13,22,23] For kth-order annihilation $kA \to \emptyset, A, \ldots, (k - 1)A$ ($k \geq 2$) the right-hand side of Eq. (5) is to be replaced with $-\lambda_k \, a(t)^k$; for diffusive propagation, dimensional analysis then yields the scaling dimension $[\lambda_k] = \kappa^{2-(k-1)\,d}$, from which one infers the upper critical dimension $d_c(k) = 2/(k-1)$. Deviations from the mean-field algebraic decay $a(t) \sim t^{-1/(k-1)}$ should only materialize for pair $(k = 2)$ reactions for $d \leq 2$, and for triplet annihilation at $d_c(3) = 1$. In accord with Smoluchowski's underlying assumption, diffusive spreading is not affected by the nonlinear annihilation processes. The resulting perturbation expansion in λ_k can be summed to all orders, and indeed recovers Eq. (16) for pair annihilation, and the corresponding logarithmic scaling at the critical dimension, which for triplet reactions becomes $a(t) \sim \left[(D\,t)^{-1} \ln(D\,t)\right]^{1/2}$.

Pair annihilation dynamics on 1D lattices with strict site exclusion can also be mapped onto non-Hermitean spin-1/2 Heisenberg models, permitting the extraction of remarkably rich non-trivial exact results.[24–27] The anomalous density decay induced by the self-generated depletion zones has been confirmed in several experiments on exciton recombination kinetics in effectively 1D molecular systems. Particularly convincing are data obtained by Allam *et al.*, at the University of Surrey in carbon nanotubes, who managed to explore the detailed crossover in the power laws from the reaction-controlled to the diffusion-limited regime (16) both in the exciton density decay and the reactivity.[28]

4.2. *Segregation in diffusive two-species annihilation*

Let us now investigate pair annihilation of particles of distinct species, $A + B \to \emptyset$, with reaction rate λ.[1,3–5] The crucial distinction to the previous situation is that alike particles do not react with each other. The associated exact time evolution as well as the coupled mean-field rate equations for the densities $a(t)$ and $b(t)$ are symmetric under species exchange $A \leftrightarrow B$,

$$\frac{\partial a(t)}{\partial t} = \frac{\partial b(t)}{\partial t} = -\lambda \langle [n_A \, n_B](t) \rangle \approx -\lambda a(t) \, b(t), \qquad (17)$$

and these binary reactions of course preserve the number difference $n_C = n_A - n_B = \text{const.}$, whence also $c(t) = a(t) - b(t) = c(0)$. One must now distinguish between two situations: If initially $n_A = n_B$

precisely, $c(t) = 0$ at all times: with identical initial conditions for their same rate equations, $a(t) = b(t)$. Equation (17) consequently reduces to Eq. (5) for pair annihilation of identical species, with the solution (6) that describes algebraic decay to zero for both populations. In the more generic case $n_A \neq n_B$, say, with majority species A, i.e., $c(0) > 0$, the B population will asymptotically go extinct, $b(\infty) = 0$, while $a(\infty) = c(0)$. At long times, we may thus replace $a(t) \approx c(0)$ in the rate equation for $b(t)$, resulting in exponential decay with time constant $c(0)\lambda$: $b(t) = a(t) - c(0) \sim e^{-c(0)\lambda t}$. The special symmetric case $c(0) = 0$ hence resembles a dynamical critical point with diverging relaxation time, and exponential density decay replaced by a power law.

In a spatial setting with diffusive transport, ultimately the stochastic pair annihilation reactions will become diffusion-limited, and the emerging depletion zones and persistent particle anti-correlations will markedly slow down the asymptotic decay of the minority species in low dimensions $d \leq d_c = 2$. Indeed, replacing $\lambda t \to \tilde{\lambda}(t) t \sim (D t)^{d/2}$ with its renormalized counterpart according to Eq. (16), one arrives at stretched exponential behavior: $\ln b(t) = \ln[a(t) - c(0)] \sim -(D t)^{d/2}$ for $d < 2$, whereas $\ln b(t) = \ln[a(t) - c(0)] \sim -(D t)/\ln(D t)$ in two dimensions.

In the special symmetric case with equal initial particle numbers, the presence of an additional conserved quantity has a profound effect on the long-time chemical kinetics: Both particle species may spatially *segregate* into inert, slowly coarsening domains, whence the reactions become confined to the contact zones separating the A- or B-rich domains.[29,30] Note that the local particle density excess satisfies a simple diffusion equation $\partial c(\vec{x}, t)/\partial t = D\nabla^2 c(\vec{x}, t)$; the associated initial value problem is then solved by means of the diffusive Green's function $G(\vec{x}, t) = \Theta(t) e^{-\vec{x}^2/4D t}/4\pi D t$ via the convolution integral $c(\vec{x}, t) = \int G(\vec{x} - \vec{x}', t) c(\vec{x}', 0) d^d x'$. If one assumes an initially random, spatially uncorrelated Poisson distribution for both A and B particles with $\overline{a(\vec{x}, 0)} = \overline{b(\vec{x}, 0)} = a(0)$, where the overbar denotes an ensemble average over initial conditions, and $\overline{a(\vec{x},0)\,a(\vec{x}',0)} = \overline{b(\vec{x},0)\,b(\vec{x}',0)} = a(0)^2 + a(0)\,\delta(\vec{x}-\vec{x}')$, whereas $\overline{a(\vec{x},0)\,b(\vec{x}',0)} = 0$, the corresponding moments for the initial density excess become $\overline{c(\vec{x},0)} = 0$ and $\overline{c(\vec{x},0)\,c(\vec{x}',0)} = 2a(0)\,\delta(\vec{x}-\vec{x}')$.

These considerations allow us to explicitly evaluate

$$\overline{c(\vec{x},t)^2} = 2\,a(0)\int G(\vec{x}-\vec{x}'t)^2\,d^d x' = \frac{2\,a(0)\,\Theta(t)}{(8\pi D\,t)^{d/2}}, \qquad (18)$$

through straightforward Gaussian integration. The distribution of the field c itself will be Gaussian as well, with zero mean and variance (18); hence we finally obtain for the average absolute value of the local density excess

$$\overline{|c(\vec{x},t)|} = \sqrt{\frac{2}{\pi}}\,\overline{c(\vec{x},t)^2} = \sqrt{\frac{4\,a(0)}{\pi}}\,\frac{\Theta(t)}{(8\pi D\,t)^{d/4}}. \qquad (19)$$

In high dimensions $d > 4$, this excess decays faster than the mean densities $a(t) \sim 1/t$, implying that the particle distribution remains largely uniform, and mean-field theory provides a satisfactory description. In contrast, for $d < d_s = 4$, the long-time behavior is dictated by the slowly decaying spatial density excess fluctuations: $a(t) \sim b(t) \sim (D\,t)^{-d/4}$. Either species accumulate in diffusively growing domains of linear size $l(t) \sim (D\,t)^{1/2}$ separated by active *reaction zones* whose width relative to $l(t)$ decreases algebraically with time.[31] Note that the borderline dimension for spatial species segregation $d_s = 4$ is not a critical dimension in the renormalization group sense; the anomalous density decay as a consequence of segregated domain formation can rather be fully described within the framework of mean-field reaction-diffusion equations.[30] The resulting power law $a(t) \sim t^{-3/4}$ in three dimensions was experimentally observed in a calcium-fluorophore system by Monson and Kopelman at the University of Michigan.[32]

The above analysis does not apply to specific, spatially correlated initial conditions. For example, if impenetrable hard-core particles are aligned in strictly alternating order $\cdots ABABABA\cdots$ on a 1D line, pair annihilation reactions will preserve this arrangement at all later times. Hence the distinction between the two species becomes in fact meaningless, and the single-species asymptotic decay $\sim (D\,t)^{-1/2}$ ensues. More generally, pair annilation processes involving q distinct species $A_i + A_j \to \emptyset$ $(1 \leq i < j \leq q)$ should eventually reduce to just a two-species system for the remaining two "strongest" particle types, as determined by their reaction rates, diffusivities, and initial concentrations. Novel, distinct behavior could thus only appear for highly symmetric situations where all reaction and diffusion rates as well as initial densities are set equal among the q species.

Yet it turns out that for any $q \geq 3$, there exist no conserved quantities in such systems. One may furthermore establish the borderline dimension for species segregation as $d_s(q) = 4/(q-1)$; for $d \geq 2$, therefore, all densities should follow the mean-field decay law $\sim 1/t$ as for a single species.[33] Only in one dimension can distinct particle species cluster into stable domains, resulting in a combination of depletion- and segregation-dominated decay:

$$a_i(t) \sim t^{-1/2} + C\,t^{-\alpha(q)}, \quad \alpha(q) = \frac{q-1}{2\,q}. \tag{20}$$

Note that $\alpha(2) = 1/4$ as established above, while the single-species decay exponent is recovered in the limit of infinitely many species, $\alpha(\infty) = 1/2$: In that situation, alike particles experience a vanishing probability to ever encounter each other, whence the distinction between the different species becomes irrelevant. Once again, correlated initial linear arrangements such as $\cdots ABCDABCD\cdots$ for four species induce special cases; here, no alike particles can ever meet, and the system behaves effectively like single-species pair annihilation again. Also, interesting cyclic variants may be constructed,[34] such as $A + B \to \emptyset, B + C \to \emptyset, C + D \to \emptyset$, and $D + A \to \emptyset$. In this case, one may collect individuals from species A and C, and similarly B and D, in just two competing "alliances", and consequently the decay kinetics is captured by the two-species pair annihilation behavior.

5. Stochastic Pattern Formation in Population Dynamics

5.1. *Activity fronts in predator–prey coexistence models*

The language and also numerical and mathematical tools developed for the investigation of chemical kinetics may be directly transferred to spatially extended stochastic population dynamics.[5,35,36] Aside from the logistic population growth with finite carrying capacity (7), another classical and paradigmatic model in ecology concerns the competition and coexistence of prey with their predator species, as first independently constructed by Lotka and Volterra.[1,3] Let B indicate the prey species, who left on their own merely undergo (asexual) reproduction $B \to B + B$ with rate σ. Their population is held in check by predators A who may either spontaneously die, $A \to \emptyset$ with rate μ, or upon encounter with a prey individual, devour it and simultaneously generate offspring: $A + B \to A + A$ with rate λ. The original deterministic Lotka–Volterra model consists of the associated coupled mean-field rate equations for the population densities $a(t)$ and $b(t)$:

$$\frac{\partial a(t)}{\partial t} = -\mu\, a(t) + \lambda\, a(t)\, b(t), \quad \frac{\partial b(t)}{\partial t} = \sigma\, b(t) - \lambda\, a(t)\, b(t). \quad (21)$$

This dynamics allows three stationary states, namely (i) complete extinction $a = b = 0$; (ii) the also absorbing pure prey state with Malthusian population explosion $a = 0$, $b \to \infty$; and (iii) a predator–prey coexistence state with finite densities $a(\infty) = \sigma/\lambda$, $b(\infty) = \mu/\lambda$. Naturally, the predators benefit from high prey fertility σ, while the prey prosper if the predators are short-lived; yet counterintuitively for the predators, both stationary population numbers decrease with enhanced predation rates λ, signaling a nonlinear feedback mechanism: If the species A too efficiently reduces the population B, they have scarce food left, whence the majority of them die.

However, this stationary coexistence state (iii) is in fact never reached under the deterministic nonlinear dynamics (21).

Indeed, eliminating time through taking the ratio $da/db = (\lambda b - \mu) a / (\sigma - \lambda a) b$, one obtains after variable separation and integration a *conserved first integral* for the mean-field dynamics: $K(t) = \lambda[a(t) + b(t)] - \sigma \ln a(t) - \mu \ln b(t) = K(0)$. The trajectories in the phase space spanned by the population numbers must therefore be strictly periodic orbits, implying undamped nonlinear population oscillations whose amplitudes and shapes are fixed by the initial values $a(0)$ and $b(0)$. For small deviations from the stationary coexistence center $\delta a(t) = a(t) - a(\infty)$, $\delta b(t) = b(t) - b(\infty)$, straightforward linearization of Eqs. (21) yields $\partial \delta a(t)/\partial t \approx \sigma \, \delta b(t)$, $\partial \delta b(t)/\partial t \approx -\mu \, \delta a(t)$, which are then readily combined to the simple harmonic oscillator differential equation $\partial^2 \delta a(t)/\partial t^2 \approx -\omega^2 \, a(t)$ with (linear) oscillation frequency $\omega = \sqrt{\mu \sigma}$. Equivalently, we may construct the linear stability matrix $\mathbf{L}$ that governs the temporal evolution of the fluctuation vector $\mathbf{v} = (\delta a \; \delta b)^T$: $\partial \mathbf{v}(t)/\partial t \approx \mathbf{L} \, \mathbf{v}(t)$, where $\mathbf{L} = \begin{pmatrix} 0 & \sigma \\ -\mu & 0 \end{pmatrix}$ with imaginary eigenvalues $\pm \, i\omega$.

Clearly the absence of any real part in the stability matrix eigenvalues represents a degenerate, atypical situation that should not be robust against even minor modifications of the model.[3] For example, in order to render the Lotka–Volterra description more realistic and prevent any population divergence, one could impose a finite carrying capacity r for species B; at the mean-field level, this alters the second differential equation in (21) to

$$\frac{\partial b(t)}{\partial t} = \sigma \, b(t) \left[1 - b(t)/r \right] - \lambda \, a(t) \, b(t), \qquad (22)$$

leading to modified stationary states (ii′) $a(\infty) = 0$, $b(\infty) = r$ and (iii′) $a(\infty) = \sigma \left(1 - \mu/\lambda r \right)/\lambda$, $b(\infty) = \mu/\lambda$. The latter two-species coexistence fixed point exists, and is linearly stable, provided the predation rate exceeds the threshold $\lambda_c = \mu/r$; for $\lambda < \lambda_c$, the predator species A is driven to extinction. At the stationary state (iii′), the linear stability matrix eigenvalues acquire negative real parts:

$$\epsilon_\pm = -\frac{\mu \, \sigma}{2 \, \lambda \, r} \left[1 \pm \sqrt{1 - \frac{4 \, \lambda \, r}{\sigma} \left(\frac{\lambda \, r}{\mu} - 1 \right)} \, \right]. \qquad (23)$$

For $\sigma > 4\lambda r(\lambda r/\mu - 1)$, these eigenvalues are both real, indicating exponential relaxation toward the stable node (iii'). For lower prey birth rates, trajectories in phase space spiral inward to reach the stationary state (iii') which now represents a stable focus, and the imaginary part of $\epsilon_\pm$ gives the frequency of the resulting damped population oscillations. One may amend this mean-field description to allow for spatial structures by replacing the population numbers with local density fields and adding diffusion terms as in Eq. (9). In one dimension, the ensuing coupled set of partial differential equations permits traveling wave solutions which describe predator invasion fronts originating in a region set in the coexistence state (iii') and moving into space occupied only by prey.[3]

Monte Carlo computer simulations with sufficiently large populations display rich dynamical features in the (quasi-)stable predator–prey coexistence regime that reflect the mean-field picture only partially.[35–37] Already in a local "urn" model without spatial degrees of freedom, intrinsic fluctuations, often termed demographic noise in this context, dominate the dynamics and induce persistent stochastic population oscillations. These can be understood through the effect of white-noise driving on a damped oscillator: On occasion, the stochastic forcing will hit the oscillator's eigenfrequency, and via this resonant amplification displace the phase space trajectories away from the stable coexistence fixed point.[38]

In stochastic Lotka–Volterra models on a lattice that permit an arbitrary number of particles per site, nearest-neighbor hopping transport generates wave-like propagation of these erratic local population oscillations.[37] As depicted in the 2D simulation snapshots of Fig. 2, prey may thus invade empty regions, followed by predators who feed on them, and in turn re-generate space devoid of particles in their wake. Surviving prey islands then act as randomly placed sources of new activity fronts; it is hence the very stochastic nature of the kinetics that both causes and stabilizes these intriguing spatio-temporal patterns.[39,40] They are moreover quite robust with respect to modification of the microscopic model implementations, and are invariably observed in the predator–prey coexistence

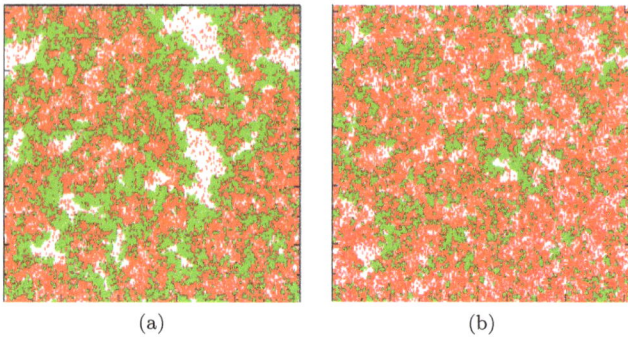

(a)   (b)

Fig. 2.   Snapshots of a stochastic Lotka–Volterra model on a square lattice with
256 × 256 sites, periodic boundary conditions, site occupation number exclusion, and random initial particle placement after (a) 500 (b) 1,000 Monte Carlo
steps. Sites occupied by predators are color-coded in red, by prey in green, and
empty spaces in white. Figures reproduced with permission from Ref. 43, copyright (2016) by IOP Publ.

phase in two and three dimensions, even when the Lotka–Volterra
process $A + B \to A + A$ is separated into independent predation and
predator birth reactions,[41] or if at most a single particle is allowed
on each lattice site. For large predation rates, one observes very distinct and prominent activity fronts; for smaller values of λ, closer to
the extinction threshold, instead there appear confined fluctuating
predator clusters immersed in a sea of abundant prey. These two different types of structures may be related to the distinct properties of
the mean-field linear stability matrix: Real eigenvalues $\epsilon_\pm$ that indicate purely relaxational kinetics correspond to isolated prey clusters
in the spatial system, whereas imaginary parts in Eq. (23) associated with spiraling trajectories and damped oscillations pertain to
spreading predator–prey fronts.

In either case, analysis of the Fourier-transformed average population densities quantitatively establishes that stochastic fluctuations
both drastically renormalize the oscillation frequencies relative to the
rate equation prediction and generate attenuation.[35,37] Based on the
Doi–Peliti mapping of the corresponding master equation to a continuum field theory,[5,13] a perturbative computation in terms of the predation rate λ qualitatively confirms these numerical observations: It

demonstrates the emergence of a noise-induced effective damping, as well as the fluctuation-driven instability toward spatially inhomogeneous structures in dimensions $d \leq 4$; the calculation also explains the strong downward renormalization for the population oscillation frequency as caused by an almost massless mode.[42]

Implementing a finite local carrying capacity through constraining the site lattice occupations, say, to at most a single particle of either type, in addition produces an extinction threshold for the predator species as in the mean-field rate equation (22). As one should indeed expect on general grounds,[19] this continuous active-to-absorbing phase transition is governed by the directed-percolation universality class. This fact can be established formally by starting from the Doi–Peliti action, and reducing it to Reggeon field theory near the predator extinction threshold.[35,42] Heuristically, the prey population almost fills the entire lattice in this situation; consequently, the presence of B particles sets no constraint on the predation process $A + B \rightarrow A + A$, which reduces to the branching reaction $A + A \rightarrow A$. Along with predator death $A \rightarrow \emptyset$ and the population-limiting pair fusion reaction $A + A \rightarrow A$ that is in fact equivalent to local site occupation restrictions, one recovers the fundamental processes of directed percolation. Extensive Monte Carlo simulations confirm directed-percolation critical exponents at the predator extinction transition,[35,36,43] including the associated critical aging scaling behavior for the two-time density autocorrelations.[5,44]

5.2. *Clusters and spiral patterns in cyclic competition games*

Extension of two-species models to systems with multiple reactants in general increases the complexity of the problem tremendously.[36] Nevertheless, simpler substructures can be amenable to full theoretical analysis. For example, on occasion, elementary competition cycles are present in food networks. As a final illustration of the often decisive role of spatial correlations in stochastic chemical reactions or population dynamics, we therefore consider the following cyclic interaction scheme involving three species subject to Lotka–Volterra predation: $A + B \rightarrow A + A$ with rate λ_A; $B + C \rightarrow B + B$ with rate λ_B; and $A + C \rightarrow C + C$ with rate λ_C, akin to the rock–paper–scissors game.[2] These elementary replacement processes conserve the total particle number $N = N_A + N_B + N_C$ and hence density $\rho = a(t) + b(t) + c(t)$; of course, this is also true on the level of the coupled rate equations

$$\frac{\partial a(t)}{\partial t} = a(t)[\lambda_A \, b(t) - \lambda_C \, c(t)], \quad \frac{\partial b(t)}{\partial t} = b(t)[\lambda_B \, c(t) - \lambda_A \, a(t)],$$
$$\frac{\partial c(t)}{\partial t} = c(t)[\lambda_B \, a(t) - \lambda_B \, b(t)]; \tag{24}$$

aside from the three absorbing states $(a, b, c) = (\rho, 0, 0)$, $(0, \rho, 0)$, and $(0, 0, \rho)$, which are linearly unstable under the mean-field dynamics (24), yet represent the sole possible final configurations in any finite stochastic realization; there exists one neutrally stable reactive coexistence state: $a(\infty) = \rho \lambda_B / (\lambda_A + \lambda_B + \lambda_C)$, $b(\infty) = \rho \lambda_C / (\lambda_A + \lambda_B + \lambda_C)$, $c(\infty) = \rho \lambda_A / (\lambda_A + \lambda_B + \lambda_C)$. The linear stability matrix at this fixed point reads

$$\mathbf{L} = \frac{\rho}{\lambda_A + \lambda_B + \lambda_C} \begin{pmatrix} 0 & \lambda_A \lambda_B & -\lambda_B \lambda_C \\ -\lambda_A \lambda_C & 0 & \lambda_B \lambda_C \\ \lambda_A \lambda_C & -\lambda_A \lambda_B & 0 \end{pmatrix};$$

one of its eigenvalues is zero, reflecting the conservation law $\rho = \text{const.}$, the other two are purely imaginary, $\epsilon_\pm = \pm i\rho \sqrt{\lambda_A \lambda_B \lambda_C / (\lambda_A + \lambda_B + \lambda_C)}$, indicating undamped population oscillations.

Akin to the Lotka–Volterra two-species competition model discussed previously, one might thus anticipate traveling wave structures. Yet Monte Carlo simulations of this rock–paper–scissors model on 2D lattices that are sufficiently large to prevent extinction events and stabilize the three-species coexistence state, one merely observes weakly fluctuating clusters containing particles of the same type,[45,46] as shown in Fig. 3(a). Upon initializing the system with a random spatial distribution, transient population oscillations appear, with a characteristic frequency that differs considerably from the mean-field prediction, but they are strongly damped, at variance

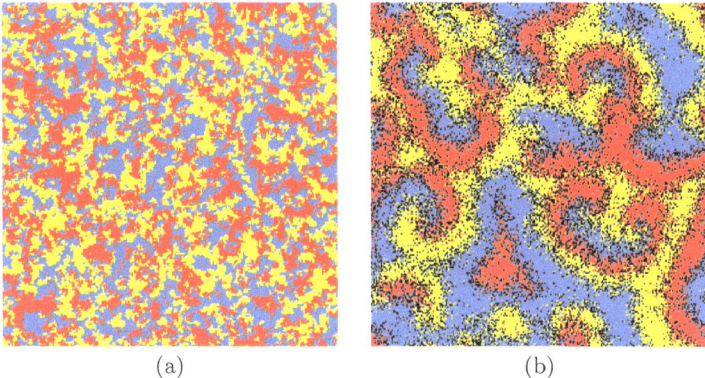

(a) (b)

Fig. 3. Snapshots of a stochastic cyclic competition models on square lattices with 256 × 256 sites, periodic boundary conditions, site occupation number exclusion, and random initial particle placement: (a) cyclic Lotka–Volterra or rock–paper–scissors model with conserved total particle number (b) May–Leonard model with separate predation and reproduction processes. Sites occupied by A, B, and C particles are, respectively, color-coded in red, yellow, and blue; empty spaces are in black. Figures reproduced with permission from Ref. 46, copyright (2010) by The American Physical Society, and Ref. 49, copyright (2011) by EDP Sciences.

with the results from the rate equation analysis. These findings are perhaps even more curious given in light of the fact that in the extreme asymmetric limit $\lambda_A \gg \lambda_B, \lambda_C$, the rock–paper–scissors system effectively reduces to the Lotka–Volterra model with predators A, prey B, and abundant, essentially saturated C population with $c(\infty) \approx \rho$. Indeed, on the mean-field level, one finds leading order $a(\infty) \approx \rho \lambda_B/\lambda_A$, $b(\infty) \approx \rho \lambda_C/\lambda_A$, and $c(\infty) \approx \rho[1 - (\lambda_B + \lambda_C)/\lambda_A]$. We may thus replace $c(t) \approx \rho$ in the rate equations (24), which takes us to Eq. (21) with effective predator death rate $\mu = c(\infty)\lambda_C$ and prey reproduction rate $\sigma = c(\infty)\lambda_B$. Lattice simulations of strongly asymmetric rock–paper–scissors model display the characteristic Lotka–Volterra spreading activity fronts and persistent oscillatory kinetics, and hence confirm this picture.[47]

In the May–Leonard model variant of cyclic competition, predation and birth reactions are explicitly separated: $A + B \to A$,

$A \rightarrow A + A$ (and cyclically permuted) with independent rates.[2] In this coupled reaction scheme, the total particle number is not conserved anymore, but still permits an active three-species coexistence state. In the absence of intraspecies competition, i.e., for infinite carrying capacities, one eigenvalue of the associated linear stability matrix **L** is real and negative, corresponding to an exponentially relaxing mode. The other two eigenvalues are purely imaginary, signifying the existence of undamped oscillations. Monte Carlo simulations of the May–Leonard model on 2D lattices show the spontaneous formation of spiral patterns with alternating A, B, C species in sequence.[36,45,48] These spiral structures become especially clear and pronounced if in addition to nearest-neighbor hopping, particle exchange $A \leftrightarrow B$ etc., is implemented, as illustrated in Fig. 3(b).[49]

On the mean-field rate equation level, one may argue the relaxing eigenmode to be irrelevant for the long-time dynamics of this system; confining the subsequent analysis to the reactive plane spanned by the remaining oscillating modes, a mapping to the complex Ginzburg–Landau equation has been achieved,[50] which is known to permit spiral spatio-temporal structures in certain parameter regimes.[51] A consistent treatment of intrinsic reaction noise for this model based again on the Doi–Peliti formalism however results in a more complex picture. Since the relaxing mode is randomly driven by nonlinear fluctuations that involve the persistent oscillatory eigenmodes, it can only be integrated out in a rather restricted parameter range that allows for distinct time scale separation. In general, therefore the May–Leonard model is aptly described by a coupled set of three Langevin equations with properly constructed multiplicative noise.[52]

6. Summary and Concluding Remarks

In this chapter, we have given an overview of crucial fluctuation and correlation effects in simple spatially extended particle reaction systems that originate from the underlying stochastic kinetics; these very same models also pertain to the basic phenomenology for infectious disease spreading in epidemiology and the dynamics of competing populations in ecology. As befits the theme of this volume, the essential dynamical features of the paradigmatic models discussed here require analysis beyond the standard textbook fare that mostly utilizes coupled deterministic rate equations, which entail a factorization of correlation functions into powers of the reactants' densities, thereby neglecting both temporal and spatial fluctuations.

Spontaneous death–birth reactions augmented with population-limiting fusion constitute the fundamental simple epidemic process. The ensuing population extinction threshold exemplifies a continuous non-equilibrium phase transition between active and absorbing states which is naturally governed by large and long-range fluctuations. It is generically characterized by the critical scaling exponents of directed percolation, which assume non-trivial values in dimensions below the upper critical dimension $d_c = 4$.

Binary (or triplet) annihilation reactions generate depletion zones in dimensions $d \leq d_c = 2$ (1); the particle anti-correlations cause a drastic slowing down of the reaction kinetics in the diffusion-limited regime. For pair annihilation involving two distinct species, spatial segregation in dimensions $d \leq d_s = 4$ confines the reactions to the interfaces separating the diffusively coarsening domains. In the special case of precisely equal species densities, this further diminishes the overall reaction activity, and considerably decelerates the asymptotic algebraic decay.

In stochastic, spatially extended variants of the classical Lotka–Volterra model for predator–prey competition, the intrinsic demographic noise causes and stabilizes spreading activity fronts in the species coexistence phase; their quantitative properties are strongly affected by the stochastic fluctuations and spatial correlations self-generated by the fundamental reaction kinetics in the system. Furthermore, the implementation of finite local carrying capacities for the prey population, mimicking limitations in their food resources, generates an extinction threshold for the predator species. Finally, we have elucidated some basic features in cyclic competition models that involve three particle species, and related the resulting spatial structures, i.e., population clusters for direct rock–paper–scissors competition and dynamical spirals in the May–Leonard model, to the presence and absence, respectively, of a conservation law for the total particle number.

In more complex systems that involve a larger number of reactants, many of these fluctuation-dominated features may of course be effectively averaged out or become inconspicuous on relevant length and/or time scales. However, even when important qualitative properties such as the topology of the phase diagram or the essential functional form of the time evolution remain adequately captured by mean-field rate equations, internal fluctuations as well as emerging spatial and temporal patterns or even short-lived correlations may well drastically modify the numerical values of effective reaction rates and transport coefficients.

Two case studies from the author's research group may illustrate this point: (I) In ligand–receptor binding kinetics on cell membranes or surface plasmon resonance devices that are commonly utilized to measure reaction affinities, repeated ligand rebinding to nearby receptor locations entails temporal correlations that persist under diffusive dynamics and even advective flow; the correct reaction rate values may consequently deviate drastically from numbers extracted by inadequate fits to straightforward rate equation kinetics.[53] (II) Circularly spreading wave fronts of "killer" as well as resistant bacteria strains induce the formation of intriguing patterns in an effectively 2D synthetic *E. coli* microecological system, emphasizing the important role of spatial inhomogeneities and local correlations.[54] A sound understanding of the potential effects of fluctuations and correlations is thus indispensable for proper theoretical modeling and analysis as well as quantitative interpretation of experimental data.

There are of course other crucial fluctuation and correlation effects on chemical reactions and population dynamics that could not be covered in this chapter.[36] Prominent examples are the strong impact of intrinsic noise on extinction probabilities and pathways in finite systems[55]; and novel phenomena related to extrinsic random influences or quenched internal disorder.[56] These topics and many more are addressed elsewhere in this book.

Acknowledgments

I would like to sincerely thank my collaborators over the past 25 years for their invaluable insights and crucial contributions to our joint research on stochastic reaction-diffusion and population dynamics, especially: Timo Aspelmeier, Nicholas Butzin, John Cardy, Jacob Carroll, Sheng Chen, Udaya Sree Datla, Oliviér Deloubrière, Ulrich Dobramysl, Kim Forsten-Williams, Erwin Frey, Ivan Georgiev, Yadin Goldschmidt, Manoj Gopalakrishnan, Qian He, Bassel Heiba, Henk Hilhorst, Haye Hinrichsen, Martin Howard, Hannes Janssen, Weigang Liu, Jerôme Magnin, William Mather, Mauro Mobilia, Michel Pleimling, Matthew Raum, Beth Reid, Gunter Schütz, Franz Schwabl (deceased), Shannon Serrao, Steffen Trimper, Ben Vollmayr-Lee, Mark Washenberger, Frédéric van Wijland, and Royce Zia.

This research was in part sponsored by the US Army Research Office and was accomplished under Grant Number W911NF-17-1-0156. The views and conclusions contained in this document are those of the author and should not be interpreted as representing the official policies, either expressed or implied, of the Army Research Office or the US Government. The US Government is authorized to reproduce and distribute reprints for Government purposes notwithstanding any copyright notation herein.

References

1. H. Haken, *Synergetics*. Springer, Berlin (1983).
2. J. Hofbauer and K. Sigmund, *Evolutionary Games and Population Dynamics*. Cambridge University Press, Cambridge (1998).
3. J. D. Murray, *Mathematical Biology*, Vols. I & II. Springer, New York, 3rd edn. (2002).
4. P. K. Krapivsky, S. Redner, and E. Ben-Naim, *A Kinetic View of Statistical Physics*. Cambridge University Press, Cambridge (2010).

5. U. C. Täuber, *Critical Dynamics — A Field Theory Approach To Equilibrium and Non-equilibrium Scaling Behavior.* Cambridge University Press, Cambridge (2014).

6. N. G. van Kampen, *Stochastic Processes in Physics and Chemistry.* North Holland, Amsterdam (1981).

7. C. M. van Vliet, *Equilibrium and Non-equilibrium Statistical Mechanics.* World Scientific, New Jersey, 2nd edn. (2010).

8. V. Kuzovkov and E. Kotomin, Kinetics of bimolecular reactions in condensed media: Critical phenomena and microscopic self-organisation, *Rep. Prog. Phys.* **51**, 1479–1523 (1988).

9. A. A. Ovchinnikov, S. F. Timashev, and A. A. Belyy, *Kinetics of Diffusion-Controlled Chemical Processes.* Nova Science, New York (1989).

10. H. Hinrichsen, Nonequilibrium critical phenomena and phase transitions into absorbing states, *Adv. Phys.* **49**, 815–958 (2001).

11. G. Ódor, Phase transition universality classes of classical, nonequilibrium systems, *Rev. Mod. Phys.* **76**, 663–724 (2004).

12. H. K. Janssen and U. C. Täuber, The field theory approach to percolation processes, *Ann. Phys.* **315**, 147–192 (2005).

13. U. C. Täuber, M. J. Howard, and B. P. Vollmayr-Lee, Applications of field-theoretic renormalization group methods to reaction-diffusion problems, *J. Phys. A.: Math. Gen.* **38**, R79–R131 (2005).

14. M. Henkel, H. Hinrichsen, and S. Lübeck, *Non-Equilibrium Phase Transitions*, Vol. 1: *Absorbing Phase Transitions.* Springer, Dordrecht (2008).

15. M. Moshe, Recent developments in Reggeon field theory, *Phys. Rep.* **37**, 255–345 (1978).

16. J. L. Cardy and R. L. Sugar, Directed percolation and Reggeon field theory, *J. Phys. A.: Math. Gen.* **13**, L423–L427 (1980).

17. H. K. Janssen, On the nonequilibrium phase transition in reaction-diffusion systems with an absorbing stationary state, *Z. Phys. B.: Cond. Matt.* **42**, 151–154 (1981).

18. P. Grassberger, On phase transitions in Schlögl's second model, *Z. Phys. B.: Cond. Matt.* **47**, 365–374 (1982).

19. H. K. Janssen, Directed percolation with colors and flavors, *J. Stat. Phys.* **103**, 801–839 (2001).

20. I. Jensen, Low-density series expansions for directed percolation on square and triangular lattices, *J. Phys. A.: Math. Gen.* **29**, 7013–7040 (1996).

21. K. A. Takeuchi, M. Kuroda, H. Chaté, and M. Sano, Experimental realization of directed percolation criticality in turbulent liquid crystals, *Phys. Rev. E.* **80**, 051116-1–12 (2009).

22. L. Peliti, Renormalisation of fluctuation effects in the $A + A \to A$ reaction, *J. Phys. A.: Math. Gen.* **19**, L365–367 (1986).

23. B. P. Lee, Renormalization group calculation for the reaction $kA \to \emptyset$, *J. Phys. A.: Math. Gen.* **27**, 2633–2652 (1994).

24. F. C. Alcaraz, M. Droz, M. Henkel, and V. Rittenberg, Reaction-diffusion processes, critical dynamics, and quantum chains, *Ann. Phys.* **230**, 250–302 (1994).

25. M. Henkel, E. Orlandini, and J. Santos, Reaction-diffusion processes from equivalent integrable quantum chains, *Ann. Phys.* **259**, 163–231 (1997).
26. G. M. Schütz, Exactly solvable models for many-body systems far from equilibrium, in: *Phase Transitions and Critical Phenomena*, Vol. 19, C. Domb and J. L. Lebowitz(eds.), Academic Press, London (2001).
27. R. Stinchcombe, Stochastic nonequilibrium systems, *Adv. Phys.* **50**, 431–496 (2001).
28. J. Allam, M. T. Sajjad, R. Sutton, K. Litvinenko, Z. Wang, S. Siddique, Q.-H. Yang, W. H. Lo, and T. Brown, Measurement of a reaction-diffusion crossover in exciton-exciton recombination inside carbon nanotubes using femtosecond optical absorption, *Phys. Rev. Lett.* **111**, 197401-1–5 (2013).
29. D. Toussaint and F. Wilczek, Particle-antiparticle annihilation in diffusive motion, *J. Chem. Phys.* **78**, 2642–2647 (1983).
30. B. P. Lee and J. Cardy, Renormalization group study of the $A + B \to \emptyset$ diffusion-limited reaction, *J. Stat. Phys.* **80**, 971–1007 (1995).
31. B. P. Lee and J. Cardy, Scaling of reaction zones in the $A + B \to \emptyset$ diffusion-limited reaction, *Phys. Rev. E.* **50**, R3287–R3290 (1994).
32. E. Monson and R. Kopelman, Nonclassical kinetics of an elementary $A + B \to C$ reaction-diffusion system showing effects of a speckled initial reactant distribution and eventual self-segregation: Experiments, *Phys. Rev. E.* **69**, 021103-1–12 (2004).
33. H. J. Hilhorst, O. Deloubrière, M. J. Washenberger, and U. C. Täuber, Segregation in diffusion-limited multispecies pair annihilation, *J. Phys. A.: Math. Gen.* **37**, 7063–7093 (2004).
34. H. J. Hilhorst, M. J. Washenberger, and U. C. Täuber, Symmetry and species segregation in diffusion-limited pair annihilation, *J. Stat. Mech.* P10002-1–19 (2004).
35. M. Mobilia, I. T. Georgiev, and U. C. Täuber, Phase transitions and spatio-temporal fluctuations in stochastic lattice Lotka–Volterra models, *J. Stat. Phys.* **128**, 447–483 (2007).
36. U. Dobramysl, M. Mobilia, M. Pleimling, and U. C. Täuber, Stochastic population dynamics in spatially extended predator–prey systems, *J. Phys. A.: Math. Theor.* **51**, 063001-1–49 (2018).
37. M. J. Washenberger, M. Mobilia, and U. C. Täuber, Influence of local carrying capacity restrictions on stochastic predator–prey models, *J. Phys. Cond. Matt.* **19**, 065139-1–14 (2007).
38. A. J. McKane and T. J. Newman, Predator–prey cycles from resonant amplification of demographic stochasticity, *Phys. Rev. Lett.* **94**, 218102-1–5 (2005).
39. T. Butler and N. Goldenfeld, Robust ecological pattern formation induced by demographic noise, *Phys. Rev. E.* **80**, 030902(R)-1–4 (2009).
40. T. Butler and N. Goldenfeld, Fluctuation-driven Turing patterns, *Phys. Rev. E.* **84**, 011112 -1–12 (2011).
41. M. Mobilia, I. T. Georgiev, and U. C. Täuber, Fluctuations and correlations in lattice models for predator–prey interaction, *Phys. Rev. E.* **73**, 040903 (R)-1–4 (2006).

42. U. C. Täuber, Population oscillations in spatial stochastic Lotka–Volterra models: A field-theoretic perturbational analysis, *J. Phys. A.: Math. Theor.* **45**, 405002-1–34 (2012).

43. S. Chen and U. C. Täuber, Non-equilibrium relaxation in a stochastic lattice Lotka–Volterra model, *Phys. Biol.* **13**, 025005-1–11 (2016).

44. M. Henkel and M. Pleimling, *Non-equilibrium Phase Transitions*, Vol. 2: *Ageing and Dynamical Scaling Far from Equilibrium.* Springer, Dordrecht (2010).

45. M. Peltomäki and M. Alava, Three-and four-state rock–paper–scissors games with diffusion. *Phys. Rev. E.* **78**, 031906-1–7 (2008).

46. Q. He, M. Mobilia, and U. C. Täuber, Spatial rock–paper–scissors models with inhomogeneous reaction rates, *Phys. Rev. E.* **82**, 051909-1–11 (2010).

47. Q. He, U. C. Täuber, and R. K. P. Zia, On the relationship between cyclic and hierarchical three-species predator–prey systems and the two-species Lotka–Volterra model, *Eur. Phys. J. B.* **85**, 141-1–13 (2012).

48. T. Reichenbach, M. Mobilia, and E. Frey, Mobility promotes and jeopardizes biodiversity in rock–paper–scissors games, *Nature* **448**, 1046–1049 (2007).

49. Q. He, M. Mobilia, and U. C. Täuber, Coexistence in the two-dimensional May–Leonard model with random rates, *Eur. Phys. J. B.* **82**, 97–105 (2011).

50. T. Reichenbach, M. Mobilia, and E. Frey, Self-organization of mobile populations in cyclic competititon, *J. Theor. Biol.* **254**, 368–383 (2008).

51. I. S. Aranson and L. Kramer, The world of the complex Ginzburg–Landau equation, *Rev. Mod. Phys.* **74**, 99–144 (2002).

52. S. R. Serrao and U .C. Täuber, A stochastic analysis of the spatially extended May–Leonard model, *J. Phys. A: Math. Theor.* **50**, 404005-1–16 (2017).

53. J. Carroll, M. Raum, K. Forsten-Williams, and U. C. Täuber, Ligand–receptor binding kinetics in surface plasmon resonance cells: A Monte Carlo analysis, *Phys. Biol.* **13**, 066010-1–12 (2016).

54. U. S. Datla, W. H. Mather, S. Chen, I. W. Shoultz, U. C. Täuber, C. N. Jones, and N. C. Butzin, The spatiotemporal system dynamics of acquired resistance in an engineered microecology, *Sci. Rep.* **7**, 16071-1–9 (2017).

55. M. Assaf and B. Meerson, WKB theory of large deviations in stochastic populations, *J. Phys. A.: Math. Theor.* **50**, 263001-1–63 (2017).

56. T. Vojta, Rare region effects at classical, quantum and nonequilibrium phase transitions, *J. Phys. A.: Math. Gen.* **39**, R143–205 (2006).

Chapter 2

Encounter Theory of Chemical Reactions in Solution: Approximate Methods of Calculating Rate Constants

Alexander B. Doktorov

V. V. Voevodsky Institute of Chemical Kinetics and Combustion, Siberian Branch of Russian Academy of Sciences, Novosibirsk, 630090, Russia

doktorov@kinetics.nsc.ru

Comparison of the approaches of the Coagulation and the Encounter Theories leading to one and the same differential kinetic equation in the consideration of one-stage irreversible chemical reactions is made. It is shown that only the concepts of the Encounter Theory admit generalization to the examination of a wide class of physicochemical processes in dilute solutions. Using the kinematic approximation in the Encounter Theory, the recipe for the calculation of steady-state constants in the general case of multistage multisite reaction is found. Based on this recipe, rate constants of irreversible one-stage reactions in the presence of reactivity anisotropy as well as rate constants of two-stage two-site reactions are calculated.

1. Introduction

Many of the bulk chemical reactions (or other physicochemical processes) in condensed medium proceed in the so-called kinetic regime, when the reaction course is completely determined by elementary chemical transformation rates rather than by reactant motion. However, the reactions, depending on reactant mobility (commonly, diffusion-influenced reactions), also include a wide class of physicochemical processes taking place in liquid solutions and solids. Among these are, for example, electron excitation energy transfer,[1] electron (or proton) migrations in photosynthetic systems,[2-4] various chemical reactions occurring in colloid or polymer solutions,[5-7] in nano- and biosystems,[8-10] or trapping and detrapping problems in semiconductors,[11] and many others. From this viewpoint, the development of the theory of reactions depending on reactant motion is important for understanding the behavior of various reacting systems. Many fundamental issues on general theories and their applications to the investigation of specific reaction systems have been addressed over many years. Such theories relied both on simple physical notions[3,5,12-17] and on many-particle considerations of the system of reactants dissolved in chemically inert continual solvent, both in the framework of simple exactly solvable models[18,19] and with the use of approximations.[20-30] A more complete list of works on the problem is available in Refs. 31–33. In the present contribution, we shall deal with the theories relying on simple physical considerations which for a long time served as a basis for the examination of diffusion-influenced chemical reactions before the development of consistent many-particle theories.

2. Coagulation Theory

The Coagulation Theory[5,12] is based on the contact model of spherical particles with total radius R and coefficient D of relative macrodiffusion. This model ("gray sphere" or "black sphere" model) was subsequently applied to the consideration of luminescence quenching processes in solutions[1] and elementary irreversible chemical reactions

$$A + B \rightarrow C + B, \tag{1}$$

proceeding with excess reactants B. Of course, strictly speaking, in this case the coefficient D cannot be treated as macroscopic diffusion coefficient, that is the proportionality coefficient between the density of the flux of particles and density gradient.

In a more general formulation, this model can take into consideration additional force interaction (along with reflection on maximum approach of particles) defined by the potential $U(r) \equiv u(r)kT$ and diffusion coefficient dependence on the distance $D(r)$ where r is the distance between two particles (A and B), $u(r)$ is the potential interaction energy in units kT, k is the Boltzmann constant, and T is the absolute temperature. According to the concepts of this theory, the change in concentration $[A]_t$ of particles A interacting with particles B, the concentration $[B]$ of which is high and thus almost time invariant, are given by the expression

$$[A]_t = [A]_0 \exp\left(-[B]\int_0^t k(\tau)d\tau\right), \qquad (2)$$

where

$$k(t) = k_r \exp(u(R))n(R,t) \equiv k_r n^+(R,t) \qquad (3)$$

and $n(r,t)$ is the pair density. Besides, the adjoint pair density $n^+(r,t)$ is introduced, which formally coincides with $n(r,t)$ at $u(r) \equiv 0$ but has a different physical meaning, as is shown below. We use the adjoint pair density $n^+(r,t)$ obeying the equation

$$\frac{\partial}{\partial t}n^+(r,t) = \frac{1}{r^2}\exp(u(r))\frac{\partial}{\partial r}D(r)r^2\exp(-u(r))\frac{\partial}{\partial r}n^+(r,t) \qquad (4)$$

with the initial condition $n^+(r,0) = 1$ and boundary conditions at $r \to \infty$ and $r = R$

$$n^+(\infty,t) = 1, 4\pi D(R)R^2\exp(-u(R))\frac{\partial}{\partial r}n^+(r,t)\bigg|_{r=R} = k_r\, n^+(R,t). \qquad (5)$$

In chemical kinetics terminology, the quantity k_r is the reaction constant.

At $k_r \to \infty$, the model corresponds to the so-called "black sphere" model and the second boundary condition (5) at $r = R$ has the form corresponding to going to zero of the pair density at the contact of reactants

$$n^+(R, t) = 0. \tag{6}$$

In the absence of additional force interaction potential $u(r) = 0$ and $D(r) = D = \text{const}$, the model corresponds to the Smoluchowski problem,[12] and, in view of additional Coulomb interaction, — to the Onsager problem.[34]

The kinetics of the process defined by expression (2) obeys the kinetic equation

$$\frac{d}{dt}[A]_t = -k(t)[B][A]_t. \tag{7}$$

In chemical kinetics terminology, $k(t)$(3) is time-dependent rate constant of the process under study. In the theory in question, its time dependence is related to non-stationary diffusion, and its characteristic time at $u(r) = 0$ and $D(r) = D = \text{const}$ is

$$\tau_D = R^2/D.$$

It determines the course of the so-called transient stage after the end of which the rate constant attains its steady-state value

$$k = \lim_{t \to 0} k(t). \tag{8}$$

Kinetic equation (7) takes the form of basic chemical kinetics[35] corresponding to the kinetic law of mass action

$$\frac{d}{dt}[A]_t = -k[B][A]_t. \tag{9}$$

Theories taking into account time dependence of kinetic coefficients are called the non-Markovian theories, while those considering just quasi-stationary stages are the Markovian theories.[32,33] In this work, in considering particular problems, we shall restrict ourselves mainly to the examination of the Markovian theories. The non-Markovian theories will be discussed elsewhere. For the calculation of the Markovian rate constant in Eq. (9), the stationary solution of Eq. (4) is sufficient, i.e., at

$$\frac{\partial}{\partial t} n^+(r,t) = 0, \quad n^+(r,t) \approx n^+(r)$$

with boundary conditions (5). With a stationary solution in Eq. (3), it gives the expression for the steady-state (Markovian) reaction rate constant

$$\frac{1}{k} = \frac{1}{k_r} + \frac{1}{k_D}, \quad k = \frac{k_D k_r}{k_D + k_r}, \tag{10}$$

which involves two different limiting regimes of the reaction course. At $k_r \ll k_D$, $k \approx k_r$ and the elementary reaction process taking place at contact is limiting. Such a stage is called a kinetic stage. At $k_D \ll k_r$, $k \approx k_D$ and diffusion of reactants resulting in their approach is limiting. Such a stage is called a diffusion-controlled stage. The quantity k_D of the diffusion-controlled reaction rate constant defined as

$$\frac{1}{k_D} = \frac{1}{4\pi} \int_R^\infty \frac{\exp(u(y))}{D(y)y^2} dy \tag{11}$$

is called diffusion constant. At $u(r) = 0$ and $D(r) = D = $ const, Eq. (11) yields the well-known expression[12]

$$k_D = 4\pi D R. \tag{12}$$

A similar formula also takes place in the case of strong attraction. However, instead of contact radius, the radius R_e determined from the condition $u(R_e) = 1$ appears. For strong Coulomb attraction, such a radius is called the Onsager radius.[34]

3. Encounter Theory

Application of the Coagulation Theory approach to the investigation of chemical reactions has a number of demerits, first of all, related to the macroscopic character of reactant motion consideration, and

it is, to a great extent, intuitive. Generalization of this approach
to the studies of reversible and multistage reactions presents significant difficulties. At the same time, dilute solutions where reactions proceed may be treated as rarefied "gas" of reactants placed in
chemically inert continual matter. In this case, to obtain kinetic equations, one can use concepts and methods of semi-classical Collision
Theory, according to which reactants are, for the most time, in the
process of "free pass" and enter into the reaction as a result of short
"collisions", the mean duration of which is much less than the mean
time between collisions. However, in dilute solutions the medium (solvent), though chemically inert (i.e., it does not participate in the reaction), has a marked effect on the course of the reaction. Just for this
reason, the dynamic character of the elementary event is replaced by
a stochastic one that is specified by the reaction's elementary event
rate $w(q)$ (q — the reaction configuration space coordinate involving all "external" parameters which determine the reaction course)
on sufficient approach of reactants. Besides, the medium has an
essential effect on the character of relative motion of reactants.
It transforms dynamic motion (realized in accordance with classical mechanics laws) into a stochastic one (executed by random
walks of reactants in "inert" medium). Thus, "free pass" gives way
to "free random walks" which, of course, in the general case, are
not macroscopic diffusion resulting from the density gradient but
microscopic molecular motion of reactants in the matter. Moreover, the "collision" character itself changes due to the so-called
cage effect.[36] The essence of this effect is that, unlike the gas
phase, where fragments, for example, of a decomposing molecule
move apart, in the liquid phase these fragments do not necessarily separate.[37] This can lead to repeat formation of chemical bond
by virtue of the fact that condensed matter keeps the partners
from moving apart by holding them together in some kind of a
"cage". It is established that collisions — *contacts* of two separate partners — occur in series. We consider that each series of
contacts corresponds to the *encounter* of reactants in condensed
matter, which begins with the first contact. So the cage ensures

a substantial increase in the characteristic encounter time τ_c as compared to duration of collisions of reactants in gas, though this time is much less than the mean time τ_f of free random walks before the encounter with other partners (quantity inverse to the mean frequency of encounters). Turning to the above contact model, we have, for example, at $u(r) = 0$ and $D(r) = D = \text{const}$

$$\frac{1}{\tau_f} = k_D[B] = 4\pi DR[B], \quad \tau_c = \frac{R^2}{3D}, \quad \eta = \frac{\tau_c}{\tau_f} = \frac{4\pi}{3}R^3[B] \ll 1,$$

$$(13)$$

i.e., just as for rarefied gases, the theory concepts require the smallness of the density parameter η.

Thus the event which is treated as collision in the Collision Theory, in reactions in solutions should be considered as an encounter consisting of a great number of contacts due to stochastic character of reactant motion in solution. Just for this reason, such a theory (analog of semi-classical Collision Theory) is called the Encounter Theory.[38,39] Note that physical notions and the corresponding terminology of the Encounter Theory differ essentially from those accepted in theories based on the Coagulation Theory. Diffusion constant multiplied by concentration $[B]$ is nothing but the frequency of encounters of reactants in solution (analog of collision frequency in gas), since in diffusion-controlled regime the first contact of reactants (beginning of encounter) gives rise to the reaction that interrupts free random walks (analog of free pass). Non-stationary diffusion time is the characteristic encounter time (analog of the mean duration of collision in gas). Such a terminology agrees more adequately with the situation when molecular motion is not diffusion.

So, by analogy with the Collision Theory, the quantity is to be calculated

$$\frac{[A]_t}{[A]_0} = \sum_{n=0}^{\infty} \left\langle \exp\left(-\int_0^t \sum_{i=1}^n w_i(q(\tau))d\tau\right)\right\rangle$$

$$= \sum_{n=0}^{\infty} \left\langle \prod_{i=1}^n \exp\left(-\int_0^t w_i(q(\tau))d\tau\right)\right\rangle, \qquad (14)$$

where broken brackets denote averaging over n encounters numbered by the index i. As in the Collision Theory, here we assume the independence of pair encounters that manifests itself in additivity of their contributions.

First, examine contact reaction. For this case, time distribution of free random walks is similar to time distribution of free pass in gas

$$dW(t) = \frac{1}{\tau_f} \exp\left(-\frac{t}{\tau_f}\right) = k_D[B] \exp(-k_D[B]t). \qquad (15)$$

As in the Collision Theory, though the reacting system evolution in the interval from 0 to t is considered, it is necessary to allow for the contributions from collisions occurred beyond the limits of this interval. However, in the Encounter Theory, in view of the fact that encounters begin from the first contact, one should take into account just the contributions of encounters occurred at times $\tau' \leq 0$ and neglect the contributions from encounters in the interval $\tau' > t$ that are considered in the Collision Theory. With this correction, the expression known from the Collision Theory (in view of Eq. (15)) is

$$[A]_t = [A]_0 \exp\left(-k_D[B] \int_{-\infty}^{t} (1 - T(t, \tau'))d\tau'\right), \qquad (16)$$

where $T(t, \tau')$ is the survival probability of two reactants by the moment of time t on the encounter that occurred at the moment of time τ'. To establish its form, introduce the quantities which are survival probabilities by the moment of time t in geminate pair $A + B$

$$n(q, t) = \left\langle \exp\left(-\int_0^t w(q(\tau))d\tau\right)\right\rangle_q,$$

$$n^+(q, t) = \left\langle \exp\left(-\int_0^t w(q(\tau))d\tau\right)\right\rangle_q. \qquad (17)$$

The first quantity is the average over stochastic trajectories $q(t)$ that pass through the point q at the moment of time t, and the second quantity — the average over stochastic trajectories that pass through the point q at the initial moment of time $t = 0$. Thus the value $n^+(q, t)$ is the survival probability by the moment of time t

on the encounter which began at the moment of time $t = 0$ at the initial position of reactants defined by the coordinate q. If stochastic trajectories $q(t)$ are a stationary Markovian stochastic process, then for a spherically symmetric problem this quantity obeys the equation[39] that differs from Eq. (4) in the presence of the sink term $w(r)n^+(r,t)$[32,33,40]

$$\frac{\partial}{\partial t}n^+(r,t) = -w(r)n^+(r,t) + \frac{1}{r^2}$$

$$\times \exp(u(r))\frac{\partial}{\partial r}D(r)r^2 \exp(-u(r))\frac{\partial}{\partial r}n^+(r,t) \quad (18)$$

with the initial condition $n^+(r,0) = 1$ and boundary conditions at $r \to \infty$ and $r = R$

$$n^+(\infty,t) = 1, \quad \frac{\partial}{\partial r}n^+(r,t)\bigg|_{r=R} = 0. \quad (19)$$

The second boundary condition corresponds to the absence of the kinematic flux on approach to the distance $r = R$. For contact reaction, we have

$$w(r) = k_0\frac{\delta(r-R)}{4\pi r R} = \frac{k_0}{4\pi r R}\lim_{\varepsilon \to 0}\frac{1}{\varepsilon}\Theta(R+\varepsilon-r),$$

$$k_0 = 4\pi\int_R^\infty W(r)r^2 dr, \quad k_r = k_0\exp(-u(R)), \quad (20)$$

where $\Theta(R+\varepsilon-r)$ is the Heaviside stepwise function and $\delta(r-R)$ — the Dirac delta function. At $r \geq R+\varepsilon$, Eq. (18) gives Eq. (4). Multiplying both sides of Eq. (18) by $4\pi r^2\exp(-u(r))$ and integrating over r between the limits from R to $R+\varepsilon$ with allowance for the second boundary condition (19) and Eqs. (20) in the limit $\varepsilon \to 0$, we obtain boundary (at $r = R$) condition (5). So, in the case of contact reaction, Eq. (18) with the reaction sink reduces to Eq. (4) without the sink but with another boundary condition defining the reaction flux into the sphere $r = R$. Now introduce the conditional probability density $\varphi(q,t|q',t') = \varphi(q,q';t-t')$ of the realization of the coordinate q at the moment of time t if at the initial moment of time t' it was q'. It depends on time difference due to stochastic process stationarity. In the specific case of spherical symmetry $(q = r)$, this function obeys the equation[39]

$$\frac{\partial}{\partial \tau}\varphi(r, r'; \tau) = \frac{1}{r^2}\frac{\partial}{\partial r}D(r)r^2 \exp(-u(r))\frac{\partial}{\partial r}\exp(u(r))\varphi(r, r'; \tau)$$

$$(21)$$

with the initial condition at $\tau = t - t' = 0$

$$\varphi(r, t'|r', t') = \varphi(r, r'; 0) = \frac{\delta(r - r')}{4\pi rr'} \tag{22}$$

and boundary conditions

$$\left.\frac{\partial}{\partial r}\right|_{r=R}\exp(u(r))\varphi(r, r'; \tau) = 0, \quad \varphi(\infty, r'; \tau) = 0. \tag{23}$$

The Laplace time transform of this probability is called a resolvent[41]

$$g(r, r'; s) = \int_0^\infty \exp(-s\tau)\varphi(r, r'; \tau)d\tau, g(q, q'; s)$$

$$= \int_0^\infty \exp(-s\tau)\varphi(q, q'; \tau)d\tau. \tag{24}$$

At $s = 0$ it is called a stationary resolvent (or the stationary Green function)

$$g(r, r') = g(r, r'; 0) = \frac{1}{4\pi}\exp(-u(r))\int_{\max\{r, r'\}}^\infty \frac{\exp(u(y))}{D(y)y^2}dy. \tag{25}$$

Now, obtain the expression for the survival probability on the encounter $T(t, \tau')$ in Eq. (16). Since encounters start from the distance $r = R$ (with the first contact) at the moment of time τ', therefore, the survival probability on the encounter $T(t, \tau')$ at $0 \leq \tau' \leq t$ is $n^+(R, t - \tau')$. At $\tau' < 0$, one should take into account that the contribution of such an encounter should be considered in the interval $0 \leq \tau \leq t$ only. In the interval $\tau' \leq \tau < 0$, averaging over trajectories starting at the contact $(r = R)$ at the moment of time $\tau' < 0$ of the encounter and ending at some value of distance r on this interval (at $\tau = 0$) gives the conditional probability density $\varphi(r, 0|R, \tau') = \varphi(r, R; -\tau')$, and on the remaining interval $0 \leq \tau \leq t$ — the value $n^+(r, t)$. Thus, eventually we have

$$T(t, \tau') = \begin{cases} n^+(R, t - \tau') & \text{at } 0 \leq \tau' \leq t, \\ 4\pi \int_R^\infty n^+(r, t)\varphi(r, R; -\tau')r^2 dr & \text{at } \tau' \leq 0. \end{cases} \quad (26)$$

Consider the argument of the exponent in Eq. (16) by dividing integration intervals from $-\infty$ to 0 and from 0 to t. We replace the integration variable $-\tau$ by τ in the first integral and $t - \tau$ by τ in the second one. With the aid of Eqs. (24), (25), and the fact that rate constant is equal to time derivative of the argument of the exponent (see Eq. (2)) divided by concentration $[B]$, we get for this constant

$$k(t) = k_D(1 - n^+(R, t)) - 4\pi k_D \int_R^\infty \frac{\partial}{\partial t} n^+(r, t)r^2 g(r, R)dr. \quad (27)$$

Using Eqs. (11), (4), and (25) and calculating the integral in Eq. (27) by parts, in view of Eq. (5), we obtain expression (3) of the Coagulation Theory. However, now it is possible to physically interpret rate constant representing it as

$$k(t) = k_r \Omega(t), \quad (28)$$

where $\Omega(t) = n^+(R, t)$ is the survival probability on the encounter of reactants, i.e., the survival probability in geminate reaction by the moment of time t, with the start from contact $r = R$ at the moment of time $t = 0$.

Now, let us generalize the result to remote reactions. For spherically symmetric problem, the difference is in that in definition (26) of $T(t, \tau')$ in Eq. (16) one should use $n^+(r, t)$ obeying Eq. (18). In the general case of the dependence on the coordinate q of the configuration space, we shall employ the Laplace transforms of quantities. The integral equation for the Laplace transform $\tilde{n}^+(q, s)$ of the quantity $n^+(q, t)$ over time is of the form[41]

$$\tilde{n}^+(q, s) = \frac{1}{s} - \int \tilde{n}^+(q', s)w(q')g(q', q; s)dq'. \quad (29)$$

Of course, in the expression for $T(t, \tau')$ (26), contact value of distance should be replaced by some distance r_0, and the value k_D — by the value $k_d(r_0)$ depending on this distance (which is one of the components of the reaction configuration space coordinate q_0). In the final calculation of the reaction rate constant $k(t)$, we put $r_0 \to \infty$. For $k_d(r_0)$, we can use the value for spherically symmetric but not contact problem that is given by the expression following from Eq. (11)

$$\frac{1}{k_d(r_0)} = \frac{1}{4\pi} \int_{r_0}^{\infty} \frac{\exp(u(y))}{D(y)y^2} dy \xrightarrow[r_0 \to \infty]{} \frac{1}{4\pi D(\infty)r_0}. \qquad (30)$$

Taking the forgoing into account, instead of Eq. (27), we have the limit $r_0 \to \infty$ of the quantity depending on q_0. Its Laplace transform is as follows:

$$\tilde{k}(s) = \lim_{r_0 \to \infty} \tilde{k}(s, q_0),$$

$$\tilde{k}(s, q_0) = k_d(r_0) \left\{ \frac{1}{s} - \tilde{n}^+(q_0, s) + s \int \left(\frac{1}{s} - \tilde{n}^+(q, s) \right) g(q, q_0) dq \right\}.$$

$$(31)$$

With Eq. (29) and resolvent relation[41] (equivalent to the Einstein–Smoluchowski relation)

$$\int g(q', q; s)g(q, q_0; 0)dq = \frac{1}{s}(g(q', q_0; 0) - g(q', q_0; s)), \qquad (32)$$

we obtain from the second equation in Eq. (31)

$$\tilde{k}(s, q_0) = k_d(r_0) \int \tilde{n}^+(q, s)w(q)g(q, q_0; 0)dq. \qquad (33)$$

So, using the original of Eq. (33) and performing the limit $r_0 \to \infty$, we arrive at the expression for rate constant

$$k(t) = \int n^+(q, t)w(q)\varphi(q)dq, \qquad (34)$$

where

$$\varphi(q) = \lim_{r_0 \to \infty} k_d(r_0)g(q, q_0) \quad (r_0 \in q_0) \qquad (35)$$

is the equilibrium distribution (the Boltzmann distribution) or static contour[41] which satisfies the detailed balancing principle and stationarity condition

$$\varphi(q, q_0; t)\varphi(q_0) = \varphi(q_0, q; t)\varphi(q), \quad g(q, q_0; s)\varphi(q_0) = g(q_0, q; s)\varphi(q),$$

$$\int \varphi(q, q_0; t)\varphi(q_0)dq_0 = \varphi(q), \quad \int g(q, q_0; s)\varphi(q_0)dq_0 = \frac{1}{s}\varphi(q).$$

(36)

Equation (35) follows from relations[41]

$$k_d(r_0)_{r_0 \to \infty} \sim 4\pi \bar{D} r_0, \quad g(q, q_0)_{r_0 \to \infty} \sim \frac{1}{4\pi \bar{D} r_0}\varphi(q) \quad (r_0 \in q_0),$$

(37)

where $\bar{D}$ is a macroscopic diffusion coefficient.

Let us introduce the function — the reaction zone shape normalized to unity with the weight $\varphi(q)$

$$\psi(q) = \frac{w(q)}{k_r}; \quad k_r = \int w(q)\varphi(q)dq; \quad \int \psi(q)\varphi(q)dq = 1. \quad (38)$$

Then reaction rate constant (34) may be represented in the previous form (28) where the survival probability $\Omega(t)$ on encounter is the survival probability in geminate pair, with the start being from the initial distribution $f(q) = \psi(q)\varphi(q)$

$$\Omega(t) = \int n^+(q, t)f(q)dq = \int n^+(q, t)\psi(q)\varphi(q)dq. \quad (39)$$

Note that the reaction on encounter is geminate process of a reaction of a pair of reactants encountering each other. However, this process should not be considered as geminate reaction of isolated pairs of reactants,[42] when isolated geminate pairs are generated at the initial moment of time (i.e., initial correlation exists). On the contrary, for bulk reactions considered in framework of the Encounter Theory, pairs of reactants are formed at different moments of time due to pair encounters of particles in solution. It is seen that the reaction rate constant dependence is related to incompleteness of pair encounters.

However, as the transient stage is completed, i.e., at $t \gg \tau_c$, the reaction rate constant reaches its steady-state value (8)

$$k = k_r \Omega(\infty) = \int w(q) n(q) dq = \int w(q) n^+(q) \varphi(q) dq. \quad (40)$$

4. Steady-State Reaction Rate Constant

Stationary equation following from Eq. (18) at $t \to \infty$ and the corresponding boundary conditions for remote reaction between hard spheres $(q = r, dq = 4\pi r^2 dr)$ with total radius R for $u(r) = 0$ and $D(r) = D = \text{const}$ $(\varphi(r) = 1, n^+(r) = n(r))$ has the form

$$-w(r)n(r) + D\frac{1}{r^2}\frac{d}{dr}r^2\frac{d}{dr}n(r) = -w(r)n(r) + D\frac{1}{r}\frac{d^2}{dr^2}rn(r) = 0,$$

$$\frac{d}{dr}n(r)|_{r=R} = 0, \quad \lim_{r \to \infty} n(r) = 1. \quad (41)$$

Steady-state reaction rate constant (40), in view of Eqs. (41), is

$$k = 4\pi \int_R^\infty w(r)n(r)r^2 dr = 4\pi D \lim_{r \to \infty} r^2 \frac{d}{dr}n(r) \equiv 4\pi D R_{\text{eff}}. \quad (42)$$

Here, R_{eff} is the effective radius that defines the reaction cross-section (which is the analog of reaction cross-section in the gas phase) depending on diffusion coefficient. In view of Eq. (41) and definition of radius in Eq. (42), the asymptote of the pair density is

$$n(r \to \infty) \sim 1 - R_{\text{eff}}/_r, \quad (43)$$

i.e., the effective radius is the asymptotic expansion coefficient (at r^{-1}) of the pair density. Calculations by Eqs. (41) and (42) for dipole–dipole (resonant)[16] and exchange[38,43] irreversible luminescence quenching,

$$\text{(a) } w(r) = ar^{-6}(\text{at } R = 0), \quad \text{(b) } w(r) = w_0 \exp((r - R)/L), \quad (44)$$

respectively, give

$$\text{(a)} \quad R_{\text{eff}} = \frac{2\sqrt{2}\pi}{\Gamma^2\left(\frac{1}{4}\right)} \left(\frac{a}{D}\right)^{1/4} \approx 0.676 \left(\frac{a}{D}\right)^{1/4},$$

$$\text{(b)} \quad R_{\text{eff}} = R + L\{\ln\gamma^2\beta + 2\vartheta(\beta,\xi)\}. \tag{45}$$

Here, $\gamma = \exp\mathbf{C}$ ($\mathbf{C} \approx 0.577$ is the Euler constant), and

$$\vartheta(\beta,\xi) = \frac{K_0(2\sqrt{\beta}) - \xi\sqrt{\beta}K_1(2\sqrt{\beta})}{I_0(2\sqrt{\beta}) + \xi\sqrt{\beta}I_1(2\sqrt{\beta})}, \tag{46}$$

where $I_\nu(x)$ and $K_\nu(x)$ are the modified Bessel functions and the MacDonald functions, respectively. Here, the parameter $\xi = R/L$ and the parameter

$$\beta = w_0\tau_c = w_0\frac{L^2}{D}.$$

The value of this parameter separates the cases of diffusion-controlled (strong) and kinetic-controlled (weak) exchange quenching.

Since for resonant quenching $w(r) \to \infty$ at $r \to 0$ the effective radius ($R_{\text{eff}} \approx R_S$) (at $R = 0$) corresponds to diffusion-controlled strong transfer, therefore, it fails to provide the attainment of the kinetic limit $k \approx k_r$ by rate constant (42) (where $R_{\text{eff}} \approx R_S$ is defined by Eq. (45a)) with increasing diffusion coefficient. The effective radius R_S (accurate to the co-factor ~ 1) can be found by the recipe

$$w(R_S)\tau_c \equiv w(R_S)\frac{R_S^2}{16D} = \frac{a}{16R_S^4 D} = 1. \tag{47}$$

The exchange diffusion-controlled strong quenching corresponds to $\beta \gg 1$ and gives[38,43]

$$R_{\text{eff}} \approx R_S = R + L\ln\gamma^2\frac{w_0L^2}{D}. \tag{48}$$

Just as in the case of dipole–dipole transfer, this radius may be found (accurate to co-factor ~ 1) from the equation[43] similar to Eq. (47)

$$w(R_S)\tau_c \equiv w(R_S)\frac{L^2}{D} = w_0\exp\left(-\frac{R_S - R}{L}\right)\frac{L^2}{D} = 1. \tag{49}$$

The condition $\beta \gg 1$ means that the radius R_S of strong exchange quenching zone exceeds the value of the total radius R of hard spheres (the distance of the closest approach), thus it is R_S that is the effective radius of quenching.

Equations (47) and (49) for the determination of effective radii of strong quenching for dipole–dipole and exchange transfer differ in the definition of residence time τ_c. By the first recipe, the extension R_S of the quenching zone itself is included in the definition. By the second recipe, the extension L of the quenching layer (the scale of the decay in the exchange quenching elementary rate $w(r)$) appears in the definition. However, this difference is not crucial from the standpoint of the universality of recipes (47) and (49). It means that in the case of resonance transfer, the extension of the quenching layer is proportional to the value of R_S. However, physically this difference is radical, since, unlike the case of resonance transfer, extension of the quenching layer of the exchange transfer may be essentially less than R and $R_S(L \ll, R_S, 1 \ll \xi)$, which is actually the case. Then the value of R_S may be considered with a good accuracy as a quenching ("black sphere") radius which increases with diffusion deceleration. This implies that as diffusion accelerates, the value of the radius R_S may become less than the total radius R of hard spheres, and remote transfer is to give way to contact reaction. General expression (45b) at $\beta \ll 1$, in view of Eq. (42) for rate constant, yields result (10) of contact spheres[5]

$$k = 4\pi D R_{\text{eff}} = 4\pi D R \frac{w_0 \tau_e}{1 + w_0 \tau_e} \equiv \frac{k_D k_r}{k_r + k_D}, \qquad (50)$$

where

$$\tau_e = \frac{RL}{D} \gg \tau_c, \quad k_r = w_0 \tau_e k_D = 4\pi R^2 L w_0 \equiv w_0 \upsilon, \qquad (51)$$

are the complete residence time in the reaction zone and reaction constant, respectively, $\upsilon = 4\pi R^2 L$ is the reaction zone volume. Note that a small value of the parameter β means that a strong quenching radius R_S is inside a hard sphere of the radius R, and a strong quenching zone is beyond reach. However, despite a small value $\beta = w_0 \tau_c$, in contrast to resonance transfer, at $k_r \gg k_D$ diffusion control of reactions typical of contact reactions is retained. The reason is that the residence time in a contact zone increases to τ_e due to a great number of recontacts. Thus in the case where the elementary event rate sharply decreases with distance, the reactions can manifest themselves both as remote reactions, and as contact ones depending on reactants mobility.[38,43]

5. Contact Reactions

Most chemical reactions are contact, i.e., they proceed on approach of reactants to the distance equal to the sum of van der Waals radii. In this section, we examine the reaction occurring near the immediate contact of reactants in a general way without contact idealization (20). This approach has received the name kinematic approximation.[44] Let us introduce the averaging of any quantity $A(q)$ over reaction zone (38)

$$\langle A(q) \rangle = \int A(q)\psi(q)\varphi(q)dq. \tag{52}$$

Averaging both sides of Eq. (29) over the reaction zone with the use of Eqs. (37) and (38), the Laplace transforms of values in Eqs. (28) and (39), and the decoupling of the averages[33]

$$\left\langle \int \psi(q')g(q',q;s)dq'\tilde{n}^+(q) \right\rangle \approx \left\langle \int \psi(q')g(q',q;s)dq' \right\rangle \langle \tilde{n}^+(q) \rangle, \tag{53}$$

which is similar to the "closure approximation" in the general theory of diffusion-controlled bimolecular reactions[40] we obtain for the Laplace transform of rate constant and steady-state constant, respectively,

$$\frac{1}{s\tilde{k}(s)} = \frac{1}{k_r} + \frac{1}{k_d(s)} \quad \text{or at} \quad s \to 0 \quad \frac{1}{k} = \frac{1}{k_r} + \frac{1}{k_d}, \quad k_d = k_d(0), \tag{54}$$

where

$$\frac{1}{k_d(s)} = \iint \psi(q')g(q',q;s)\psi(q)\varphi(q)dq'dq. \tag{55}$$

Decoupling procedure (53) is exact when contact idealization (20) is used in spherically symmetric problem, and Eqs. (54) and (55) (at $s \to 0$) reproduce results (10) for contact sphere.[5] Expressions (54) for rate constants coincide in form with the corresponding expressions for contact reactions; the only difference is in the value of the constant k_d of the reaction controlled by reactant mobility ($k_r \to \infty$). The quantity k_d does not depend on reactivity and is a pure kinematic characteristic that is a quadrature of free resolvent.

Now, employ the results obtained in the kinematic approximation for the calculation of rate constant of diffusion-controlled reaction in the presence of reactivity anisotropy that plays an important part in chemical reactions, in particular, in biological systems. Rotation of reactants is neglected. A round reaction patch on a spherical particle is the simplest model of such an anisotropy[45] (although other models also exist (see references in Ref. 33)). In this case, the geometric steric factor f (equal to the ratio between the areas of the reaction patch and the particle surface) is introduced. The reaction rate constant k_d can be represented as

$$k_d = k_D f_{\text{eff}}, \tag{56}$$

where f_{eff} is the effective steric factor. For diffusion-controlled reaction between isotropic and strongly anisotropic ($f \ll 1$) particles, the problem admits exact analytical solution[46]

$$f_{\text{eff}} = \frac{\sqrt{f}}{\pi/2 + \sqrt{f}\ln\sqrt{f}} \approx \frac{2}{\pi}\sqrt{f} \approx 0.637\sqrt{f} \quad (\sqrt{f} \ll 1). \tag{57}$$

In the kinematic approximation in the model under study at $f \ll 1$, it is necessary to average stationary resolvent at $r = r' = R$

$$g(R,\varphi,\theta;R,\varphi',\theta') = \frac{1}{k_D}\left\{\frac{\sqrt{2}}{\sqrt{1-\cos\gamma}} - \ln\left(1 + \frac{\sqrt{2}}{\sqrt{1-\cos\gamma}}\right)\right\},$$

$$\cos\gamma = \cos\theta\cos\theta' - \sin\theta\sin\theta'\cos(\varphi - \varphi'), \tag{58}$$

over azimuth angles φ and φ', and polar angles θ and θ' of spherical coordinate system

$$0 \leq \varphi, \varphi' \leq 2\pi, \quad 0 \leq \theta, \theta' \leq \theta_0 \approx 2\sqrt{f} \ll 1. \tag{59}$$

As a result, we arrive at the expression for f_{eff} that differs from Eq. (57) in that the numerical coefficient in the first term in its denominator is 8% greater. So we have[44]

$$f_{\text{eff}} = \frac{\sqrt{f}}{16/3\pi + \sqrt{f}\ln\sqrt{f}} \approx \frac{3\pi}{16}\sqrt{f} \approx 0.589\sqrt{f}. \tag{60}$$

The above expression provides a rather high accuracy in the worst situation of a small steric factor. Note that the effective steric factor is not proportional to the value of geometric steric factor and can essentially exceed it. This is due to the reactivity anisotropy averaging by translational diffusion motion with a great number of recontacts during the encounter; part of them will be favorable to the reaction.[47] Also, in the kinematic approximation one can obtain the result in the case where both reacting partners (A and B) have the reactivity anisotropy defined by geometric steric factors f_A and f_B. In this case[48,49]

$$f_{\text{eff}} = \alpha(\xi)\sqrt{f_A f_B}, \tag{61}$$

where $\alpha(\xi)$ is the numerical coefficient depending on $\xi = \sqrt{f_A/f_B}$ $(0 < \xi \leq 1)$.[48] At $\xi \to 0(f_A \ll f_B)\alpha(0) = 3\pi/16 \approx 0.589$ (as in Eq. (60)), and at $\xi = 1(f_A = f_B)\alpha(1) \approx 0.918$.

When rotation of reactants is taken into account, averaging of reactivity anisotropy occurs. Peculiarities of rate constant transformation under rotation depend on the character of translational and rotational mobilities, but rotation leads, in the end, to complete averaging of reactivity anisotropy ($f_{\text{eff}} = 1$).[48-50]

6.   Multistage Reactions

Though for the above irreversible chemical reactions the kinetic equations based on the concepts of the Coagulation and the Encounter

Theories coincide, an important advantage of the Encounter
Theory is the possibility of describing any bimolecular multistage
reaction the reversible stage of which, in the general form, may be
represented as

$$A_l + A_m \overset{k_{ik,lm}}{\underset{k_{lm,ik}}{\rightleftarrows}} A_i + A_k \quad (i, k, l, m = 1, 2, \ldots, N), \quad (62)$$

where N is the total number of species in solution. Any group of
two reacting particles $A_l + A_m$ is called the $L = (lm)$th reaction
channel. Reversible transition from the Lth reaction channel into
another channel and back (shown in Eq. (62) and occurring in a
given elementary stage) is called forward and reverse transition channels.[32,33] As the reaction rate constants are determined by the effect
of the encounter of two particles, the Encounter Theory easily admits
matrix generalization. It is because residence of reactants of species
A_l in the reacting system is considered to occur in states l of "effective" particle A and concentrations $[A_i]_t$ of species are components
of column-vectors $[\mathbf{A}]_t$ in the basis of individual states of "effective" particle.[33] Then general quasi-stationary vector equation of the
Encounter Theory is[33]

$$\frac{d}{dt}[\mathbf{A}]_t = Tr_{A'} \int \hat{\mathbf{W}}(q)\hat{\mathbf{G}}(q)dq([\mathbf{A}]_t \otimes [\mathbf{A}']_t), \quad (63)$$

where $\hat{\mathbf{W}}(q)$ is the elementary reaction rates Liouvillian, $\otimes$ denotes
the direct product of concentration vectors, $Tr_{A'}$ is the trace (summation) over states of "effective particles" of the partner. The steady-
state Liouvillian of the pair densities is the stationary solution
$\hat{\mathbf{G}}(q) = \lim_{t\to\infty} \hat{\mathbf{G}}(q,t)$ of the matrix equation defining the change in
these densities due to the encounter of reactants

$$\frac{\partial}{\partial t}\hat{\mathbf{G}}(q,t) = \hat{\mathbf{W}}(q)\hat{\mathbf{G}}(q,t) + \hat{\mathbf{L}}_q\hat{\mathbf{G}}(q,t) \quad (64)$$

with the initial conditions $\hat{\mathbf{G}}(q,0) = \hat{\varphi}_{eq}(q)$, where $\hat{\varphi}_{eq}(q)$ is a diagonal matrix with elements equal to distributions $\varphi_L(q)$ of equilibrium
coordinates q in the Lth channel.

Further, we consider the specific case of multistage reaction between non-identical particles the reversible stage of which may be represented as[51]

$$A_l + B_l \underset{k_{lk}}{\overset{k_{kl}}{\rightleftharpoons}} A_k + B_k \quad (l, k = 1, 2, \ldots, N). \tag{65}$$

Here, N is the number of channels; numbers of particles coincide with numbers of channels which are denoted by small Latin letters. Using

$$(\hat{\mathbf{W}})_{kl} = W_{kl}(q) = -\sum_m w_{ml}(q)\delta_{kl} + w_{kl}(q),$$

$$\left(\sum_k (\hat{\mathbf{W}})_{kl} = 0\right), \quad (\hat{\mathbf{L}}_q)_{kl} = \delta_{kl}\hat{\mathcal{L}}_q^{(l)}, \tag{66}$$

where δ_{kl} is the Kronecker delta, $w_{kl}(q)$ are elementary event rates (that determine the transition from the lth channel to the kth one), and $\hat{\mathcal{L}}_q^{(l)}$ is the molecular motion operator in the lth channel, we have from Eqs. (63) and (64)

$$\frac{d[A_k]_t}{dt} = \frac{d[B_k]_t}{dt} = \sum_{i,l} \int W_{ki}(q)G_{il}(q)dq[A_l]_t[B_l]_t$$

$$\equiv -k_{kk}[A_k]_t[B_k]_t + \sum_l k_{kl}[A_l]_t[B_l]_t, \tag{67}$$

$$\frac{\partial}{\partial t}G_{kl}(q,t) = -\sum_i w_{ik}(q)G_{kl}(q,t)$$

$$+ \sum_i w_{ki}(q)G_{il}(q,t) + \hat{\mathcal{L}}_q^{(k)}G_{kl}(q,t). \tag{68}$$

Introducing adjoint quantities $h_{kl}^+(q)$ by the equality $G_{kl}(q) = h_{kl}^+(q)\varphi_k(q)$, we have for rate constants in Eq. (67)

$$k_{kl} = \int \sum_{i \neq k} (w_{ki}(q)h_{il}^+(q)\varphi_i(q) - w_{ik}(q)h_{kl}^+(q)\varphi_k(q))dq \quad (k \neq l),$$

$$k_{kk} = -\int \sum_{i \neq k} (w_{ki}(q)h_{ik}^+(q)\varphi_i(q) - w_{ik}(q)h_{kk}^+(q)\varphi_k(q))dq, \tag{69}$$

and the integral equation for adjoint densities (the analog of Eq. (29) for $sn^+(s)$ at $s \to 0$) is of the form[51]

$$h_{kl}^+(q) = \delta_{kl} - \int g_k(q', q)$$

$$\sum_{i \neq k} \left\{ w_{ik}(q') h_{kl}^+(q') - w_{ki}(q') \frac{h_{il}^+(q') \varphi_i(q')}{\varphi_k(q')} \right\} dq', \quad (70)$$

where δ_{kl} is the Kronecker delta and $g_k(q', q)$ is the stationary free resolvent of the kth channel (see the second expression in Eq. (24) at $s = 0$).

The multistage reaction between two reactants on encounter proceeds at definite spatial positions of the reacting molecules or their molecular groups, i.e., from a definite site. Further we shall take that the multistage reaction under consideration is a multisite reaction, i.e., the transition from any channel, for example, from the kth one, to other channels, for example, to the lth channel, occurs from a definite site the number of which is the number of the channel the transition is accomplished to. This means that the transition from the kth channel to the lth one occurs from the lth site of the kth channel. Let us introduce the reaction zone shape of the lth site of the kth channel normalized to unity with the weight $\varphi_k(q)$ — equilibrium distribution in the kth channel

$$\psi_{lk}(q) \equiv \frac{w_{lk}(q)}{k_{lk}^0} \left(\int \psi_{lk}(q) \varphi_k(q) dq = 1 \right), \quad w_{lk}(q) = k_{lk}^0 \psi_{lk}(q),$$
$$(71)$$

where

$$k_{lk}^0 = \int w_{lk}(q) \varphi_k(q) dq \qquad (72)$$

is the reaction constant of the lth site of the kth channel. The following relations take place which express the kinetic principle of the detailed balancing, i.e., they relate elementary rates of forward and back reactions of any elementary stage of multistage reaction[32,33,51]

$$\psi_{lk}(q)\varphi_k(q) \equiv \frac{w_{lk}(q)\varphi_k(q)}{k_{lk}^0} = \psi_{kl}(q)\varphi_l(q) \equiv \frac{w_{kl}(q)\varphi_l(q)}{k_{kl}^0}. \quad (73)$$

Now, we introduce (by analogy with Eq. (52)) the averaging of some quantity $A_k(q)$ over the reaction zones $\psi_{nk}(q)$

$$\langle A_k(q)\rangle_n = \int A_k(q)\psi_{nk}(q)\varphi_k(q)dq, \langle h_{kl}^+\rangle_n = \int h_{kl}^+(q)\psi_{nk}(q)\varphi_k(q)dq. \quad (74)$$

Averaging both sides of Eq. (70) $(A_k(q) = h_{kl}^+(q))$, in view of Eq. (73), and using (in the kinematic approximation) the decoupling of averages in the right-hand side, just as in Section 5, we obtain the set of equations

$$\langle h_{kl}^+\rangle_n = \delta_{kl} - \sum_{i\neq k} g_{ni}^{(k)}(k_{ik}^0\langle h_{kl}^+\rangle_i - k_{ki}^0\langle h_{il}^+\rangle_k) \quad (n \neq k). \quad (75)$$

Hereinafter, δ_{kl} is the Kronecker delta, and

$$g_{nl}^{(k)} = \iint \psi_{nk}(q)g_k(q,q')\psi_{lk}(q')\varphi_k(q')dqdq' = g_{ln}^{(k)} \quad (76)$$

are stationary free resolvents averaged over the reaction zones (see Eq. (55)), with the upper index corresponding to the number of a channel, and lower indices being the numbers of sites in this channel. Symmetry of this quantity on rearrangement of lower indices follows from the detailed balancing relation (see Eq. (36)). In Eqs. (75) and (76), each of the functions $g_k(q,q')$ describes the free motion in the configuration space of the corresponding pair in the kth channel (without reactions), either translational motion or rotational one.

From Eqs. (69) and (74) we have the expressions for the determination of rate constants of multistage multisite reactions

$$k_{kl} = \sum_{i\neq k} (k_{ki}^0\langle h_{il}^+\rangle_k - k_{ik}^0\langle h_{kl}^+\rangle_i) \quad (k \neq l),$$

$$k_{kk} = -\sum_{i\neq k} (k_{ki}^0\langle h_{ik}^+\rangle_k - k_{ik}^0\langle h_{kk}^+\rangle_i). \quad (77)$$

Let us introduce partial rate constants

$$k_{ki}^{(l)} = k_{ki}^0 \langle h_{il}^+ \rangle_k. \tag{78}$$

Then expressions (77) for rate constants are written as

$$k_{kl} = \sum_{i \neq k} \left(k_{ki}^{(l)} - k_{ik}^{(l)} \right) \quad (k \neq l),$$

$$k_{kk} = -\sum_{i \neq k} \left(k_{ki}^{(k)} - k_{ik}^{(k)} \right) \left(k_{kk} = \sum_{i \neq k} k_{ik} \right), \tag{79}$$

and Eqs. (75) and (78), in the general case (in the presence of transition channels between any stages), give the set of $N(N-1)$ (N — the number of channels) equations for each fixed value l for the determination of $N(N-1)$ constants $k_{nk}^{(l)}$ at fixed value of the index l, i.e., all in all, N sets of equations

$$k_{nk}^{(l)} = k_{nk}^0 \delta_{kl} - k_{nk}^0 \sum_{i \neq k} g_{ni}^{(k)} \left(k_{ik}^{(l)} - k_{ki}^{(l)} \right) \quad (n \neq k). \tag{80}$$

As a particular example, consider consecutive two-stage irreversible reaction[52]

$$A_1 + B_1 \xrightarrow{k_1} A_3 + B_3 \xrightarrow{k_2} A_2 + B_2. \tag{81}$$

In this case, kinetic equation (67) involves five rate constants $k_{11}, k_{33}, k_{31}, k_{23}$, and k_{21}. Rate constants k_{11} and k_{33} correspond to the "escape" from the first and the third channels. Other rate constants correspond to transitions from the first channel to the third one to its first site, from the third channel to the second one from the second site of the third channel, and, finally, from the first channel to the second one, respectively. Note that the transition from the first channel to the second one occurs between channels not connected directly by chemical interaction, and is completely determined by diffusion motion of reactants. According to Eqs. (79),

$$k_{11} = k_{31}^{(1)}, \quad k_{31} = k_{31}^{(1)} - k_{23}^{(1)}, \quad k_{33} = k_{23}^{(3)} - k_{31}^{(3)},$$
$$k_{23} = k_{23}^{(3)}, \quad k_{21} = k_{23}^{(1)}, \tag{82}$$

and Eq. (80), yield two closed sets of equations. From these equations it follows that $k_{31}^{(3)} = 0$, so using Eq. (82), in view of designations

$$g_{22}^{(3)} = g_{22}, \quad g_{12}^{(3)} = g_{21}^{(3)} = g_{12} = g_{21}, \quad k_{31}^0 = k_1, \quad k_{23}^0 = k_2 \tag{83}$$

we get equations for the determination of rate constants

$$\begin{aligned}
k_{11} &= k_1 - k_1 g_{33}^{(1)} k_{11}, & k_{23} &= k_2 - k_2 g_{22} k_{23}, \\
k_{21} &= -k_2(g_{22}k_{21} - g_{21}k_{11}), & & \tag{84} \\
k_{31} &= k_{11} - k_{21}, & k_{33} &= k_{23},
\end{aligned}$$

which give

$$k_{11} = \frac{k_1}{1 + k_1 g_{33}^{(1)}}, \quad k_{33} = k_{23} = \frac{k_2}{1 + k_2 g_{22}},$$

$$k_{31} = \frac{k_1}{1 + k_1 g_{33}^{(1)}} \left(\frac{1 + k_2(g_{22} - g_{21})}{1 + k_2 g_{22}} \right),$$

$$k_{21} = \frac{k_1}{1 + k_1 g_{33}^{(1)}} \left(\frac{k_2 g_{21}}{1 + k_2 g_{22}} \right). \tag{85}$$

Expressions (85) may be represented in a different form by introducing diffusion constants for each independent irreversible transition channel $A_1 + B_1 \xrightarrow{k_1} A_3 + B_3$ and $A_3 + B_3 \xrightarrow{k_2} A_2 + B_2$, as well as efficiencies p_1 and p_2 of these reactions and the reduced efficiency $\bar{p}_2$ of the third channel

$$k_{d1} = \frac{1}{g_{33}^{(1)}}, \quad k_{d2} \equiv \frac{1}{g_{22}}, \quad p_1 = \frac{k_1 g_{33}^{(1)}}{1 + k_1 g_{33}^{(1)}} = \frac{k_1}{k_{d1} + k_1},$$

$$p_2 = \frac{k_2 g_{22}}{1 + k_2 g_{22}} = \frac{k_2}{k_{d2} + k_2}, \quad \bar{p}_2 = \frac{g_{21}}{g_{22}} p_2. \tag{86}$$

Then Eqs. (85) give the expressions corresponding to the specific case of paper[51]

$$k_{11} = k_{d1}p_1, \quad k_{33} = k_{23} = k_{d2}p_2,$$

$$k_{31} = k_{d1}p_1(1 - \bar{p}_2), \quad k_{21} = k_{d1}p_1\bar{p}_2. \tag{87}$$

Reaction rate constants k_{11} and $k_{33} = k_{23}$ correspond to independent irreversible reactions $A_1 + B_1 \xrightarrow{k_1} A_3 + B_3$ and $A_3 + B_3 \xrightarrow{k_2} A_2 + B_2$, respectively. It is seen that, if diffusion motion during the encounter of reactants A and B fails to provide the transition from one site to another, i.e., $g_{21} \to 0$, then the reduced efficiency $\bar{p}_{21} \to 0$. In this situation, the above independent irreversible reactions take place, i.e., from the first channel $A_1 + B_1$, the irreversible transition to the first site of the third channel $A_3 + B_3$ occurs, and from the second site of the third channel, the irreversible transition to the second channel $A_2 + B_2$ takes place, and $k_{11} = k_{31} = k_{d1}p_1$, $k_{33} = k_{23} = k_{d2}p_2$. Of course, in this case, $k_{21} = 0$, i.e., the transition from the first channel to the second one is absent, since these channels are not connected directly by chemical interaction. However, if during the encounter diffusion motion leads to the transition from the first site to the second one, the situation changes qualitatively. During the encounter (residence in the "cage") reactants of the third channel at the first site (arisen from transition from the first channel) reach the second site by diffusion from which further chemical transformation into the second channel occurs. This corresponds to transition channel from the first channel to the second channel during one encounter defined by the rate constant $k_{21} \neq 0$. Note that spherically symmetric approach most often considered in the literature is always a one-site approach, $g_{21} = g_{22}$ and $\bar{p}_2 = p_2$. Diffusion constants $k_{d1} = k_{D1}$ and $k_{d2} = k_{D2}$ for the first and the third channels, respectively, are given by expressions of the form of Eq. (12).[52]

7. Summary

Though differential kinetic equations of one-stage irreversible reactions derived in the framework of the Coagulation Theory and the Encounter Theory coincide, physical notions underlying these theories differ essentially. The Coagulation Theory makes use of the concepts of macroscopic flux proportional to density gradient, while the Encounter Theory treats solutions as "gas" of reactants dissolved in chemically inert continual solvent. It is the analog of semi-classical Collision Theory in gases and is based on the idea that most of the time reactants are in the process of free random walks, and the reaction proceeds on their encounters, the characteristic time of which is considerably less than the mean time between successive encounters. Accordingly, the terminology is different. The notion called non-stationary diffusion time in the Coagulation Theory, in the Encounter Theory is considered to be the characteristic encounter time, and time dependence of reaction rate constant is specified by incompleteness of encounters. An important advantage of the Encounter Theory is the possibility of examining any multistage bimolecular reactions and other quasi-resonant physicochemical processes determined by quantum interactions.[38,39] A significant assumption of the Encounter Theory (and the Collision Theory as well) is independence of pair encounters. However, an encounter differs essentially from collision in that it is stochastic in character, and this gives rise to space-time correlations[32,33] manifesting themselves on long-term tails of kinetic dependencies. Investigation of these correlations is possible only in the context of the non-Markovian theories based on many-particle approaches. Nevertheless, for the present many-particle approaches[32,33] the concept of the encounter is also essential for understanding of physical picture of reactions in solutions.

The use of the kinematic approximation in the calculations of steady-state constants in the Encounter Theory has made it possible to consider more realistic models of reactions taking into account reactivity anisotropy and multisite nature of multistage reactions.

Acknowledgment

The author is grateful to Russian Federal Agency of Scientific Organization (FASO) for financial support (Project No. 44.1.5.).

References

1. V. M. Agranovich and M. D. Galanin, *Electron Excitation Energy Transfer in Condensed Matter*. North-Holland, Amsterdam (1982).
2. D. V. Dodin, A. I. Ivanov, and A. I. Burshtein, Hyperfine interaction mechanism of magnetic field effects in sequential fluorophore and exciplex fluorescence, *J. Chem. Phys.* **138**(12), 124102-1 (2013).
3. N. Agmon and M. Gutman, Bioenergetics: Proton fronts on membranes, *Nat. Chem.* **3**, 840–842 (2011).
4. A. V. Popov, E. Gould, M. A. Salvitti, R. Hernandez, and K. M. Solntsev, Diffusional effects on the reversible excited-state proton transfer. From experiments to Brownian dynamics simulations, *Phys. Chem. Chem. Phys.* **13**(33), 14914–14927 (2011).
5. F. C. Collins and G. E. Kimball, Diffusion-controlled reaction rates, *J. Colloid Interface Sci.* **4**(4), 425–437 (1949).
6. G. Wilemski and M. Fixman, Diffusion-controlled intrachain reactions of polymers. I Theory, *J. Chem. Phys.* **60**(3), 866–877 (1974).
7. M. Doi and S. F. Edwards, *Theory of Polymer Dynamics*. Clarendon Press, Oxford (1986).
8. J. M. Berg, J. L. Tymoczko, and L. Stryer, *Biochemistry*, W. H. Freeman and Company, New-York, 5th edn. (2002).
9. A. Szabo and H.-X. Zhou, Diffusion-controlled intrachain reactions of polymers. I. Theory, *Bull. Korean Chem. Soc.* **33**(3), 925–928 (2012).
10. K. Park, T. Kim, and H. Kim, Facilitated protein-DNA binding: Theory and Monte Carlo Simulation, *Bull. Korean Chem. Soc.* **33**(3), 971–974 (2012).
11. K. Falkowski, W. Stampor, P. Grigiel, and W. Tomaszewicz, Sano–Tachiya–Noolandi–Hong versus Onsager modelling of charge photogeneration in organic solids, *Chem. Phys.* **392**(1–2), 122–129 (2012).
12. M. von Smoluchowski, Versucheiner Mathematischen Theorie der Koagulations Kinetic Kolloider Lousungen, *Z. Phys. Chem.* **92**, 129–168 (1917).
13. N. N. Tunitski and Kh. S. Bagdasarian, On the Resonance Intermolecular Excitation Making Allowance for Diffusion, *Opt. Spectrosc.* **15**(1), 100–106 (1963).
14. S. F. Kilin, M. S. Mikhelashvili, and I. M. Rozman, On the radio luminescence of organic substances, *Opt. Spectrosc.* **16**(4), 663–673 (1964).
15. I. I. Vasil'ev, B. P. Kirsanov, and V. A. Krongaus, Investigation of the energy transfer mechanism in liquid organic scintillators, *Kinet. Kataliz* **5**, 792–801 (1964).
16. M. Yokota and O. Tonimoto, effects of diffusion on energy transfer by resonance, *J. Phys. Soc. Japan* **22**, 779–784 (1967).
17. A. Szabo, Theoretical approaches to reversible diffusion-influenced reactions: Monomer–excimer kinetics, *J. Chem. Phys.* **95**(4), 2481–2490 (1991).
18. I. Z. Steinberg and E. Katchalski, theoretical analysis of the role of diffusion in chemical reactions, fluorescence quenching, and nonradiative energy transfer, *J. Chem. Phys.* **48**(6), 2404–2410 (1968).

19. M. Tachiya, Theory of diffusion-controlled reactions: Formulation of the bulk reaction rate in terms of the pair probability, *Radiat. Phys. Chem.* **21**(1–2), 167–175 (1983).

20. T. R. Waite, Theoretical treatment of the kinetics of diffusion-limited reactions, *Phys. Rev.* **107**, 463–470 (1957).

21. T. R. Waite, General theory of bimolecular reaction rates in solids and liquids, *J. Chem. Phys.* **28**(1), 103–106 (1958).

22. V. N. Kuzovkov and E. A. Kotomin, Generalized theory of diffusion-controlled defect annealing, *J. Phys. C.* **13**, 499–502 (1980).

23. V. N. Kuzovkov and E.A. Kotomin, Kinetics of defect accumulation and recombination. 2. Diffusion-controlled defect annihilation, *Phys. State Sol. B.* **108**, 37–44 (1981).

24. S. Lee and M. Karplus, Kinetics of diffusion-influenced bimolecular reactions in solution. I. General formalism and relaxation kinetics of fast reversible reactions, *J. Chem. Phys.* **86**(4), 1883–1903 (1987).

25. I. V. Gopich and A. B. Doktorov, Kinetics of diffusion-influenced reversible reaction A + B ⇔ C in solutions, *J. Chem. Phys.* **105**(6), 2320–2332 (1996).

26. M. Yang, S. Lee, and K. J. Shin, Kinetic theory of bimolecular reactions in liquid. I. Steady-state fluorescence quenching kinetics, *J. Chem. Phys.* **108**(1), 117–133, (1998).

27. O. A. Igoshin, A. A. Kipriyanov, and A. B. Doktorov, Many-particle treatment of nonuniform reacting systems A + B ⇒ C and A + B ⇒ C + D in liquid solutions, *Chem. Phys.* **244**, 371–385 (1999).

28. J. Sung and S. Lee, Non-equilibrium distribution function formalism for diffusion-influenced bimolecular reactions: Beyond the superposition approximation, *J. Chem. Phys.* **111**(3), 796–803 (1999).

29. A. B. Doktorov, A. A. Kadetov, and A. A. Kipriyanov, General kinetic laws of monomolecular-bimolecular reaction A + B ⇔ C in solutions, *J. Chem. Phys.* **120**(12), 8662–8670 (2004).

30. A. B. Doktorov, A-der. A. Kipriyanov, and A. A. Kipriyanov, Accumulation and decay of macroscopic correlations in elementary reactions kinetics, *Bull. Korean Chem. Sci.* **33**, 941–952 (2012).

31. S. A. Rice, *Diffusion-limited Reaction,* in: *Comprehensive Chemical Kinetics,* Vol. 25, C. H. Bamford, C. F. H. Tripper and R. G. Kompton (eds.), Elsevier, Amsterdam (1985).

32. A. B. Doktorov and A. A. Kipriyanov, Deviation from the kinetic law of mass action for reactions induced by binary encounters in liquid solutions, *J. Phys. Cond. Matt.* **19**, 065136 (2007).

33. A. B. Doktorov, *Development of the kinetic theory of physicochemical processes induced by binary encounters of reactants in solutions,* in: *Recent Research Development in Chemical Physics,* vol. 6, Chapter 6, S. G. Pandalai (ed.), Transworld Research Network, Kerala, India (2012), pp. 135–192.

34. L. Onsager, Initial recombination of ions, *Phys. Rev.* **54**, 554–557 (1938).

35. H. Eyring, S. H. Lin, and S. M. Lin, *Basic Chemical Kinetics,* Wiley, New York (1980).

36. J. Frank and E. Rabinowich, Some remarks about free radicals and the photochemistry of solutions, *Trans. Far. Soc.* **30**, 120–130 (1934).

37. R. M. Noyes, Models Relating Molecular Reactivity and Diffusion in Liquids, *J. Amer. Chem. Soc.* **78**(21), 5486–5490 (1956); in: *Progress of Reaction Kinetics*, G. Porter (ed.), Pergamon Press (1961).

38. A. B. Doktorov and A. I. Burshtein, Quantum theory of diffusion-accelerated remote transfer, *Sov. Fiz. JETP.* **41**, 671–677 (1976).

39. A. B. Doktorov, The impact approximation in the theory of bimolecular quasi-resonant processes, *Physica A.* **90**, 109–136 (1978).

40. G. Wilemsky and M. Fixman, General theory of diffusion-controlled reactions *J. Chem. Phys.* **58**(9), 4009–4019 (1973).

41. A. A. Kipriyanov and A. B. Doktorov, T-matrix representation and long time behavior of observables in the theory of migration-influenced irreversible reactions in liquid solutions, *Physica A* **230**, 75–117 (1996).

42. A. B. Doktorov and A. A. Kipriyanov, General theory of multistage geminate reactions of isolated pairs of reactants. I. Kinetic equations, *J. Chem. Phys.* **140**(18), 184104 (2014).

43. E. A. Kotomin and A. B. Doktorov, Theory of tunneling recombination of defects stimulated by their motion II. Three recombination Mechanisms, *Phys. Stat. Sol. B.* **114**, 287–318 (1982).

44. V. M. Berdnikov and A. B. Doktorov, Steric factor in diffusion-controlled chemical reactions, *Chem. Phys.* **69**, 205–212 (1982).

45. K. Solc and W. H. Stokmayer, Kinetics of diffusion-controlled reaction between chemically asymmetric molecules. I. general theory, *J. Chem. Phys.* **54**(7), 2981–2988 (1971).

46. A. B. Doktorov and N. N. Lukzen, Diffusion-controlled reactions on an active site, *Chem. Phys. Lett.* **79**, 498–502 (1981).

47. A. B. Doktorov and B. I. Yakobson, Averaging of the reactivity anisotropy by the reactant translation motion, *Chem. Phys.* **60**, 223–230 (1981).

48. A. B. Doktorov, Averaging of the reactivity anisotropy by rotation of reactants, *Sov. Chem. Phys.* **4**, 800–808 (1985).

49. S. I. Temkin and B. I. Yakobson, Diffusion-controlled reactions of chemically anisotropic molecules, *J. Phys. Chem.* **88**, 2679–2682 (1984).

50. A. I. Burshtein, A. B. Doktorov, and V. A. Morozov, Contact reactions of randomly walking particles. Rotational averaging of chemical anisotropy, *Chem. Phys.* **104**, 1–18 (1986).

51. A. B. Doktorov, The influence of the "cage" effect on the mechanism of reversible bimolecular multistage chemical reactions proceeding from different sites in solutions, *J. Chem. Phys.* **145**(8), 084114 (2016).

52. A. B. Doktorov and S. G. Fedorenko, *The influence of the cage effect on the mechanism of multistage chemical reactions in solutions*, in: *Chemistry for Sustainable Development*, Chapter 2, M. G. Bhowon, S. Jhaumeer-Laulloo, H. Li Kam Wah, and P. Ramasami (eds.), Springer, Dordrecht, Heidelberg, London, New-York (2011), pp. 11–34.

Chapter 3

Non-Markovian Kinetics
of Reactions in Solutions

Konstantin L. Ivanov[*,§], Nikita N. Lukzen[†],
and Alexander B. Doktorov[‡]

*International Tomography Center, Siberian Branch of the Russian
Academy of Sciences, Novosibirsk, 630090, Russia
†Novosibirsk State University, Novosibirsk, 630090, Russia
‡V. V. Voevodsky Institute of Chemical Kinetics and Combustion,
Siberian Branch of the Russian Academy of Sciences, Novosibirsk,
630090, Russia
§ivanov@tomo.nsc.ru

In this contribution, we give a brief overview of theoretical
approaches to diffusion-influenced reactions, focusing on non-stationary
(non-Markovian) effects in chemical kinetics. We discuss the origin of
such effects and their manifestations. We also introduce the idea of
"encounter theory", an approach that treats the reacting system as a
"gas" of reactants diluted in a chemically inert solvent, and discuss
the variants of this theory — known as integral encounter theory and
modified encounter theory.

1. Introduction

Fast reactions in solutions play an important role in many areas of
physical chemistry. Such reactions are, for instance, energy transfer,[1–3] electron transfer,[4,5] proton transfer,[6,7] and proton-coupled

electron transfer[8] giving rise to recombination of free radicals, luminescence quenching,[9] etc. Fast processes are frequently encountered in chemistry,[6,8,10–14] biochemistry,[15–20] polymer systems,[21–25] photovoltaics,[26,27] etc. In many cases, such reactions are affected not only by chemical reactivity itself, but also by mobility of the reacting particles. In the case of diffusional motion, such reactions are therefore termed diffusion-influenced or diffusion-controlled reactions. Due to the fact that such reactions are relevant in so many branches of science, development of a consistent theory of diffusion-influenced processes is of great importance for understanding the behavior of various reacting systems.

Theoretical treatment of diffusion-influenced reactions has drawn much attention from researchers over many years; there are a vast number of original research papers and reviews dedicated to this subject.[28–32] Early works in this field of research were dealing with irreversible reactions, using concepts from coagulation theory,[33,34] which were developed further in later works.[35,36] At the same time, generalization of the theory to complex reversible reactions and multistage processes, in particular, to processes involving particles with internal quasi-resonant quantum states, remained challenging for a long time. The approach which is free from this short-coming is the encounter theory,[37–39] which treats reactions in solutions using the concept of a "gas" of reactants, whereas the solvent is considered as an inert continuous medium. For this reason, the encounter theory is analogous to the semi-classical collision theory in gases.[39] At the same time, one should note that the chemically "inert" medium has a strong effect on chemical reactions in liquids, despite the fact that the solvent molecules do not react with any solute molecules. This effect has several important manifestations. First, the medium influences the rates of chemical events, due to fast relaxation over internal degrees of freedom of reactants conditioned by interactions with the solvent molecules. Second, interaction of the reacting chemical system with the medium can modify the potential energy surface, as is the case, for instance, when solvation takes place. Third, in some chemical reactions motion of solvent molecules plays an important role, e.g., in electron transfer reactions where the reaction coordinate is often the "solvent coordinate",[4,5] which describes the positions and orientations of the solvent molecules around the electron donor and acceptor. Fourth, the medium strongly affects relative motion of reactants by turning it from deterministic dynamic motion into stochastic random walks. For this reason, in solutions particles move via random walks instead of the free passes seen in gases. Consequently, the concept of a "collision" in the gas phase needs to be modified due to the presence of the *"cage" effect.*[40]

The essence of the cage effect is as follows. In liquids, the reactants do not separate after a collision: solvent molecules keep them close together preventing from immediate separation, see Fig. 1. In fact, the distribution of collisions of a pair of particles is non-uniform in time; due to the cage effect, the collisions occur in series.[28] As a consequence, when the first collision does not lead to a chemical reaction, the two particles have a finite probability to collide again and react. Hence, instead of single *collisions* in the gas phase, we need to consider *encounters* of two diffusing particles, which comprise multiple collisions (repeated contacts) between them. Multiple collisions significantly increase the encounter duration, which becomes much

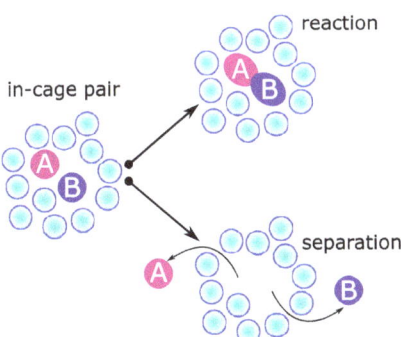

Fig. 1.   Essense of the cage effect. A pair of reactants, A and B, trapped in the solvent "cage" (the solvent molecules are symbolically shown by balls of light color) cannot separate immediately. As a consequence, there is a probability that A and B react to form a product AB. This process competes with diffusional separation.

longer than the duration of a single collision, and affect the reaction rate coefficients $\mathcal{K}(t)$, rendering them time-dependent. Thus, before the rate coefficients reach their steady-state values, which correspond to the kinetic law of mass action according to basic chemical kinetics,[41] there is a so-called *transient stage* where $\mathcal{K}$ strongly depends on time. Theories which do not treat such a time dependence are termed Markovian, while those that consider the $\mathcal{K}(t)$ dependence are essentially non-Markovian theories.

The encounter theory, which is analogous to the semi-classical collision theory in gases, makes use of the concept of independent pair encounters of reactants. One should note, however, that this approximation does not hold at long times, because the distribution of the encounter durations is non-Poissonian, having a long-time power-law "tail".[31] As a consequence, the mean encounter duration diverges, i.e., the encounter duration is literally infinite and only the characteristic encounter duration has a physical meaning.[31] This effect gives rise to long-range space-time correlations, which become manifest at long times. In addition, at long times the so-called fluctuation kinetics reveals itself,[42–45] which is due to fluctuations of concentrations when most particles have already reacted. Due to such complex effects, developing a consistent non-Markovian theory is possible only by using many-particle consideration of an ensemble of reactants in a chemically inert medium.

There are a number of many-particle approaches to reaction kinetics described in the literature. Some of them are based on *exactly solvable models*. A widely used model of such kind considers an irreversible reaction (or luminescence quenching) of immobile A particles (donors) with point-like B particles (acceptors), which are present in great excess. Since the distance between A and any of B particles changes independently, one can perform averaging of the kinetics over the ensemble of B particles. By using this model, static luminescence quenching (quenching by immobile Bs) has been considered,[2] as well as irreversible quenching by diffusing B particles (known as target model, see Fig. 2, top).[46,47] The most general exactly solvable model has been considered by Kipriyanov *et al.*:[48] in this

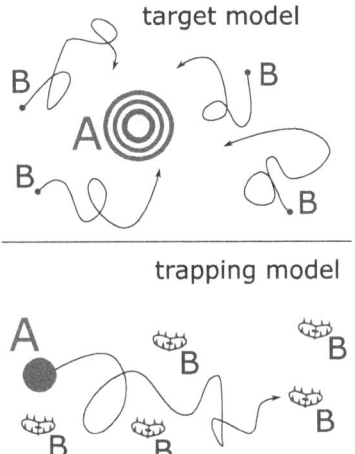

Fig. 2. Pictorial description of the target and trapping models. (top) Target model: an immobile particle, A, symbolically shown as a target, is reacting with point-like independently diffusing B particles. (bottom) Trapping model: A particle diffuses among immobile B particles, symbolically shown as traps.

model also, the hopping model of reactant motion has been treated. One should note that when A particle diffuses and B particles are immobile (trapping problem, see Fig. 2, bottom) there is no exact solution for the kinetics. A further disadvantage of this exactly solvable model is that it cannot be extended to reversible reactions.

Besides using exactly solvable models, one can exploit *many-particle approaches* to the derivation of non-Markovian kinetic equations. Such approaches exist in two variants. One of them is based on introducing the irreducible evolution operator (mass operator),[49–57] which can be calculated by diagram summation[54] or by expansion into a series in concentrations.[52,53] This method becomes relatively complex, when applied to multistage chemical processes. However, it is very useful for testing simpler methods of deriving non-Markovian kinetic equations. The other possibility is construction and closure (truncation) of hierarchies for reduced distribution functions (RDFs). The simplest truncation method is provided by the superposition decoupling,[58,59] although total superposition decoupling was also used.[60] The simplest decoupling method applied to irreversible reactions in spatially uniform reacting systems yields the differential equations of the coagulation theory and the encounter theory. However, for reversible reactions the superposition decoupling yields kinetic equations,[61] which do not reproduce those obtained by numerical modeling.[53,62] For this reason, alternative approaches have been proposed, such as the system size expansion,[63] the theory based on non-equilibrium statistical thermodynamics,[64] the kinetic theory of reactions in liquids,[53] the integral encounter theory (IET),[65] and the modified encounter theory (MET).[66,67] Originally, all these approaches have been applied to elementary reactions. However, later on substantial progress has been achieved by generalizing methods of non-equilibrium statistical mechanics to chemical systems, giving rise to general and consistent approaches to truncating hierarchies for RDFs.[68] These methods allow one to generalize the IET and MET approaches even to complex multistage processes.[69–71]

Here, we focus on the non-Markovian kinetics of diffusion-influenced processes. In Section 2, we compare various theoretical approaches to simple irreversible reactions and discuss effects of diffusion on chemical kinetics. When introducing different approaches, we explain the essense of the *binary* approximation and binary kinetics, which depends only on relative mobility of reactants. This will bring us to the concept of the encounter theory: Here, we present the IET and MET and discuss their accuracy. In Section 3, we briefly introduce many-particle approaches to the description of reacting systems. Such approaches lead to hierarchies for RDFs and correlation patterns, which can be truncated to derive kinetics equations. In this contribution, we discuss two truncation methods which allow one to obtain the IET and MET equations.

As pedagogical values of this chapter, we would like to highlight the following issues:

— we discuss the origin of non-Markovian effects in chemical kinetics;
— we describe in detail the concept of binary kinetics;
— we provide the concept of encounter theory and a detailed description of different formulations of this theory;
— we describe general many-particle approaches to derivation of kinetic equations.

2. Kinetics of Diffusion-Influenced Reactions

2.1. *Differential kinetics equations*

Let us consider the simplest reaction

$$A + B \rightarrow \text{Product} \tag{1}$$

by using various kinetic approaches assuming that the reactants, A and B, are hard spheres of radii R_A and R_B, respectively. For the sake of simplicity, we treat only pseudo-monomolecular reactions, that is, the concentration c_B of B reactant is time-independent. This is the case either when $c_A \ll c_B$, so that one can neglect reduction of the c_B concentration in the chemical process. Alternatively, one can consider the $A + B \rightarrow C + B$ reaction, in which A particles are consumed whereas the number of B particles remains unchanged. Several approaches, namely, the coagulation theory[33,34] and its modification,[36] encounter theory,[37] exactly solvable "target" model[46,47] and superposition approximation,[59] yield the same differential kinetic equation for the concentration c_A of A particles

$$\frac{d}{dt}c_A(t) = -\mathcal{K}(t)c_B c_A(t), \quad \frac{c_A(t)}{c_A(0)} = \exp\left\{-c_B \int_0^t \mathcal{K}(\tau)d\tau\right\}. \tag{2}$$

The time-dependent (non-Markovian) rate "constant" is as follows:

$$\mathcal{K}(t) = 4\pi \int_R^\infty w(r)n(r,t)r^2 dr. \tag{3}$$

Here, for the sake of simplicity and clarity, we always consider a spherically symmetric diffusional problem with $R = R_A + R_B$ being the closest approach distance and $w(r)$ being the rate of the elementary chemical event (sink term). We also introduce the pair density, which satisfies the following equation:

$$\frac{\partial}{\partial t}n(r,t) = -w(r)n(r,t) + \hat{\mathcal{L}}_r n(r,t). \tag{4}$$

For solving this equation, one should use the following initial and boundary conditions, respectively:

$$n(r,0) = 1, \quad \frac{\partial}{\partial r}n(r,t)|_{r=R} = 0. \tag{5}$$

These conditions stand for the uniform spatial distribution at $t = 0$ and for the absence of particle flux at the closest approach, $r = R$. The $\hat{\mathcal{L}}_r$ operator describes relative motion of particles; in the case of free diffusion $\hat{\mathcal{L}}_r = D\Delta_r$ with $D = D_A + D_B$ being the relative diffusion coefficient and Δ_r being the Laplace operator.

The coagulation theory has been developed[34] for contact reactivity

$$w(r) = k_r \frac{\delta(r - R)}{4\pi R^2},\tag{6}$$

where δ is the Dirac delta-function and

$$k_r = 4\pi \int_R^\infty w(r)r^2 dr\tag{7}$$

is the reaction constant (intrinsic constant). It is worth noting that in contact approximation one can set $w(r) = 0$ in Eq. (4) replacing the reflecting boundary condition (5) by an absorbing boundary condition

$$4\pi DR^2 \frac{\partial}{\partial r}n(r,t)|_{r=R} = k_r n(R,t),\tag{8}$$

which defines the reaction flux at the contact distance. The recipe given by Eq. (3) using Eq. (6) yields the following time-dependent (non-Markovian) rate "constant", see Fig. 3:

$$\mathcal{K}(t) = k_r n(R,t) = k\{1 + \alpha \exp(u^2 t)\mathrm{erfc}(u\sqrt{t})\},\tag{9}$$

where $\alpha = k_r/k_D$ (with $k_D = 4\pi RD$ being the diffusional rate constant), $u = (1 + \alpha)/\sqrt{\tau_D}$, $\tau_D = R^2/D$ (characteristic duration of diffusional encounter), erfc is the complimentary error function and k is the steady-state reaction rate constant

$$k = 4\pi DR_{\mathrm{eff}} = \lim_{t\to\infty} \mathcal{K}(t) = \frac{k_r k_D}{k_r + k_D}.\tag{10}$$

Hence, the reaction is clearly diffusion-influenced, as follows from the k value: at $k_D \gg k_r$ we obtain $k \approx k_r$, i.e., once diffusional transport of particles to the reaction zone is very fast, the reaction rate is

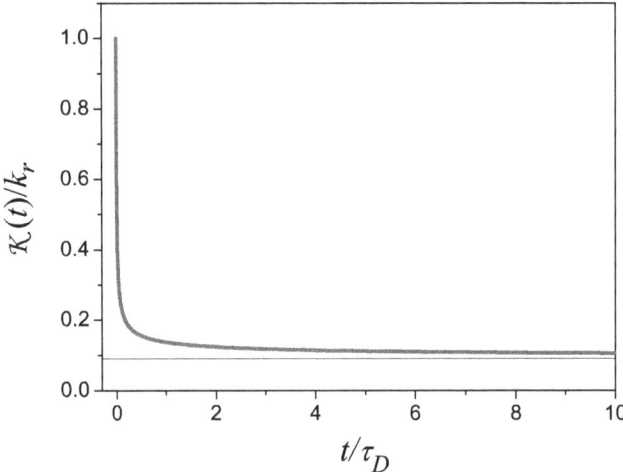

Fig. 3. Time dependence of the $\mathcal{K}(t)$ rate coefficient. In calculations, we used parameters $\alpha = k_r/k_D = 10$; time is given in units of τ_D, the rate constant is given in units of k_r. The horizontal line gives the value of the steady-state rate constant, here k/k_r.

controlled by the chemical reactivity at contact. In the opposite case $k_D \ll k_r$, every contact of reactants leads to the reaction and the reaction rate is determined solely by the efficiency of reactant transport to the reaction zone, i.e., $k \approx k_D$. As far as the time dependence of $\mathcal{K}(t)$ is concerned, at $t = 0$ it is equal to k_r because the reaction takes place from the uniform particle distribution. As the reaction occurs, $n(R, t)$ decreases and $\mathcal{K}(t)$ also decreases reaching the steady-state value at $t \to \infty$. Since the rate coefficient changes in the range given by the limiting values k_r and $k \leq k_r$, the time dependence of $\mathcal{K}(t)$ is essential only in the case $k_r \gtrsim k_D$, i.e., for diffusion-controlled processes. Indeed, once $k_r \ll k_D$, effects of diffusion are of no importance as $n(r, t) \approx 1$ at all distances and at all times. Hence, the time dependence of $\mathcal{K}(t)$ as presented in Fig. 3 is due to the evolution of the pair density $n(r, t)$. The distribution function calculated at different instants of time is shown in Fig. 4.

When $k_r \to \infty$, the problem is reduced to Smoluchowski's problem,[33] which imposes the boundary condition $n(R, t) = 0$ instead of

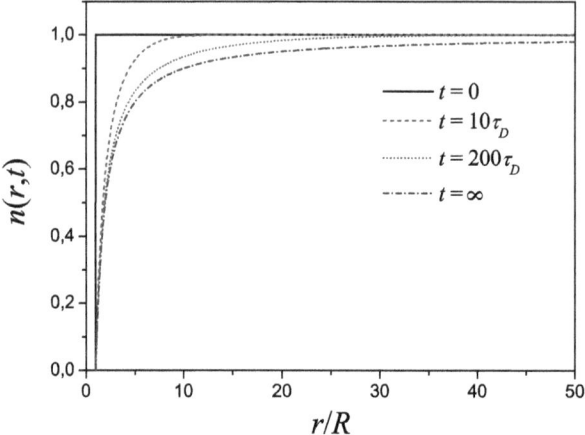

Fig. 4. Pair density distribution $n(r,t)$ at different instants of time: $t = 0$ (solid line), $t = 10\tau_D$ (dashed line), $t = 200\tau_D$ (dotted line), $t = \infty$ (dashed-dotted line). At $t = 0$, the distribution is uniform, $n = 1$, as the time goes by it is depleted near the contact radius $r = R$. The distribution becomes time-independent at long times, as the reaction sink is balanced by the diffusion flux to the reaction zone. Here, the distance is given in units of R; the reactivity at contact is taken as infinite.

Eq. (5). In this situation, from Eqs. (9) and (10) we obtain

$$\mathcal{K}(t) = k_D \left(1 + \frac{R}{\sqrt{\pi Dt}} \right) \tag{11}$$

and the reaction kinetics is given by the expression

$$\frac{c_A(t)}{c_A(0)} = \exp \left\{ -c_B k_D \left(t + 2R\sqrt{\frac{t}{\pi D}} \right) \right\}$$

$$= \exp \left\{ -2\sqrt{\frac{3\eta}{\pi} c_B k_D t} \right\} \exp\{-c_B k_D t\}. \tag{12}$$

In Eq. (12), we introduce the density parameter

$$\eta = \frac{4\pi}{3} R^3 c_B = \frac{4\pi}{3} R_{\text{eff}}^3 c_B \ll 1. \tag{13}$$

This parameter needs to be small to apply the encounter theory, which treats only *pair encounters* of reactants in solution, neglecting encounters of three or more particles. Hence, the encounter theory is essentially a *binary* theory, requiring that the encounter duration, τ_c, is much smaller than the mean time between subsequent encounters τ_f

$$\tau_c \approx \frac{R^2}{3D} = \frac{1}{3}\tau_D \ll \tau_f = \frac{1}{4\pi D R c_B}. \tag{14}$$

Here, the time between encounters is evaluated as the inverse rate of the reaction on a "black" sphere. Expression (14) is mathematically equivalent to Eq. (13). The dependence of the kinetics solely on the *relative* diffusion coefficient and inter-particle distance in $w(r)$ (i.e., on characteristics of a pair of reactants) is also an indication that the theory is a binary theory.

According to Eq. (12), the reaction kinetics at $t \leq \tau_c$, corresponding to the so-called transient stage, is essentially *non-Markovian*: the kinetics strongly deviates from the exponential kinetics $\exp\{-kc_Bt\}$, predicted by the basic chemical kinetics (kinetic law of mass action).[41] The depth of chemical transformation at $t \leq \tau_c$ is negligible when $\eta \ll 1$. However, one should note that even when $t \gg \tau_c$, i.e., when the "rate constant" $\mathcal{K}(t)$ reaches its steady-state value k, the kinetics (12) is not purely exponential. Indeed, the exponentially decreasing Markovian kinetics is not the true asymptote of the kinetics, see Fig. 5. This effect holds not only for the diffusion-controlled reaction considered here. As follows from Eq. (9), a more general asymptotic law is valid

$$\mathcal{K}(t) \xrightarrow{t\to\infty} k\left(1 + \frac{R_{\text{eff}}}{\sqrt{\pi D t}}\right) = 4\pi D R_{\text{eff}}\left(1 + \frac{R_{\text{eff}}}{\sqrt{\pi D t}}\right). \tag{15}$$

Interestingly, this law is universal,[72] since it holds for an arbitrary irreversible reaction (in this case, $R_{\text{eff}} = k/4\pi D$ is a general definition, which does not depend on whether reaction takes place only at the closest approach). As a consequence, we conclude that the long-time asymptote of *any* irreversible reaction is described by a universal long-term asymptote, which obviously deviates from the

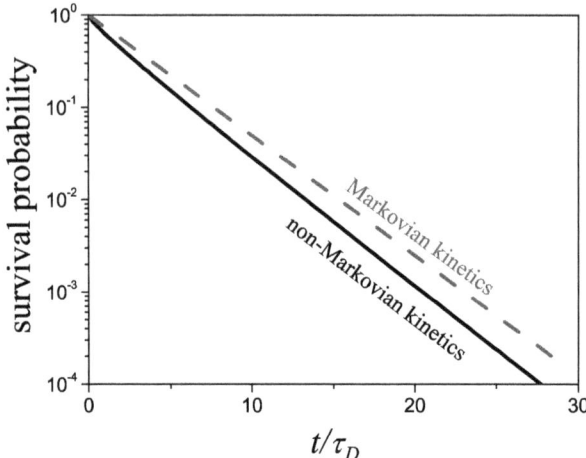

Fig. 5.   Kinetics of the $A + B \rightarrow$ Product reaction in the case of great excess of B particles (pseudo-monomolecular reaction); here we show the survival probability $c_A(t)/c_A(0)$, time is given in units of τ_D. In calculations, we used the following parameters: $k_r \gg k_D$, $c_B k_D \tau_D = 3\eta = 0.3$. Solid line shows the non-Markovian kinetics taking into account the transient stage with the rate constant taken from Eq. (11); dashed line shows the Markovian kinetics, assuming that the rate constant is time-independent and equal to k_D.

kinetic law of mass action

$$\frac{c_A(t)}{c_A(0)} \xrightarrow{t \to \infty} \exp\left\{ -2\sqrt{\frac{3\eta}{\pi}} c_B k t \right\} \exp\{-c_B k t\}. \qquad (16)$$

Hence, the non-Markovian contribution to the kinetics is of importance not only within the transient stage, but at all times as shown in Fig. 5. In this context, one should also keep in mind that the long-time asymptote given by Eq. (16) is only an intermediate asymptote but not the fluctuation asymptote, which cannot be described by any binary theory. The situation is different when the "target model"[46,47] (immobile A particles and diffusing point-like B particles) is considered. In this case, the concentration of the point particles can be arbitrary and the density parameter η is not necessarily small. In this case, the universal kinetic law (16) also coincides with the fluctuation asymptote.

A particularly interesting special case is given by static reactions between particles.[2] Of course, such reactions can only occur for non-contact reaction rates $w(r)$. In this situation, Eq. (3) for $\mathcal{K}(t)$ should be used but $n(r,t)$ should be calculated from Eq. (4) assuming $\hat{\mathcal{L}}_r = 0$

$$\frac{\partial}{\partial t} n(r,t) = -w(r)n(r,t) \Rightarrow n(r,t) = \exp[-w(r)t]. \qquad (17)$$

Using this expression, we can obtain the following result for the rate coefficient:

$$\mathcal{K}(t) = 4\pi \int_R^\infty w(r) \exp[-w(r)t] r^2 dr, \qquad (18)$$

which does not provide convergence of $\mathcal{K}(t)$ to a constant value at long times. For a given $w(r)$ rate, one should calculate the integral in Eq. (18) to evaluate the rate coefficient. However, one can also use an approximate solution, which is based on intuitive considerations. In fact, the concentration of A particles follows the law[73]

$$\frac{c_A(t)}{c_A(0)} = \exp\{-c_B V(t)\}, \quad V(t) = \frac{4\pi}{3} R_S^3. \qquad (19)$$

Hence, the ratio $c_A(t)/c_A(0)$, that is, the survival probability of A particles, is equal to the probability that there are no B particles in the volume $V(t)$ of the sphere of the radius $R_S(t)$. Hence, $R_S(t)$ can be interpreted as the distance where the reaction has occurred within the time interval from zero to t.[73] The R_S value can be estimated as follows:

$$w(R_s(t))t \approx 1. \qquad (20)$$

Rigorous calculation gives essentially the same result (which differs only by a coefficient of the order of unity). For instance, for dipolar and exchange quenching, the R_S radius is as follows:[73]

$$R_S(t) = (\pi a t)^{1/6}, \quad \text{for} \quad w(r) = \frac{a}{r^6} \quad \text{(dipolar quenching)},$$

$$R_S(t) = L \ln[w_0 t], \quad \text{for} \quad w(r) = w_0 \exp\left(-\frac{r}{L}\right)$$

$$\text{(exchange quenching)}. \qquad (21)$$

In these cases, $\mathcal{K}(t)$ behaves as follows[73]:

$$\mathcal{K}(t) = \frac{2\pi}{3}\sqrt{\frac{\pi a}{t}} \quad \text{(dipolar quenching)},$$

$$\mathcal{K}(t) = 4\pi L^3 \frac{\ln^2[w_0 t]}{t} \quad \text{(exchange quenching).} \tag{22}$$

The second expression is valid when $w_0 t \gg 1$. Hence, one can see that the time dependence of the rate coefficient is non-stationary at all times; likewise, the distribution $n(r,t)$ does not reach a stationary value at long times but gradually decays to zero, see Fig. 6. When diffusional motion comes into play, the kinetic behavior becomes even more complex: At short times static quenching dominates, whereas at long times the diffusional flux stabilizes the reaction sink and $\mathcal{K}(t)$ goes to its steady-state value k. A comparison of the rate coefficients calculated for static and diffusion-influenced exchange quenching is presented in Fig. 6.

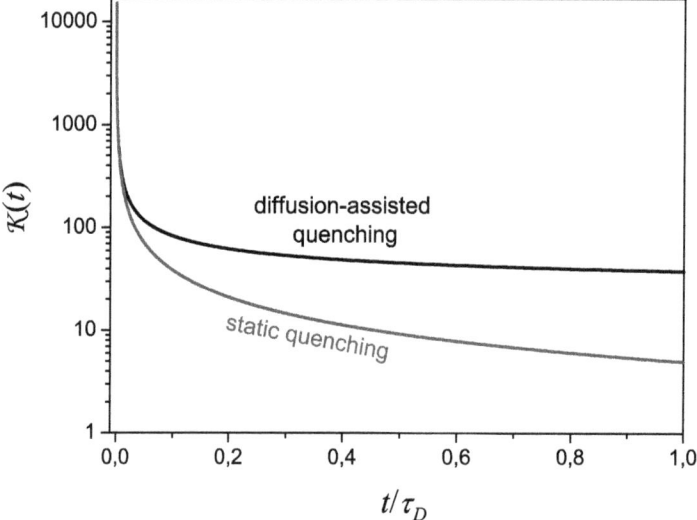

Fig. 6. Rate coefficient calculated for static quenching and diffusion-influenced quenching. Here, exchange quenching is considered. Calculation parameters: $R = 1$, $L = 1$, $w_0 = 10^4$ (in units of τ_D^{-1}), time is given in units of τ_D.

2.2. *Integro-differential kinetic equations*

Many-particle methods to the derivation of binary kinetic equations yield integro-differential (retarded) kinetic equations, in full agreement with the basic principles of non-equilibrium statistical mechanics

$$\frac{d}{dt}c_A(t) = -c_B \int_{-0}^{t} \Sigma(t-\tau)c_A(\tau)d\tau = -c_B \int_{-0}^{t} \Sigma(\tau)c_A(t-\tau)d\tau.$$
(23)

The precise expression for the integral kernel $\Sigma(t)$ (memory function) depends on the method of truncating the hierarchies for RDFs or correlation patterns. Here, we focus on the cases of IET and MET. The IET equations can be derived by keeping only two-particle correlations. The IET kernel can be evaluated as follows[68]:

$$\Sigma(\tau) = \Sigma_{\text{iet}}(\tau) = \int d\mathbf{r} \left\{ w(r)\delta(t) - w(r) \int d\mathbf{r}_0 g(r|r_0, \tau)w(r_0) \right\},$$
(24)

where $g(r|r_0, \tau)$ is the propagator (Green function) obeying the following equation:

$$\left(\frac{\partial}{\partial \tau} - \hat{\mathcal{L}}_r + w(r) \right) g(r|r_0, \tau) = \frac{\delta(r-r_0)}{4\pi r r_0}\delta(\tau).$$
(25)

The integro-differential equation (23) can be reduced to a more customary differential equation with a non-Markovian or Markovian (steady-state) rate constant, defined as follows:

$$\mathcal{K}(t) = \int_0^t \Sigma_{\text{iet}}(\tau)d\tau, \quad k = \mathcal{K}(\infty) = \int_0^{\infty} \Sigma_{\text{iet}}(\tau)d\tau.$$
(26)

Such a reduction of the IET equation to its differential form can be obtained by taking into account that the characteristic decay time of the kernel is given by τ_c and by assuming (as it is done in the collision theory in gases) that $c_A(t-\tau) \approx c_A(t)$. One should note, however, that such an approximation is justified only within a limited time interval[74]

$$t \leq \frac{\ln[1/\eta]}{kc_B}. \tag{27}$$

The reason is that the IET *does not* keep all contributions from binary encounters. Indeed, because of the repeated contacts of reactants during encounters, the kernel Σ_{iet} is not an exponentially decaying function of time having a long-time power-law tail (in contrast to gas phase reactions with rapidly decaying rate kernels). For this reason, the procedure (26) is not valid at long times.

In order to obtain the kernel, which takes into account all contributions from binary encounters, one needs to consider three-particle correlations and extract essentially binary channels in the three-particle problem.[68] In this situation, the kinetic equation also takes the form given by Eq. (23) but with the MET kernel,

$$\Sigma(\tau) = \Sigma_{\mathrm{met}}(\tau) = \Sigma_{\mathrm{iet}}(\tau) \exp(-c_B k\tau). \tag{28}$$

Here, the steady-state rate constant is defined through the IET kernel, see Eq. (26). Thus, for calculating the MET kernel one needs to know the IET kernel. Reduction of the MET equations to the differential form can be done using the "time-shift" rule[31,32]

$$c_A(t - \tau) \approx c_A(t) \exp(c_B k\tau) \tag{29}$$

in the right-hand side of Eq. (23). This recipe immediately yields the differential kinetic equation (2) with the rate constant $\mathcal{K}(t)$ given by Eq. (26). We would like to stress that the kinetics given by the original integro-differential MET equation and differential equation obtained using the procedure (29) coincide within the entire binary interval[74]

$$t \leq t_b = \frac{1}{\eta k c_B}, \quad \eta \ll 1. \tag{30}$$

It is worth noting that the integro-differential form of the equations is convenient for evaluating the integrals of the $c_A(t)$ kinetics, e.g., quantum yields as demonstrated below, because the integral term is a convolution of the kernel with the kinetics. In this case, solution using the Laplace transform becomes possible. By using such solutions, one can also estimate the deviation of the c_A kinetics calculated by IET and MET. For the Laplace transform of the kinetics (hereafter, Laplace transforms are denoted by tilde with s being the Laplace variable) using Eq. (23) we obtain

$$\tilde{c}_A(s) = \frac{c_A(0)}{s + c_B \tilde{\Sigma}(s)}. \tag{31}$$

As usual, a Laplace-transformed quantity, $\tilde{f}(s)$, and its original, $f(t)$, are related as follows:

$$\tilde{f}(s) = \int_0^\infty f(t)e^{-ts}dt. \tag{32}$$

Assuming contact reactivity, the IET kernel can be calculated as follows:

$$\tilde{\Sigma}_{\text{iet}}(s) = \frac{k_r k_D}{k_D + k_r \mathcal{G}(s)}, \quad \mathcal{G}(s) = \frac{1}{1 + \sqrt{s\tau_D}}. \tag{33}$$

According to Eq. (28), for the MET kernel we obtain

$$\tilde{\Sigma}_{\text{met}}(s) = \tilde{\Sigma}_{\text{iet}}(s + kc_B). \tag{34}$$

The $c_A(t)$ kinetics can be calculated numerically using Eq. (31) and performing the inverse Laplace transformation. A more detailed explanation on how to evaluate IET and MET kernels is given in Section 3.

Both IET and MET correctly reproduce the behavior of the kinetic at short times, where the rate coefficient $\mathcal{K}(t)$ in the differential theories considerably changes with time. As the time increases, the IET deviates from differential theories as it does not take into account all binary contributions to chemical kinetics. The MET result is consistent with that of the encounter theory within a much wider time window: The times $t \geq t_b$ where these two approaches give different results are essentially beyond the range of binary kinetics. This comparison clearly shows that both MET and IET describe the peculiarities of non-Markovian effects associated with diffusional

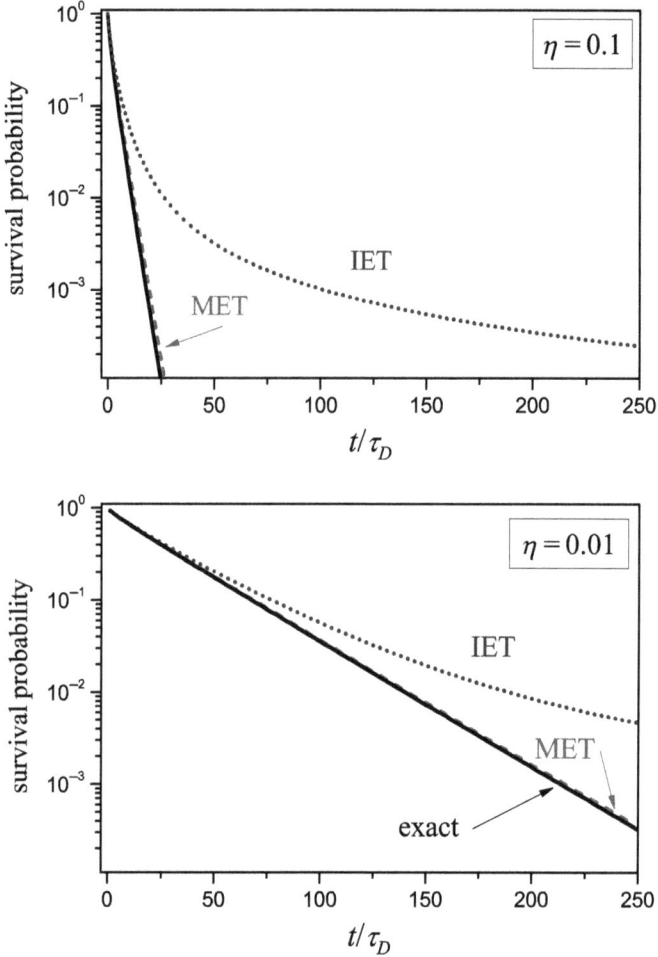

Fig. 7. Comparison of the kinetics of the pseudo-unimolecular reaction $A + B \rightarrow C + B$ obtained from the exact solution (solid line), IET (dotted line) and MET (dashed line). Here, contact reaction on a black sphere is considered; the kinetics is expressed as the survival probability of A particles equal to $c_A(t)/c_A(0)$, time is expressed in units of τ_D; the binary parameter $\eta = \frac{4\pi}{3}R^3 c_B = \tau_c/\tau_f$ is equal to 0.1 (top) and 0.01 (bottom).

motion of particles. At the same time, IET has a significantly narrower applicability time interval. Comparison of the IET, MET and Smoluchowski's theory is shown in Fig. 7 for two values of the binary parameter.

2.3. *Fluorescence quenching*

The precision of various approaches can also be compared by analyzing the quantum yield of luminescence. In this case, the kinetic equation should be modified as by adding a term, which describes unimolecular decay with a characteristic time τ_A

$$\frac{dc_A}{dt} = -\frac{c_A}{\tau_A} - \mathcal{K}(t)c_B c_A. \tag{35}$$

Likewise, the IET equation should be modified as well[75]

$$\frac{dc_A}{dt} = -\frac{c_A}{\tau_A} - c_B \int_0^t \Sigma'_{\text{iet}}(t-\tau)c_A(\tau)d\tau,$$

$$\text{where } \Sigma'_{\text{iet}}(t) = \Sigma_{\text{iet}}(t)e^{-t/\tau_A}. \tag{36}$$

In the MET case, modification of the kernel is essentially the same: the term describing unimolecular decay should be added to the right-hand side of the kinetic equation and the kernel should be modified as $\Sigma_{\text{met}}(t) \to \Sigma'_{\text{met}}(t) = \Sigma_{\text{met}}(t)e^{-t/\tau_A}$.

The luminescence quantum yield can be calculated as follows:

$$\Psi = \frac{1}{\tau_A} \int_0^\infty \frac{c_A(t)}{c_A(0)}dt = \frac{1}{1 + c_B \tau_A k_Q}, \tag{37}$$

with k_Q being the quenching rate constant. This expression for Ψ is consistent with the Stern-Volmer law.[76] In the absence of the quenching process, we obtain $\Psi = 1$, whereas the chemical reaction reduces the Ψ value, so that the actual yield depends on the competition between unimolecular decay and biomolecular quenching reaction.

In any Markovian theory, the quenching constant is simply equal to k. However, in reality the Ψ value can be strongly affected by non-Markovian effects when τ_A is short. In fact, τ_A gives the time window where quenching occurs. When at $t < \tau_A$, the actual rate coefficient of the reaction, $\mathcal{K}(t)$, differs from k. Specifically, $\mathcal{K}(t) > k$; consequently, k_Q is also *greater* than k. This effect is reproduced by the abovementioned non-Markovian theories.[77,78] For instance, the

IET result for k_Q is as follows[75,79]:

$$k_Q^{\text{iet}} = \tilde{\Sigma}'_{\text{iet}}\left(s = \frac{1}{\tau_A}\right) \xrightarrow{k_r \gg k_D} k_D\left(1 + \sqrt{\frac{\tau_D}{\tau_A}}\right). \qquad (38)$$

This value is obviously greater than k once $\tau_A \leq \tau_D$, i.e., once quenching occurs at times where the rate coefficient is greater than k_D and the kinetics is essentially non-Markovian. Corrections to the IET result come from quenching at longer times, where the IET is no longer accurate. These corrections have been derived in earlier works[77] for Smoluchowski's theory with the time-dependent $\mathcal{K}(t)$ rate coefficient and for the MET. Here, for the sake of simplicity, we present the results of both theories in the limiting case $k_r \gg k_D$ and keep only terms proportional to the density parameter η:

$$\frac{k_Q^{Sm}}{k_D} = 1 + \sqrt{\frac{\tau_D}{\tau_A}} + 3\eta\left\{\frac{1}{2}\sqrt{\frac{\tau_A}{\tau_D}} + 1 - \frac{2}{\pi}\right\} + \cdots,$$

$$\frac{k_Q^{\text{met}}}{k_D} = 1 + \sqrt{\frac{\tau_D}{\tau_A}} + 3\eta\left\{\frac{1}{2}\sqrt{\frac{\tau_A}{\tau_D}}\right\} + \cdots. \qquad (39)$$

Hence, both theories reproduce the IET result at $\eta \to 0$ and provide an additional correcting term. Of course, the correcting terms in eq. (39) are correct in the limit where they are much smaller than the IET result given by eq. (38). This correction is of no importance when $\tau_D \gg \tau_A$, i.e., when quenching occurs at very short times where the IET is still accurate. In the opposite case, the term proportional to $3\eta\sqrt{\tau_A/\tau_D}$ gives the main correction to the IET result. The two theories differ in their estimates of other terms, which are, however, unimportant in this limit. In the present case, the fully renormalized kinetic theory gives exactly the same result as MET for the linear correction of k_Q.[53]

3. Many-particle Methods for Deriving Kinetic Equations

With the aim to give an idea on how the IET and MET follow from a many-particle consideration of the problem, in this section we present a derivation[68] of the IET and MET equations. In the derivation, we exploit the formalisms of RDFs and correlation patterns to give a flavor of modern methods of non-equilibrium statistical mechanics. For the sake of simplicity and clarity, we consider only the simplest pseudo-monomolecular reaction (1), i.e., we assume that $c_A \ll c_B$.

3.1. *Reduced distribution functions and correlation patterns*

First of all, we introduce RDFs, denoted as

$$\varphi_{p,q} = \varphi_{p,q}(\mathbf{r}_{A_1}, \ldots, \mathbf{r}_{A_p}, \mathbf{r}_{B_1}, \ldots, \mathbf{r}_{B_q}, t). \tag{40}$$

The $\varphi_{p,q}$ function gives the distribution functions for p particles A and q particles B at time instant t, located at coordinates $\{\mathbf{r}_{A_1}, \ldots, \mathbf{r}_{A_p}, \mathbf{r}_{B_1} \ldots, \mathbf{r}_{B_q}\}$. The single-particle RDFs are simply the local concentrations of the corresponding particles

$$\varphi_{1,0}(\mathbf{r}_A, t) = c_A(\mathbf{r}_A, t), \qquad \varphi_{0,1}(\mathbf{r}_B, t) = c_B(\mathbf{r}_B, t). \tag{41}$$

The two-particle RDF $\varphi_{1,1}$ gives the pair distribution function. This function is analogous to the $n(r, t)$ function introduced above, differing only in normalization. At large distances $r = |\mathbf{r}_A - \mathbf{r}_B|$ between A and B, we obtain $n(r, t) \to 1$, whereas $\varphi_{1,1} \to c_A(\mathbf{r}_A) \cdot c_B(\mathbf{r}_B)$. This property, i.e., factorization of the RDF, is termed the *correlation loss* property. Due to this property, we can obtain a lower-order distribution function from a higher-order RDF, for instance

$$\varphi_{1,0}(\mathbf{r}_A, t) = \int d\mathbf{r}_B \varphi_{1,1}(\mathbf{r}_A, \mathbf{r}_B, t). \tag{42}$$

In the same way, one can obtain $\varphi_{r,s}$ from $\varphi_{p,q}$ (for $r < p$, $s < q$) by integration over extra-coordinates.

One can write down a set of coupled differential equations for the RDFs, termed "hierarchies", which are analogous to the BBGKY hierarchies in statistical physics.[80] For instance, the evolution of $\varphi_{1,0}$, i.e., of the concentration c_A, is due to reactions with B particles and free motion. Hence, we can write down the following equation:

$$\left(\frac{\partial}{\partial t} - \hat{\mathcal{L}}_A\right) \varphi_{1,0}(\mathbf{r}_A, t) = -\int d\mathbf{r}_B w(\mathbf{r}_A - \mathbf{r}_B)\varphi_{1,1}(\mathbf{r}_A, \mathbf{r}_B, t). \quad (43)$$

In the right-hand side, we take into account the decay of A particles due to the chemical reaction occurring at a rate $w(\mathbf{r}_A - \mathbf{r}_B)$ from the two-particle distribution $\varphi_{1,1}$ (integration over $\mathbf{r}_B$ is performed to remove extra-coordinates). Equation for c_B can be deduced in the same way, but here we neglect the effect of chemical reaction on B particles, which are in great excess, and take $c_B(\mathbf{r}_B) = \text{const}$.

Thus, one can see that the evolution of the c_A concentration is coupled to the two-particle RDF. To proceed further, we write down the equation for this RDF

$$\left(\frac{\partial}{\partial t} - \hat{\mathcal{L}}_A - \hat{\mathcal{L}}_B\right) \varphi_{1,1}(\mathbf{r}_A, \mathbf{r}_B, t) = -w(\mathbf{r}_A - \mathbf{r}_B)\varphi_{1,1}(\mathbf{r}_A, \mathbf{r}_B, t)$$

$$- \int d\mathbf{r}_{B'} w(\mathbf{r}_A - \mathbf{r}_{B'})\varphi_{1,2}(\mathbf{r}_A, \mathbf{r}_B, \mathbf{r}_{B'}, t). \quad (44)$$

The three-particle term arises from reaction of A with a second B particle with a coordinate $\mathbf{r}_{B'}$. A similar term, which arises from $\varphi_{2,1}$, is omitted here: Because of the condition $c_A \ll c_B$, it is much smaller than $\varphi_{1,2}$.

The hierarchy can be continued further by writing down equations for RDFs with increasing indices. However, solution of the entire hierarchy is not practical; for this reason, approximate methods for *truncating* hierarchies are required. We would like to stress that the formalism of RDFs is not always suited for such a truncation. The reason is that higher-order RDFs are not negligibly small: from the correlation loss property one can see that even at large inter-particle distances they do not tend to zero but rather to products of particle concentrations. Therefore, it is convenient to introduce so-called correlation patterns instead of RDFs.

The correlation patterns $\pi_{p,q}$ are introduced as follows. The single-particle patterns are equal to the corresponding RDFs

$$\pi_{1,0}(\mathbf{r}_A, t) = \varphi_{1,0}(\mathbf{r}_A, t) = c_A(\mathbf{r}_A, t),$$

$$\pi_{0,1}(\mathbf{r}_B, t) = \varphi_{0,1}(\mathbf{r}_B, t) = c_B(\mathbf{r}_B, t). \quad (45)$$

To introduce the two-particle pattern, we present the RDF $\varphi_{1,1}$ in the following way (as a combination of correlated and uncorrelated distributions):

$$\varphi_{1,1}(\mathbf{r}_A, \mathbf{r}_B, t) = \pi_{1,1}(\mathbf{r}_A, \mathbf{r}_B, t) + \pi_{1,0}(\mathbf{r}_A, t) \cdot \pi_{0,1}(\mathbf{r}_B, t)$$

$$\Rightarrow \pi_{1,1}(\mathbf{r}_A, \mathbf{r}_B, t) = \varphi_{1,1}(\mathbf{r}_A, \mathbf{r}_B, t) - \pi_{1,0}(\mathbf{r}_A, t) \cdot \pi_{0,1}(\mathbf{r}_B, t).$$

(46)

At large inter-particle distances, the correlation pattern decreases to zero, in contrast to the RDF. Likewise, higher-order correlation patterns can be introduced in a similar manner. For instance, the three-particle RDF can be recast as follows:

$$\varphi_{1,2}(\mathbf{r}_A, \mathbf{r}_B, \mathbf{r}_{B'}, t) = c_A(\mathbf{r}_A, t) c_B(\mathbf{r}_B) c_B(\mathbf{r}_{B'})$$

$$+ \pi_{1,1}(\mathbf{r}_A, \mathbf{r}_B, t) c_B(\mathbf{r}_{B'}) + \pi_{1,1}(\mathbf{r}_A, \mathbf{r}_{B'}, t) c_B(\mathbf{r}_B)$$

$$+ c_A(\mathbf{r}_A, t) \pi_{2,0}(\mathbf{r}_B, \mathbf{r}_{B'}) + \pi_{1,2}(\mathbf{r}_A, \mathbf{r}_B, \mathbf{r}_{B'}, t).$$

(47)

Thus, knowing all single-particle and two-particle RDFs, we can evaluate the three-particle correlation pattern $\pi_{1,2}$ from this equation. All higher-order correlation patterns can be evaluated in the same way.

The equations for correlation patterns can be deduced from those for the RDFs. For instance, the equation for $\pi_{1,1}$ is as follows:

$$\left(\frac{\partial}{\partial t} - \hat{\mathcal{L}}_A - \hat{\mathcal{L}}_B \right) \pi_{1,1}(\mathbf{r}_A, \mathbf{r}_B, t)$$

$$= -w(\mathbf{r}_A - \mathbf{r}_B)\{\pi_{1,1}(\mathbf{r}_A, \mathbf{r}_B, t) + c_A(\mathbf{r}_A, t) \cdot c_B(\mathbf{r}_B, t)\} - \mathcal{X}.$$

(48)

Here, $\mathcal{X}$ stands for three-particle terms; their explicit form is presented below. Hence, one can see that two-particle correlations are generated by chemical reactions from the non-correlated initial distribution, as described by the $w(r)c_A c_B$ source-term, and also by three-particle effects.

3.2. *Derivation of IET*

Using the many-particle description in terms of correlation patterns, one can derive the IET and MET equations.[68] To derive the

IET equation, one should consider only two-particle correlations and neglect all higher order terms in Eq. (48) by assuming $\mathcal{X} = 0$. The idea of this approximation is that A particle after an encounter with B particle completely loses correlation with it before it meets another B particle. To simplify the IET derivation, we assume that the reacting system is spatially uniform, meaning the c_A and c_B do not depend on the coordinates $\mathbf{r}_A$ and $\mathbf{r}_B$, respectively, and that $\pi_{1,1}$ depends only on $\mathbf{r} = \mathbf{r}_A - \mathbf{r}_B$. We also rewrite the two-particle motional operator as $\hat{\mathcal{L}}_r$. Hence, Eq. (48) is recast as follows:

$$\left(\frac{\partial}{\partial t} - \hat{\mathcal{L}}_r + w(r) \right) \pi_{1,1}(r, t) = -w(r)c_A(t) \cdot c_B. \qquad (49)$$

This equation can be solved by using Green function methods. To do so, we make use of the Green function defined in Eq. (25) and express the two-particle correlation

$$\pi_{1,1}(r, t) = c_B \int_0^t d\tau \int d\mathbf{r}_0 g(r|r_0, t - \tau) w(r_0) c_A(\tau). \qquad (50)$$

Substituting this expression into Eq. (46) and then to Eq. (43), we immediately obtain the IET kernel, see Eq. (24).

The expression for the IET kernel can be greatly simplified in contact approximation. Indeed, integration over spatial coordinates in Eq. (24) becomes simple

$$\Sigma_{\text{iet}}(\tau) = k_r \left(1 - k_r g(R|R, \tau) \right). \qquad (51)$$

Hence, we need to calculate the contact value of the Green function. To do so, it is convenient to evaluate first the "free" Green function, which describes the evolution of the two-particle distribution function in the absence of the reaction sink

$$\left(\frac{\partial}{\partial \tau} - \hat{\mathcal{L}}_r \right) g_0(r|r_0, \tau) = \frac{\delta(r - r_0)}{4\pi r r_0} \delta(\tau). \qquad (52)$$

In the Laplace domain, the relation between the two Green functions is as follows:

$$\tilde{g}(R|R,s) = \frac{\tilde{g}_0(R|R,s)}{1 + k_r \tilde{g}_0(R|R,s)}, \quad \text{where} \quad \tilde{g}_0(R|R,s) = \frac{1}{k_D}\mathcal{G}(s).$$

(53)

Using expressions (51), (53) we obtain the IET kernel given by Eq. (33).

3.3. Derivation of MET

In order to derive the MET equation, one needs to consider the $\mathcal{X}$ term in Eq. (48) explicitly and to find an approximate expression for this term. The $\mathcal{X}$ term can be written as follows[68]:

$$\mathcal{X} = \int d\mathbf{r}_B w(\mathbf{r}_A - \mathbf{r}_B) \{\pi_{1,2}(\mathbf{r}_A, \mathbf{r}_B, \mathbf{r}_{B'}, t)$$

$$+ \pi_{1,1}(\mathbf{r}_A, \mathbf{r}_B, t)c_B(\mathbf{r}_{B'}) + c_A(\mathbf{r}_A, t)\pi_{0,2}(\mathbf{r}_B, \mathbf{r}_{B'})\}. \quad (54)$$

To evaluate this term, one should first neglect $\pi_{0,2}$, which can be omitted in the binary approximation.[68] As far as the three-particle correlation is concerned (which is given by the first term in Eq. (54)), this contribution arises from two processes (termed binary channels): when A is correlated with B but reacts with B' and *vice versa*. Such a consideration follows from the equation for $\pi_{1,2}$, which is of the form

$$\left(\frac{\partial}{\partial t} - \hat{\mathcal{L}}_r + w(\mathbf{r}_A - \mathbf{r}_B) + w(\mathbf{r}_A - \mathbf{r}_{B'})\right)\pi_{1,2}(\mathbf{r}_A, \mathbf{r}_B, \mathbf{r}_{B'}, t)$$

$$= -w(\mathbf{r}_A - \mathbf{r}_{B'})\pi_{1,1}(\mathbf{r}_A, \mathbf{r}_B, t)c_B(\mathbf{r}_{B'})$$

$$-w(\mathbf{r}_A - \mathbf{r}_B)\pi_{1,1}(\mathbf{r}_A, \mathbf{r}_{B'}, t)c_B(\mathbf{r}_B). \quad (55)$$

In the right-hand side of this equation, we have omitted all four-particle terms. The idea of the MET is to neglect the second channel as well as interference of the two channels; such an approximation has been justified previously.[68] As a result, the second term in the right-hand side of Eq. (55) can be neglected and a simpler expression for $\pi_{1,2}$ can be obtained. As a result, $\pi_{1,2}$ expressed via $\pi_{1,1}(\mathbf{r}_A, \mathbf{r}_B, t)$

is a way analogous to Eq. (50), which expresses $\pi_{1,1}$ via c_A. To keep the description concise, we omit the mathematically strict derivation of the expression for the $\mathcal{X}$ term and only provide the final result

$$\mathcal{X} \approx -c_B \int_0^t \Sigma_{\text{iet}}(t - \tau)\pi_{1,1}(\mathbf{r}_A, \mathbf{r}_B, \tau)d\tau \xrightarrow{t \to \infty} -kc_B\pi_{1,1}(\mathbf{r}_A, \mathbf{r}_B, t).$$
(56)

Hence, in the final expression, which yields a convolution-type expression for the $\mathcal{X}$ term, one can omit non-Markovian effects and use the Markovian expression, which is given by k. The validity of such an approximation has been verified before.[68]

Hence, after evaluation of the three-particle term, one arrives at the following equation for $\pi_{1,1}$:

$$\left(\frac{\partial}{\partial t} + kc_B - \hat{\mathcal{L}}_r + w(\mathbf{r})\right)\pi_{1,1}(\mathbf{r}, t) = -w(\mathbf{r})c_A(t) \cdot c_B.$$
(57)

This equation can be solved by Green function methods, with the need to redefine the Green function as follows:

$$g(r|r_0, t) \to g(r|r_0, t)e^{-kc_Bt}.$$
(58)

Such a redefinition of the Green function results in redefinition of the kernel given by Eq. (28), giving rise to the MET kinetic equation.

Hence, in this section we explain the strategy for deriving kinetic equations from many-particle treatment of reacting systems. Such a strategy is based on writing down hierarchical sets of equations for RDFs[61,63,81] or correlation patterns,[68] which can be truncated to obtain closed equations for particle concentrations. The approaches presented here, which give rise to the IET and MET equations, can be generalized to more complex cases, that is, to reversible reactions and multistage reactions. It is important to emphasize that treatment of such reactions can be performed using a convenient matrix formalism.[69,70,82]

4. Conclusions

In this contribution, we have introduced theoretical approaches to diffusion-influenced reactions, focusing on *non-Markovian* kinetics.

We explain the peculiarities of diffusional motion, giving rise to the cage effect and repeated contacts of diffusing particles. In turn, such contacts result in diffusional encounters of reactants in liquids, in contrast to single collision events in gases. Despite this remarkable difference, one can use concepts from gas-phase kinetics to develop theoretical approaches to reaction kinetics in liquid solutions, which treat reacting systems as a gas of reactants diluted in a chemically inert medium. By analogy with the semi-classical collision theory in gases, we can term such theory "encounter theory". Here, we focus on the variants of the encounter theory known as the IET and MET, presenting their integro-differential as well as differential forms. In addition, we present a simplified derivation of the IET and MET equations, based on the many-particle treatment of the reacting system. Such a treatment leads to sets of hierarchical equations, which can be truncated by using different approaches. We discuss two truncation methods, giving rise to the IET and MET. We would like to stress that these methods are general and can be used for obtaining the IET and MET equations of complex reactions. In this situation, the concept of "effective particles" becomes very useful, which allows one to write down kinetics IET equations in a compact matrix form. Such a matrix formulation of the theory is convenient for calculating steady-state rate constants of diffusion-influenced reactions and for considering non-Markovian kinetics. In Chapter 4 we discuss the matrix-form IET and explain its important advantages.

Acknowledgment

The authors acknowledge support from Ministry of Science and Higher Education of RF (projects No. 44.1.5 and 0333-2017-0002).

References

1. V. M. Agranovich and M. D. Galanin, *Electron Excitation Energy Transfer in Condensed Matter*. North-Holland, Amsterdam (1982).

2. T. Förster, Zwischenmolekulare Energiewanderung und Fluoreszenz, *Annalen Phys.* **437**, 55–75 (1948).

3. D. L. Dexter, A theory of sensitized luminescence in solids, *J. Chem. Phys.* **21**, 836–850 (1953).

4. R. A. Marcus, On the theory of oxidation–reduction reactions involving electron transfer, *J. Chem. Phys.* **24**, 966–989 (1956).

5. R. A. Marcus, Theory of electron-transfer reactions. VI. Unified treatment for homogeneous and electrode reactions, *J. Chem. Phys.* **43**, 679–701 (1965).

6. M. Eigen, Proton transfer, acid-base catalysis, and enzymatic hydrolysis. I. elementary processes, *Angew. Chem. Intl. Ed.* **3**, 1 (1964).

7. M. V. Basilevsky and M. V. Vener, Theoretical investigations of proton and hydrogen atom transfer in the condensed phase, *Russ. Chem. Rev.* **72**, 1 (2003).

8. S. Hammers-Schiffer, Proton-coupled electron transfer: Moving together and charging forward, *J. Amer. Chem. Soc.* **137**, 8860 (2015).

9. J. R. Lakowicz, *Principles of Fluorescence Spectroscopy*. Springer, Amsterdam (1982).

10. P. F. Barbara, T. J. Meyer, and M. A. Ratner, Contemporary issues in electron transfer research, *J. Phys. Chem.* **100**, 13148–13168 (1996).

11. M. Terazima, Is the translational diffusion of organic radicals different from that of closed-shell molecules? *Acc. Chem. Res.* **33**, 687–694 (2000).

12. C. Turro, J. M. Zaleski, Y. M. Karabatsos, and D. G. Nocera, Bimolecular electron transfer in the Marcus inverted region, *J. Amer. Chem. Soc.* **118**, 6060–6067 (1996).

13. A. Rosspeintner, D. R. Kattnig, G. Angulo, S. Landgraf, and G. Grampp, The Rehm-Weller experiment in view of distant electron transfer, *Chem. Eur. J.* **14**, 6213–6221 (2008).

14. A. Rosspeintner and E. Vauthey, Bimolecular photoinduced electron transfer reactions in liquids under the gaze of ultrafast spectroscopy, *Phys. Chem. Chem. Phys.* **16**, 25741–25754 (2014).

15. O. G. Berg, R. B. Winter, and P. H. von Hippel, Diffusion-driven mechanisms of protein translocation on nucleic acids. 1. Models and theory, *Biochem.* **20**, 6929–6948 (1981).

16. A. B. Kolomeisky, Physics of protein-dna interactions: Mechanisms of facilitated target search, *Phys. Chem. Chem. Phys.* **13**, 2088–2095 (2011).

17. L. Mirny, M. Slutsky, Z. Wunderlich, A. Tafvizi, J. Leith, and A. Kosmrlj, How a protein searches for its site on dna: The mechanism of facilitated diffusion, *J. Phys. A. Math. Theor.* **42**, 434013 (2009).

18. H.-X. Zhou, Rapid search for specific sites on dna through conformational switch of nonspecifically bound proteins, *Proc. Natl. Acad. Sci. USA.* **108**, 8651–8656 (2011).

19. H.-X. Zhou, G. Rivas, and A. P. Minton, Macromolecular crowding and confinement: biochemical, biophysical, and potential physiological consequences, *Annu. Rev. Biophys.* **37**, 375–397 (2008).

20. S. Park and N. Agmon, Theory and simulation of diffusion-controlled michaelis-menten kinetics for a static enzyme in solution, *J. Phys. Chem. B.* **112**, 5977–5987 (2008).
21. G. Wilemski and M. Fixman, Diffusion-controlled intrachain reactions of polymers. I. Theory, *J. Chem. Phys.* **60**, 866–877 (1974).
22. I. M. Sokolov, J. Mai, and A. Blumen, Paradoxical diffusion in chemical space for nearest-neighbor walks over polymer chains, *Phys. Rev. Lett.* **79**, 857–860 (1997).
23. S. F. Burlatsky and G. S. Oshanin, Diffusion-controlled reactions with polymers, *Phys. Lett. A.* **145**, 61–65 (1990).
24. G. Oshanin, S. Moreau, and S. Burlatsky, Models of chemical-reactions with participation of polymers, *Adv. Colloid Interf. Sci.* **49**, 1–46 (1994).
25. P. J. Park and S. Lee, Diffusion-influenced reversible energy transfer reactions between polymers, *J. Chem. Phys.* **115**, 9594–9600 (2001).
26. C. Groves and N. C. Greenham, Bimolecular recombination in polymer electronic devices, *Phys. Rev. B.* **78**, 155205 (2008).
27. M. Hilczer and M. Tachiya, Unified theory of geminate and bulk electron-hole recombination in organic solar cells, *J. Phys. Chem. C.* **114**, 6808–6813 (2010).
28. R. M. Noyes, Effects of diffusion rates on chemical kinetics, *Prog. React. Kinet.* **1**, 129–160 (1961).
29. S. A. Rice, *Diffusion-Limited Reactions Comprehensive Chemical Kinetics*. Vol. 25, in: (C. H. Bamford, C. F. H. Tipper, and R. G. Compton), Elsevier Science Publishers, Amsterdam (1985).
30. A. B. Doktorov and E. A. Kotomin, Theory of tunnelling recombination of defects stimulated by their motion. i. general formalism, *Phys. Stat. Sol. (B)*. **113**, 9–14 (1982).
31. A. B. Doktorov, *Development of the kinetic theory of physicochemical processes induced by binary encounters of reactants in solutions*. vol. 6, Transworld Research Network, Kerala, India (2012).
32. A. B. Doktorov and A. A. Kipriyanov, Deviation from the kinetic law of mass action for reactions induced by binary encounters in liquid solutions, *J. Phys. Cond. Matt.* **19**, 065136 (2007).
33. M. Smoluchowski, Versuch einer Mathematischen Theorie der Koagulationskinetic Kolloider Lösungen, *Z. Phys. Chem.* **92**, 129 (1917).
34. F. C. Collins and G. E. Kimball, Diffusion controlled reaction rates, *J. Colloid Sci.* **4**, 425–437 (1949).
35. N. N. Tunitski and K. S. Bagdasarian, The resonance intermolecular excitation making allowance for diffusion, *Opt. Spectrosc.* **15**, 100–106 (1963).
36. G. Wilemsky and M. Fixman, General theory of diffusion-controlled reactions, *J. Chem. Phys.* **58(9)**, 4009–4019 (1973).
37. A. B. Doktorov and A. I. Burshtein, Quantum theory of diffusion-accelerated remote transfer, *Sov. Phys. JETP.* **41**, 671–677 (1976).
38. A. B. Doktorov, The impact approximation in the theory of bimolecular quasi-resonant processes, *Physica A.* **90**, 109–136 (1978).

39. A. B. Doktorov, *Encounter Theory of Chemical Reaction in Solution. Approximate Methods for Calculating Rate Constants* (2018).
40. J. Franck and E. Rabinovich, Free radicals and the photochemistry of solutions, *Trans. Fraday Soc.* **30**, 120–131 (1934).
41. H. Eyring, S. H. Lin, and S. M. Lin, *Basic Chemical Kinetics*. Wiley, New York (1980).
42. Y. B. Zeldovich and A. A. Ovchinnikov, The mass action law and the kinetics of chemical reactions with allowance for thermodynamic fluctuations of the density, *Sov. Phys. JETP Lett.* **47**, 829–834 (1977).
43. S. F. Burlatsky, G. S. Oshanin, and A. A. Ovchinnikov, Kinetics of chemical short-range ordering in liquids and diffusion-controlled reactions, *Chem. Phys.* **152**, 13–21 (1991).
44. N. Agmon and I. V. Gopich, Exact long-time asymptotics for reversible binding in three dimensions, *J. Chem. Phys.* **112**, 2863–2869 (2000).
45. I. V. Gopich, A. A. Ovchinnikov, and A. Szabo, Long-time tails in the kinetics of reversible bimolecular reactions, *Phys. Rev. Lett.* **86**, 922 (2001).
46. I. Z. Steinberg and E. Katchalsky, Theoretical analysis of the role of diffusion in chemical reactions, fluorescence quenching, and nonradiative energy transfer, *J. Chem. Phys.* **48**, 2404–2410 (1968).
47. M. Tachiya, Theory of diffusion-controlled reactions: Formulation of the bulk reaction rate in terms of the pair probability, *Radiat. Phys. Chem.* **21**, 167–175 (1983).
48. A. A. Kipriyanov, I. V. Gopich, and A. B. Doktorov, Exactly solvable models in the theory of irreversible reactions in liquids, *Physica A.* **205**, 585–622 (1994).
49. R. Zwanzig, Ensemble method in the theory of irreversibility, *J. Chem. Phys.* **33**, 1338–1341 (1960).
50. H. Mori, A continued-fraction representation of the time-correlation functions, *Progr. Theor. Phys.* **33**, 399–416 (1965).
51. H. Mori, Transport, collective motion, and brownian motion, *Progr. Theor. Phys.* **33**, 423–455 (1965).
52. V. P. Sakun, Intermolecular spin-spin interactions in liquids, *Physica A.* **80A**, 128–148 (1975).
53. M. Yang, S. Lee, and K. J. Shin, Kinetic theory of bimolecular reactions in liquid. II. Reversible association-dissociation: $A + B \leftrightarrow C + B$, *J. Chem. Phys.* **108**(1), 8557–8571 (1998).
54. A. A. Kipriyanov, I. V. Gopich, and A. B. Doktorov, A many-particle approach to the derivation of binary non-markovian kinetic equations for the reaction $A + B \rightarrow B$, *Physica A.* **255**, 347–405 (1998).
55. B. U. Felderhof and R. B. Jones, Statistical theory of time-dependent diffusion-controlled reactions in fluids and solids, *J. Chem. Phys.* **103**, 10201–10213 (1995).
56. A. S. Mikhailov and V. V. Yashin, Quantum field methods in the theory of diffusion-controlled reactions, *J. Stat. Phys.* **38**, 347–359 (1985).
57. T. Ohtsuki, Field-theoretical approach to scaling behavior of diffusion-controlled recombination, *Phys. Rev. A.* **43**, 6917–6919 (1991).

58. T. R. Waite, Theoretical treatment of the kinetics of diffusion-limited reactions, *Phys. Rev.* **107**, 463–470 (1957).

59. T. R. Waite, General theory of bimolecular reaction rates in solids and liquids, *J. Chem. Phys.* **28**, 103–106 (1958).

60. V. N. Kuzovkov and E. A. Kotomin, Generalised theory of diffusion-controlled defect annealing, *J. Phys. C.* **13**, 499–502 (1980).

61. S. Lee and M. Karplus, Kinetics of diffusion-influenced bimolecular reactions in solutions. I. General formalism and relaxation kinetics of fast reversible reactions, *J. Chem. Phys.* **86**, 1883–1903 (1987).

62. S. Lee, J. J. Lee, and K. J. Shin, Theory of diffusion-influenced radical recombination: interplay between geminate and bulk recombination, *Bull. Kor. Chem. Soc.* **15**, 311–320 (1994).

63. W. Naumann, The reversible reaction $A + B = C$ in solution. A system-size expansion approach on the base of reactive many-particle diffusion equations, *J. Chem. Phys.* **98**, 2353–2365 (1993).

64. A. Molski and J. Keizer, Kinetics of nonstationary, diffusion-influenced reversible reactions in solution, *J. Chem. Phys.* **96**, 1391–1398 (1992).

65. A. A. Kipriyanov, A. B. Doktorov, and A. I. Burshtein, Binary theory of dephasing in liquid solutions. I. the non-markovian theory of encounters, *Chem. Phys.* **76**, 149–162 (1983).

66. A. A. Kipriyanov, I. V. Gopich, and A. B. Doktorov, A modification of the non-Markovian encounter theory. I. Markovian description in non-Markovian theories, *Chem. Phys.* **187**, 241–249 (1995).

67. I. V. Gopich and A. B. Doktorov, Kinetics of diffusion-influenced reversible reaction $A + B \rightleftharpoons C$ in solutions, *J. Chem. Phys.* **105**, 2320–2332 (1996).

68. A. A. Kipriyanov, O. A. Igoshin, and A. B. Doktorov, A new approach to the derivation of binary non-markovian kinetic equations, *Physica A.* **268**, 567–606 (1999).

69. K. L. Ivanov, N. N. Lukzen, A. B. Doktorov, and A. I. Burshtein, Integral encounter theories of multistage reactions. I. Kinetic equations, *J. Chem. Phys.* **114**, 1754–1762 (2001).

70. K. L. Ivanov, N. N. Lukzen, A. A. Kipriyanov, and A. B. Doktorov, The integral encounter theory of multistage reactions containing association-dissociation reaction stages. I. Kinetic equations, *Phys. Chem. Chem. Phys.* **6**, 1706–1718 (2004).

71. K. L. Ivanov, N. N. Lukzen, and A. B. Doktorov, The integral encounter theory of multistage reactions containing association-dissociation reaction stages. II. The kinetics of reversible excitation binding, *Phys. Chem. Chem. Phys.* **114**, 1719–1724 (2004).

72. A. A. Kipriyanov and A. B. Doktorov, Long-time behaviour of the observables in irreversible reactions in liquid solutions, *Chem. Phys. Lett.* **246**, 359–363 (1995).

73. A. B. Doktorov, R. F. Khairutdinov, and K. I. Zamaraev, Analysis of kinetic models for the tunnel electron transfer reactions. Reaction kinetics for various radical and angular dependences of the tunneling probability, *Chem. Phys.* **61**, 351–364 (1981).

74. I. V. Gopich, A. A. Kipriyanov, and A. B. Doktorov, A many-particle treatment of the reversible reaction $A + B \leftrightarrow C + B$, *J. Chem. Phys.* **110**(22), 10888–10898 (1999).

75. N. N. Lukzen, A. B. Doktorov, and A. I. Burshtein, Non-markovian theory of diffusion-controlled excitation transfer, *Chem. Phys.* **102**(3), 289–304 (1986).

76. O. Stern and M. Volmer, über die Abklingzeit der Fluoreszenz, *Zeitschrift für Physik*, **20**, 183–188 (1919).

77. A. I. Burshtein, I. V. Gopich, and P. A. Frantsuzov, Accumulation and distribution of energy quenching products, *Chem. Phys. Lett.* **289**, 60–66 (1998).

78. S. Murata, M. Nishimura, S. Y. Matsuzaki, and M. Tachiya, Transient effect in fluorescence quenching induced by electron transfer. I. Analysis by the Collins-Kimball model of diffusion-controlled reactions, *Chem. Phys. Lett.* **219**, 200–206 (1994).

79. A. I. Burshtein and N. N. Lukzen, Reversible reactions of metastable reactants, *J. Chem. Phys.* **103**, 9631–9641 (1995).

80. R. C. Balescu, *Equilibrium and Non-Equilibrium Statistical Mechanics*. John Wiley & Sons, Chichester, New York, Sydney, Toronto (1975).

81. J. Sung and S. Lee, Nonequilibrium distribution function formalism for diffusion-influenced bimolecular reactions: Beyond the superposition approximation, *J. Chem. Phys.* **111**(3), 796–803 (1999).

82. K. L. Ivanov, N. N. Lukzen, and A. B. Doktorov, The integral encounter theory of multistage reactions containing association-dissociation reaction stages. III. Taking account of quantum states of reactants, *J. Chem. Phys.* **121**, 5115–5124 (2004).

Chapter 4

Integral Encounter Theory: A Universal Method for Kinetic Description of Multistage Reactions in Solutions

Konstantin L. Ivanov[*,‡] and Nikita N. Lukzen[†]

*International Tomography Center, Siberian Branch of the Russian
Academy of Sciences, Novosibirsk, 630090, Russia
†Novosibirsk State University, Novosibirsk, 630090, Russia
‡ivanov@tomo.nsc.ru

In this contribution, we provide a detailed overview of the Integral Encounter Theory (IET), which is currently the most general approach to non-Markovian kinetics in liquid solutions. We introduce the IET of multistage reactions of arbitrary complexity and explain how the kinetic IET equations can be derived and written in a compact matrix form. A method of calculating rate constants of diffusion-influenced processes by using the IET is described; applications of the IET to reactions involving metastable particles are also discussed.

1. Introduction

Theoretical treatment of diffusion-influenced reactions is a relevant scientific problem. Such reactions play an important role in many areas of chemical physics,[1–7] biophysics,[8] biochemistry,[9–13] polymer chemistry,[14–18] photovoltaics,[19,20] etc. Within the last decades, strong progress in this field has been achieved,[21–37] initiated by early theories of coagulation processes.[38,39] These early studies have revealed that liquid-state reactions can be *diffusion-influenced*, in the sense that their rates are determined not only by chemical reactivity parameters but also by diffusional transport of particles to the reaction zone. For instance, the rate constant of a bimolecular reaction on a perfectly absorbing "black sphere" is equal to $k_D = 4\pi RD$ with R being the contact distance (equal to the sum of reactant's radii) and D being the relative diffusion coefficient (equal to the sum of reactant's diffusion coefficients). Hence, the rate of the reaction is conditioned by reactant mobility rather than by chemical reactivity. It is worth noting that diffusion-influenced reactions exhibit non-Markovian kinetic behavior meaning that their "rate constants" $\mathcal{K}(t)$ are time-dependent: for this reason, it is more appropriate to term $\mathcal{K}(t)$ *rate coefficient*. The main concepts of non-Markovian kinetics have been described in Chapter 3 of this book.[40]

Despite the strong progress in the theory of diffusion-influenced reactions, consistent description of complex multistage reactions taking into account non-Markovian effects is often challenging. A method which allows one to perform generalization of the theory to such reactions is based on deriving sets of hierarchical equations for reduced distribution functions (RDFs). The hierarchies can be truncated to obtain closed kinetics equations for reactant concentrations. There are a number of such theories, for instance, the superposition approximation,[22,23] many-particle kernel formalism,[41] fully renormalized theory,[42] integral encounter theory (IET)[43] and its modified version.[44,45] The last two theories can be generalized[46–48] to describe multistage processes of arbitrary complexity by using a convenient matrix form of the hierarchical equations and kinetic equations for particle concentrations. At the same time, both approaches are capable of describing non-Markovian kinetics. For this reason, the focus of this review is the IET; we do not dwell on the more complex modified encounter theory. In this contribution, we discuss the physical background of the IET, introduce the concept of "effective particles" that allows one to write down matrix-form equations, and present a derivation of the IET as well as examples showing the IET in action. These examples demonstrate that the IET is a very general approach to calculating rate constants of complex chemical processes in solutions. Finally, we discuss applications of the IET to reactions of metastable particles, demonstrating that the approach presented here is perfectly suited for describing non-Markovian kinetics.

2. Physical Background of the IET

The IET is an approach analogous to the semi-classical collision theory in gases,[49] since it treats a "gas" of reactants in a chemically inert medium. Although the medium is non-reactive, interaction of reactants with solvent molecules is significant as it changes the type of motion. In liquids, reactants move by means of random walks; in the limiting case where the length of such walks tends to zero, the motion is given by continuous diffusion.

Diffusional motion has important consequences. In contrast to gases, after a collision of two reactants they do not separate immediately as they are kept together by solvent molecules.[50] This is the essence of the "cage" effect: the reactants are trapped in the solvent cage for a finite period of time. The time which they stay close together until diffusional separation can be estimated as $\tau_c = R^2/3D = \tau_D/3$. As we see below, the τ_D parameter comes into play when the IET kernels and kinetic coefficients are calculated.

Furthermore, we would like to emphasize that collisions of diffusing particles occur in series: this is an inherent property of diffusional motion.[21] Interestingly, in the case of 3D-diffusion the probability of at least one collision of two particles separated by the distance $r_0 > R$ is finite, being equal to $p = R/r_0$; in the case of 2D-diffusion, this probability is equal to unity.[51] As a consequence, when two particles *encounter* each other, they stay close to each other over an extended period of time in the solvent "cage" and collide multiple times.

Hence, instead of single *collisions* in the gas phase, we need to consider *encounters* of two diffusing particles, A and B, which consist of multiple collisions between them, see Fig. 1. This effect has important manifestations in reaction kinetics in solutions. For instance,

Fig. 1.   Diffusional encounter of reactants A and B in solution: Due to random walks caused by interaction with the solvent molecules, the encounter consists of multiple contacts of reactants (here the solvent molecules are symbolically shown by light color balls).

when the first collision does not lead to a chemical reaction, the two particles have a finite probability to collide again and react. The cage effect also gives rise to non-Markovian kinetics. Hence, the peculiarities of diffusional motion are of great importance for reactions in solution, as the rate constants of chemical reactions are strongly affected by the cage effect, as demonstrated in this contribution.

In this work, we focus on the IET, which is a general method to consider non-Markovian kinetics of diffusion-influenced processes. The IET takes into account non-Markovian effects assuming that pair encounters of reactants are completely independent of each other. This assumption implies that when two reactants, A and B, encounter each other, no third particle interferes until they separate by diffusion. This means that the IET is essentially a *binary* theory, which is valid when the encounter duration, τ_c, is much smaller than the mean time between subsequent encounters of reactant A with reactant B τ_f

$$\tau_c \approx \frac{R^2}{3D} = \frac{1}{3}\tau_D \ll \tau_f = \frac{1}{4\pi DRN_B}. \tag{1}$$

This expression is mathematically equivalent to the condition that the density parameter

$$\eta = \frac{4\pi}{3} R^3 N_B \tag{2}$$

is much smaller than unity. Hereafter, unless otherwise stated, concentrations are denoted as $N_X = [X]$ (different notations for concentrations are used only in Subsection 4.4).

The approximation of independent pair encounters is valid only within a limited time interval because the encounter duration is formally infinite.[36] Hence, as correlations between reactants accumulate with time, the IET description becomes inaccurate at long times. In such cases, modification of the IET may become necessary, which extends its applicability time interval. One should note that the IET and its modification are essentially binary approaches, meaning that they express the rate coefficients via *relative* rather than absolute mobility. For the simplest pseudo-monomolecular reactions $A + B \rightarrow C + B$, the IET applicability time interval is as follows:[52]

$$t \leq \frac{\ln[1/\eta]}{kN_B}. \tag{3}$$

The modification extends this interval to[52]

$$t \leq t_b = \frac{1}{\eta k N_B} \tag{4}$$

because its takes into account all binary contributions to the kinetics, as has been demonstrated before.[52] The time t_b thus gives the time interval where binary approximation is valid. One should note that at long times the kinetics is no longer given by the relative mobility, but rather depends on mobility of both partners of a bimolecular reaction. Such a kinetic behavior, known as fluctuation kinetics,[53–56] becomes manifest at $t > t_b$ when most particles have already reacted. Discussion of the fluctuation kinetics is beyond the scope of this contribution. Likewise, we do not discuss the modified encounter theory, which is more complex than the IET.

3. Integral Encounter Theory of Multistage Reactions

3.1. *Reaction scheme*

In this section, we present a derivation of the IET in the case of multistage chemical processes. To make it simpler, we start from the approach of "effective particles",[46,47,57] which is very useful for theoretical treatment of multistage reactions, as it dramatically simplifies derivation of kinetic equations.

By a multistage reaction we mean a process comprising an arbitrary number of unimolecular and bimolecular stages, as well as association-dissociation stages. Hence, the reaction comprises the following unimolecular stages

$$A_i \leftrightarrow A_p, \quad B_j \leftrightarrow B_q, \quad C_m \leftrightarrow C_r, \tag{5}$$

bimolecular stages

$$A_i + B_j \leftrightarrow A_k + B_l \tag{6}$$

and association–dissociation stages

$$A_i + B_j \leftrightarrow C_m. \tag{7}$$

We assume that there are I, J and M kinds of particles A_i, B_j and C_m, respectively; hence, $i, k, p = 1, \ldots, I$, $j, l, q = 1, \ldots, J$ and $m, r = 1, \ldots, M$. The scheme given by Eqs. (5–7) represents the most general chemical reaction.

To go on further, we need to introduce the rates of the elementary processes (5–7).

For the unimolecular reactions (5), we introduce the rate constants $k_{A_i \to A_p}$, $k_{B_j \to B_q}$ and $k_{C_m \to C_r}$.

For the bimolecular processes (6), we define the rates $W_{A_i + B_j \to A_k + B_l}(\mathbf{r})$. These rates depend on the relative positions of the reactants, i.e., on the distance r between them (here $\mathbf{r}$ is the vector connecting the reactants). The IET can also be formulated in a more general way,[47,58] by introducing three-center rates of processes (7) and four-center rates of the bimolecular reactions (6), which depend on positions of the initial reactants as well as of the reactions' products. For the sake of clarity, here we avoid such complications.

Finally, we introduce the rates of the association–dissociation reactions (7) as $W_{A_i + B_j \to C_m}(\mathbf{r})$ and $W_{C_m \to A_i + B_j}(\mathbf{r})$. The r-dependence of $W_{A_i + B_j \to C_m}(\mathbf{r})$ means that the chemical rate depends on the distance between A_i and B_j, which associate. The backward process, dissociation, is described by introducing the rate $W_{C_m \to A_i + B_j}(\mathbf{r})$.

Here, the r-dependence reflects the fact that the probability of forming A_i and B_j by dissociation of C_m depends on the distance between them (predominantly, the products of dissociation are formed at small distances, which are close to R). By integrating over r, we can obtain from the rate $W_{C_m \to A_i + B_j}(\mathbf{r})$ the rate constant of the dissociation process

$$k_{C_m \to A_i + B_j} = \int d\mathbf{r} W_{C_m \to A_i + B_j}(\mathbf{r}). \tag{8}$$

This is the rate constant averaged over all possible relative positions of A_i and B_j.

We also assume that the reactants diffuse. Since we are interested only in the binary kinetics, we take into account only *relative* mobility of the A_i and B_j reactants, which is represented by operator $\mathcal{L}_{A_i B_j}$. In the simplest case of free diffusion, it is equal to $\mathcal{L}_{A_i B_j} = D_{ij} \Delta_r = (D_{A_i} + D_{B_j}) \Delta_r$ (for simplicity, we consider diffusion in three dimensions and isotropic reactivity). When D_{ij} depends on r and force interaction between A_i and B_j is present, described by the potential $V_{ij}(r)$, this operator takes the form (assuming spherical symmetry)

$$\mathcal{L}_{A_i B_j} = \frac{1}{r^2} \frac{\partial}{\partial r} r^2 D_{ij}(r) e^{-\beta V_{ij}(r)} \frac{\partial}{\partial r} e^{\beta V_{ij}(r)}, \tag{9}$$

with $\beta = 1/k_B T$ (here k_B is the Boltzmann constant and T is the absolute temperature).

Having introduced the rates of the elementary stages and operators describing relative reactant motion, let us now present the approach of "effective particles", which enables dramatic simplification of hierarchical equations and resulting kinetic equations for reactant concentration.

3.2. *Effective particles*

To go on further, we introduce the *effective particles* $\mathbb{A}$, $\mathbb{B}$ and $\mathbb{C}$ such that A_i, B_j and C_m are treated as their *internal states*

$$\mathbb{A} = \{A_i\} = \{A_1, A_2, \ldots, A_I\}, \quad \mathbb{B} = \{B_j\} = \{B_1, B_2, \ldots, B_J\},$$

$$\mathbb{C} = \{C_m\} = \{C_1, C_2, \ldots, C_M\}. \tag{10}$$

In some cases, reactants A_i, B_j, C_m are true internal states, e.g., excited states or ionic states of $\mathbb{A}$, $\mathbb{B}$, $\mathbb{C}$. However, one can as well consider introducing the "effective particles" as a formal mathematical trick, which allows one to simplify the form of kinetic equations. Simplification comes from the fact that the unimolecular reactions (5) and bimolecular reactions (6) only change the "states" of the "effective particles" keeping their number unchanged, while only the association–dissociation processes (7) change the number of particles of each kind. We would like to mention that by using the concept of effective particles, one can generalize the reaction scheme even further and include into consideration stages $A_i + A_i \rightarrow A_k + B_l$: this can be done by assuming that A_i is the "state" of both $\mathbb{A}$ and $\mathbb{B}$.[47]

By introducing the effective particles, one can greatly simplify the form of equations by writing them in a compact matrix form. For instance, to describe the effect of the unimolecular processes $A_i \leftrightarrow A_p$, we use a single matrix-form equation,

$$\frac{d\sigma_{\mathbb{A}}}{dt} = \hat{Q}_{\mathbb{A}}\sigma_{\mathbb{A}}. \tag{11}$$

Here, $\sigma_{\mathbb{A}}$ is the column-vector composed of concentrations of all A_is, that is, $\{\sigma_{\mathbb{A}}\}_i = [A_i]$; the sum of all elements of $\sigma_{\mathbb{A}}$ gives the total concentration of the $\mathbb{A}$ particle, $\sum_i \sigma_{\mathbb{A}i} = [\mathbb{A}]$. Hence, we introduce the basis of states $|A_i\rangle$ and introduce the operator matrix $\hat{Q}_{\mathbb{A}}$, which describes all unimolecular processes $A_i \leftrightarrow A_p$. Hereafter, all matrices are denoted by capital letters written with operator hats. In this basis, the matrix $\hat{Q}_{\mathbb{A}}$ has the following elements:

$$\langle A_p|\hat{Q}_{\mathbb{A}}|A_i\rangle = k_{A_i \rightarrow A_p}, \quad \langle A_i|\hat{Q}_{\mathbb{A}}|A_i\rangle = -\sum_{p \neq i}\langle A_p|\hat{Q}_{\mathbb{A}}|A_i\rangle. \tag{12}$$

The concentration vectors $\sigma_{\mathbb{B}}$ and $\sigma_{\mathbb{C}}$ are introduced in the same way in the bases of states $|B_j\rangle$ and $|C_m\rangle$, respectively. Likewise, the matrices $\hat{Q}_{\mathbb{B}}$ and $\hat{Q}_{\mathbb{C}}$ can be introduced, which describe the processes $B_j \leftrightarrow B_q$ and $C_m \leftrightarrow C_r$. The convenient matrix form thus reduces dramatically the number of kinetic equations, since the unimolecular processes only lead to transitions between the "states" of the effective particles.

In order to take into account the bimolecular processes, we need to introduce two-particle distribution functions and reaction operators. The two-particle distribution function of $\mathbb{A}$ and $\mathbb{B}$ is introduced as a column-vector $\varphi_{\mathbb{AB}}(\mathbf{r})$ written in the basis $|A_iB_j\rangle = |A_i\rangle|B_j\rangle$. This vector has the following components:

$$\{\varphi_{\mathbb{AB}}(\mathbf{r})\}_{ij} = \varphi_{A_iB_j}(\mathbf{r}). \tag{13}$$

Here, $\varphi_{A_iB_j}(\mathbf{r})$ is the pair distribution function of A_i and B_j. An important operation, which we need to introduce in order to reduce dimensionality required to obtain the single-particle distribution function, i.e., concentration, from a two-particle distribution function is the trace operator. For example, the jth component of $\mathrm{Tr}_{\mathbb{A}}\{\varphi_{\mathbb{AB}}\}$ is as follows:

$$\mathrm{Tr}_{\mathbb{A}}\{\varphi_{\mathbb{AB}}\}_j = \sum_{i=1}^{I} \varphi_{A_iB_j}(\mathbf{r}). \tag{14}$$

Trace over the $\mathbb{B}$ subspace can be taken in the same way.

Now, we can introduce the relevant operators describing the reactions (5–7) and free motion of particles. The unimolecular reactions changing the "states" of the $\mathbb{AB}$ pair are described by the $\hat{Q}_{\mathbb{AB}} = \hat{Q}_{\mathbb{A}} \oplus \hat{Q}_{\mathbb{B}}$ operator with the following elements:

$$\langle A_kB_l|\hat{Q}_{\mathbb{AB}}|A_iB_j\rangle = \delta_{jl}\langle A_k|\hat{Q}_{\mathbb{A}}|A_i\rangle + \delta_{ki}\langle B_l|\hat{Q}_{\mathbb{B}}|B_j\rangle. \tag{15}$$

To describe the bimolecular reactions (6), we introduce the operator $\hat{U}_{\mathbb{AB}}(\mathbf{r})$ with the following elements:

$$\langle A_kB_l|\hat{U}_{\mathbb{AB}}(\mathbf{r})|A_iB_j\rangle = W_{A_i+B_j\to A_k+B_l}(\mathbf{r}),$$

$$\langle A_iB_j|\hat{U}_{\mathbb{AB}}(\mathbf{r})|A_iB_j\rangle = -\sum_{k,l\neq i,j}\langle A_kB_l|\hat{U}_{\mathbb{AB}}(\mathbf{r})|A_iB_j\rangle. \tag{16}$$

The association–dissociation reactions (7) can also be described using matrix-form equations. The decay of the $\mathbb{AB}$ pair due to association can be described by means of the operator $\hat{V}_{\mathbb{AB}}(\mathbf{r})$ with the following elements:

$$\langle A_k B_l | \hat{V}_{\mathbb{AB}} | A_i B_j \rangle = -\delta_{ik} \delta_{jl} \sum_{m=1}^{M} W_{A_i + B_j \rightarrow C_m}(\mathbf{r}). \qquad (17)$$

In turn, generation of the C_m particles by association can be described by the following "association matrix" $\hat{V}_a$:

$$\langle C_m | \hat{V}_a | A_i B_j \rangle = W_{A_i + B_j \rightarrow C_m}(\mathbf{r}). \qquad (18)$$

In the general case, this matrix is non-square (when $M \neq I \cdot J$) as it performs conversion from the $|A_i B_j\rangle$ basis to the $|C_m\rangle$ basis. Hence, we can introduce a "dissociation matrix" $\hat{V}_d$ with the following elements:

$$\langle A_i B_j | \hat{V}_d | C_m \rangle = W_{C_m \rightarrow A_i + B_j}(\mathbf{r}). \qquad (19)$$

This matrix is also non-square as it performs conversion from the $|C_m\rangle$ basis to the $|A_i B_j\rangle$ basis. The rate constants introduced in Eq. (8) constitute the matrix that describes decay of $\mathbb{C}$ due to dissociation

$$\langle C_n | \hat{V}_{\mathbb{C}} | C_m \rangle = -\delta_{mn} \sum_{i,j=1}^{I,J} k_{C_m \rightarrow A_i + B_j}. \qquad (20)$$

Finally, we introduce the operator $\hat{\mathcal{L}}_{\mathbb{AB}}$ describing the relative motion of particles. This operator has the following matrix elements:

$$\langle A_k B_l | \hat{\mathcal{L}}_{\mathbb{AB}} | A_i B_j \rangle = \mathcal{L}_{A_i B_j} \delta_{ik} \delta_{jl}. \qquad (21)$$

Using the operators introduced in this subsection, we can write down the equations that describe the evolution of many-particle distribution functions of any order. To keep the description concise, we do so only for the single-particle distribution functions and two-particle distribution function of the $\mathbb{AB}$ pair. All such equations can be written in the matrix form using the reactivity operators introduced above. This is demonstrated in Subsection 3.3.

3.3.   Kinetic equations

Having introduced the effective particles and reaction operators, we can now *derive* the kinetic equations. To do so, we start with a set of *hierarchical equations*, which couple various multidimensional distribution functions. For instance, the equation for the single-particle distribution functions, which are nothing else but particle concentrations, are written in the following way (hereafter, all equations are written in the matrix form)

$$\frac{d\sigma_A}{dt} = \hat{Q}_A \sigma_A + \text{Tr}_B \int d\mathbf{r} \hat{W}_{AB}(\mathbf{r}) \varphi_{AB}(\mathbf{r}) + \text{Tr}_B \int d\mathbf{r} \hat{V}_d(\mathbf{r}) \sigma_C,$$

$$\frac{d\sigma_B}{dt} = \hat{Q}_B \sigma_B + \text{Tr}_A \int d\mathbf{r} \hat{W}_{AB}(\mathbf{r}) \varphi_{AB}(\mathbf{r}) + \text{Tr}_A \int d\mathbf{r} \hat{V}_d(\mathbf{r}) \sigma_C,$$

$$\frac{d\sigma_C}{dt} = \{\hat{Q}_C + \hat{V}_C\} \sigma_C + \int d\mathbf{r} \hat{V}_a(\mathbf{r}) \varphi_{AB}(\mathbf{r}). \tag{22}$$

Here, we used a simplified notation $\hat{W}_{AB}(\mathbf{r}) = \hat{U}_{AB}(\mathbf{r}) + \hat{V}_{AB}(\mathbf{r})$.

Hence, the changes in the concentrations of particles caused by bimolecular processes and association are given by the products of reactivities and corresponding pair distribution functions (integrated over the entire space). Dissociation leads to decay of $\mathbb{C}$ particles described by the $\hat{V}_C$ operator, giving rise to formation of $\mathbb{A}$s and $\mathbb{B}$s. In the first two equations, extra-dimensionality is reduced by taking the trace operation. The unimolecular processes (5) are treated by using the $\hat{Q}_{A,B,C}$ operators. To proceed further, we need to write down the equation for $\varphi_{AB}(\mathbf{r})$

$$\left(\frac{\partial}{\partial t} - \hat{\mathcal{L}}_{AB}\right) \varphi_{AB}(\mathbf{r}) = \{\hat{Q}_{AB} + \hat{W}_{AB}(\mathbf{r})\} \varphi_{AB}(\mathbf{r})$$

$$+ \hat{V}_d(\mathbf{r}) \sigma_C + \mathcal{X}. \tag{23}$$

Here, $\mathcal{X}$ stands for the terms that arise from three-particle contributions; for the sake of simplicity, we do not present explicit expressions for these terms. Hence, the two-particle distribution function depends on three-particle functions, which depend on four-particle functions and so on. In other words, we can write down a set of hierarchical equations for distribution functions of different order.

To proceed further, *truncation* of the hierarchy is required (hierarchy closure). The idea of the IET that allows one to make such a truncation is to neglect all correlations of particles, except for the two-particle correlations. This means that reactant A_i, which encounters particle B_j, can encounter another particle, e.g., particle

B_l, only after the encounter with B_j is accomplished so that A_i and B_j separate by diffusion and completely loose correlation with each other. To perform the truncation under these assumptions, we write down the expression for $\varphi_{AB}(\mathbf{r}, t)$ in the following way:

$$\varphi_{AB}(\mathbf{r}, t) = \sigma_A(t) \otimes \sigma_B(t) + \sigma_{AB}(\mathbf{r}, t). \tag{24}$$

Hence, we present the pair distribution as a sum of a completely uncorrelated contribution (which is simply given by the products of particle concentrations) and a contribution describing *correlations* of particles. The equation for the new function $\sigma_{AB}(\mathbf{r}, t)$ can be obtained from Eqs. (23) and (24)

$$\left(\frac{\partial}{\partial t} - \hat{\mathcal{L}}_{AB} - \hat{Q}_{AB} \right) \sigma_{AB}(\mathbf{r}, t)$$

$$\approx \hat{W}_{AB}(\mathbf{r}) \left\{ \sigma_{AB}(\mathbf{r}, t) + \sigma_A(t) \otimes \sigma_B(t) \right\} + \hat{V}_d(\mathbf{r})\sigma_C(t). \tag{25}$$

Here, the approximate equality sign means that we have completely neglected all higher-order correlations of particles. Specifically, we take into account only two-particle correlations of $\mathbb{A}$ and $\mathbb{B}$ and completely neglect all other correlations. One should note that this does not mean that higher-order distribution functions are zero: these distributions are non-zero but they are uncorrelated. At the same time, we also neglect correlations between $\mathbb{A}$ and $\mathbb{C}$ (and also between $\mathbb{B}$ and $\mathbb{C}$): One can show that these correlations give rise to higher-order corrections.

Equation (25) for $\sigma_{AB}(\mathbf{r}, t)$ can be solved using the Green function technique

$$\sigma_{AB}(\mathbf{r}, t) = \int d\mathbf{r}_0 \int_0^t d\tau \hat{G}_{AB}(\mathbf{r}|\mathbf{r}_0, t - \tau)$$

$$\times [\hat{W}_{AB}(\mathbf{r}_0)\sigma_A(\tau) \otimes \sigma_B(\tau) + \hat{V}_d(\mathbf{r}_0)\sigma_C(\tau)]. \tag{26}$$

Here, as it is usually done, the Green function is introduced as a solution of the following equation:

$$\left(\frac{\partial}{\partial t} - \hat{\mathcal{L}}_{AB} - \hat{Q}_{AB} - \hat{W}_{AB}(\mathbf{r}) \right) \hat{G}_{AB}(\mathbf{r}|\mathbf{r}_0, t)$$

$$= \frac{\delta(r - r_0)}{4\pi r r_0} \delta(t)\hat{E}, \tag{27}$$

with $\hat{E}$ being the unity matrix of a dimensionality $IJ \times IJ$. Inserting the expression (26) for σ_{AB} into Eqs. (22), we arrive at closed equations for the concentration vectors, i.e., at the IET equations

$$\frac{d\sigma_A(t)}{dt} = \hat{Q}_A \sigma_A(t) + \text{Tr}_B \int_0^t d\tau \hat{\Sigma}_{AB}(t-\tau)\sigma_A(\tau) \otimes \sigma_B(\tau)$$

$$+ \text{Tr}_B \int_0^t d\tau \hat{\Sigma}_d(t-\tau)\sigma_C(\tau),$$

$$\frac{d\sigma_B(t)}{dt} = \hat{Q}_B \sigma_B(t) + \text{Tr}_A \int_0^t d\tau \hat{\Sigma}_{AB}(t-\tau)\sigma_A(\tau) \otimes \sigma_B(\tau)$$

$$+ \text{Tr}_A \int_0^t d\tau \hat{\Sigma}_d(t-\tau)\sigma_C(\tau),$$

$$\frac{d\sigma_C(t)}{dt} = \hat{Q}_C \sigma_C(t) + \int_0^t d\tau \hat{\Sigma}_a(t-\tau)\sigma_A(\tau) \otimes \sigma_B(\tau)$$

$$+ \int_0^t d\tau \hat{\Sigma}_C(t-\tau)\sigma_C(\tau). \tag{28}$$

One can readily see that these equations are integro-differential. The kernels of the equations are expressed via chemical reactivities and parameters of relative motion of $\mathbb{A}$ and $\mathbb{B}$; hence, the IET is indeed a binary theory. The expressions for the kernels are as follows:

$$\hat{\Sigma}_{AB}(t) = \int d\mathbf{r} \hat{W}_{AB}(\mathbf{r}) \left[\delta(t) + \int d\mathbf{r}_0 \hat{G}_{AB}(\mathbf{r}|\mathbf{r}_0, t)\hat{W}_{AB}(\mathbf{r}_0) \right],$$

$$\hat{\Sigma}_d(t) = \int d\mathbf{r} \left[\hat{V}_d(\mathbf{r})\delta(t) + \hat{W}_{AB}(\mathbf{r}) \int d\mathbf{r}_0 \hat{G}_{AB}(\mathbf{r}|\mathbf{r}_0, t)\hat{V}_d(\mathbf{r}_0) \right],$$

$$\hat{\Sigma}_a(t) = \int d\mathbf{r} \hat{V}_a(\mathbf{r}) \left[\delta(t) + \int d\mathbf{r}_0 \hat{G}_{AB}(\mathbf{r}|\mathbf{r}_0, t)\hat{W}_{AB}(\mathbf{r}_0) \right],$$

$$\hat{\Sigma}_d(t) = \hat{V}_C \delta(t) + \int d\mathbf{r} \hat{V}_a(\mathbf{r}) \int d\mathbf{r}_0 \hat{G}_{AB}(\mathbf{r}|\mathbf{r}_0, t)\hat{V}_d(\mathbf{r}_0). \tag{29}$$

The physical meaning of the IET kernels is as follows. (i) $\hat{\Sigma}_{AB}$ takes into account all bimolecular processes and the decay of $\mathbb{A}$ and $\mathbb{B}$ due to association. The off-diagonal elements of this operator determine the rates of the $A_i + B_j \rightarrow A_k + B_l$ processes, whereas the diagonal elements give the total decay of the A_i and B_j reactants in the course of the biomolecular reactions and association processes. (ii) The kernel $\hat{\Sigma}_d$ describes formation of the $\mathbb{A}$ and $\mathbb{B}$ particles by dissociation of the C_m reactants. This matrix is generally a non-square matrix (except for the special case $M = I \cdot J$) and its elements determine the effective rates of processes $C_m \to A_i + B_j$. (iii) The kernel $\hat{\Sigma}_a$ describes formation of $\mathbb{C}$s by association of $\mathbb{A}$ and $\mathbb{B}$ particles; its elements give the rates of the processes $A_i + B_j \to C_m$ (this matrix is also non-square, unless $M = I \cdot J$). (iv) Finally, the kernel $\hat{\Sigma}_{\mathbb{C}}$ describes the decay of $\mathbb{C}$s upon dissociation. Interestingly, this matrix is not necessarily diagonal (in contrast to the $\hat{V}_{\mathbb{C}}$ matrix). The diagonal elements give the effective rates of decay of C_m particles upon dissociation. At the same time, there are also off-diagonal elements of $\hat{\Sigma}_{\mathbb{C}}$ corresponding to the processes $C_m \to C_r$. The physical origin of such processes is as follows. First, C_m dissociates into a pair $\{A_i + B_j\}$, which is trapped in the solvent cage and thus cannot separate immediately. Hence, this pair can recombine through the pathway $A_i + B_j \to C_r$ effectively giving rise to the process $C_m \to C_r$. There are also additional pathways, which can give rise to the same process, i.e., $C_m \to A_i + B_j \to A_k + B_l \to C_r$. It is also important to emphasize that (i) all four terms, which stand in Eqs. (28) for the bimolecular reactions and association–dissociation processes, are integral convolutions and that (ii) the rates of all processes depend on the relative mobility of $\mathbb{A}$ and $\mathbb{B}$. This holds even for the diagonal elements of $\hat{\Sigma}_{\mathbb{C}}$, which describe dissociation reactions. The reason is that after dissociation of C_m, the pair $\{A_i + B_j\}$ does not separate immediately and can recombine in the solvent cage producing the $\mathbb{C}$ particle again. For this reason, the effective rates of dissociation also depend on relative mobility of $\mathbb{A}$ and $\mathbb{B}$ as has been first demonstrated by Berg.[59] This effect is discussed below in more detail.

Further extensions of the IET can be found in the literature. For instance, one can generalize the approach to take into account[57] the presence of internal quantum states of reactants A_i, B_j and C_m, i.e., $|\lambda_{A_i}\rangle$, $|\mu_{B_j}\rangle$ and $|\nu_{C_m}\rangle$, respectively. One can also extend the treatment to the case where at $t = 0$ there are correlated pairs of reactants $\mathbb{A}$ and $\mathbb{B}$ present in the system. This can be done by formally assuming that one of the C_m reactants dissociates irreversibly at an infinite rate. In this situation, geminate pairs $\{\mathbb{A} + \mathbb{B}\}$ are produced instantaneously: recombination of these pairs and interplay between geminate and bulk reactions can also be taken into account by means of the IET.[57] Here, we will not dwell on these extensions of the IET approach and recommend an interested reader to consult with the original publication.[57]

Generally speaking, the approach of "effective particles" presented here is very useful for writing down kinetic equations and sets of hierarchical equations for distribution functions in a compact matrix form. Truncation of hierarchies written in the matrix form is also possible; for instance, methods for deriving the general formulation of modified encounter theory using matrix-form equations have been previously discussed.[46] Here, for the sake of simplicity and clarity, we do not present any detail of such derivations.

4. Examples

4.1. *Contact approximation*

To simplify calculation of the IET kernels, we restrict ourselves to contact approximation for the bimolecular processes between A_i and B_j and association–dissociation processes. Contact approximation implies that the reactivity operators are as follows:

$$\hat{W}_{\mathbb{AB}}(\mathbf{r}) = \hat{K}_{\mathbb{AB}} \frac{\delta(r - R)}{4\pi R^2}, \quad \text{where} \quad \hat{K}_{\mathbb{AB}} = \int d\mathbf{r} \hat{W}_{\mathbb{AB}}(\mathbf{r}),$$

$$\hat{V}_a(\mathbf{r}) = \hat{K}_a \frac{\delta(r - R)}{4\pi R^2}, \quad \text{where} \quad \hat{K}_a = \int d\mathbf{r} \hat{V}_a(\mathbf{r}),$$

$$\hat{V}_d(\mathbf{r}) = \hat{K}_d \frac{\delta(r - R)}{4\pi R^2}, \quad \text{where} \quad \hat{K}_d = \int d\mathbf{r} \hat{V}_d(\mathbf{r}). \tag{30}$$

Here, the contact distance R is taken the same for all $\{A_i + B_j\}$ pairs. Hence, the elements of these K-matrices are the reaction rate constants of the corresponding processes, for instance,

$$\langle A_k B_l | \hat{K}_{AB} | A_i B_j \rangle = k_{A_i + B_j \to A_k + B_l} = \int d\mathbf{r} W_{A_i + B_j \to A_k + B_l}(r).$$

(31)

Using the definitions (30) of the K-matrices, we can dramatically simplify the expressions for the kernels because integration over $\mathbf{r}$ and $\mathbf{r}_0$ in Eqs. (29) becomes straightforward. As a result, we obtain the IET kernels

$$\hat{\Sigma}_{AB}(t) = \hat{K}_{AB}[\delta(t) + \hat{G}_{AB}(R|R,t)\hat{K}_{AB}],$$
$$\hat{\Sigma}_d(t) = [\delta(t) + \hat{K}_{AB}\hat{G}_{AB}(R|R,t)]\hat{K}_d,$$
$$\hat{\Sigma}_a(t) = \hat{K}_a[\delta(t) + \hat{G}_{AB}(R|R,t)\hat{K}_{AB}],$$
$$\hat{\Sigma}_{\mathbb{C}}(t) = \hat{V}_{\mathbb{C}}\delta(t) + \hat{K}_a\hat{G}_{AB}(R|R,t)\hat{K}_d.$$

(32)

It is convenient to perform calculations in the Laplace domain. Hereafter, the Laplace transform of a function $f(t)$ is denoted as $\tilde{f}(s)$ with s being the Laplace variable. As usual, the Laplace-transformed quantity is calculated as follows:

$$\tilde{f}(s) = \int_0^\infty f(t)e^{-ts}dt.$$

(33)

Calculation of the Green function can be simplified by introducing the "free" Green function satisfying a simpler equation (with all bimolecular processes neglected)

$$\left(\frac{\partial}{\partial t} - \hat{\mathcal{L}}_{AB} - \hat{Q}_{AB}\right)\hat{\mathcal{G}}_{AB}(r|r_0, t) = \frac{\delta(r - r_0)}{4\pi r r_0}\delta(t)\hat{E},$$

(34)

and the following relations between the Laplace transforms of both Green functions:[48]

$$\hat{\tilde{G}}_{AB}(R|R,s) = \hat{\tilde{\mathcal{G}}}_{AB}(R|R,s)[\hat{E} - \hat{K}_{AB}\hat{\tilde{\mathcal{G}}}_{AB}(R|R,s)]^{-1}$$
$$= [\hat{E} - \hat{\tilde{\mathcal{G}}}_{AB}(R|R,s)\hat{K}_{AB}]^{-1}\hat{\tilde{\mathcal{G}}}_{AB}(R|R,s).$$

(35)

Calculation of $\hat{\tilde{\mathcal{G}}}_{AB}(R|R,s)$ is discussed further in this contribution.

Hence, one can see that in the Laplace domain, the problem of calculating the kernels becomes a standard linear algebra problem.

4.2. *Elementary reactions*

Before introducing applications of the IET to complex chemical processes, let us consider an example of an irreversible single-stage reaction,

$$A + B \rightarrow C. \tag{36}$$

In this case, the effective particles are $\mathbb{A} = \{A\}$, $\mathbb{B} = \{B\}$, $\mathbb{C} = \{C\}$ (each "particle" has only one "state"). Hence, we obtain that the matrices of rate constants become scalars

$$K_{\mathbb{AB}} = -k_a, \quad K_d = 0, \quad K_a = k_a, \quad V_{\mathbb{C}} = 0. \tag{37}$$

Here, k_a is the reaction rate constant of association. In this situation, from Eq. (32) we obtain that $\Sigma_{\mathbb{AB}} = -\Sigma_a = -\Sigma_{\text{iet}}$, while the other two kernels are equal to zero; all kernels also become scalars. Hence, the IET equation takes the form

$$\frac{d}{dt} N_A(t) = \frac{d}{dt} N_B(t) = -\frac{d}{dt} N_C(t)$$

$$= - \int_0^t \Sigma_{\text{iet}}(t - \tau) N_A(\tau) N_B(\tau) d\tau. \tag{38}$$

The IET kernel can be evaluated from Eqs. (32) and (35)

$$\tilde{\Sigma}_{\text{iet}}(s) = k_a[1 - k_a \tilde{G}_{\mathbb{AB}}(R|R, s)] = \frac{k_a}{1 + k_a \tilde{G}_{\mathbb{AB}}(R|R, s)}. \tag{39}$$

When reactant motion is given by free diffusion in three dimensions ($D(r) = D = \text{const}, V(r) = 0$), for the Green function we can use the following expression:[41]

$$\tilde{G}_{\mathbb{AB}}(R|R, s) = \frac{1}{k_D} \mathcal{G}(s), \quad \mathcal{G}(s) = \frac{1}{1 + \sqrt{s\tau_D}}. \tag{40}$$

The integro-differential IET equation (38) can be reduced to a differential equation,

$$\frac{d}{dt}N_A(t) = \frac{d}{dt}N_B(t) = -\mathcal{K}(t)N_A(t)N_B(t). \tag{41}$$

Here, the rate constant and its steady-state value are defined as follows:

$$\mathcal{K}(t) = \int_0^t \Sigma_{\text{iet}}(\tau)d\tau, \quad k = \mathcal{K}(\infty) = \int_0^\infty \Sigma_{\text{iet}}(\tau)d\tau. \tag{42}$$

To obtain the differential equation (41), we assume that $N_{A,B}$ $(t - \tau) \approx N_{A,B}(t)$ in the integrand of Eq. (38). This approximation is justified because the IET kernel decays on the time scale of $t \approx \tau_D$. One should note, however, that such an approximation fails at long times because Σ_{iet} is not an exponentially decaying function of time having a long-term power-law tail. As shown below, reduction of the IET equations to differential equations may also fail when fast unimolecular reactions are present.

If we assume that the reaction (36) is reversible, for calculating the IET kernels we need to keep in mind that the K_d and $V_{\mathbb{C}}$ rate constants are non-zero

$$K_{\mathbb{AB}} = -k_a, \quad K_d = k_d, \quad K_a = k_a, \quad V_{\mathbb{C}} = -k_d. \tag{43}$$

Here, k_d is the dissociation rate constant. The IET kernels remain scalars and they are as follows:

$$\Sigma_{\mathbb{AB}} = -\Sigma_a = -\Sigma_{\text{iet}}, \quad \Sigma_{\mathbb{C}} = -\Sigma_d = -\Sigma'_{\text{iet}} = -\frac{k_d}{k_a}\Sigma_{\text{iet}}. \tag{44}$$

The IET kernel Σ_{iet} remain the same as before; it is given by Eq. (39). Hence, the kinetic equations take the form

$$\frac{d}{dt}N_A(t) = \frac{d}{dt}N_B(t) = -\frac{d}{dt}N_C(t)$$

$$= -\int_0^t \Sigma_{\text{iet}}(t - \tau)N_A(\tau)N_B(\tau)d\tau$$

$$+ \int_0^t \Sigma'_{\text{iet}}(t - \tau)N_C(\tau)d\tau. \tag{45}$$

The additional term containing Σ'_{iet} describes the reverse dissociation process.

If we evaluate the steady-state rate constants of association–dissociation

$$k_a^{\text{eff}} = \tilde{\Sigma}_{\text{iet}}(s = 0) = \frac{k_a k_D}{k_a + k_D}, \qquad k_d^{\text{eff}} = \tilde{\Sigma}'_{\text{iet}}(s = 0) = k_d \frac{k_D}{k_a + k_D},$$

(46)

we notice that the effective dissociation rate constant is given not only by the intrinsic dissociation rate k_d but it also depends on k_D, i.e., on the parameters of relative mobility of A and B. This result is conditioned by the cage effect: when the geminate pair $\{A + B\}$ is formed after dissociation of C, the pair can recombine back (before A and B separate) forming the C particle, see Fig. 2. As a consequence, the effective rate of producing A and B in the solvent bulk, i.e., k_d^{eff}, is reduced as compared to k_d. The probability of pair separation is equal to $p = k_D/(k_a + k_D)$; hence, the effective dissociation rate is equal to $p \cdot k_d$, in full agreement with Eq. (46) and with earlier findings of Berg.[59] It is worth noting that the values of the rate constants given by Eq. (46) are consistent with the detailed balance principle: the ratio $k_a^{\text{eff}}/k_d^{\text{eff}} = K_{eq} = k_a/k_d$ does not depend on relative mobility of A and B, which is possible only when k_d^{eff} depends on the k_D value.

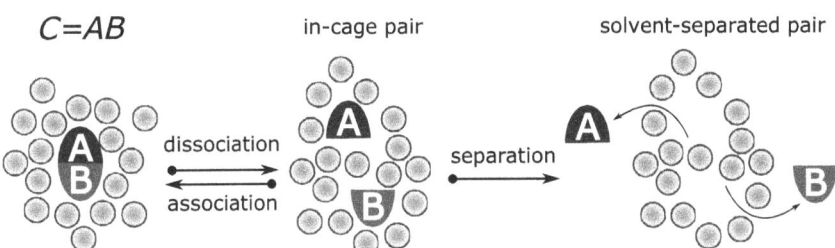

Fig. 2. Scheme illustrating the impact of the cage effect on the dissociation rate constant. Upon dissociation of $C = AB$, an in-cage pair $A + B$ is formed, which can recombine back to produce C or separate. As a consequence, the effective dissociation rate constant is smaller than the intrinsic rate constant k_d, since it is the product of k_d and the separation probability of the in-cage pair.

This simple example illustrates the importance of the cage effect and its impact on rate constants of diffusion-influenced processes. The influence of the cage effect on the rate constants of reactions in solutions can be taken into account by means of the IET, which is a universal method for calculating reaction rate constants, as demonstrated in Subsection 4.3.

4.3. *Rate constants of multistage reactions*

As has been mentioned above, the IET equations can be reduced to the differential form once the IET kernels change with time much faster than the reactant concentrations. The same idea can be used in the case of a multistage reaction. In this subsection, we exploit this idea assuming that there are no unimolecular reactions, i.e., $\hat{Q}_{A,B,C}$ are zero. For the sake of simplicity, we consider only the steady-state limit of the IET equations. Hence, the integral terms in the kinetic equations are modified as follows:

$$\int_0^t \hat{\Sigma}_{AB}(t-\tau)\sigma_A(\tau)\otimes\sigma_B(\tau)d\tau$$

$$\approx \left[\int_0^t \hat{\Sigma}_{AB}(\tau)d\tau\right]\sigma_A(t)\otimes\sigma_B(t) \approx \hat{R}_{AB}\sigma_A(t)\otimes\sigma_B(t). \quad (47)$$

Hence, the matrix of the rate constants can be expressed via the Laplace transform of the IET kernel at $s = 0$

$$\hat{R}_{AB} = \int_0^\infty \hat{\Sigma}_{AB}(t)dt = \hat{\tilde{\Sigma}}_{AB}(s=0). \quad (48)$$

The other three matrices of the rate constants can be computed in the same way, giving rise to differential equations for the concentrations

$$\frac{d\sigma_A}{dt} = \text{Tr}_B\{\hat{R}_{AB}\sigma_A\otimes\sigma_B\} + \text{Tr}_B\{R_d\sigma_C\},$$

$$\frac{d\sigma_B}{dt} = \text{Tr}_A\{\hat{R}_{AB}\sigma_A\otimes\sigma_B\} + \text{Tr}_A\{\hat{R}_d\sigma_C\},$$

$$\frac{d\sigma_C}{dt} = \hat{R}_a\sigma_A\otimes\sigma_B + \hat{R}_C\sigma_C. \quad (49)$$

Now, let us evaluate the R-matrices of interest. When $\hat{Q}_{\mathbb{A},\mathbb{B}} = 0$, the free Green function is a diagonal matrix

$$\langle A_k B_l | \hat{\tilde{\mathcal{G}}}_{\mathbb{A}\mathbb{B}}(R|R, s) | A_i B_j \rangle = \delta_{ik} \delta_{jl} \frac{1}{4\pi R D_{ij}} \mathcal{G}(s) \xrightarrow{s \to 0} \delta_{ik} \delta_{jl} \frac{1}{k_{D_{ij}}},$$

(50)

where $k_{D_{ij}} = 4\pi R D_{ij}$ is the diffusional rate constant for the $A_i + B_j$ pair of reactants. Using the expressions for the free Green function, contact value of the $\hat{\tilde{\mathcal{G}}}_{\mathbb{A}\mathbb{B}}$ Green function and definition of the IET kernels, we obtain

$$\hat{R}_{\mathbb{A}\mathbb{B}} = [\hat{E} - \hat{K}_{\mathbb{A}\mathbb{B}} \hat{\mathcal{G}}_c]^{-1} \hat{K}_{\mathbb{A}\mathbb{B}} = \hat{K}_{\mathbb{A}\mathbb{B}} [\hat{E} - \hat{\mathcal{G}}_c \hat{K}_{\mathbb{A}\mathbb{B}}]^{-1},$$

$$\hat{R}_d = [\hat{E} - \hat{K}_{\mathbb{A}\mathbb{B}} \hat{\mathcal{G}}_c]^{-1} \hat{K}_d, \quad \hat{R}_a = \hat{K}_a [\hat{E} - \hat{\mathcal{G}}_c \hat{K}_{\mathbb{A}\mathbb{B}}]^{-1},$$

$$\hat{R}_{\mathbb{C}} = \hat{K}_{\mathbb{C}} + \hat{K}_a \hat{\mathcal{G}}_c \hat{R}_d = \hat{K}_{\mathbb{C}} + \hat{R}_a \hat{\mathcal{G}}_c \hat{K}_d.$$

(51)

Here, $\hat{\mathcal{G}}_c = \hat{\tilde{\mathcal{G}}}_{\mathbb{A}\mathbb{B}}(R|R, s = 0)$, i.e., it is the contact value of the free Green function at $s = 0$.

Let us demonstrate how this method of calculating rate constants works for a specific example.[60] Here, we consider a three-stage reaction

$$A_1 + B_1 \leftrightharpoons A_2 + B_2 \leftrightharpoons C.$$

(52)

Let the reaction rates for the $A_1 + B_1 \leftrightharpoons A_2 + B_2$ process be k_f (for the forward process) and k_b (for the backward process); let the rates of the $A_2 + B_2 \leftrightharpoons C$ association–dissociation reaction be k_a and k_d; the diffusional rate constant is taken the same for the $A_1 + B_1$ pair and for the $A_2 + B_2$ pair, being equal to k_D. In the case under study, the effective particles are

$$\mathbb{A} = \begin{pmatrix} A_1 \\ A_2 \end{pmatrix}, \quad \mathbb{B} = \begin{pmatrix} B_1 \\ B_2 \end{pmatrix}, \quad \mathbb{C} = (C).$$

(53)

Hence, the concentration vectors of interest are

$$\sigma_{\mathbb{A}} = \begin{pmatrix} N_{A_1} \\ N_{A_2} \end{pmatrix}, \quad \sigma_{\mathbb{B}} = \begin{pmatrix} N_{B_1} \\ N_{B_2} \end{pmatrix},$$

$$\sigma_{\mathbb{C}} = (N_C), \quad \sigma_{\mathbb{A}} \otimes \sigma_{\mathbb{B}} = \begin{pmatrix} N_{A_1} N_{B_1} \\ N_{A_1} N_{B_2} \\ N_{A_2} N_{B_1} \\ N_{A_2} N_{B_2} \end{pmatrix}. \tag{54}$$

The relevant matrices of rate constants can be written as follows:

$$\hat{K}_{\mathbb{AB}} = \begin{pmatrix} -k_f & 0\,0 & k_b \\ 0 & 0\,0 & 0 \\ 0 & 0\,0 & 0 \\ k_f & 0\,0 & -k_b - k_a \end{pmatrix}, \quad \hat{K}_d = \begin{pmatrix} 0 \\ 0 \\ 0 \\ k_d \end{pmatrix},$$

$$\hat{K}_a = (0\ 0\ 0\ k_a), \ \hat{K}_{\mathbb{C}} = -k_d. \tag{55}$$

By using Eqs. (51), (55) and straightforward matrix multiplication, one can obtain the following matrices of reaction rates:

$$\hat{R}_{\mathbb{AB}} = \begin{pmatrix} -k_1 & 0\,0 & k_2 \\ 0 & 0\,0 & 0 \\ 0 & 0\,0 & 0 \\ k_5 & 0\,0 & -k_4 \end{pmatrix}, \quad \hat{R}_d = \begin{pmatrix} k_3 \\ 0 \\ 0 \\ k_6 \end{pmatrix},$$

$$\hat{R}_a = (k_8\ 0\ 0\ k_9), \qquad \hat{R}_{\mathbb{C}} = -k_7. \tag{56}$$

Using these matrices, we obtain the kinetic equations for concentrations

$$\frac{dN_{A_1}}{dt} = \frac{dN_{B_1}}{dt} = -k_1 N_{A_1} N_{B_1} + k_2 N_{A_2} N_{B_2} + k_3 N_C,$$

$$\frac{dN_{A_2}}{dt} = \frac{dN_{B_2}}{dt} = -k_4 N_{A_2} N_{B_2} + k_5 N_{A_1} N_{B_1} + k_6 N_C,$$

$$\frac{dN_C}{dt} = -k_7 N_C + k_8 N_{A_1} N_{B_1} + k_9 N_{A_2} N_{B_2}. \tag{57}$$

The rate constants introduced here are

$$k_1 = \frac{k_f}{Z}\left(1 + \frac{k_a}{k_D}\right), \quad k_2 = \frac{k_b}{Z}, \quad k_3 = \frac{k_d}{Z} \cdot \frac{k_b}{k_D},$$

$$k_4 = \frac{1}{Z}\left(k_a + k_b + \frac{k_a k_f}{k_D}\right), \quad k_5 = \frac{k_f}{Z}, \quad k_6 = \frac{k_d}{Z}\left(1 + \frac{k_f}{k_D}\right),$$

$$k_7 = \frac{k_d}{Z}\left(1 + \frac{k_b + k_f}{k_D}\right), \quad k_8 = \frac{k_a}{Z} \cdot \frac{k_f}{k_D}, \quad k_9 = \frac{k_a}{Z}\left(1 + \frac{k_f}{k_D}\right),$$

$$(58)$$

where

$$Z = \left(1 + \frac{k_a}{k_D}\right)\left(1 + \frac{k_f}{k_D}\right) + \frac{k_b}{k_D}.$$

Hence, the matrix-form IET gives a straightforward way to evaluate rate constants of multistage diffusion-influenced processes.

By analyzing the results, one can notice two remarkable properties of diffusion-influenced reactions.

First, the rate constant of dissociation is diffusion-influenced. The reason is the same as in the previous example:[59] After dissociation of C and formation of the pair $A_2 + B_2$ reactants do not separate immediately and react back to produce C. Hence, the dissociation rate effectively becomes lower, furthermore, it is influenced by the relative mobility of A_2 and B_2 because the probability of the in-cage reaction depends on the interplay between diffusional separation and association reaction. One can also see that the ratio of the rate constants of the corresponding processes, i.e., k_6/k_9, is equal to the ratio of the reaction rate constants, k_d/k_a being in full accordance with the detailed balance principle. Hence, since the association rate is diffusion-influenced, the dissociation rate becomes diffusion-influenced as well.

Second, one can see that a diffusion-influenced reaction has additional pathways. In the present case, we set the reaction rate of the process $A_1 + B_1 \rightleftharpoons C$ equal to zero. Nonetheless, as follows from the kinetic equations (57), such a reaction pathway exists because the effective rate constants k_3 and k_8 are non-zero. This means that the diffusion-influenced reaction has opened additional pathway. The physical reason for this is as follows. When A_1 and B_1 encounter each other, they form an in-cage pair, which can react to produce A_2 and B_2, which also do not separate immediately but can associate producing the C reactant. This means that encounters of A_1 and B_1 can give rise to formation of C; hence, the k_8 rate constant is non-zero. The origin of $k_3 \neq 0$ is similar. When C dissociates, it produces the in-cage pair of A_2 and B_2, which has a probability (before the reactants separate by diffusion) to react producing A_1 and B_1 so that k_3 is also non-zero.

This simple example demonstrates the usefulness of the IET for evaluating the rate constants of diffusion-influenced reactions. Extension of the formalism to more complex reaction schemes is straightforward, since calculation of the rate constants is a formal linear algebra problem that requires matrix inversion and multiplication. It is worth noting that the method presented here can account for unusual effects in diffusion-influenced reactions, such as the dependence of dissociation rate constants on mobility of reactants and the presence of additional reaction pathways due to in-cage reactions.

4.4. *Reactions of metastable particles*

Correct description of the Markovian stage of chemical reactions is not the only merit of the IET, which is essentially a non-Markovian theory. In this subsection, we discuss two examples that demonstrate that the Markovian description is not valid for a general chemical process even at long times. In this situation, using the IET is a remedy: the description using the integro-differential IET equations is free of any inconsistencies. To Illustrate this, here we consider reactions involving metastable particles with finite lifetimes.

We consider two examples of chemical reactions following previous works.[48,61–63]

The first example is inter-molecular energy transfer

$$A^* + B \rightleftharpoons A + B^*, \quad A^* \rightarrow A, \quad B^* \rightarrow B. \qquad (59)$$

Hence, excitation is transferred between A and B; we also assume that the excited-state molecules A^* and B^* are metastable and decay to their ground state with the characteristic times τ_A and τ_B, respectively. For simplicity, we assume that the concentrations of A and B, denoted as c_A and c_B, are much greater than those of A^* and B^*, denoted as N_A and N_B. Hence, in this subsection the notations for concentrations are different from those used in the rest of the chapter.

The second example is reversible binding involving metastable particles

$$A^* + B \rightleftharpoons C^*, \quad A^* \to A, \quad C^* \to C. \tag{60}$$

Like in the previous case, A^* and C^* are metastable particles decaying to form A and C with characteristic times τ_A and τ_C, respectively. Excited-state reversible proton transfer to the solvent from a photo-acid (i.e., a molecule that becomes more acidic after light absorption)[64–69] is a prominent example of such a reaction. Another example is given by reversible formation of exciplexes.[70–74] For reaction (60), we assume that the concentration of B, denoted as c_B, is much greater than that of A^* and C^*, which are denoted as N_A and N_C, respectively.

Let us first describe the kinetics of inter-molecular energy transfer by using the IET. In this case, as there are no association–dissociation stages, we need to introduce only two effective particles,

$$\mathbb{A} = \begin{pmatrix} A^* \\ A \end{pmatrix}, \quad \mathbb{B} = \begin{pmatrix} B^* \\ B \end{pmatrix}. \tag{61}$$

To calculate the IET kernel (in this case $\hat{\Sigma}_{\mathbb{AB}}$ as all other kernels are equal to zero) we need to specify the operators $\hat{Q}_{\mathbb{AB}}$ and $\hat{K}_{\mathbb{AB}}$ in the basis of states $|AB\rangle, |A^*B\rangle, |AB^*\rangle, |A^*B^*\rangle$:[62,63]

$$\hat{Q}_{\mathbb{AB}} = \begin{pmatrix} 0 & 1/\tau_A & 1/\tau_B & 0 \\ 0 & -1/\tau_A & 0 & 1/\tau_B \\ 0 & 0 & -1/\tau_B & 1/\tau_A \\ 0 & 0 & 0 & -1/\tau_A - 1/\tau_B \end{pmatrix}, \quad \hat{K}_{\mathbb{AB}} = \begin{pmatrix} 0 & 0 & 0 & 0 \\ 0 & -k_f & k_b & 0 \\ 0 & k_f & -k_b & 0 \\ 0 & 0 & 0 & 0 \end{pmatrix}. \tag{62}$$

Here, k_f and k_b are the reaction rate constants of the forward and backward energy transfer reactions. To make the treatment simpler, one can make all calculations for 2×2 matrices in the space spanned by the $|A^*B\rangle, |AB^*\rangle$ states. In this space, the free Green function and $\hat{K}_{AB}$ are as follows:

$$\hat{\mathcal{G}}_{AB} = \frac{1}{k_D} \begin{pmatrix} \mathcal{G}(s + 1/\tau_A) & 0 \\ 0 & \mathcal{G}(s + 1/\tau_B) \end{pmatrix}, \quad \hat{K}_{AB} = \begin{pmatrix} -k_f & k_b \\ k_f & -k_b \end{pmatrix}. \tag{63}$$

Hence, unimolecular decay alters the free Green function.[62,63] Straightforward operations with matrices yield the IET kernel

$$\hat{\Sigma}_{AB}(t) = \begin{pmatrix} -k_f & k_b \\ k_f & -k_b \end{pmatrix} F_1(t), \tag{64}$$

where the Laplace transform of the auxiliary function F_1 is as follows:[62,63]

$$\tilde{F}_1(s) = \frac{1}{1 + \frac{k_f}{k_D}\mathcal{G}(s + 1/\tau_A) + \frac{k_b}{k_D}\mathcal{G}(s + 1/\tau_B)}.$$

Using this expression, we can write down the IET equations for the concentrations $N_A = [A^*]$ and $N_B = [B^*]$

$$\frac{dN_A(t)}{dt} = -\frac{N_A(t)}{\tau_A} - k_f c_B \int_0^t F_1(t - \tau)N_A(\tau)d\tau$$

$$+ k_b c_A \int_0^t F_1(t - \tau)N_B(\tau)d\tau,$$

$$\frac{dN_B(t)}{dt} = -\frac{N_B(t)}{\tau_B} + k_f c_B \int_0^t F_1(t - \tau)N_A(\tau)d\tau$$

$$- k_b c_A \int_0^t F_1(t - \tau)N_B(\tau)d\tau. \tag{65}$$

An important question that we would like to address here is reduction of these non-Markovian equations to differential equations and, eventually, to Markovian equations. The standard recipe for such a reduction is to approximate $N_A(\tau)$ and $N_B(\tau)$ in the integrands of Eq. (65) as follows:

$$N_A(\tau) \approx N_A(t) \exp\left(\frac{t-\tau}{\tau_A}\right), \quad N_B(\tau) \approx N_B(t) \exp\left(\frac{t-\tau}{\tau_B}\right). \quad (66)$$

The simpler recipe $N_{A,B}(\tau) \approx N_{A,B}(t)$ obviously does not hold because due to unimolecular decay $N_{A,B}$ can change in time faster than the IET kernel. Using Eq. (66), we can write down the sought differential equations for concentrations

$$\frac{dN_A(t)}{dt} = -\frac{N_A(t)}{\tau_A} - \kappa_f(t)c_B N_A(t) + \kappa_b(t)c_A N_B(t),$$

$$\frac{dN_B(t)}{dt} = -\frac{N_B(t)}{\tau_B} + \kappa_f(t)c_B N_A(t) - \kappa_b(t)c_A N_B(t), \quad (67)$$

where the rate coefficients are

$$\kappa_f(t) = k_f \int_0^t F_1(\tau) \exp[\tau/\tau_A] d\tau,$$

$$\kappa_b(t) = k_b \int_0^t F_1(\tau) \exp[\tau/\tau_B] d\tau. \quad (68)$$

At a first glance, reduction to differential equations is possible. However, one should keep in mind that one of the steady-state rate constants

$$\kappa_f(\infty) = k_f \tilde{F}_1(s = -1/\tau_A), \quad \kappa_b(\infty) = k_b \tilde{F}_1(s = -1/\tau_B), \quad (69)$$

is formally complex when $\tau_A \neq \tau_B$. In reality, this means that the corresponding integral in Eq. (68) diverges at $t \to \infty$, i.e., reduction to the rate equations is not possible in this case. Hence, when the unimolecular decay reactions interfere with the in-cage processes reduction of the original integro-differential equations to rate equations is generally not possible.

The actual time dependences of $\kappa_f(t)$ and $\kappa_b(t)$ are shown in Fig. 3. Here, we assume that $\tau_B \to \infty$. One can see that the $\kappa_b(t)$ rate coefficient behaves in the same way as in the Smoluchowsky theory[38,39] (see discussion in Chapter 3[40]): it decreases from the value $\kappa_b(t = 0) = k_b$ to a positive steady-state value at $t \to \infty$. At the same

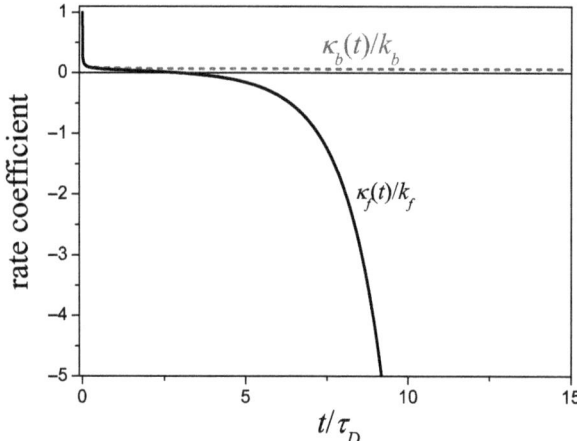

Fig. 3. Time dependence of the rate coefficients introduced in Eq. (68) divided by the corresponding reaction constants, $\kappa_f(t)/k_f$ (solid line) and $\kappa_b(t)/k_b$ (dashed line); time is given in units of τ_D. Calculation parameters: $k_f = k_b = 10k_D$, $\tau_A = \tau_D$, $\tau_B \to \infty$.

time, $\kappa_f(t)$ at $t = 0$ is also equal to the corresponding reaction rate constant k_f and then decreases. However, at longer times it goes to negative values and finally diverges at $t \to \infty$. The explanation for this behavior is as follows. When A^* reacts with B, an in-cage pair $\{A + B^*\}$ is formed, which includes the stable excitation B^* (as we assume that $\tau_B \to \infty$). Hence, excitation is accumulated in B^* particles. However, as the $\{A + B^*\}$ pair reacts back, A^* is formed again. Hence, the flux of energy changes its sign as well as the rate coefficient $\kappa_f(t)$. This effect can also be seen from the $N_B(t)$ kinetics, see Fig. 4. In this example, we assume that at $t = 0$ there are only A^* excitations; for the sake of simplicity, we set $\tau_B \to \infty$ (the B^* excitation is stable) and $c_A \to 0$ (neglecting the backward reaction $B^* + A \to B + A^*$ in the solvent bulk but not in the in-cage process). In this situation, the $N_B(t)$ concentration starts from zero and then increases reaching a constant value at $t \to \infty$ because the B^* particles do not decay to the ground state and the bulk reaction $B^* + A \to B + A^*$ is infinitely slow. Of course, this level is higher for the irreversible transfer process than for the reversible reaction;

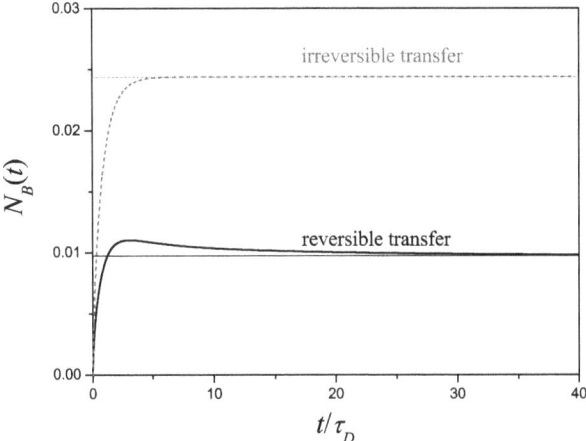

Fig. 4. Kinetics of $N_B(t)$ accumulation in the case $\tau_B \to \infty$ calculated for reversible (solid line) and irreversible (dashed line) energy transfer from A^*. One can see that in the former case the N_B value at long times is approached from above, while in the latter case it is approached from below. Calculation parameters: $k_f = 10k_D$, $k_b = 10k_D$ (reversible transfer) or $k_b = 0$ (irreversible transfer), $\tau_A = \tau_D$, $\eta = \frac{4\pi}{3}R^3 c_B = 0.01$, $c_A \to 0$. Here time is given in units of τ_D and $N_B(t)$ is given in units of $N_A(t = 0)$.

however, this is not the only difference between the two cases. When the transfer reaction is reversible, the long-time asymptotic value is approached from above, reflecting the abovementioned effect of the sign change of the energy flux. This effect is completely absent for the irreversible reaction (the asymptotic value is approached from below), in accordance with the considerations given above.

It is important to emphasize that the unusual effect of the sign change of the energy flux is of relevance when $\tau_A \neq \tau_B$ and also when the unimolecular reactions are of importance on the time scale of τ_D (i.e., $\tau_A \sim \tau_D$ or $\tau_B \sim \tau_D$) and when the bimolecular reactions are diffusion-influenced (otherwise the in-cage pairs separate too fast so that the in-cage reactions are of no relevance).

Now, let us turn to the second example and introduce the effective particles as

$$\mathbb{A} = \begin{pmatrix} A^* \\ A \end{pmatrix}, \quad \mathbb{B} = (B), \quad \mathbb{C} = \begin{pmatrix} C^* \\ C \end{pmatrix}. \tag{70}$$

The relevant operators are

$$\hat{Q}_A = \hat{Q}_{AB} = \begin{pmatrix} -1/\tau_A & 0 \\ 1/\tau_A & 0 \end{pmatrix}, \quad \hat{Q}_B = 0, \quad \hat{Q}_C = \begin{pmatrix} -1/\tau_C & 0 \\ 1/\tau_C & 0 \end{pmatrix},$$

$$\hat{K}_{AB} = \begin{pmatrix} -k_f & 0 \\ 0 & 0 \end{pmatrix}, \quad \hat{K}_a = \begin{pmatrix} k_f & 0 \\ 0 & 0 \end{pmatrix}, \quad \hat{K}_d = \begin{pmatrix} k_b & 0 \\ 0 & 0 \end{pmatrix},$$

$$\hat{K}_C = \begin{pmatrix} -k_b & 0 \\ 0 & 0 \end{pmatrix}. \tag{71}$$

The same procedure as before gives the following IET kernels (in this case all four kernels need to be computed)

$$\hat{\Sigma}_{AB} = \begin{pmatrix} -k_f & 0 \\ 0 & 0 \end{pmatrix} F_2, \quad \hat{K}_a = \begin{pmatrix} k_f & 0 \\ 0 & 0 \end{pmatrix} F_2,$$

$$\hat{K}_d = \begin{pmatrix} k_b & 0 \\ 0 & 0 \end{pmatrix} F_2, \quad \hat{K}_C = \begin{pmatrix} -k_b & 0 \\ 0 & 0 \end{pmatrix} F_2, \tag{72}$$

where

$$\tilde{F}_2(s) = \frac{1}{1 + \frac{k_f}{k_D}\mathcal{G}(s + 1/\tau_A)}.$$

The IET equations for the concentrations $[A^*] = N_A$ and $[C^*] = N_C$ are as follows:

$$\frac{dN_A(t)}{dt} = -\frac{N_A(t)}{\tau_A} - k_f c_B \int_0^t F_2(t - \tau) N_A(\tau) d\tau$$

$$+ k_b \int_0^t F_2(t - \tau) N_C(\tau) d\tau,$$

$$\frac{dN_C(t)}{dt} = -\frac{N_C(t)}{\tau_C} + k_f c_B \int_0^t F_2(t - \tau) N_A(\tau) d\tau$$

$$- k_b \int_0^t F_2(t - \tau) N_C(\tau) d\tau. \tag{73}$$

The method for reducing these equations to the differential form is the same as before. We approximate $N_A(\tau)$ and $N_C(\tau)$ in the integrands of Eq. (73) as follows:

$$N_A(\tau) \approx N_A(t) \exp\left(\frac{t-\tau}{\tau_A}\right), \quad N_C(\tau) \approx N_C(t) \exp\left(\frac{t-\tau}{\tau_C}\right), \quad (74)$$

and obtain differential equations

$$\frac{dN_A(t)}{dt} = -\frac{N_A(t)}{\tau_A} - \kappa_f(t)c_B N_A(t) + \kappa_b(t)N_C(t),$$

$$\frac{dN_C(t)}{dt} = -\frac{N_C(t)}{\tau_C} + \kappa_f(t)c_B N_A(t) - \kappa_b(t)N_C(t), \quad (75)$$

where the rate coefficients are

$$\kappa_f(t) = k_f \int_0^t F_2(\tau) \exp[\tau/\tau_A]d\tau,$$

$$\kappa_b(t) = k_b \int_0^t F_2(\tau) \exp[\tau/\tau_C]d\tau. \quad (76)$$

In contrast to the previous case, the rate coefficient $\kappa_f(t)$ never exhibits any unusual behavior, e.g., a sign change: at $t \to \infty$ it tends to a constant value $\kappa_f \to k_f k_D/(k_f + k_D)$. The behavior of $\kappa_b(t)$ depends on the relation between τ_A and τ_C. When $\tau_A > \tau_C$, dissociation of C^* leads to formation of in-cage pairs $\{A^* + B\}$, in which excitation is accumulated. Subsequent association of these pairs gives rise to formation of C^* with a consequence that the energy flux changes its sign, as well as the dissociation rate constant κ_b. This effect is absent for the association process because a single association event produces not a geminate pair of reactants but free C^* in solution. Hence, all in-cage processes are interrupted with a consequence that unimolecular decay does not interfere with any in-cage processes. Typical time dependences of the rate coefficients are shown in Fig. 5.

The behavior of the rate coefficients is also reflected by the kinetic behavior, see Fig. 6 illustrating these considerations. Here, we consider two different situations.

In the first case (see Fig. 6, top), we assume that initially the excitation is at the A^* molecules, and also imply a short τ_A time and an infinitely long τ_C time. In this situation, the $N_A(t)$ kinetics exhibits

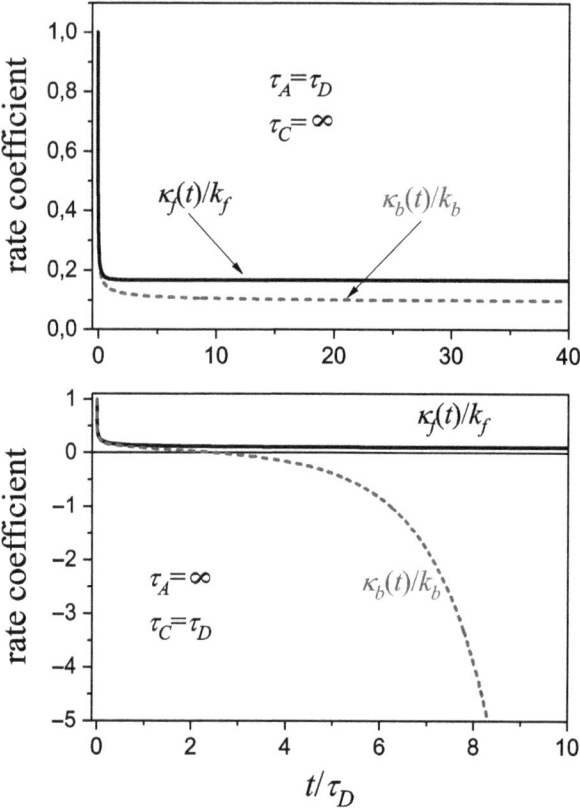

Fig. 5. Time dependence of the rate coefficiencts introduced in Eq. (76) divided by the corresponding reaction constants, $\kappa_f(t)/k_f$ (solid line) and $\kappa_b(t)/k_b$ (dashed line); time is given in units of τ_D. Calculation parameters: $k_f = 10k_D$; for the top graph $\tau_A = \tau_D$ and $\tau_C \to \infty$, for the bottom graph $\tau_A \to \infty$ and $\tau_C = \tau_D$.

fast decay at short times followed by slower decay at long times. Such a behavior is conditioned by (i) accumulation of excitation in the form of C^*, which is more stable, and (ii) the back reaction $C^* \to A^* + B$ producing the A^* particles. The $N_C(t)$ concentration increases at short times and then decays as the C^* molecules disappear in the dissociation process. The $N_C(t)$ kinetics approaches its long-time asymptote, representing decay due to dissociation, from below.

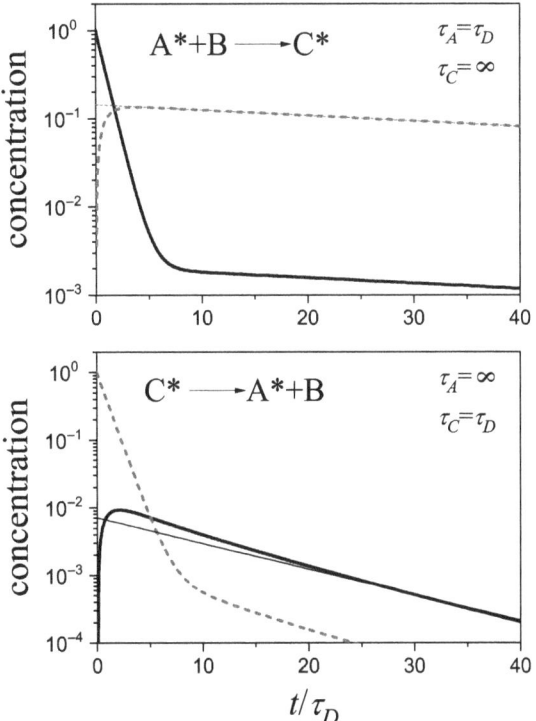

Fig. 6. $N_A(t)$ (solid line) and $N_C(t)$ (dashed line) kinetics. (top) At $t = 0$, we assume $N_C = 0$ so that all excitations exist in the form of A^*; here, $\tau_A = \tau_D$ and $\tau_C \to \infty$. One can see that the long-time asymptote of N_C (exponential decay, shown by the thin dotted line) is approached from below. (bottom) At $t = 0$, we assume $N_A = 0$ so that all excitation exist in the form of C^*; here, $\tau_A \to \infty$ and $\tau_C = \tau_D$. One can see that the long-time asymptote of N_C (exponential decay, shown by the thin solid line) is approached from above. Calculation parameters: $k_f = 10 k_D$, $\eta = \frac{4\pi}{3} R^3 c_B = 1/30$, $k_b \tau_D = 0.1$. In both subplots, time is given in units of τ_D and $N_B(t)$ is given in units of $N_A(t = 0)$ (top) or $N_C(t = 0)$ (bottom).

In the second case (see Fig. 6, bottom), we start from excited molecules in the form of C^* and assume that τ_C is short whereas τ_A is infinitely long. The concentration of the short-lived species behaves in the same way as in the previous case: fast initial decay is followed by a slower "tail" due to (i) accumulation of excitation in the form of A^*, and (ii) the back association reaction $A^* + B \to C^*$. At the same time, the kinetic behavior of the long-lived species, here A^*, exhibits an unusual behavior. As expected from qualitative considerations, the $N_A(t)$ concentration increases at short times and then decays due to association. However, the $N_A(t)$ kinetics approaches its long-time asymptote from above. The origin of the effect is the sign change of the energy flux. Hence, the change of the direction of the energy flux manifests itself not only in the rate coefficients, but also in the reaction kinetics.

5. Conclusions

In this chapter, we present the general formulation of the IET. By using the concept of "effective particles," one can dramatically simplify the form of the kinetics equations. This approach allows one to formalize derivation of the IET equations, as well as of the rate constants, for the most general reaction scheme of a multistage chemical process.

The discussion of the IET formalism is supported by several examples, starting from the simplest single-stage reaction and ending up with reversible reactions of metastable particles. For the $A + B \leftrightarrow C$ process, we derive the IET equations and evaluate the rate constants; essentially the same method can be applied to the general reaction scheme having an arbitrary number of unimolecular, bimolecular and association–dissociation stages. We also present applications of the IET to reversible reactions of metastable particles, demonstrating that Markovian description of such processes is generally not possible as the "rate constants" may diverge with time. The IET is free from this limitation, providing consistent description of the reaction kinetics. Hence, the IET is indeed a general approach to non-Markovian kinetics of complex multistage chemical reactions in liquid solutions.

Acknowledgments

The authors acknowledge support from the Ministry of Science and Higher Education of the Russian Federation (Project No. 0333-2017-0002).

References

1. P. F. Barbara, T. J. Meyer, and M. A. Ratner, Contemporary issues in electron transfer research, *J. Phys. Chem.* **100**, 13148–13168 (1996).
2. M. Terazima, Is the translational diffusion of organic radicals different from that of closed-shell molecules? *Acc. Chem. Res.* **33**, 687–694 (2000).
3. C. Turro, J. M. Zaleski, Y. M. Karabatsos, and D. G. Nocera, Bimolecular electron transfer in the Marcus inverted region, *J. Amer. Chem. Soc.* **118**, 6060–6067 (1996).
4. A. Rosspeintner, D. R. Kattnig, G. Angulo, S. Landgraf, and G. Grampp, The Rehm-Weller experiment in view of distant electron transfer, *Chem. Eur. J.* **14**, 6213–6221 (2008).
5. A. Rosspeintner and E. Vauthey, Bimolecular photoinduced electron transfer reactions in liquids under the gaze of ultrafast spectroscopy, *Phys. Chem. Chem. Phys.* **16**, 25741–25754 (2014).
6. M. Eigen, Proton transfer, acid" base catalysis, and enzymatic hydrolysis. I. elementary processes, *Angew. Chem. Intl. Ed.* **3**, 1 (1964).
7. S. Hammers-Schiffer, Proton-coupled electron transfer: Moving together and charging forward, *J. Amer. Chem. Soc.* **137**, 8860 (2015).
8. H.-X. Zhou, G. Rivas, and A. P. Minton, Macromolecular crowding and confinement: Biochemical, biophysical, and potential physiological consequences, *Annu. Rev. Biophys.* **37**, 375–397 (2008).
9. O. G. Berg, R. B. Winter, and P. H. von Hippel, Diffusion-driven mechanisms of protein translocation on nucleic acids. 1. Models and theory, *Biochem.* **20**, 6929–6948 (1981).
10. A. B. Kolomeisky, Physics of protein-DNA interactions: mechanisms of facilitated target search, *Phys. Chem. Chem. Phys.* **13**, 2088–2095 (2011).
11. L. Mirny, M. Slutsky, Z. Wunderlich, A. Tafvizi, J. Leith, and A. Kosmrlj, How a protein searches for its site on DNA: The mechanism of facilitated diffusion, *J. Phys. A: Math. Theor.* **42**, 434013 (2009).
12. H.-X. Zhou, Rapid search for specific sites on DNA through conformational switch of nonspecifically bound proteins, *Proc. Natl. Acad. Sci. USA.* **108**, 8651–8656 (2011).
13. S. Park and N. Agmon, Theory and simulation of diffusion-controlled michaelis-menten kinetics for a static enzyme in solution, *J. Phys. Chem. B.* **112**, 5977–5987 (2008).
14. G. Wilemski and M. Fixman, Diffusion-controlled intrachain reactions of polymers. I. Theory, *J. Chem. Phys.* **60**, 866–877 (1974).
15. I. M. Sokolov, J. Mai, and A. Blumen, Paradoxical diffusion in chemical space for nearest-neighbor walks over polymer chains, *Phys. Rev. Lett.* **79**, 857–860 (1997).
16. S. F. Burlatsky and G. S. Oshanin, Diffusion-controlled reactions with polymers, *Phys. Lett. A.* **145**, 61–65 (1990).
17. G. Oshanin, S. Moreau, and S. Burlatsky, Models of chemical-reactions with participation of polymers, *Adv. Colloid Interf. Sci.* **49**, 1–46 (1994).

18. P. J. Park and S. Lee, Diffusion-influenced reversible energy transfer reactions between polymers, *J. Chem. Phys.* **115**, 9594–9600 (2001).

19. C. Groves and N. C. Greenham, Bimolecular recombination in polymer electronic devices, *Phys. Rev. B.* **78**, 155205 (2008).

20. M. Hilczer and M. Tachiya, Unified theory of geminate and bulk electron-hole recombination in organic solar cells, *J. Phys. Chem. C.* **114**, 6808–6813 (2010).

21. R. M. Noyes, Effects of diffusion rates on chemical kinetics, *Progr. Reaction Kinetics.* **1**, 129–160 (1961).

22. T. R. Waite, Theoretical treatment of the kinetics of diffusion-limited reactions, *Phys. Rev.* **107**, 463–470 (1957).

23. T. R. Waite, General theory of bimolecular reaction rates in solids and liquids, *J. Chem. Phys.* **28**, 103–106 (1958).

24. R. Zwanzig, Ensemble method in the theory of irreversibility, *J. Chem. Phys.* **33**, 1338–1341 (1960).

25. I. Z. Steinberg and E. Katchalsky, Theoretical analysis of the role of diffusion in chemical reactions, fluorescence quenching, and nonradiative energy transfer, *J. Chem. Phys.* **48**, 2404–2410 (1968).

26. G. Wilemsky and M. Fixman, General theory of diffusion-controlled reactions, *J. Chem. Phys.* **58**, 4009–4019 (1973).

27. S. A. Rice, *Diffusion-Limited Reactions Comprehensive Chemical Kinetics.* Vol. 25, C. H. Bamford, C. F. H. Tipper, and R. G. Compton Elsevier Science Publishers (1985).

28. A. B. Doktorov and A. I. Burshtein, Quantum theory of diffusion-accelerated remote transfer, *Sov. Phys. JETP.* **41**, 671–677 (1976).

29. A. B. Doktorov and E. A. Kotomin, Theory of tunnelling recombination of defects stimulated by their motion. I. General formalism, *Phys. Stat. Sol. B.* **113**, 9–14 (1982).

30. A. Szabo and W. Naumann, Comparison of the smoluchowski approach with modern alternative approaches to the diffusion-influenced fluorescence quenching: the effect of intense excitation pulses, *J. Chem. Phys.* **107**, 402–407 (1997).

31. M. Tachiya, Theory of diffusion-controlled reactions: Formulation of the bulk reaction rate in terms of the pair probability, *Radiat. Phys. Chem.* **21**, 167–175 (1983).

32. A. Molski and J. Keizer, Kinetics of nonstationary, diffusion-influenced reversible reactions in solution, *J. Chem. Phys.* **96**, 1391–1398 (1992).

33. B. U. Felderhof and R. B. Jones, Statistical theory of time-dependent diffusion-controlled reactions in fluids and solids, *J. Chem. Phys.* **103**, 10201–10213 (1995).

34. A. I. Burshtein, Energy quenching kinetics beyond the rate concept, *J. Lumin.* **93**, 229–241 (2001).

35. I. V. Gopich and A. Szabo, Kinetics of reversible diffusion influenced reactions: The self-consistent relaxation time approximation, *J. Chem. Phys.* **117**(2), 507–517 (2002).

36. A. B. Doktorov, *Development of the Kinetic Theory of Physicochemical Processes Induced by Binary Encounters of Reactants in Solutions.* Vol. 6, Transworld Research Network, Kerala, India (2012).

37. A. B. Doktorov and A. A. Kipriyanov, Deviation from the kinetic law of mass action for reactions induced by binary encounters in liquid solutions, *J. Phys. Cond. Matt.* **19**, 065136 (2007).

38. M. Smoluchowski, Versuch einer Mathematischen Theorie der Koagulationskinetic Kolloider Lösungen, *Z. Phys. Chem.* **92**, 129 (1917).

39. F. C. Collins and G. E. Kimball, Diffusion controlled reaction rates, *J. Colloid Sci.* **4**, 425–437 (1949).

40. K. L. Ivanov, N. N. Lukzen, and A. B. Doktorov, Non-Markovian Kinetics of Reactions in Solutions, Chapter 3 in *Chemical Kinetics: Beyond the Textbook*, K. Linderberg, R. Metzler, and G. Oshanin (eds.), World Scientific, Singapore, forthcoming.

41. J. Sung and S. Lee, Nonequilibrium distribution function formalism for diffusion-influenced bimolecular reactions: Beyond the superposition approximation, *J. Chem. Phys.* **111**, 796–803 (1999).

42. M. Yang, S. Lee, and K. J. Shin, Kinetic theory of bimolecular reactions in liquid. II. Reversible association-dissociation: $A + B \leftrightarrow C + B$, *J. Chem. Phys.* **108**, 8557–8571 (1998).

43. A. A. Kipriyanov, A. B. Doktorov, and A. I. Burshtein, Binary theory of dephasing in liquid solutions. I. the non-markovian theory of encounters, *Chem. Phys.* **76**, 149–162 (1983).

44. A. A. Kipriyanov, I. V. Gopich, and A. B. Doktorov, A modification of the non-Markovian encounter theory. I. Markovian description in non-Markovian theories, *Chem. Phys.* **187**, 241–249 (1995).

45. I. V. Gopich and A. B. Doktorov, Kinetics of diffusion-influenced reversible reaction $A + B \rightleftharpoons C$ in solutions, *J. Chem. Phys.* **105**, 2320–2332 (1996).

46. K. L. Ivanov, N. N. Lukzen, A. B. Doktorov, and A. I. Burshtein, Integral encounter theories of multistage reactions. I. Kinetic equations, *J. Chem. Phys.* **114**, 1754–1762 (2001).

47. K. L. Ivanov, N. N. Lukzen, A. A. Kipriyanov, and A. B. Doktorov, The integral encounter theory of multistage reactions containing association–dissociation reaction stages. I. Kinetic equations, *Phys. Chem. Chem. Phys.* **6**, 1706–1718 (2004).

48. K. L. Ivanov, N. N. Lukzen, and A. B. Doktorov, The integral encounter theory of multistage reactions containing association–dissociation reaction stages. II. The kinetics of reversible excitation binding, *Phys. Chem. Chem. Phys.* **114**, 1719–1724 (2004).

49. A. B. Doktorov, Encounter Theory of Chemical Reaction in Solution. Approximate Methods for Calculating Rate Constants, Chapter 2 in *Chemical Kinetics: Beyond the Textbook*, K. Linderberg, R. Metzler, and G. Oshanin (eds.), World Scientific, Singapore, forthcoming.

50. J. Franck and E. Rabinovich, Free radicals and the photochemistry of solutions, *Trans. Fraday Soc.* **30**, 120–131 (1934).

51. J. M. Deutch, Theory of chemically induced dynamic polarization in thin films, *J. Chem. Phys.* **56**, 6076–6081 (1972).

52. I. V. Gopich, A. A. Kipriyanov, and A. B. Doktorov, A many-particle treatment of the reversible reaction $A + B \leftrightarrow C + B$, *J. Chem. Phys.* **110**, 10888–10898 (1999).

53. Y. B. Zeldovich and A. A. Ovchinnikov, The mass action law and the kinetics of chemical reactions with allowance for thermodynamic fluctuations of the density, *Sov. Phys. JETP Lett.* **47**, 829–834 (1977).

54. S. F. Burlatsky, G. S. Oshanin, and A. A. Ovchinnikov, Kinetics of chemical short-range ordering in liquids and diffusion-controlled reactions, *Chem. Phys.* **152**, 13–21 (1991).

55. N. Agmon and I. V. Gopich, Exact long-time asymptotics for reversible binding in three dimensions, *J. Chem. Phys.* **112**, 2863–2869 (2000).

56. I. V. Gopich, A. A. Ovchinnikov, and A. Szabo, Long-time tails in the kinetics of reversible bimolecular reactions, *Phys. Rev. Lett.* **86**, 922 (2001).

57. K. L. Ivanov, N. N. Lukzen, and A. B. Doktorov, The integral encounter theory of multistage reactions containing association–dissociation reaction stages. III. Taking account of quantum states of reactants, *J. Chem. Phys.* **121**, 5115–5124 (2004).

58. A. A. Kipriyanov, O. A. Igoshin, and A. B. Doktorov, The effect of chemical displacement of B species in the reaction $A + B \rightarrow B$, *Physica A.* **275**, 99–133 (2000).

59. O. G. Berg, On diffusion-controlled dissociation, *Chem. Phys.* **31**, 47–57 (1978).

60. K. L. Ivanov and N. N. Lukzen, A novel method for calculating rate constants of diffusion-influenced reactions, *J. Chem. Phys.* **121**, 5109–5114 (2004).

61. N. N. Lukzen, A. B. Doktorov, and A. I. Burshtein, Non-markovian theory of diffusion-controlled excitation transfer, *Chem. Phys.* **102**(3), 289–304 (1986).

62. A. I. Burshtein and N. N. Lukzen, Reversible reactions of metastable reactants, *J. Chem. Phys.* **103**, 9631–9641 (1995).

63. A. I. Burshtein and N. N. Lukzen, Excitation trapping in liquid solutions, *J. Chem. Phys.* **105**, 9588–9596 (1996).

64. E. Pines and D. Huppert, Observation of geminate recombination in excited state proton transfer, *J. Chem. Phys.* **84**, 3576–3577 (1986).

65. N. Agmon, E. Pines, and D. Huppert, Geminate recombination in proton-transfer reactions. II. Comparison of diffusional and kinetic schemes, *J. Chem. Phys.* **88**, 5631–5638 (1988).

66. D. Huppert, S. Y. Goldberg, A. Masad, and N. Agmon, Experimental determination of the long-time behavior in reversible binary chemical reactions, *Phys. Rev. Lett.* **68**, 3932–3935 (1992).

67. I. V. Gopich, K. M. Solntsev, and N. Agmon, Excited-state reversible geminate reaction. I. Two different lifetimes, *J. Chem. Phys.* **110**, 2164–2174 (1999).

68. N. Agmon, Excited-state reversible geminate reaction. II. Contact geminate quenching, *J. Chem. Phys.* **110**, 2175–2180 (1999).

69. I. V. Gopich and N. Agmon, Excited-state reversible geminate reaction. III. Exact solution for noninteracting partners, *J. Chem. Phys.* **110**, 10433–10444 (1999).

70. A. Weller, Mechanism and spin dynamics of photoinduced electron transfer reactions, *Z. Phys. Chem.* **130**, 129–138 (1982).

71. T. Förster, *The Exiplex, M. Gordon and W. R. Ware* (eds.), Academic, New York (1971).

72. T. Asahi and N. Mataga, Charge recombination process of ion pair state produced by excitation of charge-transfer complex in acetonitrile solution. Essentially different character of its energy gap dependence from that of geminate ion pair formed by encounter between fluorescer and quencher, *J. Phys. Chem.* **93**, 6578 (1989).

73. L. Fodor, A. Horvath, K. A. Hötzer, S. Walbert, and U. E. Steiner, Enhancement of magnetic field effect in $Ru(bpy)_3^{2+}/MV_2^+$ system by $Ru(bpy)_3^{2+}$-Ag^+ exciplex formation, *Chem. Phys. Lett.* **316**, 411–418 (2000).

74. T. Tachikawa, Y. Kobori, K. Akiyama, A. Katsuki, U. E. Steiner, and S. Tero-Kubota, Spin dynamics and zero-field splitting constants of the triplet exciplex generated by photoinduced electron transfer reaction between Erythrosin B and duroquinone, *Chem. Phys. Lett.* **360**, 13–21 (2002).

Chapter 5

A New Method of Solution for the Fredholm Integral Equation and Its Application to the Diffusion-Influenced Reaction Kinetics

Sangyoub Lee

Department of Chemistry, Seoul National University,
Seoul 08826, South Korea

sangyoub@snu.ac.kr

The solution to the Fredholm integral equation of the second kind is often represented by the Liouville–Neumann series. We present a new type of solution, which may be considered as resumming the series. Except that the kernel function must not be zero, the method is general and provides an approximate analytic solution that can be improved systematically. It gives numerically more accurate results than the Padé approximation method for the same computational cost. This new solution method has been found to be very useful in the problems of diffusion-influenced chemical kinetics, and some examples are given.

1. Introduction

Boundary value problems arise in many branches of physical sciences and engineering. In many cases, one may put the differential equation with the associated boundary conditions into an equivalent Fredholm integral equation of the second kind whose standard form is given by[1]

$$f(x) = g(x) + \int_a^b dy \, K(x,y)f(y), \tag{1}$$

where $g(x)$ and the kernel $K(x,y)$ are known functions and $f(x)$ is the function to be determined.

When an exact solution to Eq. (1) is not available and the kernel contains a smallness parameter as $K(x,y) = \varepsilon k(x,y)$, one may resort to a power series solution that can be generated by the method of successive approximations as

$$f(x) = g(x) + \varepsilon \int_a^b dy_1 k(x,y_1)g(y_1)$$

$$+ \varepsilon^2 \int_a^b dy_1 k(x,y_1) \int_a^b dy_2 k(y_1,y_2)g(y_2) + \cdots . \tag{2}$$

If the kernel is indeed small, Eq. (2) called the Liouville–Neumann series may give a convergent result with a small number of expansion terms. Otherwise, when the series converges slowly or even diverges, one must resort to a systematic method for resumming the series. For example, the methods of Padé approximants, Borel transform, and Euler summation have been used extensively to deal with the divergent or slowly convergent series.[2,3] Usually, this procedure provides a numerically easier route than dealing with the resolvent function expressed in terms of Fredholm determinants.[1]

In Section 2, we will present another simple solution method to Eq. (1). Except that the kernel function must not be zero, the method is general. Furthermore, as explained below, one can easily incorporate additional physical approximations into the method. In many cases, it gives numerically more accurate results than the widely used Padé approximation method for the same computational cost. In particular, our new solution method has been found to be very useful in the problems of diffusion-influenced chemical kinetics.[4–8] Several illustrative examples are given in Sections 3 and 4.

2. A New Solution Method for the Fredholm Integral Equation

We have proposed a new method of solution for Eq. (1) that gives very accurate results if $f(x)$ is a slowly varying non-zero function.[4,6] The idea is very simple. We first rewrite Eq. (1) as

$$f(x) = g(x) + f(x) \int_a^b dy\, K(x,y)[f(y)/f(x)], \qquad (3)$$

where we have simply multiplied $f(x)$ to the integral term and then divided it by $f(x)$. This equation is solved formally to give

$$f(x) = g(x) \left\{ 1 - \int_a^b dy\, K(x,y)[f(y)/f(x)] \right\}^{-1}. \qquad (4)$$

To get an explicit result, we then replace the ratio $f(y)/f(x)$ by $f_a(y)/f_a(x)$ where $f_a(x)$ is an appropriate approximation for $f(x)$. We thus have

$$f(x) \cong g(x) \left\{ 1 - \int_a^b dy\, K(x,y)[f_a(y)/f_a(x)] \right\}^{-1}. \qquad (5)$$

When $K(x,y) = \varepsilon k(x,y)$, an immediate choice for $f_a(x)$ is given by the first few terms in the perturbation series in Eq. (2). With higher order approximations for $f_a(x)$, Eq. (5) will give systematically improved approximation of $f(x)$ for small ε values. We have found that this procedure gives more accurate results than the Padé approximation method for the same computational cost.[5,6] The reason is that (i) Eq. (5) has the same formal structure as the exact relation in Eq. (4) and (ii) the error caused by using the approximate function $f_a(x)$ would be partially offset because it enters as a ratio.

Furthermore, we have an additional flexibility in using Eq. (5). For example, we may find an approximate form of $f(x)$ in the large ε limit by an asymptotic analysis or physical reasoning. If we take such an approximation for $f_a(x)$, Eq. (5) produces a solution that is applicable to a wider range of ε values.[6] Note that such a solution gives an exact result also in the opposite $\varepsilon \to 0$ limit as can be seen from the structure of Eq. (5). An example of this line of approach is given in Section 4.

3. Diffusive Propagator for an Interacting Particle Pair

We have applied the above solution method for the Fredholm integral equation of the second kind to obtain an approximate expression of the propagator for diffusive dynamics of a pair of particles interacting via an arbitrary central potential and hydrodynamic interaction.[4] The propagator $G(r, t | r_0)$, which represents a probability density of finding a pair of particles at a separation r at time t, given that their initial separation was r_0, is one of the most important dynamic quantities in the liquid state physics.[9] Its knowledge enables the calculation of various dynamic quantities such as time correlation functions of the physical variables that depend on the particle separation, the first encounter time distribution of a pair of particles, the time-dependent survival probability of reactive molecules, and so on.

When the medium surrounding the particles are very viscous, the time evolution of $G(r, t | r_0)$ is described by the Smoluchowski equation,[10-12]

$$\frac{\partial G(r, t | r_0)}{\partial t} = D \frac{1}{r^2} \frac{\partial}{\partial r} r^2 h(r) e^{-U(r)} \frac{\partial}{\partial r} e^{U(r)} G(r, t | r_0). \qquad (6)$$

Here, $U(r)$ denotes an arbitrary central interaction potential in units of $k_B T$, with k_B and T denoting the Boltzmann constant and the absolute temperature. The relative diffusion coefficient at large separation is D, but its value decreases with decreasing r due to hydrodynamic interaction, which is described by the function $h(r)$. When the hydrodynamic interaction can be described by the Oseen tensor and the hydrodynamic radii of the particles are given by $\sigma/2$ with σ denoting the contact distance, we have $h(r) = 1 - 3\sigma/4r$.[13] However, other forms of $h(r)$ may also be used.[10] The initial and boundary conditions associated with Eq. (6) are

$$G(r, t = 0 | r_0) = \delta(r - r_0)/(4\pi r_0^2), \tag{7}$$

$$\lim_{r \to \infty} G(r, t | r_0) = 0 \tag{8}$$

and

$$\partial[e^{U(r)} G(r, t | r_0)]/\partial r|_{r=\sigma} = 0. \tag{9}$$

By making the generalized Flannery's transformation[14] of variables given by

$$\tilde{r}(r) = \left\{ \int_r^\infty dr_1 \, [r_1^2 h(r_1) e^{-U(r_1)}]^{-1} \right\}^{-1}, \tag{10}$$

$$F(\tilde{r}, t | \tilde{r}_0) = e^{U(r)} \varphi(r_0)^{-1} G(r, t | r_0), \tag{11}$$

with $\varphi(r) = (\tilde{r}^4/r^4) e^{2U(r)}/h(r)$, it can be shown that the Laplace transform of $F(\tilde{r}, t | \tilde{r}_0)$ satisfies the integral equation,[4]

$$\hat{F}(\tilde{r}, s | \tilde{r}_0) = \varphi(\tilde{r}_0)^{-1} \left[\hat{F}_0(\tilde{r}, s | \tilde{r}_0) \right.$$

$$\left. - s \int d\tilde{\mathbf{r}}_1 \, d(\tilde{r}_1) \hat{F}(\tilde{r}, s | \tilde{r}_1) \hat{F}_0(\tilde{r}_1, s | \tilde{r}_0) \right] \tag{12}$$

with $d(\tilde{r}) = 1 - \varphi(\tilde{r})$. For notational convenience, we will not distinguish the dimensionless function of $\tilde{r}$ from that of r representing the same quantity; that is, $\varphi(\tilde{r}(r)) = \varphi(r)$ and so on. $\hat{F}_0(\tilde{r}, s | \tilde{r}_0)$ is a seemingly interaction-free propagator in the $\tilde{r}$ space,[15]

$$\hat{F}_0(\tilde{r}, s | \tilde{r}_0) = \frac{1}{8\pi D \zeta \tilde{r} \tilde{r}_0} \left[e^{-\zeta |\tilde{r} - \tilde{r}_0|} + \frac{\tilde{\sigma}\zeta - 1}{\tilde{\sigma}\zeta + 1} e^{-\zeta(\tilde{r} + \tilde{r}_0 - 2\tilde{\sigma})} \right] \tag{13}$$

with $\tilde{r}_0 = \tilde{r}(r_0)$, $\tilde{\sigma} = \tilde{r}(\sigma)$, and $\zeta = (s/D)^{1/2}$.

The usual iterative method for solving the Fredholm integral equation of the second kind gives the following series solution to

Eq. (12),

$$\hat{F}(\tilde{r},s|\tilde{r}_0) = \varphi(\tilde{r}_0)^{-1}\left[\hat{F}_0(\tilde{r},s|\tilde{r}_0) - s\int d\tilde{\mathbf{r}}_1\,\psi(\tilde{r}_1)\hat{F}_0(\tilde{r},s|\tilde{r}_1)\hat{F}_0(\tilde{r}_1,s|\tilde{r}_0)\right.$$

$$\left. + s^2\int d\tilde{\mathbf{r}}_2\,\psi(\tilde{r}_2)\hat{F}_0(\tilde{r},s|\tilde{r}_2)\int d\tilde{\mathbf{r}}_1\,\psi(\tilde{r}_1)\hat{F}_0(\tilde{r}_2,s|\tilde{r}_1)\hat{F}_0(\tilde{r}_1,s|\tilde{r}_0) + \cdots\right],$$

$$(14)$$

where $\psi(\tilde{r}) = d(\tilde{r})/\varphi(\tilde{r})$. It is obvious that this solution would be useful only in the small-s limit. For large s, the series will diverge. With the lowest-order approximation from Eq. (14), we have

$$\frac{\hat{F}(\tilde{r},s|\tilde{r}_1)}{\hat{F}(\tilde{r},s|\tilde{r}_0)} \simeq \frac{\varphi(\tilde{r}_1)^{-1}\hat{F}_0(\tilde{r},s|\tilde{r}_1)}{\varphi(\tilde{r}_0)^{-1}\hat{F}_0(\tilde{r},s|\tilde{r}_0)} \qquad (15)$$

and with Eq. (15) the new solution method described in Section 2 gives the following small-s approximation for $\hat{F}(\tilde{r},s|\tilde{r}_0)$,

$$\hat{F}(\tilde{r},s|\tilde{r}_0) \simeq \frac{\hat{F}_0(\tilde{r},s|\tilde{r}_0)}{\varphi(\tilde{r}_0)}$$

$$\times \left[1 + s\int d\tilde{\mathbf{r}}_1\,\psi(\tilde{r}_1)\frac{\hat{F}_0(\tilde{r},s|\tilde{r}_1)\hat{F}_0(\tilde{r}_1,s|\tilde{r}_0)}{\hat{F}_0(\tilde{r},s|\tilde{r}_0)}\right]^{-1}. \quad (16)$$

Equation (16) provides the propagator expression that becomes exact in the small-s limit and remains quite accurate up to intermediate s values, as will be seen below.

As an application of the reaction-free propagator expression given by Eqs. (11) and (16), we now consider the time-dependent rate coefficient for diffusion-influenced irreversible bimolecular reactions.[4,5] We will assume that the reactants are distributed in equilibrium at $t = 0$ and that the reaction can occur only at the contact distance σ with an inherent rate constant κ, so that the effect of reaction can be incorporated into the Smoluchowski equation as a δ-function sink, $\kappa\delta(r - \sigma)/(4\pi\sigma^2)$.[16] Then, in the low reactant concentration limit, the Laplace transform of the bimolecular rate coefficient can be expressed in terms of the reaction-free propagator as[11]

$$\hat{k}_f(s) = \frac{s^{-1}\kappa e^{-U(\sigma)}}{1 + \kappa \hat{G}(\sigma, s\,|\sigma)} = \frac{s^{-1}\kappa e^{-U(\sigma)}}{1 + \kappa e^{-U(\sigma)}\varphi(\sigma)\hat{F}(\tilde{\sigma}, s\,|\tilde{\sigma})}. \quad (17)$$

In the diffusion-controlled limit ($\kappa \to \infty$), we have from Eqs. (13), (16), and (17)

$$\hat{k}_f(s) = k_D\left[\frac{1 + \tilde{\sigma}\zeta}{s} + \frac{\tilde{\sigma}}{D}\int_{\tilde{\sigma}}^{\infty} d\tilde{r}\,\psi(\tilde{r})e^{-2\zeta(\tilde{r}-\tilde{\sigma})}\right], \quad (18)$$

whose inverse Laplace transformation gives

$$k_f(t) = k_D\left\{1 + \frac{\tilde{\sigma}}{(\pi Dt)^{1/2}} + \frac{\pi\tilde{\sigma}}{(\pi Dt)^{3/2}}\right.$$
$$\left.\times \int_{\sigma}^{\infty} dr\,\frac{1 - \varphi(r)}{[h(r)\varphi(r)]^{1/2}}(\tilde{r} - \tilde{\sigma})\exp\left[-\frac{(\tilde{r} - \tilde{\sigma})^2}{Dt}\right]\right\}. \quad (19)$$

Here, k_D is the steady-state rate constant attained at long times and is given by

$$k_D \equiv k_f(\infty) = 4\pi D\tilde{\sigma} = 4\pi D\left\{\int_{\sigma}^{\infty} dr_1\,[r_1^2 h(r_1)e^{-U(r_1)}]^{-1}\right\}^{-1}. \quad (20)$$

With the biexponential approximation to $\psi(\tilde{r})$,

$$\psi(\tilde{r})/\psi(\tilde{\sigma}) \cong c_1 e^{-\alpha_1(\tilde{r}-\tilde{\sigma})} + c_2 e^{-\alpha_2(\tilde{r}-\tilde{\sigma})}, \quad (21)$$

the radial integral in Eq. (18) can be evaluated to give

$$\frac{k_f(t)}{k_D} = 1 + \frac{\tilde{\sigma}}{(\pi Dt)^{1/2}} + \tilde{\sigma}\psi(\tilde{\sigma})\left[\frac{1}{2(\pi Dt)^{1/2}}\right.$$
$$\left. - \frac{c_1\alpha_1}{4}\Omega\left(\frac{\alpha_1\sqrt{Dt}}{2}\right) - \frac{c_2\alpha_2}{4}\Omega\left(\frac{\alpha_2\sqrt{Dt}}{2}\right)\right]. \quad (22)$$

Here, the coefficients c_1, c_2, α_1, and α_2 can be determined explicitly as described in Refs. 4 and 5, and $\Omega(x) \equiv \exp(x^2)\mathrm{erfc}(x)$.

The accuracy of the rate coefficient expressions given by Eqs. (19) and (22) have been checked against the exact results obtained by solving the Smoluchowski equation numerically. Figure 1 displays the effects of the Coulomb interaction, $U(r) = r_C/r$, on the rate coefficients. r_C is the Onsager distance defined by $r_C/\sigma = z_1 z_2 e^2/(4\pi\varepsilon_0\varepsilon_r\sigma k_B T)$ where $z_1 e$ and $z_2 e$ are the charges of the reactants, ε_0 is the permittivity of vacuum, and ε_r is the relative permittivity of the medium. Figure 1(a) displays the results for $r_C/\sigma = -8$ while Fig. 1(b) for $r_C/\sigma = -16$. In both cases, we neglect the effects of hydrodynamic interaction; that is, $h(r) = 1$. The exact numerical results are represented by filled squares, and the results obtained from Eqs. (19) and (22) are represented by the solid and dotted curves, respectively.

As for the effects of the interaction potential on $k_f(t)$, Dudko and Szabo constructed a quite simple and accurate time-domain rate expression by the Pade approximation using two exact terms in both the short- and long-time expansions of $k_f(t)$.[17] In Fig. 1, the dashed curves, denoted as DS, are the results of their rate expression. On the other hand, Zharikov and Shokhirev derived an approximate expression for $\hat{G}(\sigma, s|\sigma.)$,[18] which is given in our notations by

$$\hat{G}(\sigma, s\,|\sigma) = \frac{e^{-U(\sigma)}}{4\pi D\tilde{\sigma}}\left[1 + \tilde{\sigma}\zeta^2 \int_\sigma^\infty dr\, \varphi(\tilde{r})^{-1/2}e^{-\zeta(r-\sigma)}\right]^{-1}. \quad (23)$$

This expression is closely related to ours given by

$$\hat{G}(\sigma, s\,|\sigma) = \frac{e^{-U(\sigma)}}{4\pi D\tilde{\sigma}}\left[1 + \tilde{\sigma}\zeta + \tilde{\sigma}\zeta^2 \int_{\tilde{\sigma}}^\infty d\tilde{r}\, \psi(\tilde{r})e^{-2\zeta(\tilde{r}-\tilde{\sigma})}\right]^{-1}. \quad (24)$$

It is obvious that the Zharikov–Shokhirev expression for $\hat{G}(\sigma, s|\sigma)$ in Eq. (23), missing the $\tilde{\sigma}\zeta$ term in the square bracket of the denominator, would predict an inaccurate long-time behavior. However, they introduced some additional approximation which gives

$$\hat{G}(\sigma, s\,|\sigma) = \frac{e^{-U(\sigma)}}{4\pi D\tilde{\sigma}}\left[1 + \tilde{\sigma}\zeta + \tilde{\sigma}\zeta^2 \frac{\varphi(\tilde{\sigma})^{-1/2} - 1}{\zeta + \alpha(\sigma)}\right]^{-1},$$

$$\alpha(\sigma) = \left[\varphi(\tilde{\sigma})^{-1/2} - 1\right]\left\{\int_\sigma^\infty dr\left[\varphi(\tilde{r})^{-1/2} - 1\right]\right\}^{-1}. \quad (25)$$

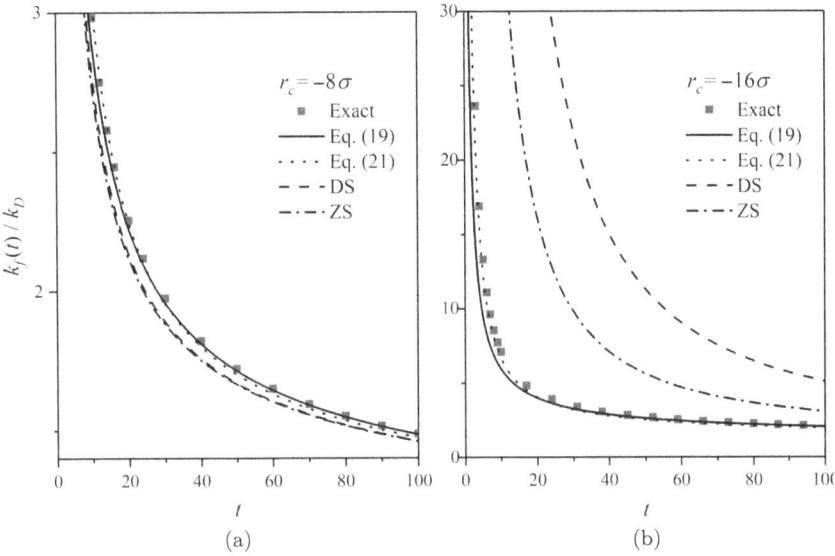

Fig. 1. Effects of attractive Coulomb interaction on the bimolecular rate coefficient in the diffusion-controlled limit. The unit of time t is σ^2/D.

In fact, this expression for $\hat{G}(\sigma, s \,|\, \sigma)$, put into Eq. (17), gives better results for the rate coefficient than that of Eq. (23) due to error cancellation. In the numerical comparisons in Fig. 1, the Zharikov-Shokhirev (ZS) results, represented by dot-dashed curves, are those based on Eq. (25).

From Fig. 1, we see that both of our rate coefficient expressions in Eqs. (19) and (21) give results in excellent agreement with the exact results. In comparison, the Dudko–Szabo and the Zharikov–Shokhirev expressions underestimate the rate coefficient slightly when the strength of attractive Coulomb potential is moderate ($r_C/\sigma = -8$), and overestimate the rate to an unreliable extent at large potential strength ($r_C/\sigma = -16$).

We have displayed the results only for the attractive interaction potentials. As proved rigorously by Traytak,[19] when the interaction potential is repulsive, the rate coefficient converges much faster to the steady-state value $k_f(\infty)$ than in the attractive potential cases. Hence all the theories considered here provide better results for the repulsive potentials than for the attractive potentials with the same $|r_C|$ value.

4. Steady-State Rates of Reactions with Long-Range Reactivity

The transfer of energy or electron occurs at a range of separation between donor and acceptor molecules.[11] When the reaction medium is very viscous or the inherent transfer process is very fast, the overall rate of the transfer reaction depends on the relative diffusion rate of the reactant molecules. Then the steady-state rate constant k_D can be determined by solving the Smoluchowski equation,[16]

$$Dr^{-2}\frac{\partial}{\partial r}r^2 h(r)e^{-U(r)}\frac{\partial}{\partial r}e^{U(r)}\rho(r) = S(r)\rho(r). \qquad (26)$$

$\rho(r)$ denotes the non-equilibrium pair correlation function; if the bulk number density of the acceptor molecules is [A], $4\pi r^2 \rho(r)[A]dr$ gives the number of A molecules around a donor molecule D located at a distance between r and $r + dr$. The sink function $S(r)$ denotes the energy or electron transfer rate when the distance between D and A is r. The steady-state rate constant k_D is related to $\rho(r)$ as[20]

$$k_D = \int_\sigma^\infty dr\, 4\pi r^2 S(r)\rho(r). \qquad (27)$$

The exact solution of Eq. (26) is known for certain types of $S(r)$ in the case of non-interacting reactants with $U(r) = 0$ and $h(r) = 1$, and for $S(r) \propto r^{-6}$ in the case when $U(r) = r_C/r$ and $h(r) = 1$.[21] Zharikov and Shokhirev derived very narrow upper and lower bounds to k_D in the case when $h(r) = 1$.[22] For the general case involving arbitrary $U(r)$, $h(r)$, and $S(r)$, one may carry out a usual perturbative analysis which can provide a solution that is accurate in either small or large reactivity limit.[11] We now illustrate the utility of the new solution method of Fredholm integral equations in obtaining a solution which becomes exact in the two opposite limits of small and large reactivity and also performs well in the intermediate regime.[6]

We will consider the general case in which the reaction sink function contains the contact reaction term as well as the long-range reaction term

$$S(r) = \kappa \frac{\delta(r-\sigma)}{4\pi\sigma^2} + S_L(r). \tag{28}$$

By introducing the generalized Flannery's transformation given by Eq. (10) with the relation, $\tilde{\rho}(\tilde{r}) = e^{U(r)}\rho(r)$, it can be shown that $\tilde{\rho}(\tilde{r})$ satisfies the following integral equation,[6]

$$\tilde{\rho}(\tilde{r}) = \left(1 - \frac{\nu_\sigma\tilde{\sigma}}{\tilde{r}}\right) - \int_{\tilde{\sigma}}^{\infty} d\tilde{r}_0 \left(\frac{\tilde{r}_0}{\max(\tilde{r},\tilde{r}_0)} - \frac{\nu_\sigma\tilde{\sigma}}{\tilde{r}}\right) \frac{\tilde{r}_0 S_L(\tilde{r}_0)}{D\varphi(\tilde{r}_0)} \tilde{\rho}(\tilde{r}_0), \tag{29}$$

where $\tilde{r}_0$, $\tilde{\sigma}$, and $\varphi(\tilde{r}_0)$ are those defined in Section 3, and ν_σ is defined as

$$\nu_\sigma \equiv \frac{\kappa e^{-U(\sigma)}}{4\pi D\tilde{\sigma} + \kappa e^{-U(\sigma)}}. \tag{30}$$

The solution method described in Section 2 gives

$$\tilde{\rho}(\tilde{r}) \cong \left(1 - \frac{\nu_\sigma\tilde{\sigma}}{\tilde{r}}\right) \times \left[1 + \int_{\tilde{\sigma}}^{\infty} d\tilde{r}_0 \right.$$

$$\left. \times \left(\frac{\tilde{r}_0}{\max(\tilde{r},\tilde{r}_0)} - \frac{\nu_\sigma\tilde{\sigma}}{\tilde{r}}\right) \frac{\tilde{r}_0 S_L(\tilde{r}_0)}{D\varphi(\tilde{r}_0)} \frac{\tilde{\rho}_a(\tilde{r}_0)}{\tilde{\rho}_a(\tilde{r})}\right]^{-1}, \tag{31}$$

where $\tilde{\rho}_a(\tilde{r})$ is an approximate form of $\tilde{\rho}(\tilde{r})$.

In the small reactivity limit with small $S_L(r)$ and large D, we may take the following expression for $\tilde{\rho}_a(\tilde{r})$

$$\tilde{\rho}_a(\tilde{r}) = 1 - \nu_\sigma\tilde{\sigma}/\tilde{r}. \tag{32}$$

By substituting Eq. (32) into Eq. (31), we obtain

$$\tilde{\rho}(\tilde{r}) \cong \tilde{\rho}_1(\tilde{r}) = \left(1 - \frac{\nu_\sigma\tilde{\sigma}}{\tilde{r}}\right) \left\{1 + \frac{1}{D} \left[\int_{\tilde{\sigma}}^{\tilde{r}} d\tilde{r}_0 \frac{(\tilde{r}_0 - \nu_\sigma\tilde{\sigma})^2}{(\tilde{r} - \nu_\sigma\tilde{\sigma})} \frac{S_L(\tilde{r}_0)}{\varphi(\tilde{r}_0)} \right.\right.$$

$$\left.\left. + \int_{\tilde{r}}^{\infty} d\tilde{r}_0 (\tilde{r}_0 - \nu_\sigma\tilde{\sigma}) \frac{S_L(\tilde{r}_0)}{\varphi(\tilde{r}_0)}\right]\right\}^{-1}. \tag{33}$$

From Eqs. (27), (28), and (33), we have

$$k_D \cong \kappa e^{-U(\sigma)} \tilde{\rho}_1(\tilde{\sigma}) + \int_{\tilde{\sigma}}^{\infty} d\tilde{r} \, 4\pi \tilde{r}^2 \varphi(\tilde{r})^{-1} S_L(\tilde{r}) \tilde{\rho}_1(\tilde{r}). \qquad (34)$$

To get a solution that is applicable to the case of large reactivity [with large $S_L(r)$ and small D], we first note that the asymptotic form of $\tilde{\rho}(\tilde{r})$ at large distance is given by $\tilde{\rho}(\tilde{r}) = 1 - k_D/(4\pi D\tilde{r})$, because the steady-state inward flux is equal to k_D. We thus assume an expression for $\tilde{\rho}(\tilde{r})$ given as

$$\tilde{\rho}(\tilde{r}) = \left(1 - \frac{\nu \tilde{\sigma}}{\tilde{r}}\right) \exp\left[-\int_{\tilde{r}}^{\infty} d\tilde{r} \, \chi(\tilde{r})\right]. \qquad (35)$$

It can be shown that in the limit of large reactivity $\chi(\tilde{r})$ is approximately given by $\chi(\tilde{r}) \cong [S_L(\tilde{r})/D\varphi(\tilde{r})]^{1/2}$.[6] We thus take the following expression for $\tilde{\rho}_a(\tilde{r})$ to calculate the ratio in Eq. (31)

$$\tilde{\rho}_a(\tilde{r}) = \tilde{\rho}^{(0)}(\tilde{r}) = \left(1 - \frac{\nu^{(0)} \tilde{\sigma}}{\tilde{r}}\right) \exp\left[-\int_{\tilde{r}}^{\infty} d\tilde{r}_1 \left(\frac{S_L(\tilde{r}_1)}{D\varphi(\tilde{r}_1)}\right)^{1/2}\right], \qquad (36)$$

where $\nu^{(0)}$ is an approximation to the unknown constant ν given by

$$\nu^{(0)} = \frac{k_{eq}}{k_{eq} + 4\pi \tilde{\sigma} D}; \quad k_{eq} = \int_{\sigma}^{\infty} dr \, 4\pi r^2 S(r) e^{-U(r)}. \qquad (37)$$

With Eq. (36), Eqs. (27), (28), and (31) give

$$k_D \cong \kappa e^{-U(\sigma)} \tilde{\rho}_2(\tilde{\sigma}) + \int_{\tilde{\sigma}}^{\infty} d\tilde{r} \, 4\pi \tilde{r}^2 \varphi(\tilde{r})^{-1} S_L(\tilde{r}) \tilde{\rho}_2(\tilde{r}), \qquad (38)$$

$$\tilde{\rho}_2(\tilde{r}) = \frac{1 - \frac{\nu_{\sigma} \tilde{\sigma}}{\tilde{r}}}{1 + \int_{\tilde{\sigma}}^{\infty} d\tilde{r}_0 \left(\frac{\tilde{r}_0}{\max(\tilde{r}, \tilde{r}_0)} - \frac{\nu_{\sigma} \tilde{\sigma}}{\tilde{r}}\right) \frac{\tilde{r}_0 S_L(\tilde{r}_0)}{D\varphi(\tilde{r}_0)} \frac{\tilde{\rho}^{(0)}(\tilde{r}_0)}{\tilde{\rho}^{(0)}(\tilde{r})}}. \qquad (39)$$

We have evaluated the accuracy of the approximate rate constant expressions in Eqs. (34) and (38) against the exact results. The necessary numerical integrations can be carried out by noting that

$$\int_{\tilde{\sigma}}^{\infty} d\tilde{r}(\cdots) = \int_{\sigma}^{\infty} dr\, h(r)^{-1/2} \varphi(r)^{1/2}(\cdots),$$

$$\int_{\tilde{\sigma}}^{\infty} d\tilde{r}\, 4\pi\tilde{r}^2(\cdots) = \int_{\sigma}^{\infty} dr\, 4\pi r^2 \varphi(r) e^{-U(r)}(\cdots).$$

(40)

We also compared the results with those obtained from the Wilemski–Fixman (WF) rate theory[16] and the upper and lower bound expressions given by Zharikov and Shokhirev (ZS).[22]

The distance-dependent reactivity of electron transfer reactions is often described by an exponential sink function, $S(r) = \kappa_e e^{-2\beta r}$.[11] When $U(r) = 0$ and $h(r) = 1$, the exact expression for k_D is given by[23–26]

$$\frac{k_D}{4\pi D\sigma} = (\beta\sigma)^{-1} \left[\frac{K_0(x) - \beta\sigma x K_1(x)}{I_0(x) + \beta\sigma x I_1(x)} + \gamma - \ln 2\mu \right], \quad (41)$$

where $I_\nu(x)$ and $K_\nu(x)$ are the νth-order modified Bessel functions of the first and the second kind, respectively.[27] $x = \mu^{-1} e^{-\beta\sigma}$, $\mu = (D\beta^2/\kappa_e)^{1/2}$, and $\gamma\,(= 0.5772\ldots)$ is the Euler's constant. Figure 2 displays the results of the present, WF, and ZS theories compared to the exact result. The abscissa λ is a dimensionless reactivity parameter defined by $\lambda = k_{eq} t_L/\sigma^3$ with $k_{eq} = \kappa_e \pi \beta^{-3} e^{-2\beta\sigma}(2\beta^2\sigma^2 + 2\beta\sigma + 1)$ and $t_L = \sigma^2/D$; k_{eq} is the equilibrium rate constant defined by Eq. (37). For the electron transfer case, the value of λ may range up to 100.[11] We set the value of $\beta\sigma$ to 0.1. It can be seen that both Eqs. (34) and (38) of the present theory, as well as the WF theory, give very accurate results for the whole range of λ considered; indeed, the resulting three curves are hardly distinguishable. The ZS theory also provides quite narrow upper and lower bounds to k_D; see the curves designated as ZS1 and ZS2, respectively.

When the energy transfer occurs via dipole-dipole interaction, the orientation-averaged reactivity is usually modeled by a sink function given by $S(r) = \kappa_F (r_F/r)^6 \equiv \alpha r^{-6}$.[11] Here κ_F is the natural decay rate of an excited donor molecule in the absence of acceptor molecules

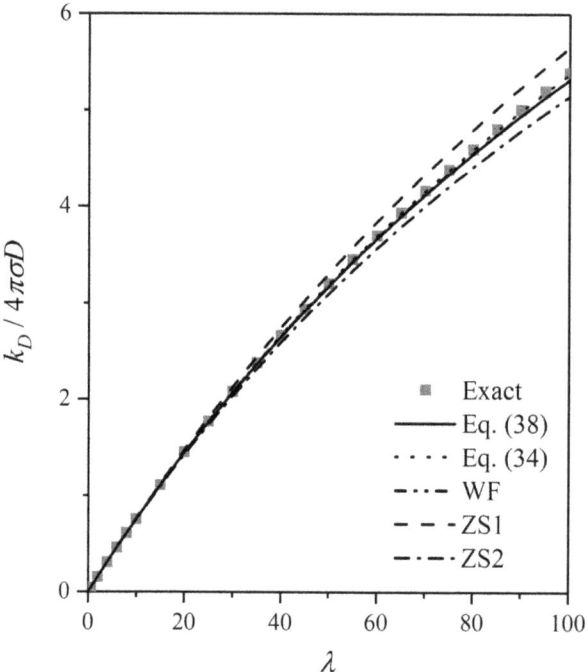

Fig. 2.   The steady-state rate constant k_D of diffusion-influenced electron transfer reaction as a function of the dimensionless reaction strength parameter λ.
(Reproduced from Ref. 6 with the permission of AIP Publishing.)

and r_F is the Förster radius at which κ_F and the rate of energy transfer are of equal magnitude. For the purpose of present calculations,
we need not specify separate values of κ_F and r_F, and will denote
$\kappa_F r_F^6$ as α. In the interaction free case with $U(r) = 0$ and $h(r) = 1$
for $r \geq \sigma$, the exact expression for k_D is given by[26,28,29]

$$\frac{k_D}{4\pi D\sigma} = \frac{2(3\pi^3\lambda)^{1/4}}{[\Gamma(1/4)]^2} \frac{I_{3/4}(u_0)}{I_{-3/4}(u_0)}. \tag{42}$$

Here $\Gamma(x)$ is the gamma function,[27] $\lambda = k_{eq}t_L/\sigma^3$ with $k_{eq} = 4\pi\alpha/(3\sigma^3)$, and $u_0 = [3\lambda/(16\pi)]^{1/2}$. Figure 3 displays the results
of the present, WF, and ZS theories compared to the exact result.
For the Förster energy transfer, the value of λ may range over
10^4.[11] Again, our general rate expression given by Eq. (38) provides

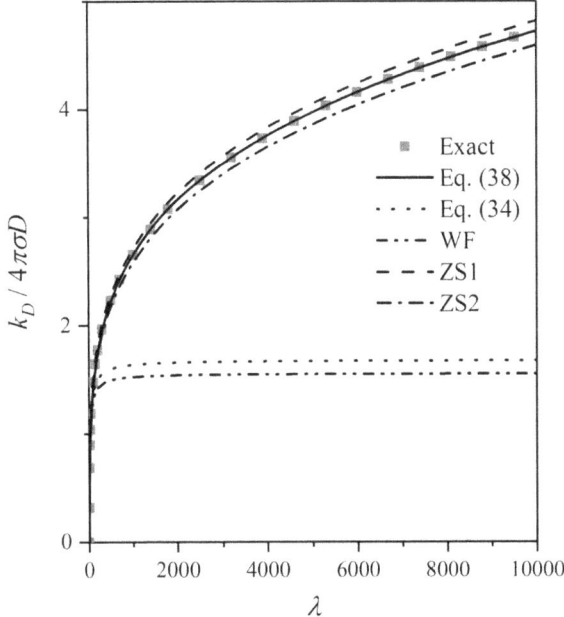

Fig. 3. The steady-state rate constant k_D of the diffusion-influenced Förster energy transfer as a function of the dimensionless reaction strength parameter λ. (Reproduced from Ref. 6 with the permission of AIP Publishing.)

excellent results for the whole range of λ, and the ZS theory provides very narrow lower and upper bounds. Both Eq. (34) and the WF theory give reliable results only for very small λ, but still the former is a little better than the latter. The performance of the solution given by Eq. (38) is remarkable. Equation (38) is based on the approximate $\tilde{\rho}_a(\tilde{r})$ in Eq. (36) that is valid only in the large λ limit. However, $\tilde{\rho}_a(\tilde{r})$ enters as a ratio in the formal solution in Eq. (31), which has the form that becomes exact in the $\lambda \to 0$ limit. This makes the approximate solution in Eq. (38) applicable over the whole range of λ.

5. Conclusion

We have presented a new solution method for the Fredholm integral equation of the second kind and showed its utility in dealing with the reaction kinetic problems occurring in viscous media. As described in Section 3, the new method provided an accurate expression of the propagator for diffusive dynamics of a pair of particles interacting via an arbitrary central potential and hydrodynamic interaction. In Section 4, we derived a general expression for the steady-state rate constant of diffusion-influenced reactions involving long-range reactivity and arbitrary interactions between reactants. The rate expression becomes exact in the two opposite limits of small and large reactivity, and also performs very well in the intermediate regime. In addition to the problems described in Sections 3 and 4, we applied the method in investigating the many-particle effects on the rates of diffusion-influenced reactions[7] and calculating the separation probability of geminate charge pairs in the presence of an external electric field.[8] We believe that the new procedure for solving the integral equations can be applied to diverse problems; it may provide a convergent solution even in the case where the direct iterative solution method leads to a divergent perturbation series. Finally, it is worthy of mentioning that a similar procedure has been used in deriving the resummed Brillouin–Wigner perturbation theory for quantum mechanical problems.[30]

Acknowledgments

The work described here was carried out in collaboration with C. Y. Son, J. Sung, S.-H. Chong, J. Kim, J.-H. Kim, and J. S. Kim. This work has been supported by a grant from by the Samsung Science and Technology Foundation (SSTF-BA1701-12).

References

1. A. D. Polyanin and A. V. Manzhirov, *Handbook of Integral Equations*. Chapman & Hall/CRC, Boca Raton, 2nd edn. (2008).
2. G. A. Baker, Jr., *Essentials of Pade Approximants*. Academic Press, New York (1975).
3. G. A. Arteca, F. M. Fernández, and E. A. Castro, *Large Order Perturbation Theory and Summation Methods in Quantum Mechanics*. Springer, Berlin (1990).

4. S. Lee, C. Y. Son, J. Sung, and S. Chong, *J. Chem. Phys.* **134**, 121102 (2011).
5. C. Y. Son and S. Lee, *J. Chem. Phys.* **135**, 224512 (2011).
6. C. Y. Son, J. Kim, J.-H. Kim, J. S. Kim, and S. Lee, *J. Chem. Phys.* **138**, 164123 (2013).
7. M. Kim, S. Lee, and J.-H. Kim, *J. Chem. Phys.* **141**, 084101 (2014).
8. K. Lee, S. Lee, C. H. Choi, and S. Lee, *J. Chem. Phys.* **147**, 144111 (2017).
9. U. Balucani and M. Zoppi, *Dynamics of the Liquid State.* Oxford University Press, Oxford (1994).
10. S. H. Northrup and J. T. Hynes, *J. Chem. Phys.* **71**, 871 (1979).
11. S. A. Rice, *Comprehensive Chemical Kinetics* **25**, *Diffusion-Limited Reactions*, C. H. Bamford, C. F. H. Tipper, and R. G. Compton, (eds.), Elsevier, Amsterdam (1985).
12. S.-H. Chong, C. Y. Son, and S. Lee, *Phys. Rev. E.* **83**, 041201 (2011).
13. P. G. Wolynes and J. M. Deutch, *J. Chem. Phys.* **65**, 450 (1976).
14. M. R. Flannery, *Phys. Rev. Lett.* **47**, 163 (1981).
15. H. S. Carslaw and J. C. Jaeger, *Conduction of Heat in Solids.* Oxford Universty Press, Oxford, 2nd edn. (1959).
16. G. Wilemski and M. Fixman, *J. Chem. Phys.* **58**, 4009 (1973).
17. O. K. Dudko and A. Szabo, *J. Phys. Chem. B.* **109**, 5891 (2005).
18. A. A. Zharikov and N. V. Shokhirev, *Chem. Phys. Lett.* **186**, 253 (1991).
19. S. D. Traytak, *Chem. Phys.* **140**, 281 (1990); **150**, 1 (1991); **154**, 263 (1991).
20. S. Lee and M. Karplus, *J. Chem. Phys.* **86**, 1883 (1987); **86**, 1904 (1987).
21. A. B. Doktorov, A. A. Kiprijanov, and A. I. Burshtein, *Opt. Spectrosc.* **45**, 279, 640 (1978).
22. A. A. Zharikov and N. V. Shokhirev, *Chem. Phys. Lett.* **190**, 423 (1992).
23. M. J. Pilling and S. A. Rice, *J. Chem. Soc. Faraday Trans.* **71**, 1555 (1975); **72**, 792 (1976).
24. A. B. Doktorov and A. I. Burshtein, *Sov. Phys. JETP.* **41**, 671 (1976).
25. E. A. Kotomin and A. B. Doktorov, *Phys. Stat. Sol. B.* **114**, 287 (1982).
26. F. Heisel and J. A. Miehe, *J. Chem. Phys.* **77**, 2558 (1982).
27. M. Abramowitz and I. A. Stegun, *Handbook of Mathematical Functions.* Dover, New York (1972).
28. M. Yokota and O. Tanimoto, *J. Phys. Soc. Jpn.* **22**, 779 (1967).
29. U. Gösele, M. Hauser, U. K. A. Klein, and R. Frey, *Chem. Phys. Lett.* **34**, 519 (1975).
30. S. Lee, C. H. Choi, and S. Lee, *Bull. Korean Chem. Soc.* **39**, 347 (2018).

Chapter 6

Geminate Electron–Hole Recombination in Homogeneous and Heterojunction Systems

Mariusz Wojcik

Institute of Applied Radiation Chemistry,
Lodz University of Technology,
Wroblewskiego 15, 93-590 Lodz, Poland
mariusz.wojcik@p.lodz.pl

Understanding the recombination kinetics of geminate pairs of oppositely charged particles is important for the construction of efficient solar cells and detectors of ionizing radiation. This chapter presents both the classical theory of geminate charge recombination, formulated by Onsager, and its extensions which make it more applicable to real systems. Special attention is given to the recent theoretical results on geminate electron–hole recombination in the presence of a donor–acceptor heterojunction, which are particularly useful for applications in photovoltaics.

1. Introduction

The physical problem which we deal with in this chapter is, at least at first glance, quite straightforward. We consider a pair of oppositely charged particles, initially separated by distance r_0, in a condensed medium (see Fig. 1). We assume that these particles perform a random motion affected by their mutual Coulomb interaction and have a chance to react to form a neutral product when they approach

M. Wojcik

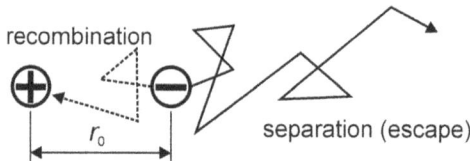

Fig. 1.   Illustration of the problem of geminate charge recombination.

each other to a close distance. We ask the following questions: how the survival probability of this pair depends on time and, in consequence, what is the ultimate probability that these particles avoid mutual reaction and separate to become free charge carriers.

The significance of this theoretical problem is related to its practical applications. First of all, this problem is crucial for our understanding of the operation of various types of solar cells. A quantum of light absorbed in a solar cell usually produces an electronic excitation, which soon dissociates into a pair of charged particles — an electron and a hole. For the efficient operation of a solar cell, it is required that these particles have a high chance to get separated rather than react with each other or, in other words, recombine. Because the considered electron and hole originate from the same dissociation, we call them a geminate electron–hole pair, and their reaction is called geminate recombination.

A similar situation appears in systems exposed to ionizing radiation, for example in the elementary particle detectors. In a collision with an ionizing particle, an electron is ejected from an atom (or a molecule) of the irradiated medium. This electron quickly dissipates its initially high kinetic energy and gets thermalized at some distance r_0 from its parent cation. Thus a geminate electron–ion pair is created, which subsequently may either recombine (often emitting luminescence) or escape from the mutual attraction. An electric current formed by the escaped charges carries a signal of the original ionization, which can be measured in the radiation detectors.

It is not clear why the theory of geminate charge recombination, despite its practical importance, is rarely discussed in the chemical kinetics textbooks. As you will see, this theory not only has its elegant classical works but is still a field of active research with interesting problems waiting to be solved.

Because of the space limitations of this chapter, we consider here only the theories in which the Coulomb interaction between the geminate particles is explicitly taken into account. However, a reader should be aware that a vast field of important theoretical research exists which is devoted to geminate reactions between uncharged particles or those whose long-range interaction is neglected. We will also not discuss here the non-geminate, bulk recombination processes which involve spatial distributions of positively and negatively charged particles and are usually covered by the physical chemistry textbooks as diffusion-controlled reactions between ions.

2. The Onsager Theory

The problem of geminate charge recombination can be treated by considering the probability density $w(\mathbf{r}, t)$ that the electron–hole pair is separated by distance $\mathbf{r}$ at time t. If the random motion of both particles is an ideal diffusion, then the time evolution of $w(\mathbf{r}, t)$ may be described by the Smoluchowski equation[1]

$$\frac{\partial w(\mathbf{r}, t)}{\partial t} = D\left[\nabla^2 w(\mathbf{r}, t) + \frac{1}{k_B T}\nabla w(\mathbf{r}, t)\nabla V(\mathbf{r})\right]. \tag{1}$$

Here, D is the sum of diffusion coefficients of the electron and the hole, k_B is the Boltzmann constant, and T is temperature. The second term in the square brackets is due to the drift of both particles in the mutual interaction potential $V(\mathbf{r})$. Without this term, Eq. (1) has the form of the classical diffusion equation.

For the pair of particles initially separated by $\mathbf{r}_0$, the initial condition is

$$w(\mathbf{r}, 0) = \frac{1}{4\pi r_0^2}\delta(\mathbf{r} - \mathbf{r}_0), \tag{2}$$

and the boundary conditions are given by

$$w(0, t) = 0, \tag{3a}$$

$$w(\infty, t) = 0. \tag{3b}$$

While the probability density $w(\mathbf{r}, t)$ contains the complete information about the fate of the electron–hole pair, it is more relevant from the point of view of chemical kinetics to concentrate on the time-dependent pair survival probability $W(t)$, which is related to $w(\mathbf{r}, t)$ by

$$W(t) = \int w\,(\mathbf{r}, t)\,d\mathbf{r}. \tag{4}$$

Interestingly, $W(t)$ satisfies the differential equation adjoint to Eq. (1), namely[1,2]

$$\frac{\partial W}{\partial t} = D\left(\nabla_{r_0}^2 W - \frac{1}{k_B T}\nabla_{r_0} W \nabla_{r_0} V\right). \tag{5}$$

Note that the space derivatives are here calculated with respect to the initial separation r_0 (as indicated by the subscripts).

By solving Eq. (5), one can directly determine the kinetics of the recombination process and thus bypass the much more complicated calculations that involve the probability density $w(\mathbf{r}, t)$. This possibility was first noted by Lars Onsager,[1] but with a very brief explanation. However, the derivation of Eq. (5) turns out to be not so simple. We refer the reader to other sources[2,3] where this issue is described in more detail.

The long-time limit of $W(t)$,

$$\varphi = \lim_{t \to \infty} W(t), \tag{6}$$

gives us the probability that the electron–hole pair ultimately avoids the mutual reaction. φ is often called the escape probability and has great practical significance, as discussed earlier. The escape probability can be directly determined by solving the following differential equation:

$$\nabla_{r_0}^2 \varphi - \frac{1}{k_B T}\nabla_{r_0}\varphi\,\nabla_{r_0} V = 0, \tag{7}$$

which is readily derived from Eq. (5) by considering $t \to \infty$. The appropriate boundary conditions are

$$\varphi(r_0 = 0) = 0, \tag{8a}$$

$$\varphi(r_0 \to \infty) = 1. \tag{8b}$$

Onsager[1] solved Eq. (7) for the case where the geminate particles not only interact Coulombically but are also under the influence of an external electric field F. This is an important situation because many experiments on charge recombination are performed in an applied field. The interaction potential is then given by

$$V(r, \vartheta) = -\frac{e^2}{4\pi\varepsilon_0\varepsilon r} - eFr\cos\vartheta, \tag{9}$$

where e is the elementary charge, ε_0 is the vacuum permittivity, ε is the dielectric constant, and ϑ is the angle that describes the orientation of a charge pair with respect to the external field.

Onsager's result is given by

$$\varphi(r_0, \vartheta, F) = \exp\left[-\beta r(1 + \cos\vartheta)\right]$$
$$\times \int_{r_c/r}^{\infty} J_0\{2\left[-\beta r(1 + \cos\vartheta)s\right]^{1/2}\}e^{-s}ds, \tag{10}$$

where $\beta = eF/2k_BT$ and J_0 is a zeroth-order Bessel function of the first kind. r_c is defined as

$$r_c = \frac{e^2}{4\pi\varepsilon_0\varepsilon k_BT} \tag{11}$$

and represents the distance at which the energy of Coulomb attraction is balanced by the thermal energy k_BT (r_c is often called the Onsager radius).

Because the initial orientation of the geminate pair is usually unknown, Eq. (10) is not very useful for practical applications. Instead, it is often more convenient to consider the angle-averaged escape probability, which can be calculated as[4]

$$\varphi(r_0, F) = \exp(-a - c)\frac{1}{c}\sum_{l=1}^{\infty} l\left(\frac{c}{a}\right)^{1/2} I_l\left(2\sqrt{ac}\right), \tag{12}$$

where $a = r_c/r_0$, $c = eFr_0/k_BT$, and I_l are the modified Bessel functions. Different expressions of $\varphi(r_0, F)$ are also available in the literature.[5,6]

In the absence of an external field, the escape probability of a geminate charge pair reduces to the simple and celebrated Onsager formula

$$\varphi(r_0) = \exp\left(-\frac{r_c}{r_0}\right). \tag{13}$$

In comparison with the calculations of the escape probability, it is much more complicated to determine the time dependence of the pair survival probability $W(t)$ or, in other words, the recombination kinetics. While the formal analytical solution of $W(t)$ has been obtained by Hong and Noolandi,[7] their treatment is so complicated that numerical solutions of Eq. (5) seem to be more convenient for some applications. A useful analytical result can only be obtained for the asymptotic behavior of $W(t)$ at long times, which takes the form

$$W(t) \underset{t\to\infty}{=} \exp\left(-\frac{r_c}{r_0}\right)\left(1 + \frac{r_c}{\sqrt{\pi Dt}}\right). \tag{14}$$

The theoretical framework described in this section and the analytical results presented above form the classical theory of geminate charge recombination, also known as the Onsager theory. While this theory is quite elegant and mathematically strict, it has serious limitations which become apparent when the Onsager theory is applied to interpret the experimental data. The source of these limitations is the very simplified picture of real physico-chemical processes that the equations of the Onsager theory provide. For example, the boundary conditions (3a) or (8a) imply that the chemical reaction which occurs when the electron and hole approach each other is infinitely fast. This is not a realistic assumption for many systems. Moreover, the motion of electrons and holes in various media cannot be represented as an ideal diffusion. Finally, many photophysical systems of practical importance are not homogeneous or not isotropic. A lot of research has been done over the last decades to improve or extend the Onsager theory. In the remaining part of this chapter, we will try to present a bird's eye view of those works which address important and interesting issues related to the Onsager theory and, more generally, to the kinetics of the geminate charge recombination process.

3.   Finite Rate of the Contact Reaction

From the experiments on charge carrier photogeneration in photo-conductive organic glasses, it was found that the escape probability of geminate electron–hole pairs is often $\sim 10^{-2}$. The dielectric constant in these non-polar systems is $\varepsilon = 3 - 4$, so the Onsager radius (see Eq. (11)) at room temperature is $r_c = 14 - 19\,\text{nm}$. Using the Onsager formula, Eq. (13), one can estimate the initial distance between the electron and hole in this case as $r_0 = 3-4\,\text{nm}$. It is rather difficult to rationalize such a large value of r_0 in a photoconductive organic system. Most probably, the photoseparated electron and hole initially reside on neighboring molecules of the medium, so the distance between them is only $\sim 1\,\text{nm}$. On the other hand, when we take $r_0 = 1\,\text{nm}$ and calculate the escape probability using Eq. (13), we obtain $\varphi \sim 10^{-7}$ in contrast to the experimental observations.

This problem with the Onsager theory is a consequence of its unphysical assumption of the inner boundary condition given by Eq. (3a) or, equivalently, Eq. (8a). It is clear that, first, the geminate electron and hole cannot approach each other to the distance equal to zero and, second, their contact reaction is never infinitely fast so the probability density $w(\mathbf{r}, t)$ should not vanish at their contact distance. While the rate constant of the contact electron–hole reaction can be measured as $k_r \sim 10^8\,\text{1/s}$ and the intermolecular separation of $R \sim 1\,\text{nm}$ can be taken as the contact distance, building this information into the Onsager theory turns out to be not so straightforward.

Using the formalism presented in Section 2, based on the continuous probability density $w(\mathbf{r}, t)$, one cannot conveniently describe the first-order electron–hole reaction which takes place only at a given distance R. Instead, one can assume that this reaction takes place over a range of distances and is characterized by a distance-dependent rate constant $k(\mathbf{r})$. Then, in order to include the electron–hole reaction, Eq. (1) can be modified in the following way[8,9]

$$\frac{\partial w(\mathbf{r}, t)}{\partial t} = D\left[\nabla^2 w(\mathbf{r}, t) + \frac{1}{k_B T}\nabla w(\mathbf{r}, t)\nabla V(\mathbf{r})\right] - k(\mathbf{r})w(\mathbf{r}, t).$$

$$(15)$$

Because the reaction is now represented by a sink term $-k(\mathbf{r})w(\mathbf{r}, t)$, the boundary condition (3a) needs to be replaced by the reflective boundary condition

$$\left.\frac{\partial w(r, t)}{\partial r}\right|_{r=R} = 0. \tag{16}$$

Although the model of electron–hole recombination given by Eq. (15) is in principle quite realistic, it does not allow us to obtain useful analytical expressions of the escape probability and practically can only be used in numerical calculations.[9] It is also a disadvantage that the real form of the distance dependence $k(r)$ is unknown and can only be approximated.

A much more convenient method of including the finite intrinsic reactivity of the electron–hole pair in the Onsager theory is by using the so-called radiation boundary condition, in analogy with the Collins–Kimball theory of diffusion controlled reactions between ions.[10] This boundary condition has the form

$$D\left[\frac{\partial w(r, t)}{\partial r} + \frac{w(r, t)}{k_B T}\frac{\partial V(r)}{\partial r}\right]_{r=R} = pw(R, t), \tag{17}$$

where the left-hand side represents the reactive flux through the surface $r = R$ and p is the reactivity parameter which has the dimension of velocity. Instead of using this parameter, some authors prefer to describe the intrinsic electron–hole reactivity by a second-order rate constant $k_r^{(II)}$ which is related to p by

$$k_r^{(II)} = 4\pi R^2 p. \tag{18}$$

Although the quantity $k_r^{(II)}$ can be regarded as more intuitive than parameter p, a careful reader should note that the use of a second-order rate constant to describe a reaction between two stationary particles (it is assumed here that the electron and hole are in contact at $r = R$) is not mathematically strict.

The advantage of introducing the radiation boundary condition (17) into the Onsager theory is that this approach yields a simple expression of the escape probability[2]

$$\varphi(r_0) = \frac{\exp(-r_c/r_0) - (1 - Dr_c/pR^2)\exp(-r_c/R)}{1 - (1 - Dr_c/pR^2)\exp(-r_c/R)}. \qquad (19)$$

One can see that at $R \to 0$ and $p \to \infty$, this equation correctly reduces to the Onsager formula (Eq. (13)). In the practically important situation where an electron–hole pair is initially separated by $r_0 = R$, Eq. (19) takes the form

$$\varphi(R) = \frac{1}{1 + (pR^2/Dr_c)[\exp(r_c/R) - 1]}. \qquad (20)$$

Although Eqs. (19) and (20) are quite convenient, their use is complicated by the fact that the reactivity parameter p cannot be uniquely related with the first-order recombination rate constant k_r. However, by considering the appropriate reaction volume, an approximate relation between p and k_r can be derived in the form[11]

$$p \approx k_r R. \qquad (21)$$

The theoretical model of geminate recombination based on the solutions of the Smoluchowski equation (1) with the boundary condition given by Eq. (17) can also be used to determine the time dependence of the pair survival probability $W(t)$. Unfortunately, the formal analytical solutions of $W(t)$ obtained by Noolandi and Hong[7] are too complicated for practical use, similarly as in the case of infinite electron–hole intrinsic reactivity discussed in Section 2.

Also the dependence of the escape probability on the applied electric field, analytically derived using the radiation boundary condition (17) by Noolandi and Hong,[12] is quite complicated. The calculations of $\varphi(F)$ based on their methodology are presented in an accessible way in Ref. 11. Moreover, the first- and second-order expansions of $\varphi(F)$ can be expressed by more convenient formulas, as described further below.

In the recombination theories presented so far, we have considered only continuous models of the particle motion. However, the discreteness of real molecular systems suggests that the lattice models of the medium may become useful in describing the geminate recombination process. Such models were applied by Scher and Rackovsky[13] in their analytical calculations of the escape probability. They considered the particle motion by nearest-neighbor hopping between the lattice sites, with the transition rates proportional to the Boltzmann factors $\exp(-\Delta V/k_B T)$, and evaluated for this transport mechanism the lattice Green's functions. They also assumed a finite rate of the ultimate electron–hole reaction. Unfortunately, their calculations involved various approximations and did not provide useful expressions of the escape probability.

The discrete nature of the medium can be partially incorporated into a theoretical model of geminate recombination using a different, mixed approach.[14,15] Consider a hole placed at the lattice origin and an electron initially located at one of its nearest-neighbor lattice sites. When we assume that the electron hopping between the lattice sites (except the origin) may be described by a continuous diffusion model, while a hop to the origin results in the first-order reaction with the hole, we can arrive at the following expression of the escape probability[15]

$$\varphi(d) = \frac{1}{1 + (N_c V_c k_r / 4\pi D r_c)[\exp(r_c/d) - 1]}, \tag{22}$$

where N_c denotes the lattice coordination number and V_c is the primitive cell volume (volume per site). The results obtained using Eq. (22) are in a rather good agreement with the corresponding simulation results calculated for different lattice types,[15] including the close-packed fcc structure ($N_c = 12$, $V_c = d^3/\sqrt{2}$) which is quite a good representation of various real systems.

4. The Braun Model — A Controversy

As already explained, the problem of geminate electron–hole recombination concerns a competition between the reaction of the geminate particles and their ultimate separation. The latter effect is a result of a random walk of the particles, during which they move forth and

Fig. 2.   Illustration of the Braun model and its controversy.

back visiting many sites. An assumption of the continuous diffusion model for this random walk leads to the construction of the theories described in Sections 2 and 3. Although these theories yield some useful analytical results of the escape probability, the mathematical picture is generally quite complicated, especially if one is interested in the kinetics of recombination or the electric field effect on the escape probability.

Trying to provide a simpler analytical description of geminate recombination, Braun[16] made use of the model in which separation of the geminate particles is treated as a first-order process and described by a "dissociation" rate constant k_d (see Fig. 2). With k_r being the first-order rate constant of the electron–hole reaction, this approach readily yields

$$\varphi = \frac{k_d}{k_r + k_d}. \tag{23}$$

The rate constant k_d was derived by Braun from the theoretical result of the equilibrium constant that describes the dissociation of ions in a weak electrolyte solution. The expression he proposed has the form

$$k_d = \frac{3Dr_c}{R^3} \exp(-r_c/R). \tag{24}$$

Moreover, Braun made use of the Onsager result[17] which describes the dependence of the considered equilibrium constant on the applied electric field, and proposed the field dependence of k_d in the form

$$k_d(F) = k_d(0) J_1 [2\sqrt{2}\,(-b)^{1/2}]/\sqrt{2}\,(-b)^{1/2}$$
$$= k_d(0)\left(1 + b + b^2/3 + \cdots\right), \tag{25}$$

where $b = er_cF/2k_BT$ and J_1 is the Bessel function of order one. Eqs. (23)–(25) allow us to calculate the electric field dependence of the geminate electron–hole escape probability taking into account the finite rate constant of the electron–hole reaction. As these equations are also not too complicated, they have been very often used to analyze the experimental data on charge carrier photogeneration.[a]

Despite its frequent use, the theoretical approach described above, sometimes referred to as the Braun model, is the matter of controversy.[11] As already pointed out, the separation process of geminate particles has the character of a random walk, so it cannot be treated as a simple first-order process. This is reflected, for example, in the kinetics of geminate recombination. The strict theory described in Section 3 predicts for the survival probability $W(t)$ the following asymptotic form[2]

$$W(t) \underset{t\to\infty}{=} \varphi(R) \left[1 + \frac{1}{1 - e^{-r_c/R} + \left(Dr_c/pR^2\right) e^{-r_c/R}} \frac{r_c}{(\pi Dt)^{1/2}} \right],$$

(26)

which is clearly different from the exponential kinetics that characterizes a competition between two first-order reactions.

The above arguments raise concerns about applicability of the electric field dependence of escape probability given by Eqs. (23)–(25). In fact, as shown in Ref. 11, the dependence $\varphi(F)$ obtained from these equations has a different shape than that calculated using the analytical results of Noolandi and Hong.[12] However, the discrepancy is not large at low values of k_r.

It is also interesting to note that for non-polar systems where $\exp(r_c/R) \gg 1$, the first- and second-order field expansions of the escape probability obtained from Eqs. (23)–(25) and from the Noolandi–Hong theory have the same form[15]

$$\varphi(b) = \varphi_0 \left[1 + (1 - \varphi_0)\, b + \frac{1}{3}\, (1 - \varphi_0)\, (1 - 3\varphi_0)\, b^2 \right], \qquad (27)$$

where $\varphi_0 = \varphi(0)$.

---

[a] See the references 8–22 in Ref. 11.

Although the theoretical background of the Braun model is definitely questionable, this fact is not widely understood. The simplicity of the Braun model and its convenient analytical expressions stand behind its frequent applications. Despite the controversy, it seems that the Braun model may be treated as a useful empirical theory, especially when we are not concerned about the recombination kinetics and interested only in the escape probability. Moreover, new analytical results still appear[18,19] which modify Braun's approach and provide better approximations of the real $\varphi(F)$ dependence than that given by Eqs. (23)–(25).

5. Effects of Non-Ideal Diffusion

In the theories of geminate recombination presented so far, it is generally assumed that the motion of geminate particles can be described as ideal diffusion. However, in many photoconductive materials, the transport of charge carriers deviates from this idealized picture. A major reason for this deviation is the fact that the energy of the hopping sites visited by electrons or holes during the recombination process often has a wide statistical distribution, which reflects an amorphous structure of the medium or other forms of disorder. As a consequence, the hopping times may also have specific distributions, so that the description of the particle motion merely by a diffusion coefficient (or mobility) becomes, at least at short times, rather unrealistic.

The problem of geminate electron–hole recombination in an energetically random system is difficult to treat analytically. In the case where the energy disorder can be described using a Gaussian distribution, the electron–hole escape probability was calculated by simulations using lattice models. Albrecht and Bässler[20] found that the escape probability is in this case larger than that given by the Onsager formula (Eq. (13)). However, in other simulations,[21] where the finite rate of the electron–hole reaction was properly taken into account, the opposite effect was observed.

When the energy of hopping sites has an exponential or other long-tailed distribution, the motion of charge carriers may take the form of subdiffusion, which means that the mean square displacement of a particle depends on time as $\langle r^2(t) \rangle \sim t^\alpha$, where $\alpha < 1$. A lot of theoretical research has been done over the last decades to understand the mechanisms of this anomalous charge carrier diffusion. Within this field, interesting results are also available that describe reactions between geminate particles, for example the results obtained using the fractional diffusion formalism and the fractional reaction-diffusion equation.[22] Unfortunately, these results are mostly obtained for neutral particles or by neglecting their long-range interactions. An attempt to determine the recombination kinetics for Coulombically interacting particles has also been undertaken[23] but the developed methodology, while elegant, is not so convenient for practical applications.

6. Geminate Recombination in Heterojunction Systems

As already pointed out, an important application of the theory of geminate charge recombination is that in explaining the operation of the solar cells. These devices are often constructed as binary systems composed of an electron donor and electron acceptor materials. There are two main advantages of using such two-phase structures. First, energy differences between the corresponding electronic states (or bands) in the donor and acceptor materials lead to very efficient dissociation of electron photoexcitations near the donor–acceptor boundary, and thus to efficient creation of geminate electron–hole pairs. Second, a heterojunction which is formed at this boundary restricts the motion of electrons and holes to the acceptor and donor phases, respectively, which in turn causes a significant increase in the electron–hole escape probability.

While the charge separation processes in the solar cells in the presence of a donor–acceptor heterojunction are qualitatively rather well understood, their quantitative description based on an appropriate analytical theory is much more complicated. If we denote the hole and electron positions by (x_1, y_1, z_1) and (x_2, y_2, z_2), respectively, and

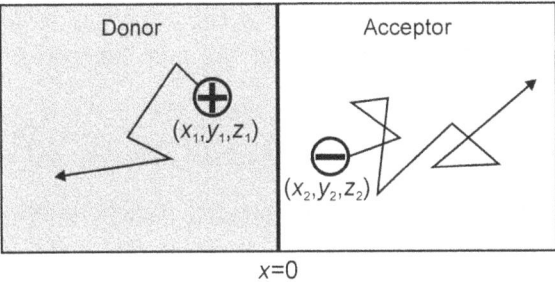

Fig. 3.   Illustration of geminate electron–hole recombination in a donor–acceptor heterojunction system.

assume that a plane heterojunction is located at $x = 0$, so that the donor and acceptor materials are at $x < 0$ and $x > 0$, respectively (see Fig. 3), then the Smoluchowski equation (see Eq. (1)) can be written as[24]

$$\frac{\partial w}{\partial t} = D_1 \frac{\partial^2 w}{\partial x_1^2} + D_2 \frac{\partial^2 w}{\partial x_2^2} + (D_1 + D_2) \left(\frac{\partial^2 w}{\partial y^2} + \frac{\partial^2 w}{\partial z^2} \right)$$

$$+ \beta D_1 \frac{\partial}{\partial x_1} \left(\frac{\partial V}{\partial x_1} w \right) + \beta D_2 \frac{\partial}{\partial x_2} \left(\frac{\partial V}{\partial x_2} w \right)$$

$$+ \beta (D_1 + D_2) \left[\frac{\partial}{\partial y} \left(\frac{\partial V}{\partial y} w \right) + \frac{\partial}{\partial z} \left(\frac{\partial V}{\partial z} w \right) \right], \qquad (28)$$

where $w = w(x_1, x_2, y, z, t)$, $y = y_2 - y_1$, $z = z_2 - z_1$, and $\beta = (k_B T)^{-1}$. D_1 and D_2 denote the diffusion coefficients of the electron and hole, respectively. The initial and boundary conditions can also be appropriately expressed.[24] Equation (28) is quite complex, and there is little hope to find its exact analytical solution that describes the geminate-electron hole recombination in a heterojunction system.

It is interesting to note that Eq. (28) can also be looked at as an equation which describes geminate recombination in a 4D homogeneous system with a diffusion anisotropy. Then, D_1 and D_2 can be interpreted as the diffusion coefficients along x_1 and x_2 coordinates, respectively, while $D_1 + D_2$ as that along both y and z coordinates. The above observation allows one to predict the asymptotic behavior of the recombination kinetics at long times. Using the theoretical result of Barzykin and Tachiya,[25] one can expect in a 4D system the long-time asymptotic dependence of the pair survival probability in the form

$$W(t) \underset{t\to\infty}{=} \varphi\left(1 + Bt^{-1}\right), \tag{29}$$

where B is a constant. This dependence is clearly different than that determined for homogeneous 3D systems (see Eq. (14)). The validity of Eq. (29) has been confirmed by comparison with the corresponding simulation results obtained for a heterojunction system.[24]

An attempt to find an approximate analytical solution of Eq. (28) has recently been reported by Lukin.[26] He used the so-called pre-scribed diffusion approach, in which the probability density that describes the electron-hole pair is approximated as

$$w(x_1, x_2, y, z, t) = w_1(x_1, t)w_2(x_2, t)w_{yz}(y, z, t)P(t), \tag{30}$$

where $P(t)$ is independent of the spatial coordinates and w_{yz} is assumed to satisfy a 2D diffusion equation. Although the obtained analytical results of the escape probability and its dependence on an applied electric field were found to differ from more accurate simulation results, the reported discrepancy is not very large.

In view of the complications in the analytical treatment, geminate electron-hole recombination in heterojunction systems was extensively studied by computer simulations.[15,21,27,28] By analyzing a large set of the simulation results, a semi-empirical equation was proposed which describes the escape probability as[15]

$$\varphi = \left[1 + \frac{k_r R^3}{ADr_c} \exp\left(\frac{r_c}{R}\right)\right]^{-1}, \tag{31}$$

where A is a parameter. For the case of a plane heterojunction, this parameter is approximately expressed as $A \approx 90/\varepsilon$, where ε denotes the dielectric constant. Equation (31) was also found to be valid for homogeneous systems (without a heterojunction). In this case, A is independent of ε and can usually be taken as $A \approx 1.5$.[15]

It is important to note that in heterojunction systems, the kinetics of geminate recombination depends not only on the sum of the

diffusion coefficients of electron and hole, $D = D_1 + D_2$, but also on
their relative values. These diffusion coefficients, which are related to
the electron and hole mobilities, are usually different in the donor and
acceptor materials. Finding their optimum combination is crucial in
the design of efficient solar cells. Recently, a controversy appeared
in the literature after it was reported that a mismatch between the
electron and hole mobilities in heterojunction systems may increase
the charge carrier separation probability.[29] This conclusion was later
shown to be not true and result from a wrong interpretation of the
simulation data. As finally demonstrated,[28] the escape probability is
the highest in the case of equal electron and hole mobilities. A mobility mismatch decreases the charge separation probability, although
this effect is not strong unless the electron and hole mobilities differ
by more than an order of magnitude.

The theoretical results presented in this section are directly applicable to real systems when the shape of the heterojunction can be well
approximated by a plane. Although this assumption is fulfilled for
some types of photovoltaic devices (for example, the bi-layer organic
solar cells), there are also other cell architectures where the heterojunction has a more complicated structure. An important example
is the bulk heterojunction solar cell, where the donor and acceptor materials are mixed and form an interpenetrating network (see
Fig. 4). While the geminate electron–hole recombination in bulk heterojunction cells was studied by simulations,[29] no analytical theory

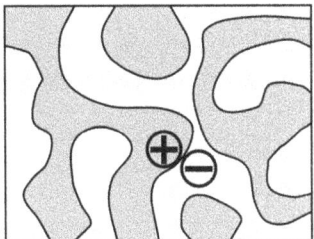

Fig. 4.   Illustration of the structure of a bulk heterojunction solar cell, where the
donor (dark shaded) and acceptor materials form an interpenetrating network.

is available that describes the effect of the cell architecture on the charge separation probability.

References

1. L. Onsager, *Phys. Rev.* **54**, 554 (1938).
2. H. Sano and M. Tachiya, *J. Chem. Phys.* **71**, 1276 (1979).
3. M. Tachiya, *J. Chem. Phys.* **69**, 2375 (1978).
4. W. Que and J. A. Rowlands, *Phys. Rev. B* **51**, 10500 (1995).
5. G. C. Abell and K. Funabashi, *J. Chem. Phys.* **58**, 1079 (1973).
6. A. Mozumder, *J. Chem. Phys.* **60**, 4300 (1974).
7. K. M. Hong and J. Noolandi, *J. Chem. Phys.* **68**, 5163 (1978).
8. M. Tachiya, *Radiat. Phys. Chem.* **21**, 167 (1983).
9. M. Wojcik and M. Tachiya, *Radiat. Phys. Chem.* **74**, 132 (2005).
10. F. C. Collins and G. E. Kimball, *J. Colloid Sci.* **4**, 425 (1949).
11. M. Wojcik and M. Tachiya, *J. Chem. Phys.* **130**, 104107 (2009).
12. J. Noolandi and K. M. Hong, *J. Chem. Phys.* **70**, 3230 (1979).
13. H. Scher and S. Rackovsky, *J. Chem. Phys.* **81**, 1994 (1984).
14. P. Giazitzidis, P. Argyrakis, J. Bisquert, and V. S. Vikhrenko, *Org. Electron.* **15**, 1043 (2014).
15. M. Wojcik, A. Nowak, and K. Seki, *J. Chem. Phys.* **146**, 054101 (2017).
16. C. L. Braun, *J. Chem. Phys.* **80**, 4157 (1984).
17. L. Onsager, *J. Chem. Phys.* **2**, 599 (1934).
18. K. Seki and M. Wojcik, *J. Phys. Chem. C.* **121**, 3632 (2017).
19. K. Lee, S. Lee, C. H. Choi, and S. Lee, *J. Chem. Phys.* **147**, 144111 (2017).
20. U. Albrecht and H. Bässler, *Chem. Phys. Lett.* **235**, 389 (1995).
21. M. Wojcik, P. Michalak, and M. Tachiya, *Bull. Korean Chem. Soc.* **33**, 795 (2012).
22. K. Seki, M. Wojcik, and M. Tachiya, *J. Chem. Phys.* **119**, 2165 (2003).
23. A. I. Shushin, *J. Chem. Phys.* **129**, 114509 (2008).
24. M. Wojcik and M. Tachiya, *Chem. Phys. Lett.* **537**, 58 (2012).
25. A. V. Barzykin and M. Tachiya, *J. Chem. Phys.* **99**, 9591 (1993).
26. L. V. Lukin, *Chem. Phys.* **491**, 102 (2017).
27. P. Peumanns and S. R. Forrest, *Chem. Phys. Lett.* **398**, 27 (2004).
28. M. Wojcik, P. Michalak, and M. Tachiya, *Appl. Phys. Lett.* **96**, 162102 (2010).
29. C. Groves, R. A. Marsh, and N. C. Greenham, *J. Chem. Phys.* **129**, 114903 (2008).

Chapter 7

Anomalous Kinetics of Catalytic Conversion Reactions in Linear Nanopores Mediated by Inhibited Transport: Multiscale Modeling

Andrés Garcia[*,†], Chi-Jen Wang[¶], David M. Ackerman[*],
Mark S. Gordon[*,‡], Igor I. Slowing[*,‡], and James W. Evans[*,†,§,‖]

[*]*Ames Laboratory — USDOE, Iowa State University,
Ames, Iowa 50011, USA*
[†]*Department of Physics & Astronomy, Iowa State University,
Ames, Iowa 50011, USA*
[‡]*Department of Chemistry, Iowa State University,
Ames, Iowa 50011, USA*
[§]*Department of Mathematics,
Iowa State University, Ames, Iowa 50011, USA*
[¶]*Department of Mathematics,
National Chung Cheng University, Chiayi 62102, Taiwan*
[‖]*evans@ameslab.gov*

The yield of irreversible conversion reactions, $A \to B$, can be enhanced through the use of catalytically active nanoporous materials due to the high surface area of exposed catalytic sites. Reactants, A, enter the pore openings and diffuse through a solvent within the pore to catalytic sites where they convert to product, B. However, a restricted propensity, P, for passing of A and B (in the extreme case, single-file diffusion, SFD) inhibits removal of product from the pores, greatly reducing yield. Multiscale modeling of this process utilizes Langevin simulation (with implicit treatment of solvent) to assess P versus pore diameter; then P provides input to coarse-grained (CG) spatially discrete stochastic modeling. For

173

the latter, the pore is divided into a linear array of cells. A adsorbs, and both A and B desorb from the end cells. Diffusion involves hopping of A and B to adjacent empty cells, and also exchange with probability P_{ex} (related to P) of A and B in adjacent cells. Yield strongly decreases as P or $P_{ex} \to 0$ (SFD). Behavior is not described by mean-field kinetics due to the development of strong spatial correlations reflecting the interplay between SFD and reaction. However, behavior is captured by a suitable generalized hydrodynamic formalism.

1. Introduction

Catalytic reaction-diffusion processes wherein reactants are irreversibly introduced and products are irreversibly removed from a catalytic system generally operate in non-equilibrium steady states, although more complex oscillatory behavior sometimes occurs.[1,2] Examples of different catalytic systems where such considerations apply are: (i) arrays of 1D linear nanopores in zeolites[3] or catalytically functionalized mesoporous silica nanoparticles (MSN)[4]; and (ii) 2D low-index surfaces of crystalline catalytic metals or oxides.[5]

Two general scenarios apply. First, if the catalytic system is "well-stirred", then the spatial distribution of reactant and product species is equilibrated with concentrations controlled by addition (adsorption) and removal (reaction, desorption) kinetics. This situation applies for traditional ultra-high vacuum studies of catalytic surface reactions under low-pressure conditions where adsorbed species are highly mobile.[6,7] If interactions between mobile species are neglected, which implies that they are randomly distributed, then behavior can be described by mean-field kinetics (although interactions are typically significant and accurate treatment must account for equilibrium spatial correlations[6]).

The second scenario is where the system is not well-stirred due to transport limitations. One such example is provided by 1D nanoporous systems with inhibited passing of reactants and product species (and in the extreme case SFD) in narrow nanopores.[7-9] Another example is provided by 2D surface reactions under high-pressure conditions, which result in crowding of adsorbed reactant and product species on the surface at near-jamming coverages. In these cases, spatial correlations between reactant and product species, which impact reaction kinetics, result from the interplay between adsorption, desorption, reaction, and limited transport. Clearly mean-field kinetics will not apply, and in fact there is no general formulation to reliably describe these correlations. Nonetheless, recently there have been advances, e.g., in analysis of realistic models for high-pressure CO-oxidation.[10–12]

In this chapter, however, we analyze irreversible catalytic conversion reactions, $A \to B$, occurring in arrays of 1D narrow linear nanopores of zeolites or MSN. Our modeling is geared to solution-phase reactions although we note many reactions studies in zeolites involve the gas phase. See Fig. 1(a) illustrating neopentane conversion to isobutene in a zeolite[13] and Fig. 1(b) illustrating PNB conversion to an aldol product in amine-functionalized MSN.[14] In both cases, SFD can apply. Ideally, many-particle molecular dynamics (MD) with explicit treatment of all species (including the solvent) could provide the most complete description of the system, although many-particle Langevin simulation with implicit treatment of solvent might provide a reasonable alternative. However, neither approach could efficiently access the relevant time- and length-scales. Thus, instead, CG spatially discrete stochastic modeling has traditionally

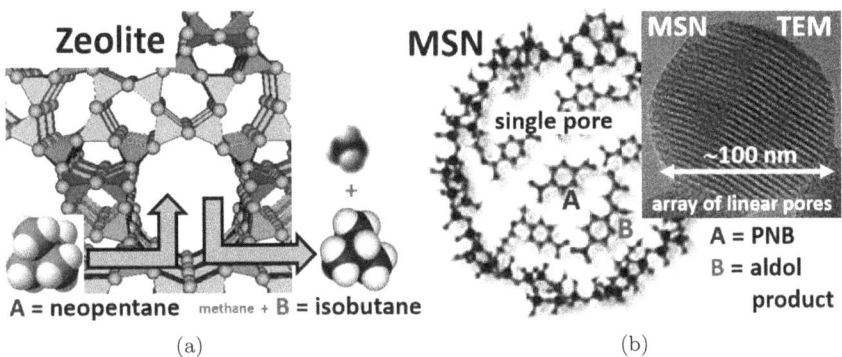

Fig. 1. Schematic of: (a) neopentane conversion to isobutene in a zeolites, (b) p-nitrobenzaldehyde (PNB) conversion to an aldol product in amine-functionalized MSN.

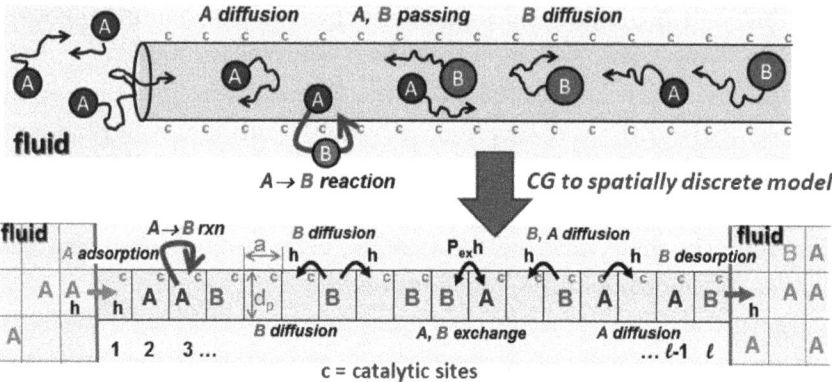

Fig. 2.   Schematic of CG from continuum Langevin to spatially discrete stochastic model. Copyright (2015) by American Chemical Society.

been applied to analyze the overall reaction–diffusion process in these systems[7–9,15–18] (see Fig. 2). Here, the pore is divided into a linear array of cells, the width of which corresponds to the typical linear dimension of the reactant, A, and product, B, species of ~1 nm. A adsorbs, and both A and B desorb from the end cells. Diffusion involves hopping of A and B to adjacent empty cells, and also exchange with suitably defined probability P_{ex} of A and B in adjacent cells. Irreversible conversion $A \rightarrow B$ occurs in catalytically active cells.

This presentation adopts a multiscale modeling strategy wherein first we implement targeted two-particle Langevin simulations to assess the passing propensity of reactant and product species, P, as a function of pore diameter. P provides input to CG stochastic modeling of the type described above, specifically determining the exchange probability, P_{ex}. As indicated above, mean-field reaction-diffusion equations fail to describe model behavior for small P (or P_{ex}) due to the development of strong spatial correlations determined by the interplay of reaction and near-SFD transport. Kinetic Monte Carlo simulations can provide a precise characterization of behavior. However, for deeper insight, and also for more efficient model analysis, we will show that a generalized hydrodynamic formulation proves effective.

In Section 2, we describe Langevin and equivalent Fokker–Planck analysis of molecular passing propensity in narrow pores. In Section 3, the CG stochastic model is described in detail, and Kinetic Monte Carlo results for model behavior are presented. Section 4 develops exact hierarchical master equations for the CG model, elucidates the failure of mean-field-type truncation approximations, and provides an alternative and effective generalized hydrodynamic analysis. Conclusions are presented in Section 5.

2. Passing Propensity for Reactants and Products in Pores

The passing propensity, P, is a key parameter controlling reaction yield. Ideally, high-fidelity MD studies with explicit molecular-level description of the pore, the solvent, as well as a reactant (A) and product (B) pair, could be used to assess P versus pore diameter. However, no such studies exist to date, although recent developments in fragmentation methods open this possibility even for the complex MSN systems.[19] Here, however, we will consider only simpler analysis based on strongly damped Langevin (i.e., Brownian) dynamics of A and B through an implicit solvent considering only steric (non-overlap) interactions between A and B and the pore wall (treated by a simple cylindrical geometry). Other formulations may be appropriate in different experimental regimes.

One basic issue is the appropriate definition of P which should be crafted from the perspective that this parameter is to be passed to the CG modeling. In the simplest case, suppose that A and B have similar characteristic linear size, L_c, which will also correspond to the cell width in the CG modeling. Let z denote the coordinate along the pore axis. Then, to match exchange processes in the CG modeling, it is natural to start with the centers of the reactant and product species, A and B, separated by a distance, $\delta z = +L_c$ in the direction of the pore axis, with A (B) on the left (right), say. Then, one runs Brownian dynamics until δz reaches for the first time either $\delta z = -L_c$ (passing, with now B on the left of A) or $\delta z = +2L_c$

(separation without passing with A still on the left of B). Then, P is simply defined as the fraction of passing outcomes after running many trials.

For a CG model with rates h_A (h_B) of hopping to adjacent empty cells for A (B), we will naturally define the rate of exchange of adjacent A and B as $\frac{1}{2}(h_A + h_B)P_{ex}$. Then, comparing this rate with that separation of $h_A + h_B$ (A hops left, or B hops right), it follows that $P = P_{ex}/(2 + P_{ex})$. For wide pores where A and B do not interact, since $|\delta z|$ changes by $2L_c$ for passing versus $1L_c$ for separation, first-passage analysis shows that $P = 1/3$ and $P_{ex} = 1$. At the other extreme, for SFD, one has that $P = P_{ex} = 0$.

2.1. *Langevin analysis of passing propensity*

For our models with steric (non-overlap) interactions between A and B and the pore wall, there is a well-defined critical value of pore diameter $d = d_c$ below which passing is not possible; d_c is simply determined by system geometry. It is natural to consider the scaling $P \sim (d - d_c)^\sigma$, as $d \to d_c$ from above. A simple transition state theory (TST) analysis[20,21] identifies $P_{TST} \sim \exp[-\delta F/(k_B T)]$, where δF is the entropic free energy barrier for passing. Let $\Omega(\delta z)$ denote the phase space volume for a specific separation δz of A and B centers along the pore (which considers molecular positions orthogonal to the pore axis and orientation angles for non-spherical molecules). Then, one has $\delta F \sim -k_B T \ln(\Omega_{min})$ where Ω_{min} denotes the minimum Ω at the transition state (TS) for passing. It follows that $P_{TST} \sim \Omega_{min}$ which can be directly calculated given molecular sizes and shapes.[20] We will find, however, that the corresponding exponent, σ_{TST}, fails to recover the correct value of σ obtained from Langevin simulation.

It is natural to consider the canonical problem where A and B are represented by spheres of equal radii, r, with equal diffusivities, D_A and D_B, where $D_{A,B} = (L_c)^2 h_{A,B}$, in a cylindrical pore of radius $R \geq 2r$. See Fig. 3(a). Thus, the critical pore diameter is $d_c = 4r$. It is also instructive to define a gap size $g = 2(R - 2r) = d - d_c$, so that $P \sim g^{\sigma}$, as $g \to 0$. See Fig. 3(b). Langevin simulation results for P versus g are shown in Fig. 3(b). Simulations are demanding as $10^4/P$ trials are required to determine P with 1% uncertainty. Thus, for reliable estimate of the scaling exponent, we supplement Langevin simulation with a Fokker–Planck equation (FPE) analysis for small g (and P), as described below. For two spheres passing in a cylinder, we conclude that $\sigma \approx 1.4$, which should be contrasted with $\sigma_{\mathrm{TST}} = 2.5$.[20]

It should be emphasized that the above analysis for spherical molecules (where dynamics is described by a single translational diffusion coefficient, and P is independent of its value) is far simpler than for arbitrarily shaped molecules. For the latter, each molecule has 3 translational and 3 rotational degrees of freedom. In the simplest scenario, different translational and rotational diffusion degrees of freedom would be uncoupled and described by a total of 6 diffusion coefficients. However, according to a hydrodynamic formulation of diffusion, this only occurs for higher-symmetry molecules in 3D. In general, diffusional degrees of freedom are coupled, and diffusion must be described by a 6×6 diffusion tensor.[22] However, Langevin simulation can still be performed with appropriate random forces and torques satisfying a fluctuation–dissipation relation compatible with the 6×6 diffusion tensor.[23] See Fig. 4 for a schematic of this procedure including our geometric models for PNB and aldol molecules

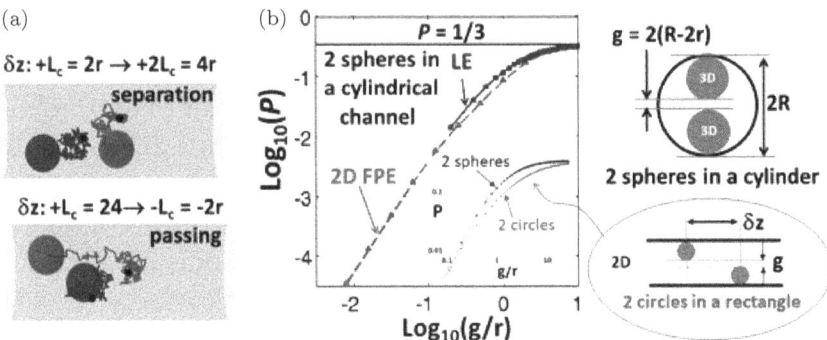

Fig. 3.   (a) Schematic of Langevin simulation of passing for 2 spheres in a cylindrical pore, (b) results for passing propensity, P, versus gap size, g.

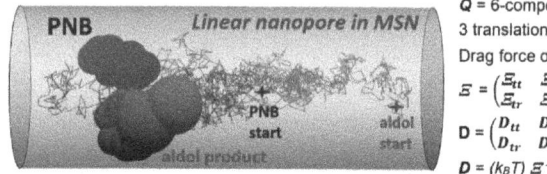

Q = 6-component state vector for each molecule
3 translation (t) coordinates; 3 rotation (r) angles

Drag force on molecule: $F^{drag} = -\Xi\, d/dt\, Q$

$\Xi = \begin{pmatrix} \Xi_{tt} & \Xi_{rt} \\ \Xi_{tr} & \Xi_{rr} \end{pmatrix}$ 6 x 6 grand resistance tensor

$D = \begin{pmatrix} D_{tt} & D_{rt} \\ D_{tr} & D_{rr} \end{pmatrix}$ 6x6 diffusion tensor

$D = (k_B T)\, \Xi^{-1} = \mu^2$ and $d/dt\, Q = \sqrt{2}\,\mu\, W$

W entries mean 0, variance 1 random numbers

Langevin analysis: passing of complex shaped molecules

Fig. 4.   Schematic of the Langevin simulation formalism general shapes molecules with application to passing of PNB and an aldol product.

which reasonably capture their actual size and elongated shapes. For these models, one has that $d_c = 9.3$ nm, and for pore diameter d just above d_c, molecules must align with each other and the pore to pass, so reliable treatment of both rotational and translational diffusion is key. Our analysis shows P varying from $P = 0.028$ for $d = 1.5$ nm (close to SFD) to $P = 0.24$ for $d = 3.0$ nm.[24]

2.2.   Equivalent Fokker–Planck equation (FPE) analysis

The Langevin analysis above for passing of molecules in a cylindrical pore can be reformulated as a boundary value problem for an equivalent FPE.[20] The passing problem corresponds to an initial value problem (IVP) for diffusion of probability in a high-dimensional channel with axis δz and where the lateral size of the channel equals the volume of phase space, $|\Omega(\delta z)|$. Probability is initially located at $\delta z = +L_c$ and either diffuses through constriction at the TS (often at $\delta z = 0$) to reach a trap at $\delta z = -L_c$ (corresponding to passing) or diffuses without constriction to reach a trap as $\delta z = +2L_c$ (corresponding to separation). This is illustrated in a 2D schematic in Fig. 5 (a) for a channel with axis δz and width $|\Omega|$.

Rather than solve this IVP, it is more convenient to solve an equivalent steady-state problem[20] where probability flux is injected at a constant rate J at $\delta z = +L_c$, diffuses in the constricted channel, and absorbs at traps located at $\delta z = -L_c$ with flux J_{pass}, and $\delta z = +2L_c$ with flux J_{sep} (where $J = J_{pass} + J_{sep}$). See Fig. 5 (b) for a schematic of the corresponding steady-state probability distribution, $f^{ss}(\delta z)$. It follows that $J_{pass} : J_{sep} = P : 1 - P$ allowing determination of P.

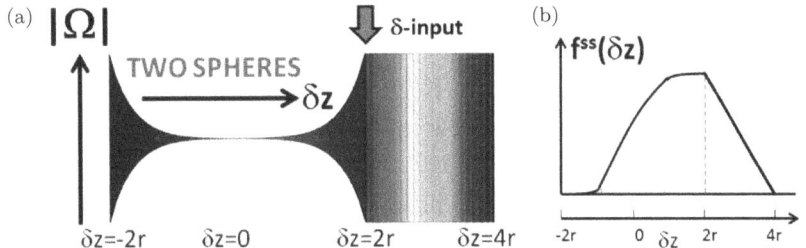

Fig. 5. (a) 2D schematic of FPE analysis of passing of two spheres in a cylindrical channel, (b) steady-state probability distribution where probability is injected at $+L_c = 2r$ and is absorbed trapped at $-L_c$ (passing) or $+2L_c$ (separation).

The potential advantage of formulating passing as a steady-state diffusion problem is that one might exploit adaptive mesh numerical methods to accurately treat behavior around the narrow constriction (for small g and P).[20] Unfortunately, such methods are only well developed for systems with 2 or 3 spatial variables, but in the simplest case of passing of 2 spheres has 5 variables (δz and 2 lateral positions for each sphere). Thus, we instead use 2 variable representation of the constricted channel illustrated in Fig. 5(a) where the channel width is just taken as $|\Omega|$. FPE results are shown in Fig. 3(b). This approach is validated for passing of 2 circles in a rectangle where the full 3 variable FPE analysis can be compared with our reduced 2 variable problem.[20]

3. Coarse-Grained Spatially Discrete Stochastic Model

3.1. Model formulation

Here, we provide more details of the key ingredients of the CG model[7,18] illustrated in Fig. 2. As noted previously, pores are divided into a linear array of cells which we label $n = 1, 2, \ldots, n_{\max}$, from left to right, and which can be populated by at most one reactant, A, or product, B, species. Cells populated by neither A or B are described as empty, E (although they are still populated by solvent). It is convenient to extend the 1D array of cells into a 3D array in the external fluid outside the pore. The external fluid is regarded as well-stirred and populated with randomly distributed reactant of concentration $\langle A_0 \rangle$ per cell, i.e., any external cell is populated with (at most one) reactant with probability $\langle A_0 \rangle$. Here, we just consider the early stages of the reaction where this concentration in the large external fluid volume does not change significantly due to conversion $A \to B$, and any product reaching the external fluid is rapidly dispersed and diluted to negligible concentration (and thus does not reenter the pore). We let $\langle A_n \rangle (\langle B_n \rangle)$ denote the concentration, i.e., occupation probability, of $A(B)$ at cell n. Reactant just outside the pore adsorbs at empty end cells at rate h_A. A (B) within the pore hops to adjacent empty cells with rate $h_A(h_B)$, where this can include desorption where A or B at end cell hops outside the pore. A and B at adjacent cells within the pore exchange at rate $h_{ex} = \frac{1}{2}(h_A + h_B)P_{ex}$, so $P_{ex} = 0$ corresponds to SFD, and $P_{ex} = 1$ corresponds to uninhibited passing. Exchange at this rate can also occur between B just inside the pore and A just outside. We assume all cells are catalytically active and that conversion $A \to B$ occurs irreversibly at rate k.

One anticipates that this model supports a non-equilibrium steady-state where the concentration $\langle A_n \rangle$ decays from the pore opening to its center, and the concentration $\langle B_n \rangle$ increases. For SFD, the decay of $\langle A_n \rangle$ is strongest as after A converts to B inside the pore, it is difficult for B to be removed from the pore and the pore center tends to be exclusively populated by this quasi-trapped product, B. In this chapter, we focus on the simplest scenario where $h_A = h_B = h$. Then, in a "color blind" analysis of the model where one cannot distinguish between A or B, one just sees particles $X = A$ or B diffusing into the pore at rate h, until an equilibrium state is achieved with uniform concentration, $\langle X_n \rangle = \langle A_n \rangle + \langle B_n \rangle = \langle A_0 \rangle \equiv \langle X_0 \rangle$.

3.2.   Kinetic Monte Carlo (KMC) simulation analysis

Precise assessment of model behavior is provided by KMC simulation which simply implements various processes in the model with probabilities proportional to their assigned rates. KMC can track time evolution, but here we just focus on (non-equilibrium) steady-state behavior. First, we consider behavior for the case of SFD ($P_{ex} = 0$), where the steady-state profile is determined by the ratio of hop rate to reaction rate, h/k (and by the external reactant or total concentration, $\langle A_0 \rangle = \langle X_0 \rangle$). See Fig. 6(a). For $k = h$, there is very limited penetration of reactant into the pore. However, as k decreases relative to h (so h/k increases), the depth of reactant penetration, $L_p \sim (h/k)^\zeta$, increases as expected since there is more time for A to diffuse into the pore before converting to B. A mean-field (MF) reaction-diffusion equation analysis not surprisingly indicates that $\zeta_{MF} = 1/2$. However, simulation results reveal quite different behavior with $\zeta \approx 1/4 - 1/3$ reflecting the interplay of SFD with reaction.

Second, for fixed $k/h = 0.001$, we explore the dependence of the steady-state concentration profiles on P_{ex}. See Fig. 6(b). There is a strong increase in reactant penetration depth, L_p, as P_{ex} increases from $P_{ex} = 0$ (SFD) to $P_{ex} = 1$. Increasing P_{ex} enhances diffusion in the pore, and particularly diffusion of product out of the pore (facilitating diffusion of reactant in the pore). MF behavior is achieved for $P_{ex} = 1$.

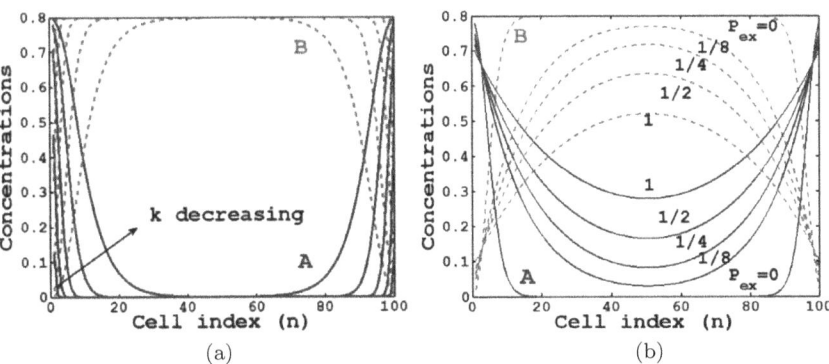

Fig. 6. KMC simulation results for steady-state concentration profiles with pore length $n_{max} = 100$, external concentration $\langle X_0 \rangle = 0.8$, and h = 1: (a) dependence on $k = 1, 0.1, 0.01, 0.001, 0.0001$ for $P_{ex} = 0$ (SFD), (b) dependence on P_{ex} for $k = 0.001$. Copyright (2015) by American Chemical Society.

4. Analysis of the CG Spatially Discrete Stochastic Model

4.1. Exact hierarchical master equations — Heterogeneous states

The evolution of concentrations is described by the master equations for the Markovian model which can be written in an instructive hierarchical form as[7,18]

$$d/dt\langle A_n\rangle = -k\langle A_n\rangle - \nabla J_A(n \to n+1),$$

$$d/dt\langle B_n\rangle = +k\langle A_n\rangle - \nabla J_B(n \to n+1),$$

where $\nabla K_n = K_n - K_{n-1}$ is the discrete derivative, and

$$J_A(n \to n+1) = h[\langle A_n E_{n+1}\rangle - \langle E_n A_{n+1}\rangle]$$

$$+ h_{\text{ex}}[\langle A_n B_{n+1}\rangle - \langle B_n A_{n+1}\rangle]$$

is the net flux of A from cell n to $n + 1$. The expression for $J_B(n \to n + 1)$ is analogous. In these expressions, $\langle A_n E_{n+1}\rangle(\langle A_n B_{n+1}\rangle)$ denotes the probability that cell n is occupied by A, and cell $n + 1$ is empty (occupied by B), etc. Separate equations are required for concentrations at the end cells.[7,18] This set of equations for single-cell concentrations is not closed as the fluxes involve pair probabilities for configurations of adjacent pairs of cells. Equations can be developed for the latter which couple to probabilities of triplets of cells, etc., so one generates a hierarchy of equations. For practical analysis, this hierarchy must be truncated.

The simplest MF truncation neglects all spatial correlations, so that $\langle A_n E_{n+1}\rangle = \langle A_n\rangle\langle B_{n+1}\rangle$, $\langle A_n\rangle\langle B_{n+1}\rangle = \langle A_n\rangle\langle B_{n+1}\rangle$, etc. Then, in the steady state where $\langle A_n\rangle + \langle B_n\rangle = \langle X_0\rangle$ so that $\nabla\langle A_n\rangle + \nabla\langle B_n\rangle = 0$, the above expression for J_A reduces to[7,25]

$$J_A(n \to n+1) = -D_{\text{tr}}(\text{MF})\nabla\langle A_n\rangle,$$

$$\text{where } D_{\text{tr}}(\text{MF}) = h[(1 - \langle X_0\rangle) + P_{\text{ex}}\langle X_0\rangle].$$

$D_{\text{tr}}(\text{MF})$ corresponds to the MF version of the tracer (not collective) diffusion coefficient, for reasons clarified below. Analysis of the closed single-cell concentrations resulting from this MF approximation for $P_{ex} = 0$ results in concentration profiles for which reactant penetration into the pore is far greater than the actual penetration.[7,18] This is because the above $|J_A(\text{MF})| \gg |J_A(\text{exact})|$, i.e., the MF approximation does not account for the impact of SFD in restricting diffusion, at least for small k/h. One naturally considers higher-order pair, triplet, etc., truncation approximations. These yield improvement with predictions somewhat closer to exact behavior, but still fail fundamentally for $P_{ex} = 0$.[7,18]

The origin of this failure is the presence of strong correlations between occupancy of adjacent populated and empty cells. The exact $\langle A_n E_{n+1} \rangle$ and $\langle E_n A_{n+1} \rangle$ are much closer to each other than the MF estimates (corresponding to $|J_A(\text{exact})| \ll |J_A(\text{MF})|$). See Fig. 7(a). Similarly, $\langle A_n E_{n+1} E_{n+1} \rangle$, $\langle E_n A_{n+1} E_{n+2} \rangle$, and $\langle E_n E_{n+1} A_{n+2} \rangle$ are also much closer than the MF and pair estimates explaining the failure of the pair approximation. This behavior can be directly tied to the SFD dynamics: an $A_n E_{n+1}$ configuration can readily transform to $E_n A_{n+1}$ by A hopping right, and the reverse transformation is also facile. A could be corralled on sites n and $n + 1$ due to population of sites $n - 1$ and $n + 2$ resulting in an equalization of $\langle A_n E_{n+1} \rangle$ and $\langle E_n A_{n+1} \rangle$. The same argument explains the equalization of the triplet probabilities listed above and shown in Fig. 7(b).

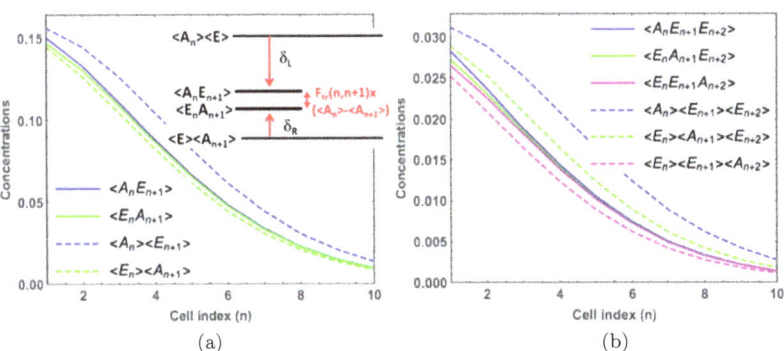

Fig. 7. Spatial correlations near the pore opening for $k/h = 0.001$, $\langle X_0 \rangle = 0.8$, $n_{max} = 100$, $P_{ex} = 0$ (SFD): (a) occupied-empty pair probabilities, (b) occupied-empty triplet probabilities. Copyright (2015) by American Chemical Society.

4.2. Generalized hydrodynamic formulation & generalized D_{tr}

For lattice-gas systems in the class of our CG model with a mixture of two distinguishable species which otherwise have identical diffusive dynamics, an exact analysis is possible for collective or chemical diffusion in the hydrodynamic regime of slowly varying concentrations.[26,27] For the case of constant total concentration, $\langle X_n \rangle = \langle X_0 \rangle$, of relevance for our study, the key result translated to our notation is

$$J_A(n \to n+1) = -D_{tr}(\langle X_0 \rangle)\nabla\langle A_n \rangle,$$

with an analogous expression for $J_B(n \to n+1)$. Here, $D_{tr}(\langle X_0 \rangle)$ denotes the exact hydrodynamic tracer diffusion coefficient for the lattice gas in an infinite system with concentration $\langle X_0 \rangle$. For SFD, $D_{tr}(\langle X_0 \rangle) \equiv 0$ in an infinite system for any $\langle X_0 \rangle > 0$,[28] and it has been shown that $D_{tr}(\langle X_0 \rangle) \sim 1/n_{max}$ for a finite open system.[7,25,29] D_{tr} values will also increase monotonically with increasing P_{ex}.

Using the above hydrodynamic formulation for J_A and J_B with very small hydrodynamic $D_{tr} \sim 1/n_{max}$ also fails to recover model behavior for $P_{ex} = 0$ since diffusion fluxes are underestimated.[7] Thus, our strategy is to replace hydrodynamic D_{tr} with "generalized hydrodynamic" (GH) $D_{tr}(n, n+1)$ which are enhanced relative to hydrodynamic values near the pore openings. This enhancement reflects fluctuation effects associated with adsorption-desorption near pore openings which are not present in an infinite SFD system. We have used various strategies to systematically determine the GH $D_{tr}(n, n+1)$. Here, we just describe a "tracer counter permeation" (TCP) approach.[7,25,29]

In TCP, non-reactive species $A(B)$ adsorbs at the right (left) end of a finite pore which are connected to separated reservoirs of A and B, respectively. See Fig. 8(a). The diffusive dynamics inside the pore is identical to our reaction model. The TCP system reaches a steady state with quasi-linear gradients of the species A and B across the pore when averaged over time. There is a flux, $J_B(TCP)$, of B through the pore from left to right, which for SFD is associated with

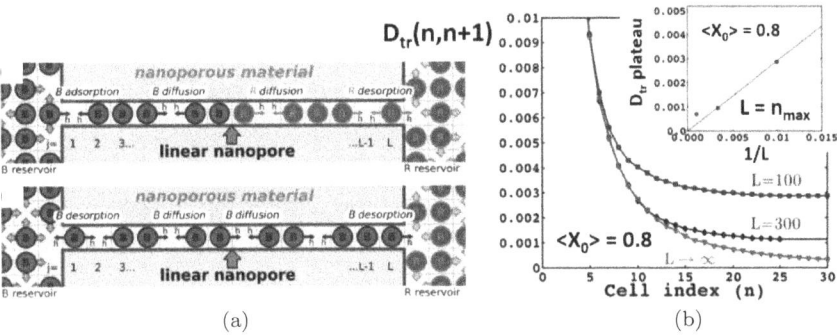

Fig. 8. (a) TCP for SFD. Copyright (2017) American Physical Society, (b) $D_{tr}(n, n + 1)$ for SFD with $\langle X_0 \rangle = 0.8$. Copyright (2012) by American Physical Society.

occasional rare configurations where the pore is completely populated with B (Fig. 8(a) bottom). Similarly, there is an equal but opposite flux, $J_A(\text{TCP})$, of A from right to left. After measuring these fluxes in the simulation as well as the concentration gradients, and since $J_A(\text{TCP}) = -D_{tr}(n, n + 1)\nabla\langle A_n \rangle$, we can extract $D_{tr}(n, n + 1) = |J_A(\text{TCP})|/|\nabla\langle A_n \rangle|$. Results are shown in Fig. 8(b) for SFD for $\langle X_0 \rangle = 0.8$ and for various pore lengths. The behavior $D_{tr} \sim 1/n_{\max}$ in the pore center is also clear from the inset.

Using these $D_{tr}(n, n + 1)$ to determine $J_A(n \rightarrow n + 1) = -D_{tr}(n, n + 1)\nabla\langle A_n \rangle$ in the master equations for cell concentrations in the reaction-diffusion model, we find that this analytic GH theory essentially exactly recovers KMC results for profiles shown in Fig. 6.[7,18] Furthermore, the observed decay of $D_{tr}(n, n + 1) \sim 1/n^p$ for long pores can be shown to produce a scaling of the reactant penetration depth $L_p \sim (h/k)^\zeta$, with $\zeta = 1/(p+2)$. Values of $p \approx 1 - 2$ from the TCP analysis are consistent with value of $\zeta \approx 1/4 - 1/3$ reported in Section 3.2 for steady-state behavior of the model.

5. Discussion and Conclusions

For catalytic conversion reactions in linear nanopores, inhibited passing of reactants and products is a key factor controlling reaction kinetics, thus motivating our analysis of the associated passing propensity, P. Our CG stochastic modeling reveals a severe failure of mean-field treatments of kinetics for small P (most dramatically for SFD with $P = 0$) due to the development of strong spatial correlations. This failure cannot be fully resolved by implementation of higher-level mean-field-type truncation approximations which attempt but still fail to accurately estimate spatial correlations. However, a generalized hydrodynamic formulation proves capable of capturing the subtle interplay between restricted passing and fluctuations and adsorption and desorption at pore openings. Many extensions are possible of the modeling presented here, e.g., to treat kinetics of the overall reaction during which reactant in the external fluid is slowly converted to product, to allow different hopping rates for reactants and products, to treat non-uniform distributions of catalytic sites, to allow reversible conversion reactions,[7] to include interactions between species,[30] and to incorporate finer-grained spatially discrete modeling.[31]

Acknowledgments

AG, DMA, MSG, IIS, and JWE were supported by the USDOE BES Division of Chemical Sciences, Geosciences, and Biosciences. Research was performed at the Ames Laboratory, which is operated by Iowa State University under contract number DE-AC02-07CH11358. CJW performed the FPE analysis for which he was also partially supported by the Ministry of Science and Technology (MOST) of Taiwan 105-2115-M-194-011-MY2.

References

1. K. Lindenberg, G. Oshanin, and M. Tachiya (eds.), Chemical kinetics beyond the textbook: Fluctuations, many-particle-effects and anomalous dynamics. *J. Phys. Cond. Matt.* **19** (2007).
2. P. Gray and S. K. Scott, *Chemical Oscillations and Instabilities: Non-linear Chemical Kinetics.* Clarendon, Oxford (1994).

3. G. Ohlmann, H. Pfeifer, and G. Fricke, (eds.), *Catalysis and Adsorption in Zeolites.* Elsevier, Amsterdam (1991).

4. T. Maschmeyer, F. Rey, G. Sankar, and J. M. Thomas, Heterogeneous catalysts obtained by grafting metallocene complexes onto mesoporous silica. *Nature* **378**, 159–162 (1995).

5. R. Imbihl and G. Ertl, Oscillatory kinetics in heterogeneous catalysis. *Chem. Rev.* **95**, 697–733 (1995).

6. D.-J. Liu and J. W. Evans, Realistic multisite lattice-gas modeling and KMC simulation of catalytic surface reactions: Kinetics and multiscale spatial behavior for CO-oxidation on metal (100) surfaces. *Prog. Surf. Sci.* **88**, 393–521 (2013).

7. D.-J. Liu, A. Garcia, J. Wang, D. M. Ackerman, C.-J. Wang, and J.W. Evans, Kinetic Monte Carlo simulation of statistical mechanical models and coarse-grained mesoscale descriptions of catalytic reaction-diffusion processes: 1D nanoporous and 2D surface systems. *Chem. Rev.* **115**, 5979–6050 (2015).

8. J. Karger, M. Petzold, H. Pfeiffer, S. Ernst, and J. Weitkamp, Single-file diffusion and reaction in zeolites. *J. Catal.* **136**, 283–299 (1992).

9. C. Rodenbeck, J. Karger, and K. Han, Tracer exchange and catalytic reaction in single-file systems. *J. Catal.* **157**, 656–664 (1995).

10. B. Temel, H. Meskine, K. Reuter, M. Scheffler, and H. Metiu, Does phenomenological kinetics provide an adequate description of heterogeneous catalytic reactions? *J. Chem. Phys.* **126**, 204711 (2007).

11. S. Matera, H. Meskine, and K. Reuter, Adlayer inhomogeneity without lateral interactions: Rationalizing correlation effects in CO oxidation at $RuO2(110)$ with first-principles kinetic Monte Carlo. *J. Chem. Phys.* **134**, 064713 (2011).

12. D.-J. Liu and J.W. Evans, Transitions between strongly correlated and random steady-states for catalytic CO-oxidation on surfaces at high-pressure. *J. Chem. Phys.* **142**, 134703 (2015).

13. Z. Karpinski, S. N. Gandhi, and W. M. H. Sachtler, Neopentane conversion catalyzed by Pd in L-Zeolite: Effects of protons, ions, and zeolite structure. *J. Catal.* **141**, 337–346 (1993).

14. K. Kandel, S. M. Althaus, C. Peeraphatdit, T. Kobayashi, B. G. Trewyn, M. Pruski, and I. I. Slowing, Substrate inhibition in heterogeneous catalyzed aldol condensation: A mechanistic study of supported organocatalysts. *J. Catal.* **129**, 63 (2012).

15. M. S. Okino, R. Q. Snurr, H. H. Kung, J. E. Ochs, and M. L. Mavrovouniotis, A consistent correlation approach to single-file diffusion with reaction. *J. Chem. Phys.* **111**, 2210–2221 (1999).

16. S. V. Nedea, A. P. J. Jansen, J. J. Lukkien, and P. A. Hilbers, Steady-states of single-file systems with conversion. *J. Phys. Rev. E.* **65**, 066701 (2002).

17. D. M. Ackerman, J. Wang, J. H. Wendel, D.-J. Liu, M. Pruski, and J. W. Evans, Catalytic conversion reactions mediated by single-file diffusion in linear nano-pores: Hydrodynamic vs stochastic behavior. *J. Chem. Phys.* **134**, 114107 (2011).

18. D. M. Ackerman, J. Wang, and J. W. Evans, Generalized hydrodynamic treatment of the interplay between restricted transport and catalytic reactions in nanoporous materials. *Phys. Rev. Lett.* **108**, 228301 (2012).

19. A. P. D. de L. Batista, F. Zahariev, I. I. Slowing, A. A. C. Braga, F. R. Ornellas, and M. S. Gordon, Silanol-assisted carbinolamine formation in an amine-functionalized mesoporous silica surface: Theoretical investigation by fragmentation methods. *J. Phys. Chem. B.* **120**, 1660–1669 (2016).

20. C.-J. Wang, D. M. Ackerman, I. I. Slowing, and J. W. Evans, Langevin and Fokker–Planck analyses of inhibited molecular passing processes controlling transport and reactivity in nanoporous materials. *Phys. Rev. Lett.* **113**, 038301 (2014).

21. D. S. Sholl, Characterizing adsorbate passage in molecular sieve pores. *Chem. Eng. J.* **74**, 25–32 (1999).

22. B. Carrasco and J. Garcia de la Torre, Hydrodynamical properties of rigid particles: Comparison of different modeling and computational procedures. *Biophys. J.* **75**, 3044–3057 (1999).

23. S. Delong, F. B. Usabiaga, and A. Donev, Brownian dynamics of confined rigid bodies. *J. Chem. Phys.* **143**, 144107 (2015).

24. A. Garcia, I. I. Slowing, and J. W. Evans, Pore diameter dependence of catalytic activity: p-nitrobenzaldehyde conversion to an aldol product in amine-functionalized mesoporous silica. *J. Chem. Phys.* **149**, 024101 (2018).

25. D. M. Ackerman and James W. Evans, Tracer counter-permeation analysis of diffusivity in finite-length nanopores with and without single-file dynamics. *Phys. Rev. E.* **95**, 012132 (2017).

26. H. Spohn, *Large Scale Dynamics of Interacting Particles*. Springer, Berlin (1991).

27. J. Quastel, Diffusion of color in the simple exclusion process. *Commun. Pure Appl. Math.* **45**, 623–679 (1992).

28. T. E. Harris, Diffusion with "collisions" between particles. *J. Appl. Probab.* **2**, 323–338 (1965).

29. P. H. Nelson and S. M. Auerbach, Self-diffusion in single-file zeolite membranes is Fickian at long times. *J. Chem. Phys.* **110**, 9235–9243 (1999).

30. A. Garcia and J. W. Evans, Catalytic conversion in nanoporous materials: Concentration oscillations and spatial correlations due to inhibited transport and intermolecular interactions. *J. Chem. Phys.* **145**, 174705 (2016).

31. A. Garcia and J. W. Evans, Boundary conditions for diffusion-mediated processes within linear nanopores: Exact treatment of coupling to an equilibrated external fluid. *J. Phys. Chem. C.* **121**, 8873–8888 (2017).

Chapter 8

Imperfect Diffusion-Controlled Reactions

Denis S. Grebenkov

Laboratoire de Physique de la Matière Condensée,
CNRS – Ecole Polytechnique, 91128 Palaiseau, France
denis.grebenkov@polytechnique.edu

This chapter aims at emphasizing the crucial role of partial reactivity of a catalytic surface or a target molecule in diffusion-controlled reactions. We discuss various microscopic mechanisms that lead to imperfect reactions, the Robin boundary condition accounting for eventual failed reaction events, and the construction of the underlying stochastic process, the so-called partially reflected Brownian motion. We show that the random path to the reaction event can naturally be separated into the transport step toward the target and the exploration step near the target surface until reaction. While most studies are focused exclusively on the transport step (describing perfect reactions), the exploration step, consisting of an intricate combination of diffusion-mediated jumps between boundary points, and its consequences for chemical reactions, remain poorly understood. We discuss the related mathematical difficulties and recent achievements. In particular, we derive a general representation of the propagator, show its relation to the Dirichlet-to-Neumann operator, and illustrate its properties in the case of a flat surface.

1. Introduction[a]

Classical reaction kinetics describes the time evolution of spatially homogeneous concentrations of reacting species via a set of coupled nonlinear ordinary differential equations. This description relies on the assumption that the produced species move fast enough to maintain homogeneous concentrations, and the limiting factor is the chemical kinetics itself. Marian von Smoluchowski was the first who emphasized on the importance of the transport step by computing the macroscopic reaction rate of small molecules A diffusing toward a spherical target B of radius R.[1] If the reaction occurs immediately upon the first encounter between the two molecules, the rate is equal to the diffusive flux which is obtained by solving the underlying diffusion equation on the concentration $c(\boldsymbol{x}, t)$ of molecules A

$$\frac{\partial c(\boldsymbol{x}, t)}{\partial t} = D\Delta c(\boldsymbol{x}, t) \qquad \text{diffusion in the bulk } \Omega, \tag{1a}$$

$$c(\boldsymbol{x}, t = 0) = c_0 \qquad \text{uniform initial condition,} \tag{1b}$$

$$c(\boldsymbol{x}, t)|_{\partial\Omega} = 0 \qquad \text{vanishing on the target surface } \partial\Omega, \tag{1c}$$

$$c(\boldsymbol{x}, t)|_{|\boldsymbol{x}|\to\infty} = c_0 \qquad \text{regularity condition at infinity,} \tag{1d}$$

where D is the diffusion coefficient of molecules A (in units m^2/s), c_0 is their initial concentration (in units mol/m^3), the confining domain $\Omega = \{\boldsymbol{x} \in \mathbb{R}^3 \ : \ |\boldsymbol{x}| > R\}$ is the bulk space outside the target, $\partial\Omega = \{\boldsymbol{x} \in \mathbb{R}^3 \ : \ |\boldsymbol{x}| = R\}$ is the reactive surface of the target, and $\Delta = \partial^2/\partial x^2 + \partial^2/\partial y^2 + \partial^2/\partial z^2$ is the Laplace operator governing diffusion. Here and throughout the text, we assume that the target B is immobile while the molecules A can be treated as independent and point-like. Although Smoluchowski considered both molecules to be mobile and of finite size (in which case D and R are, respectively, the sum of diffusivities and of radii of two molecules), the equivalence between two settings is not valid in general.

---

[a]Disclaimer: Apart from the general introduction and motivation, the chapter summarizes former results obtained by the author and his co-workers. For this reason, the chapter presents the author's personal view on the topic, whereas the bibliography is biased.

The diffusion equation (1a) states that the time evolution of the concentration (the left-hand side) is caused exclusively by diffusive motion (the right-hand side), whereas the Dirichlet boundary condition (1c) ensures that any molecule A arriving at the boundary of the target B immediately reacts, i.e., it is transformed into another molecule A',

$$A + B \longrightarrow A' + B, \tag{2}$$

which diffuses away. This is the basic catalytic reaction, in which the target molecule B is a catalyst needed to initiate the chemical transformation of A into A'. Since the molecule A, being transformed into A' upon the contact, disappears at the target surface, the concentration $c(\boldsymbol{x}, t)$ vanishes on $\partial \Omega$. Finally, the regularity condition (1d) claims that the concentration (infinitely) far away from the target remains unaffected and equal to c_0.

Smoluchowski provided the exact solution of the partial differential equation (1) and deduced the diffusive flux of molecules A onto the target

$$J_{\mathrm{S}}(t) = \int_{|\boldsymbol{x}|=R} d\boldsymbol{s} \left(-D \frac{\partial c(\boldsymbol{x}, t)}{\partial n} \right) \Big|_{\partial \Omega} = 4\pi c_0 D R \left(1 + \frac{R}{\sqrt{\pi D t}} \right), \tag{3}$$

where $\partial / \partial n$ is the normal derivative at the boundary, oriented outward the domain (i.e., toward to center of the target in this case). In the long-time limit, the second term vanishes, and one gets the steady-state reaction rate, $J_{\mathrm{S}}(\infty) = 4\pi c_0 D R$, which is proportional to the diffusion coefficient D and to the radius R of the target. The seminal work by Smoluchowski focused exclusively onto the transport step as a limiting factor, considering *perfect* immediate reactions upon the first encounter. This setting was later called "diffusion-limited reactions", in contrast to conventional "kinetics-limited reactions".[2] In numerous following studies, the basic diffusion problem (1) was extended in various directions, in particular, by replacing the exterior of a spherical target by an arbitrary Euclidean domain $\Omega \subset \mathbb{R}^d$,[3,4] by considering one or multiple targets on the otherwise inert impenetrable boundary,[5–8] by replacing the Laplace operator (ordinary diffusion) by a general second-order elliptic differential operator[9] or a general Fokker–Planck operator,[10] by introducing bulk reactivity,[11-13] trapping events,[14,15] or intermittence.[16-19] However, the focus on the transport step till the first encounter with the target, expressed via Eq. (1c), still remains the dominant paradigm nowadays.

In practice, chemical reactions always combine the transport step until an encounter and the reaction step. While in some situations one of these steps can be much longer than the other, it is important to consider them together, as the impact of a seemingly "irrelevant" step may still be crucial, as we discuss below. Collins and Kimball proposed to describe *imperfect* reactions on the target surface $\partial\Omega$ by replacing the Dirichlet boundary condition (1c) by the Robin boundary condition (also known as Fourier, radiation or third boundary condition)

$$\left(-D\frac{\partial c(\boldsymbol{x},t)}{\partial n}\right)\Bigg|_{\partial\Omega} = \kappa\, c(\boldsymbol{x},t)|_{\partial\Omega}, \qquad (4)$$

where κ (in units m/s) is called the *reactivity*.[20] Although κ can in general be any non-negative function of a boundary point, we focus on the practically relevant case of a constant κ (we also discuss below the mixed Robin–Neumann boundary condition when κ is a piecewise constant function).

The relation (4) states that the diffusive flux density of molecules at the target surface (the left-hand side) is proportional to the concentration of the molecules at the target (the right-hand side), κ being the proportionality coefficient. It is important to stress that the diffusive flux density is a macroscopic quantity that describes the average difference between the diffusive flux toward the target and the diffusive flux from the target back to the bulk. In the limit $\kappa = 0$, the diffusive flux density vanishes at the boundary $\partial\Omega$, meaning that, on average, the number of molecules A arriving from the bulk onto the target surface is equal to the number of molecules diffusing from the target back to the bulk. As a consequence, none of these molecules actually react on the target. So, the limit $\kappa = 0$ and the corresponding Neumann boundary condition describe a chemically inactive (inert) boundary that just confines the molecules in a prescribed spatial region, with no reaction. When $\kappa > 0$, there is a net difference between the diffusive flux toward the target and that from the target, and this difference is precisely the flux of reacted molecules. In the limit $\kappa \to \infty$, one retrieves the Dirichlet boundary condition (1c) by dividing Eq. (4) by κ and taking the limit. Varying κ from zero to infinity allows one to change the chemical reactivity of the target from inert to perfectly reactive.

Collins and Kimball solved the diffusion problem (1) with the Robin boundary condition (4) and found the macroscopic reaction rate[20]

$$J_{\text{CK}}(t) = \frac{4\pi c_0 D R}{1 + \frac{D}{\kappa R}}\left(1 + \frac{\kappa R}{D}\operatorname{erfcx}\big((1/R + \kappa/D)\sqrt{Dt}\big)\right), \quad (5)$$

where $\operatorname{erfcx}(z)$ is the scaled complementary error function (see Ref. 21 for an extension to distinct reactivities on two complementary caps of a sphere and Ref. 22 for a perturbative approach for a dilute suspension of multiple partially reactive spheres). In the diffusion-limited case $\kappa \to \infty$, one recovers the Smoluchowski rate $J_S(t)$, whereas the reaction-limited case $\kappa \to 0$ gives the classical limit $J_{\text{reac}} = 4\pi R^2 c_0 \kappa$, i.e., the rate is proportional to the surface area of the target and to the reactivity κ but is independent of the diffusion coefficient D. One can combine these limiting expressions to rewrite the steady-state limit of Eq. (5) as

$$\frac{1}{J_{\text{CK}}(\infty)} = \frac{1}{J_S(\infty)} + \frac{1}{J_{\text{reac}}}. \quad (6)$$

Understanding the inverse of the flux as a "resistance", one gets a simple electric interpretation of the Collins–Kimball rate $J_{\text{CK}}(\infty)$ as the serial connection of two "resistances" characterizing the "difficulty" to access the target (the transport step) and the "difficulty" to react on the target (the reaction step).[23] Alternatively, one can think of this relation as the sum of two characteristic times of the underlying steps. Note that a similar additivity law was established for the mean reaction time.[24,25] Chemical reactions for which both the transport and reaction steps are relevant are sometimes called *diffusion-influenced, diffusion-mediated,* or *diffusion-controlled.*

The macroscopic boundary condition (4) can describe various *microscopic* mechanisms of imperfect reactions.

(i) A molecule needs to overcome an energetic activation barrier for reaction to occur[26]; in this setting, the reaction (2) can be written more accurately as

$$A + B \xrightleftharpoons[k_{\mathrm{off}}]{k_{\mathrm{on}}} AB \xrightarrow{k_{\mathrm{reac}}} A' + B, \qquad (7)$$

where AB is a metastable complex which can either result in the production of A' (if the activation barrier is overpassed, with the rate k_{reac}), or be dissociated back to A and B (with the dissociation rate k_{off}). The reactivity is thus related to the height of the activation barrier. For instance, the reactivity κ of a spherical target can be directly related to the association rate k_{on} (in units $\mathrm{m}^3/\mathrm{s}/\mathrm{mol}$) as $k_{\mathrm{on}} = 4\pi R^2 \kappa N_{\mathrm{av}}$, where $N_{\mathrm{av}} \simeq 6 \cdot 10^{23}$ $1/\mathrm{mol}$ is the Avogadro number, and $4\pi R^2$ is the surface area of the target.[27,28] This description was shown to be important for partial diffusion-controlled recombination.[29,30] The reversible binding in (7) is also the key element for coupling bulk diffusion to other diffusion-reaction processes on the target. For instance, such a coupling was employed to describe the flux of receptors across the boundary between the dendrite and its spines.[31]

(ii) A molecule needs to overcome an entropic barrier[32,33] if the reaction is understood as an escape from a confining domain through a small opening region on the boundary (e.g., ion channels or aquaporins on the membrane of a living cell). In this case, the reactivity can be related to the geometric shape of the opening region (see Ref. 24 for further discussion).

(iii) There is a stochastic gating when the target can randomly switch between "open" and "closed" states, implying that the diffusing molecule can either go through the open gate, or be stopped at the closed gate and thus reflected back to resume its diffusion;[34–36] the same stochastic mechanism is relevant for enzymatic reactions when an enzyme can randomly change its conformational state to be active or passive. In both cases, the reactivity is related to the switching rates or to the probability of finding the gate open or the enzyme active.

(iv) When the target is an inert surface covered by small reactive patches, the diffusing molecule can either hit a patch and react immediately, or be reflected on an inert part and resume its diffusion until the next arrival on the target. Homogenizing such microscopic reactivity heterogeneities by setting an effective finite reactivity κ uniformly on the target is often used to facilitate the analysis (given that the diffusion equation with multiple targets is a much more complicated problem, for both analytical and numerical computations). For instance, for a spherical target covered uniformly by N disks of radius a, Berg and Purcell obtained[37] $\kappa = DNa/(\pi R^2)$ (see further extensions in Refs. 38 and 39).

(v) For larger scale problems (e.g., animal foraging), the finite reactivity can model a "recognition" step when a particle or a species may need to recognize the target.[40]

The above (incomplete) list of microscopic mechanisms urges for considering imperfect reactions. We also mention that the Robin boundary condition (4) can describe permeation across a membrane (in which case κ is called permeability),[41–43] impedance of an electrode,[23,44,45] surface relaxation in nuclear magnetic resonance,[46–48] etc.

According to Eq. (6), the macroscopic reaction rate results from a balance between the transport and the reaction steps, whereas the diffusion coefficient D and the reactivity κ play the roles of respective weights in the Robin boundary condition (4). The ratio of these quantities naturally defines the *reaction length*[23]

$$\Lambda = D/\kappa, \tag{8}$$

which has to be compared to *geometric length scales* of the problem (e.g., the radius of the target R in our spherical example). In particular, the diffusion-limited and reaction-limited cases correspond to conditions $\Lambda \ll R$ and $\Lambda \gg R$, respectively.

When molecular diffusion occurs in a structurally complex environment (e.g., an overcrowded cytoplasm or a multiscale porous medium), both the transport and reaction steps strongly depend on the geometrical structure of the environment. While the transport step is relatively well understood (see Refs. 49–51 and references therein), the more sophisticated reaction step was most often just ignored. In the next section, we summarize the main steps of a mathematical construction of the stochastic process allowing one to investigate the reaction step.

2.   Partially Reflected Brownian Motion

The macroscopic formulation of diffusion-reaction processes in terms of partial differential equations such as Eq. (1) admits an equivalent microscopic probabilistic interpretation in terms of random trajectories of appropriate stochastic processes.[9,52–54] From the mathematical point of view, the Dirichlet boundary condition (1c) is the easiest to deal with. In fact, for a given continuous stochastic process X_t, it is sufficient to define the first passage time τ_0 to the target surface $\partial\Omega$

$$\tau_0 = \inf\{t > 0 \; : \; X_t \in \partial\Omega\}, \qquad (9)$$

and then to stop the process at this time. For instance, the solution of the steady-state diffusion equation, $\Delta u(\boldsymbol{x}) = 0$, subject to the Dirichlet boundary condition, $u(\boldsymbol{x})|_{\partial\Omega} = \psi(\boldsymbol{x})$, is $u(\boldsymbol{x}) = \mathbb{E}\{\psi(X_{\tau_0})|X_0 = x\}$, where $\mathbb{E}$ denotes the (conditional) expectation and $\psi(\boldsymbol{x})$ is a prescribed function on the boundary. In other words, one can think of the solution $u(\boldsymbol{x})$ as the average of the function $\psi(\boldsymbol{x})$ evaluated at random boundary points X_{τ_0} at which the stochastic process X_t (started from x) has arrived onto the boundary $\partial\Omega$ for the first time. In this way, the presence of a perfectly absorbing boundary can be accounted for via the first passage time τ_0, *with no modification to the stochastic process itself.*

In contrast, accounting for an inert or partially reactive boundary requires to modify the stochastic process. In fact, when a molecule arrives onto an inert boundary, its motion across this boundary should be prohibited. Physically, one can think of applying a very strong potential, localized in an infinitesimal vicinity of the boundary, to force the molecule moving back into the bulk. In mathematical terms, the so-called *reflected Brownian motion* can be defined as the solution of the stochastic Skorokhod equation.[52] This is a standard probabilistic way of implementing the Neumann boundary condition for an inert boundary. Once the reflected Brownian motion is constructed, one can implement eventual reactions to treat the Robin boundary condition.[55–59] While the related mathematical theory is well developed, its details are beyond the scope of this chapter.

The easier and physically more appealing approach consists in modeling the diffusion process as a discrete-time random walk on a lattice $\mathbb{Z}^d$ with the inter-site distance a. In a bulk site, the molecule jumps with the probability $1/(2d)$ to one of the neighboring sites. When the molecule sits at a boundary site, it can either react with the probability

$$q_a = (1 + \Lambda/a)^{-1}, \tag{10}$$

or move back to a neighboring bulk site with the probability $1 - q_a$ and continue the random walk until the next arrival onto the boundary.[44,60] The trajectory of a molecule near a partially reactive boundary can thus be seen as a sequence of independent diffusion-mediated jumps between boundary points, each jump being a random walk in the bulk that starts from the closest bulk neighbor of a boundary site and terminates at another boundary site upon the first arrival onto the boundary (Fig. 1). This process is called *partially reflected random walk*, whereas the statistical properties of the jumps are described by the so-called Brownian self-transport operator.[44,60] One naturally recovers the two limiting cases: certain reaction ($q_a = 1$) for a perfectly reactive surface ($\Lambda = 0$ or $\kappa = \infty$) and certain reflection ($q_a = 0$) for an inert surface ($\Lambda = \infty$ or $\kappa = 0$).

Since the reaction event and jumps are independent, the random number η of jumps until reaction follows a geometric distribution: $\mathbb{P}\{\eta = n\} = q_a(1 - q_a)^n$ $(n = 0, 1, 2, \ldots)$. In particular, the mean

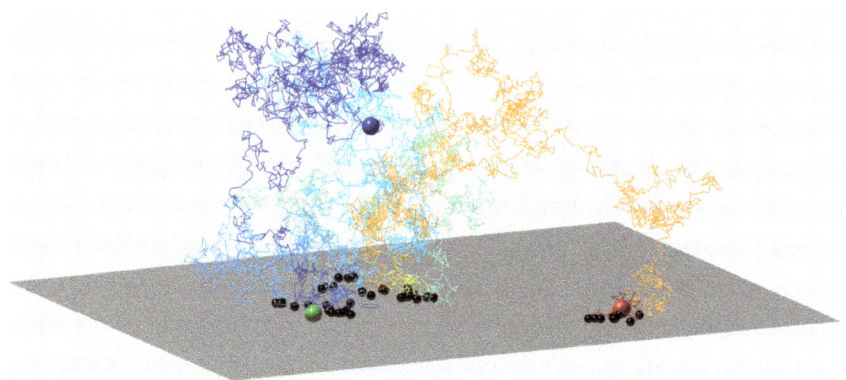

Fig. 1. A simulated trajectory of partially reflected Brownian motion with $\Lambda = 10$ and $a = 0.05$. Blue, green, and red balls show, respectively, the starting point $(0,0,2)$, the first arrival point, and the reaction point, whereas small black balls indicate reflection events. Diffusion-mediated jumps are drawn by different colors. Length units are arbitrary. We chose this continuous-space simulation instead of a lattice random walk for a better visualization.

number of jumps, $\mathbb{E}\{\eta\} = (1 - q_a)/q_a = \Lambda/a$, is proportional to the reaction length Λ and inversely proportional to the lattice mesh size a. When a is small, it is thus convenient to consider a rescaled variable $\chi = \eta a$ which follows an exponential distribution

$$\mathbb{P}\{\chi \geq x\} = \mathbb{P}\{\eta \geq x/a\} = (1 - q_a)^{x/a} = (1 + a/\Lambda)^{-x/a} \approx e^{-x/\Lambda}.$$
(11)

From the probabilistic point of view, consecutive individual trials after each jump with probability q_a are fully equivalent to saying that the trajectory is stopped when the number of realized jumps, multiplied by a, exceeds an independent exponentially distributed random variable χ.

Each jump started at the distance a from the boundary has high chances to return to the boundary within the distance a to the starting point. In other words, each jump explores the boundary at a typical distance of the order of a (we emphasize that the *mean* exploration distance can be infinite, see Section 4). As a consequence, the whole trajectory from the first arrival onto the boundary until reaction, being a sequence of Λ/a jumps on average, explores the surface up to a typical distance Λ. These qualitative arguments provide the geometric meaning of the reaction length Λ as a typical size of the surface region around the first arrival point explored until reaction (see Section 4 for precise statements).

The above lattice-based construction of partially reflected random walks employs the mesh distance a. In the limit $a \to 0$, each jump starts closer to the boundary and thus explores shorter distances, but the number of jumps increases (given that $q_a \to 0$) so that these effects compensate each other. One can thus expect that such partially reflected trajectories on a lattice converge to a well-defined continuous limit as $a \to 0$ (in the same way as ordinary random walks converge to Brownian motion). This limit that we call partially reflected Brownian motion (PRBM),[61] is defined as reflected Brownian motion stopped at the random time

$$\tau_\Lambda = \inf\{t > 0 \; : \; \ell_t > \chi\}, \tag{12}$$

where ℓ_t is the local time process of the reflected Brownian motion on the boundary, which is a continuous analog of the rescaled number of jumps on the boundary (ηa) up to time t, and an independent random variable χ was defined by Eq. (11). The reaction length Λ appears in the exponential law (11) and thus parameterizes the family of PRBMs. In the limit $\Lambda = 0$, the exponential distribution for χ degenerates to the Dirac distribution at 0, so that Eq. (12) is reduced to $\tau_0 = \inf\{t > 0 \; : \; \ell_t > 0\}$, i.e., τ_0 is the first moment when the local time process becomes strictly positive that occurs on the first arrival onto the boundary, and one retrieves Eq. (9).

In spite of apparent similarities between discrete-space and continuous-space partially reflected diffusions, there is a peculiar difference in their construction. In the discrete-space approach, one considers a sequence of diffusion-mediated jumps, in which each jump terminates upon the first arrival on the boundary. In other words, the boundary is treated as perfectly absorbing for each jump, whereas the partial reactivity is implemented and controlled by the number of jumps. In contrast, in the continuous-space approach, the underlying process is the reflected Brownian motion so that the boundary is treated as fully reflecting, whereas the partial reactivity is again controlled by the (rescaled) number of jumps (i.e., the local time process).

The introduction of a small distance a to restart each jump was crucial to avoid an immediate return to the boundary. In contrast, discrete-space and discrete-time random walks on a lattice can be easily replaced by continuous trajectories of Brownian motion.[61,62] While we focused on PRBM governed by the Laplace operator, one can construct much more general partially reflected diffusions governed by elliptic second-order differential operators.[59] In particular, one can consider Markovian jump processes generated by the Euler discretization scheme of a more general stochastic process in the bulk.[63] Once the generated trajectory crosses the boundary, it can be either terminated with probability q_a, or reflected back. In our notations, the probability q_a obtained by Singer *et al.* reads

$$q_a = \sqrt{\pi}\kappa\sqrt{\Delta t}/\sqrt{2D} = (\sqrt{\pi}/2)\kappa a/D = (\sqrt{\pi}/2)a/\Lambda, \qquad (13)$$

with $a = \sqrt{2D\Delta t}$, where Δt is the time step. This expression is close to the formula (10), which for small a/Λ yield $q_a \approx a/\Lambda$. The extra factor $\sqrt{\pi}/2 \approx 0.89$ is related to the use of Gaussian jumps in the Euler scheme instead of discrete-space random walks. In contrast to Eq. (13), the formula (10) is applicable for any reactivity, i.e., a/Λ does not need to be a small parameter.

In the next section, we discuss the propagator of PRBM and show the impact of the reactivity onto various statistics relevant to diffusion-controlled reactions.

3. General Representation of the Propagator

In this section, we derive a general representation of the propagator $G_\Lambda(\boldsymbol{x}_0, \boldsymbol{x}; t)$ of PRBM, i.e., the probability density of finding a molecule started at a point $\boldsymbol{x}_0$ at time 0 in a vicinity of a point $\boldsymbol{x}$ at time t in the presence of a partially reactive boundary $\partial\Omega$ characterized by the reaction length Λ. The propagator satisfies the following equations

$$\frac{\partial G_\Lambda(\boldsymbol{x}_0, \boldsymbol{x}; t)}{\partial t} - D\Delta G_\Lambda(\boldsymbol{x}_0, \boldsymbol{x}; t) = 0 \qquad (\boldsymbol{x} \in \Omega), \qquad (14a)$$

$$G_\Lambda(\boldsymbol{x}_0, \boldsymbol{x}; t = 0) = \delta(\boldsymbol{x} - \boldsymbol{x}_0), \qquad (14b)$$

$$\left(\Lambda\frac{\partial}{\partial n} + 1\right) G_\Lambda(\boldsymbol{x}_0, \boldsymbol{x}; t) = 0 \qquad (\boldsymbol{x} \in \partial\Omega), \qquad (14c)$$

where $\delta(\boldsymbol{x} - \boldsymbol{x}_0)$ is the Dirac distribution (if the domain Ω is unbounded, these equations should be completed by the regularity condition at infinity: $G_\Lambda(\boldsymbol{x}_0, \boldsymbol{x}; t) \to 0$ as $|\boldsymbol{x}| \to \infty$). To avoid technical details, we assume that the boundary $\partial\Omega$ is smooth.

The propagator can be decomposed into two parts: the contribution of direct trajectories from $\boldsymbol{x}_0$ to $\boldsymbol{x}$ without touching the reactive surface, $G_0(\boldsymbol{x}_0, \boldsymbol{x}; t)$, and the contribution of trajectories that hit this surface at a point $\boldsymbol{s}_1 \in \partial\Omega$ at an intermediate time $0 < t_1 < t$

$$G_\Lambda(\boldsymbol{x}_0, \boldsymbol{x}; t) = G_0(\boldsymbol{x}_0, \boldsymbol{x}; t) + \int_{\partial\Omega} d\boldsymbol{s}_1 \int_0^t dt_1\, j_0(\boldsymbol{x}_0, \boldsymbol{s}_1; t_1)$$
$$\times G_\Lambda(\boldsymbol{s}_1, \boldsymbol{x}; t - t_1), \qquad (15)$$

where

$$j_0(\boldsymbol{x}_0, \boldsymbol{s}_1; t_1) = \left(-D\frac{\partial}{\partial n} G_0(\boldsymbol{x}_0, \boldsymbol{x}; t_1)\right)_{\boldsymbol{x}=\boldsymbol{s}_1} \qquad (16)$$

is the diffusive flux density at time t_1 at a point $\boldsymbol{s}_1$ of the *perfectly reactive surface* (i.e., the probability density of the first arrival in a vicinity of $\boldsymbol{s}_1$ at time t_1). This density describes the transport step and is independent of Λ. Next, employing the reversal symmetry of the propagator, $G_\Lambda(\boldsymbol{x}_0, \boldsymbol{x}; t) = G_\Lambda(\boldsymbol{x}, \boldsymbol{x}_0; t)$, which is also valid when $\boldsymbol{x}_0$ is a boundary point for $\Lambda > 0$, one can represent $G_\Lambda(\boldsymbol{s}_1, \boldsymbol{x}; t)$ in Eq. (15) using again Eq. (15) to obtain

$$G_\Lambda(\boldsymbol{x}_0, \boldsymbol{x}; t) = G_0(\boldsymbol{x}_0, \boldsymbol{x}; t) + \int_{\partial\Omega} d\boldsymbol{s}_1 \int_{\partial\Omega} d\boldsymbol{s}_2 \int_0^t dt_1$$
$$\int_{t_1}^t dt_2\, j_0(\boldsymbol{x}_0, \boldsymbol{s}_1; t_1) G_\Lambda(\boldsymbol{s}_1, \boldsymbol{s}_2; t_2 - t_1) j_0(\boldsymbol{x}, \boldsymbol{s}_2; t - t_2).$$
$$(17)$$

This relation expresses the propagator $G_\Lambda(\boldsymbol{x}_0, \boldsymbol{x}; t)$ in the whole domain in terms of the propagator $G_\Lambda(\boldsymbol{s}_1, \boldsymbol{s}_2; t)$ from one boundary point to another via bulk diffusion. The second term in Eq. (17) has a simple probabilistic interpretation: a molecule reaches the boundary for the first time at t_1, performs PRBM (with eventual reaction) over time $t_2 - t_1$ and, if not reacted during this time, diffuses to the bulk point $\boldsymbol{x}$ during time $t - t_2$ without hitting the reactive surface. We emphasize that the last step would not be possible in the perfectly reactive case because the propagator $G_0(\boldsymbol{x}_0, \boldsymbol{x}; t)$ is zero when $\boldsymbol{x}$ is a boundary point. We also stress that Eq. (17) is valid if $\boldsymbol{x} = \boldsymbol{s} \in \partial\Omega$ is a boundary point. In this case, $G_0(\boldsymbol{x}_0, \boldsymbol{s}; t) = 0$, while $j_0(\boldsymbol{s}, \boldsymbol{s}_2; t - t_2) = \delta(\boldsymbol{s} - \boldsymbol{s}_2)\delta(t - t_2)$, so that the integrals over $\boldsymbol{s}_2$ and t_2 are removed, reducing Eq. (17) to

$$G_\Lambda(\boldsymbol{x}_0, \boldsymbol{s}; t) = \int_{\partial\Omega} d\boldsymbol{s}_1 \int_0^t dt_1\, j_0(\boldsymbol{x}_0, \boldsymbol{s}_1; t_1)\, G_\Lambda(\boldsymbol{s}_1, \boldsymbol{s}; t - t_1). \quad (18)$$

This relation justifies the qualitative separation of the diffusion-reaction process into two steps: the transport step (described by $j_0(\boldsymbol{x}_0, \boldsymbol{s}_1; t_1)$) and the reaction step (described by $G_\Lambda(\boldsymbol{s}_1, \boldsymbol{s}; t - t_1)$ or related quantities, see below). We stress, however, that the reaction step involves intricate diffusion process near the partially reactive surface. In addition to the new conceptual view onto PRBM, the representations (17, 18) can be helpful for a numerical computation of the propagator because only the boundary-to-boundary transport $G_\Lambda(\boldsymbol{s}_1, \boldsymbol{s}_2; t)$ needs to be evaluated. Note also that the time convolution can be removed by passing to the Laplace domain, in which Eq. (17) becomes

$$\tilde{G}_\Lambda(\boldsymbol{x}_0, \boldsymbol{x}; p) = \tilde{G}_0(\boldsymbol{x}_0, \boldsymbol{x}; p)$$
$$+ \int_{\partial\Omega} d\boldsymbol{s}_1 \int_{\partial\Omega} d\boldsymbol{s}_2\, \tilde{j}_0(\boldsymbol{x}_0, \boldsymbol{s}_1; p)\, \tilde{G}_\Lambda(\boldsymbol{s}_1, \boldsymbol{s}_2; p)\, \tilde{j}_0(\boldsymbol{x}, \boldsymbol{s}_2; p),$$
$$(19)$$

where tilde denotes the Laplace transform: $\tilde{f}(p) = \int_0^\infty dt\, e^{-pt}\, f(t)$.

3.1. *Relation to the Dirichlet-to-Neumann operator*

The Laplace-transformed propagator $\tilde{G}_\Lambda(\boldsymbol{s}_1, \boldsymbol{s}_2; p)$ turns out to be the resolvent of the *Dirichlet-to-Neumann operator* $\mathcal{M}_p$ for the modified Helmholtz equation: To a given function f on the boundary $\partial\Omega$ of a confining domain $\Omega \subset \mathbb{R}^d$, the operator $\mathcal{M}_p$ associates another function on the boundary,

$$\mathcal{M}_p f = \left(\frac{\partial u(\boldsymbol{x}; p)}{\partial n} \right)_{|\boldsymbol{x} \in \partial\Omega}, \tag{20}$$

where $u(\boldsymbol{x}; p)$ is the solution of the Dirichlet boundary value problem

$$(p - D\Delta)u(\boldsymbol{x}; p) = 0 \qquad (\boldsymbol{x} \in \Omega), \tag{21a}$$

$$u(\boldsymbol{x}; p)_{|\partial\Omega} = f(\boldsymbol{x}; p) \quad (\boldsymbol{x} \in \partial\Omega). \tag{21b}$$

For instance, if $f(\boldsymbol{x}; p)$ is understood as a source of molecules on the boundary $\partial\Omega$ emitted into the reactive bulk, then the operator $\mathcal{M}_p$ gives their flux density on that boundary. Note that there is a family of operators parameterized by p (or p/D). While the Dirichlet-to-Neumann operator was conventionally studied for the Laplace equation (i.e., $p = 0$), the above extension to the modified Helmholtz equation is straightforward.

One can use the Dirichlet-to-Neumann operator to represent the Laplace-transformed propagator $\tilde{G}_\Lambda(\boldsymbol{x}_0, \boldsymbol{x}; p)$ which satisfies for each fixed $\boldsymbol{x}_0 \in \Omega$ the modified Helmholtz equation

$$(p - D\Delta)\tilde{G}_\Lambda(\boldsymbol{x}_0, \boldsymbol{x}; p) = \delta(\boldsymbol{x} - \boldsymbol{x}_0) \quad (\boldsymbol{x} \in \Omega), \tag{22a}$$

$$\left(\Lambda \frac{\partial}{\partial n} + 1\right)\tilde{G}_\Lambda(\boldsymbol{x}_0, \boldsymbol{x}; p) = 0 \quad (\boldsymbol{x} \in \partial\Omega). \tag{22b}$$

To get rid off $\delta(\boldsymbol{x} - \boldsymbol{x}_0)$, one can search the solution in the form

$$\tilde{G}_\Lambda(\boldsymbol{x}_0, \boldsymbol{x}; p) = \tilde{G}_0(\boldsymbol{x}_0, \boldsymbol{x}; p) + \tilde{g}_\Lambda(\boldsymbol{x}_0, \boldsymbol{x}; p), \tag{23}$$

where the unknown regular part $\tilde{g}_\Lambda(\boldsymbol{x}_0, \boldsymbol{x}; p)$ satisfies

$$(p - D\Delta)\tilde{g}_\Lambda(\boldsymbol{x}_0, \boldsymbol{x}; p) = 0 \quad (\boldsymbol{x} \in \Omega), \tag{24a}$$

$$\left(\Lambda\frac{\partial}{\partial n}+1\right)\tilde{g}_\Lambda(\boldsymbol{x}_0,\boldsymbol{x};p)=\underbrace{-\Lambda\left(\frac{\partial}{\partial n}\tilde{G}_0(\boldsymbol{x}_0,\boldsymbol{x};p)\right)}_{=(\Lambda/D)\tilde{j}_0(\boldsymbol{x}_0,\boldsymbol{x};p)}\qquad(\boldsymbol{x}\in\partial\Omega).\quad(24b)$$

Suppose that we have solved this problem and found that the solution $\tilde{g}_\Lambda(\boldsymbol{x}_0,\boldsymbol{x};p)$ on the boundary $\partial\Omega$ is equal to some function $f(\boldsymbol{x};p)$. Once we know $f(\boldsymbol{x};p)$, we can simply reconstruct $\tilde{g}_\Lambda(\boldsymbol{x}_0,\boldsymbol{x};p)$ as the solution of the corresponding Dirichlet problem. Applying then the Dirichlet-to-Neumann operator to $f(\boldsymbol{x};p)$, we can express the normal derivative of $\tilde{g}_\Lambda(\boldsymbol{x}_0,\boldsymbol{x};p)$. Summarizing these steps, one can express the restriction of the solution $\tilde{g}_\Lambda(\boldsymbol{x}_0,\boldsymbol{x};p)$ onto the boundary as

$$\tilde{g}_\Lambda(\boldsymbol{x}_0,\boldsymbol{s};p)=\left(\Lambda\mathcal{M}_p+I\right)^{-1}\frac{\Lambda}{D}\tilde{j}_0(\boldsymbol{x}_0,\boldsymbol{s};p)\qquad(\boldsymbol{s}\in\partial\Omega),\qquad(25)$$

where I is the identity operator. Finally, if the starting point $\boldsymbol{x}_0=\boldsymbol{s}_0$ lies on the boundary, one has $\tilde{j}_0(\boldsymbol{s}_0,\boldsymbol{s};p)=\delta(\boldsymbol{s}-\boldsymbol{s}_0)$, the Dirichlet propagator $\tilde{G}_0(\boldsymbol{s}_0,\boldsymbol{x};p)$ vanishes in Eq. (23), and we get

$$D\tilde{G}_\Lambda(\boldsymbol{s}_0,\boldsymbol{s};p)=(\mathcal{M}_p+I/\Lambda)^{-1}\delta(\boldsymbol{s}-\boldsymbol{s}_0)\qquad(\boldsymbol{s}_0,\boldsymbol{s}\in\partial\Omega).\quad(26)$$

We conclude that $D\tilde{G}_\Lambda(\boldsymbol{s}_0,\boldsymbol{s};p)$ is the resolvent of the Dirichlet-to-Neumann operator $\mathcal{M}_p$. In particular, one can rewrite Eq. (19) in the form of a scalar product between two functions on the boundary

$$\tilde{G}_\Lambda(\boldsymbol{x}_0,\boldsymbol{x};p)=\tilde{G}_0(\boldsymbol{x}_0,\boldsymbol{x};p)+\frac{\Lambda}{D}$$
$$\times\,(\tilde{j}_0(\boldsymbol{x}_0,\cdot;p)\cdot(I+\Lambda\mathcal{M}_p)^{-1}\tilde{j}_0(\boldsymbol{x},\cdot;p))_{L_2(\partial\Omega)}.\quad(27)$$

Remarkably, all the "ingredients" of this formula correspond to the Dirichlet condition on a perfectly reactive boundary, and only the reaction length Λ keeps track of partial reactivity. Note that this is a significant extension of the formula for the total steady-state flux derived in Ref. 45. This profound connection (that was earlier established for the conventional setting of the Laplace equation[45,61]) allows one to express many probabilistic quantities through the spectral properties of the Dirichlet-to-Neumann operator.

3.2. Reaction time distribution and spread harmonic measure

The propagator determines many quantities often considered in the context of diffusion-controlled reactions: the survival probability up to time t, the reaction time distribution, the distribution of the reaction point (at which reaction occurs), etc. For instance, the diffusive flux density at a partially reactive point $s \in \partial\Omega$ is

$$j_\Lambda(\boldsymbol{x}_0, \boldsymbol{s}; t) = \left(-D\frac{\partial}{\partial n}G_\Lambda(\boldsymbol{x}_0, \boldsymbol{x}; t)\right)_{\boldsymbol{x}=\boldsymbol{s}} = \frac{D}{\Lambda}G_\Lambda(\boldsymbol{x}_0, \boldsymbol{s}; t), \quad (28)$$

where we used the Robin boundary condition (14c). This is the joint probability density for the reaction time and reaction point on the boundary (from the starting point $\boldsymbol{x}_0$). The integral over $\boldsymbol{s}$ yields the marginal probability density of reaction times,

$$\rho_\Lambda(t; \boldsymbol{x}_0) = \int_{\partial\Omega} d\boldsymbol{s}\, j_\Lambda(\boldsymbol{x}_0, \boldsymbol{s}; t) = \frac{D}{\Lambda}\int_{\partial\Omega} d\boldsymbol{s}\, G_\Lambda(\boldsymbol{x}_0, \boldsymbol{s}; t), \quad (29)$$

whereas the integral over time t yields the marginal probability density of reaction points

$$\omega_\Lambda(\boldsymbol{s}; \boldsymbol{x}_0) = \int_0^\infty dt\, j_\Lambda(\boldsymbol{x}_0, \boldsymbol{s}; t) = \frac{D}{\Lambda}\tilde{G}_\Lambda(\boldsymbol{x}_0, \boldsymbol{s}; p = 0). \quad (30)$$

The latter was called the *spread harmonic measure*.[45,61,64,65] This is a natural extension of the harmonic measure[66] that characterizes the first arrival onto the perfectly reactive boundary.[67,68]

In general, the reactive surface does not need to be the whole boundary of the confining domain Ω. The above analysis remains applicable when only a part of the boundary is partially reactive, whereas the remaining part is inert. In this case, the Neumann boundary condition is imposed on the inert part so that one faces the mixed Robin–Neumann boundary condition. In this setting, the propagator $G_0(\boldsymbol{x}_0, \boldsymbol{x}; t)$ corresponds to the respective Dirichlet–Neumann problem, with a perfectly reactive part. In other words, the presence of the inert part does not change the above arguments, once the involved quantities are treated accordingly.

4. An Example: The Half-Space[b]

As a basic example, we consider the PRBM in an upper half-space, $\Omega = \{x = (y, z) \in \mathbb{R}^d : z > 0\}$, with partially reactive hyperplane $\partial\Omega = \mathbb{R}^{d-1}$. For convenience, we represent each point x as (y, z), where $y \in \mathbb{R}^{d-1}$ are lateral coordinates, and $z \geq 0$ is the transverse coordinate. In this example, the transverse and lateral motions are independent, and the propagator $G_\Lambda(x_0, x; t)$ is the product of the Gaussian propagator in lateral directions and the propagator on the transverse positive semi-axis with the Robin boundary condition at the zero endpoint. The latter propagator is well known (see e.g., the book[69]), and we get

$$G_\Lambda(x_0, x; t) = \frac{\exp\left(-\frac{|y-y_0|^2}{4Dt}\right)}{(4\pi Dt)^{(d-1)/2}} \left\{ \frac{\exp\left(-\frac{(z-z_0)^2}{4Dt}\right) + \exp\left(-\frac{(z+z_0)^2}{4Dt}\right)}{\sqrt{4\pi Dt}} \right.$$
$$\left. -\frac{1}{\Lambda} \exp\left(\frac{z+z_0}{\Lambda} + \frac{Dt}{\Lambda^2}\right) \text{erfc}\left(\frac{z+z_0}{\sqrt{4Dt}} + \frac{\sqrt{Dt}}{\Lambda}\right) \right\}, \quad (31)$$

where $\text{erfc}(z)$ is the complementary error function. This is a rare example when the propagator can be obtained in a fully explicit form. In the limit $\Lambda = 0$, Eq. (31) is reduced to the Dirichlet propagator in the upper half-space

$$G_0(x_0, x; t) = \underbrace{\frac{\exp\left(-\frac{|y-y_0|^2}{4Dt}\right)}{(4\pi Dt)^{(d-1)/2}}}_{\text{lateral}} \underbrace{\frac{\exp\left(-\frac{(z-z_0)^2}{4Dt}\right) - \exp\left(-\frac{(z+z_0)^2}{4Dt}\right)}{\sqrt{4\pi Dt}}}_{\text{transverse}}$$

$$(32)$$

from which

$$j_0(x_0, s; t) = \frac{\exp\left(-\frac{|s-y_0|^2}{4Dt}\right)}{(4\pi Dt)^{(d-1)/2}} \frac{z_0 \exp\left(-\frac{z_0^2}{4Dt}\right)}{\sqrt{4\pi Dt^3}}. \quad (33)$$

The above formulas help to compute other quantities of interest. For instance, the probability density of the reaction time follows

---

[b]This section is rather technical and can be skipped at first reading.

from Eq. (29)

$$\rho_\Lambda(t; z_0) = \frac{D\exp\left(-\frac{z_0^2}{4Dt}\right)}{\Lambda}\left\{\frac{1}{\sqrt{\pi Dt}} - \frac{1}{\Lambda}\,\text{erfcx}\left(\frac{z_0}{\sqrt{4Dt}} + \frac{\sqrt{Dt}}{\Lambda}\right)\right\},$$
(34)

(note that the density depends only on the height z_0 of the starting point $\boldsymbol{x}_0 = (\boldsymbol{y}_0, z_0)$ above the boundary). In the limit $\Lambda = 0$, one retrieves the classical expression $\rho_0(t; z_0) = \frac{z_0}{\sqrt{4\pi Dt^3}}e^{-z_0^2/(4Dt)}$. In the long-time limit, one gets $\rho_\Lambda(t; z_0) \simeq \frac{z_0+\Lambda}{\sqrt{4\pi Dt^3}}$, as though the height of the starting point is increased by the reaction length Λ. In the short-time limit, one gets either a very fast vanishing as $t \to 0$ for $z_0 > 0$, or a power law divergence $\rho_\Lambda(t; z_0) \simeq \frac{D/\Lambda}{\sqrt{\pi Dt}}$ for $z_0 = 0$. This behavior is illustrated in Fig. 2.

The spread harmonic measure density is obtained from Eq. (30)

$$\omega_\Lambda(\boldsymbol{s}; \boldsymbol{x}_0) = \int_{\mathbb{R}^{d-1}} \frac{d\boldsymbol{q}}{(2\pi)^{d-1}} \frac{e^{-i(\boldsymbol{q}(\boldsymbol{s}-\boldsymbol{y}_0))}}{1 + \Lambda|\boldsymbol{q}|}$$

$$= \int_0^\infty dz \frac{e^{-z/\Lambda}}{\Lambda} \omega_0(\boldsymbol{s}; (\boldsymbol{y}, z_0 + z)),$$
(35)

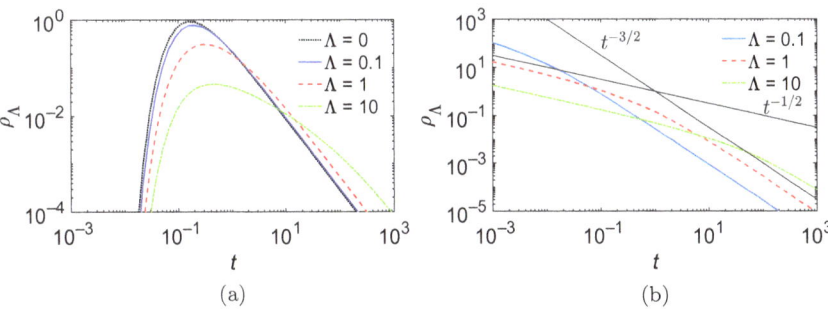

Fig. 2. The probability density $\rho_\Lambda(t; z_0)$ of the reaction time for a molecule started at height z_0 above a flat surface, with $z_0 = 1$ (a) and $z_0 = 0$ (b), and four values of the reaction length Λ (arbitrary units are used, with $D = 1$). In the second plot, the density $\rho_0(t; z_0 = 0) = \delta(t)$ for $\Lambda = 0$ is not shown. Gray straight lines indicate power law asymptotics $t^{-1/2}$ and $t^{-3/2}$.

where

$$\omega_0(\boldsymbol{s};(\boldsymbol{y_0},z_0)) = \frac{\Gamma(d/2)}{\pi^{d/2}} \frac{z_0}{[z_0^2 + |\boldsymbol{s} - \boldsymbol{y_0}|^2]^{d/2}} \qquad (36)$$

is the harmonic measure density (see Ref. 65 for the derivation of the last equality in Eq. (35)). In other words, the effect of partial reflections on the boundary can be seen as an effective increase the height z_0 of the starting point above the boundary. This increase is randomly distributed with an exponential law determined by the reaction length Λ. As expected, the density ω_Λ is radially symmetric with respect to the line which is perpendicular to the reactive surface and passes through the starting point $\boldsymbol{x_0}$. In other words, it depends only on the height z_0 and the radial distance $r = |\boldsymbol{s} - \boldsymbol{y_0}|$. The behavior of the spread harmonic measure is illustrated in Fig. 3. One can easily check that the *mean* exploration distance, defined as the standard deviation of explored distance from the first arrival point to the reaction point, is infinite (see also Ref. 70). This is related to the fact that each jump can go unlimitedly far away from the reactive boundary, and such rare but anomalously long trajectories dominate in the second moment, due to the power-law asymptotic decay of the harmonic measure density in Eq. (36). In this way, the successive arrival points of PRBM onto the reactive surface can be seen as Lévy flights on that surface.[71–75]

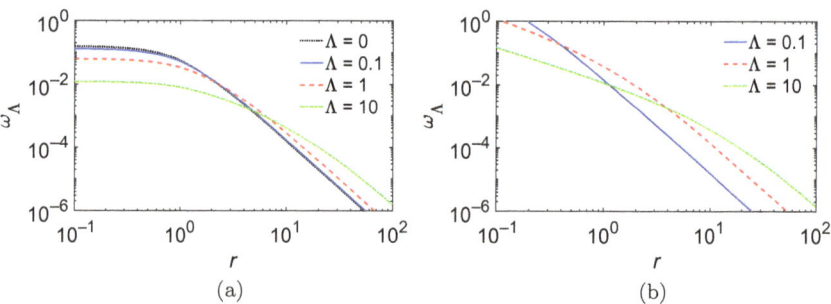

Fig. 3.   The spread harmonic measure density $\omega_\Lambda(\boldsymbol{s};(\boldsymbol{y_0},z_0))$ of the reaction point $\boldsymbol{s}$ as a function of the radial distance $r = |\boldsymbol{s} - \boldsymbol{y_0}|$ for a molecule started at height z_0 above the plane in the 3D space ($d = 3$), with $z_0 = 1$ (a) and $z_0 = 0$ (b), and four values of the reaction length Λ (arbitrary units are used, with $D = 1$). In the second plot, the harmonic measure density $\omega_0(\boldsymbol{s};(\boldsymbol{y_0},0)) = \delta(\boldsymbol{s} - \boldsymbol{y_0})$ for $\Lambda = 0$ is not shown.

Integrating the spread harmonic measure over s with $|s - y_0| > R$, one gets the probability of the reaction event at boundary points whose distance to the first arrival point is greater than R

$$
P_\Lambda(R) = \int_{|s-y_0|>R} ds\,\omega_\Lambda(s;(y_0,0))
$$

$$
= \frac{2\Gamma(d/2)}{\sqrt{\pi}\,\Gamma((d-1)/2)} \int_0^\infty dx\,\frac{e^{-xR/\Lambda}}{(1+x^2)^{d/2}}, \tag{37}
$$

(the last equality was derived in Ref. 65). This probability characterizes the exploration of a flat partially reactive surface after the first arrival and till the reaction. This is a function of R/Λ, which monotonously decreases from 1 to 0. The condition $P_\Lambda(R_m) = 0.5$ defines the median radius R_m of the spherical domain around the first arrival point, at which half of molecules react. In three dimensions, one gets $R_m \approx 1.17\,\Lambda$, providing a geometric interpretation of the reaction length Λ. The behavior of the probability $P_\Lambda(R)$ is illustrated in Fig. 4.

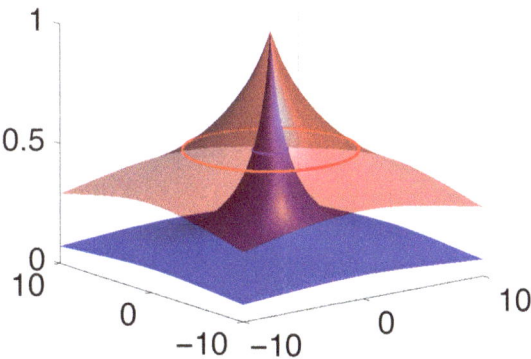

Fig. 4. The probability $P_\Lambda(R)$ of the reaction event at a boundary point s at distance greater than R to the first arrival point (here, the origin) on the plane in the three-dimensional space ($d = 3$), for two values of the reaction length: $\Lambda = 1$ (blue narrower surface) and $\Lambda = 5$ (red broader surface). Two circles show the median reaction radius $R_m \approx 1.17\,\Lambda$ at which this probability is equal to 0.5. Arbitrary length units are used.

5. Conclusion: Is the Finite Reactivity Important?

As discussed in Section 1, there are various microscopic mechanisms that naturally lead to a finite reactivity of the target surface. However, the mathematical description of imperfect reactions involves more sophisticated partially reflected Brownian motion and while most of the results known for perfect reactions can be generalized to imperfect ones, mathematical derivations are usually much more involved. It is probably for this reason that the majority of theoretical studies remains focused on perfect reactions. It is therefore natural to ask whether the finite reactivity is indeed important?

The reactivity of the target introduces the reaction length $\Lambda = D/\kappa$ which controls the balance between the transport step toward the target and the reaction step on the target. The latter step involves the intricate exploration of the partially reactive surface until the reaction occurs. This process mixes bulk diffusion from surface to surface points and eventual reaction trials. Since the process is strongly coupled to the geometry of the environment, the dependence of diffusion characteristics (such as the total flux) on Λ can reveal geometric structure of the target surface.[76]

When the reaction length is the smallest length scale of the problem, the effects of finite reactivity can be neglected. In practice, however, there is a conceptual difference between setting $\Lambda = 0$ (the idealized case of perfect reactions with Dirichlet boundary condition, $\kappa = \infty$), and $\Lambda > 0$, even if Λ remains small. For instance, in the analysis of the short-time asymptotic behavior of a diffusion problem, the diffusion length $\sqrt{Dt}$ can become smaller than the reaction length $\Lambda > 0$ as $t \to 0$. As a consequence, the short-time asymptotic behavior for perfect and imperfect reactions is different.[40,77] For example, the probability density of the reaction time on a spherical target of radius R reads[40]

$$\rho_\Lambda(t; \boldsymbol{x}_0) \simeq \frac{R}{|\boldsymbol{x}_0|} \frac{\exp\left(-\frac{(|\boldsymbol{x}_0|-R)^2}{4Dt}\right)}{t\sqrt{\pi}} \times \begin{cases} \dfrac{|\boldsymbol{x}_0| - R}{\sqrt{4Dt}} & (\Lambda = 0) \\[2mm] \dfrac{\sqrt{Dt}}{\Lambda} & (\Lambda > 0) \end{cases}. \tag{38}$$

(with the starting point x_0 outside the target: $|x_0| > R$). One can see that the power law prefactor is different in two cases. Moreover, when Λ is small, one can also expect the transition from the short-time regime at $\sqrt{Dt} \ll \Lambda \ll |x_0| - R$ (given by the second expression) to an intermediate regime at $\Lambda \ll \sqrt{Dt} \ll |x_0| - R$ (given by the first expression).

Another important example is the so-called narrow escape problem from an Euclidean domain Ω through a small hole Γ on the boundary $\partial\Omega$ (see the review[6] and references therein). In this setting, the escape through a hole Γ can be understood as reaction on the target Γ upon the first arrival. When the (non-dimensional) size ε of the escape region goes to zero, the mean exit time diverges as $\log(1/\varepsilon)$ in two dimensions and as ε^{-1} in three dimensions.[6,78] However, an escape through a small opening requires overpassing either an entropic or an energetic barrier, or both. Once the escape region is considered as partially reactive ($\Lambda > 0$), the asymptotic behavior of the mean exit time changes drastically.[24] In the narrow escape limit $\varepsilon \to 0$, the dominant contribution to the mean exit time comes from the finite reactivity and scales as ε^{-1}/κ in two dimensions and ε^{-2}/κ in three dimensions. This contribution does not exist for idealized perfect escape ($\kappa = \infty$) but becomes dominant for a finite reactivity κ. Once again, in the narrow escape limit, the size of the hole becomes the smallest length scale, and thus the finite reactivity cannot be ignored. The impact of κ on the whole distribution of the reaction time was recently investigated.[79]

These two examples illustrate how an idealized assumption of perfect reactions can be misleading. Even though the analysis of imperfect reactions is mathematically more involved, the related technical difficulties are generally not insurmountable. As a matter of fact, the finite reactivity often "regularizes" solutions and helps to resolve some apparent "paradoxical" properties of Brownian motion. For instance, the Smoluchowski flux (3) toward a perfectly reactive spherical target diverges as $t \to 0$, meaning that too many molecules in a close vicinity of the target react at first time instances. This unrealistic divergence does not appear in the Collins–Kimball flux (5) toward a partially reactive target. Moreover, PRBM can be started right at the partially reactive boundary whereas ordinary Brownian motion started on an absorbing boundary would immediately react. As a consequence, the Robin boundary condition is necessary for a natural implementation of reversible association of a diffusing molecule with the target. In fact, when the formed metastable complex AB in (7) dissociates (with some rate k_{off}), the just dissociated molecule is released *on* the target surface. If the latter was perfectly reactive, the molecule would *immediately* react again, making such a dissociation process impossible.[80,81] A standard trick to overcome this problem consists in ejecting the just dissociated molecule to a finite distance a from the target, from which its diffusion is released.[16–19,44,60] While the introduction of such an ejection distance can be rationalized from the physical point of view (e.g., a finite size of molecules, the presence of a specific surface layer, a finite range binding interaction, etc.), this is precisely an intermediate step of the construction of PRBM as described in Section 2. In other words, even when one tries to avoid using partial reactivity and the associated PRBM, one often uses it implicitly via a "regularization" by a finite ejection distance a.

The mathematical theory of partially reflected diffusions is rather well developed.[55–59] In particular, the fundamental interconnection between the stochastic process, the diffusion equation with Robin boundary condition, and the spectral properties of the underlying diffusion operator and the Dirichlet-to-Neumann operator provides the solid mathematical foundation for investigating imperfect diffusion-controlled reactions. In turn, the intricate exploration of the partially reactive surface via diffusion-mediated jumps in complicated structures such as multiscale porous media or domains with irregular or fractal boundaries, remains poorly understood. In this light, efficient numerical techniques such as fast Monte Carlo methods[67,68,82,83] or semi-analytical solutions[84,85] become particularly important.

References

1. M. Smoluchowski, Versuch einer Mathematischen Theorie der Koagulations Kinetic Kolloider Lösungen, *Z. Phys. Chem.* **129**, 129–168 (1917).
2. S. Rice, *Diffusion-Limited Reactions.* Elsevier, Amsterdam (1985).
3. H. S. Carslaw and J. C. Jaeger, *Conduction of Heat in Solids*, Oxford University Press, Oxford (2nd edn.) (1959).
4. J. Crank, *The Mathematics of Diffusion.* Oxford University Press, Oxford (1956).
5. S. Redner, *A Guide to First Passage Processes.* Cambridge University Press, Cambridge, England (2001).
6. D. Holcman and Z. Schuss, The narrow escape problem, *SIAM Rev.* **56**, 213–257 (2014).
7. D. Holcman and Z. Schuss, *Stochastic Narrow Escape in Molecular and Cellular Biology.* Springer, New York (2015).
8. D. S. Grebenkov, Universal formula for the mean first passage time in planar domains, *Phys. Rev. Lett.* **117**, 260201 (2016).
9. R. F. Bass, *Diffusions and Elliptic Operators.* Springer, New York (1998).
10. H. Risken, *The Fokker–Planck equation: Methods of Solution and Applications*, Berlin: Springer (3rd Edn.) (1996).
11. S. B. Yuste, E. Abad, and K. Lindenberg, Exploration and trapping of mortal random walkers, *Phys. Rev. Lett.* **110**, 220603 (2013).
12. B. Meerson and S. Redner, Mortality, redundancy, and diversity in stochastic search, *Phys. Rev. Lett.* **114**, 198101 (2015).
13. D. S. Grebenkov and J.-F. Rupprecht, The escape problem for mortal walkers, *J. Chem. Phys.* **146**, 084106 (2017).
14. J.-P. Bouchaud and A. Georges, Anomalous diffusion in disordered media: Statistical mechanisms, models and physical applications, *Phys. Rep.* **195**, 127–293 (1990).
15. R. Metzler and J. Klafter, The random walk's guide to anomalous diffusion: A fractional dynamics approach, *Phys. Rep.* **339**, 1–77 (2000).
16. O. Bénichou, D. S. Grebenkov, P. Levitz, C. Loverdo, and R. Voituriez, Optimal reaction time for surface-mediated diffusion, *Phys. Rev. Lett.* **105**, 150606 (2010).
17. O. Bénichou, D. S. Grebenkov, P. Levitz, C. Loverdo, and R. Voituriez, Mean first-passage time of surface-mediated diffusion in spherical domains, *J. Stat. Phys.* **142**, 657–685 (2011).
18. J.-F. Rupprecht, O. Bénichou, D. S. Grebenkov, and R. Voituriez, Kinetics of active surface-mediated diffusion in spherically symmetric domains, *J. Stat. Phys.* **147**, 891–918 (2012).
19. J.-F. Rupprecht, O. Bénichou, D. S. Grebenkov, and R. Voituriez, Exact mean exit time for surface-mediated diffusion, *Phys. Rev. E.* **86**, 041135 (2012).
20. F. C. Collins and G. E. Kimball, Diffusion-controlled reaction rates, *J. Colloid Sci.* **4**, 425–437 (1949).

21. S. D. Traytak and W. Price, Exact solution for anisotropic diffusion-controlled reactions with partially reflecting conditions, *J. Chem. Phys.* **127**, 184508 (2007).

22. J. Qian and P. N. Sen, Time dependent diffusion in a disordered medium with partially absorbing walls: A perturbative approach, *J. Chem. Phys.* **125**, 194508 (2006).

23. B. Sapoval, General formulation of Laplacian transfer across irregular surfaces, *Phys. Rev. Lett.* **73**, 3314–3317 (1994).

24. D. S. Grebenkov and G. Oshanin, Diffusive escape through a narrow opening: New insights into a classic problem, *Phys. Chem. Chem. Phys.* **19**, 2723–2739 (2017).

25. D. S. Grebenkov, R. Metzler, and G. Oshanin, Effects of the target aspect ratio and intrinsic reactivity onto diffusive search in bounded domains, *New J. Phys.* **19**, 103025 (2017).

26. P. Hänggi, P. Talkner, and M. Borkovec, Reaction-rate theory: Fifty years after Kramers, *Rev. Mod. Phys.* **62**, 251–341 (1990).

27. D. Shoup and A. Szabo, Role of diffusion in ligand binding to macromolecules and cell-bound receptors, *Biophys. J.* **40**, 33–39 (1982).

28. D. A. Lauffenburger and J. Linderman, *Receptors: Models for binding, trafficking, and signaling.* Oxford University Press (1993).

29. H. Sano and M. Tachiya, Partially diffusion-controlled recombination, *J. Chem. Phys.* **71**, 1276–1282 (1979).

30. H. Sano and M. Tachiya, Theory of diffusion-controlled reactions on spherical surfaces and its application to reactions on micellar surfaces, *J. Chem. Phys.* **75**, 2870–2878 (1981).

31. P. C. Bressloff, B. A. Earnshaw, and M. J. Ward, Diffusion of protein receptors on a cylindrical dendritic membrane with partially absorbing traps, *SIAM J. Appl. Math.* **68**, 1223–1246 (2008).

32. H.-X. Zhou and R. Zwanzig, A rate process with an entropy barrier, *J. Chem. Phys.* **94**, 6147–6152 (1991).

33. D. Reguera, G. Schmid, P. S. Burada, J. M. Rubí, P. Reimann, and P. Hänggi, Entropic transport: Kinetics, scaling, and control mechanisms, *Phys. Rev. Lett.* **96**, 130603 (2006).

34. O. Bénichou, M. Moreau, and G. Oshanin, Kinetics of stochastically gated diffusion-limited reactions and geometry of random walk trajectories, *Phys. Rev. E.* **61**, 3388–3406 (2000).

35. J. Reingruber and D. Holcman, Gated narrow escape time for molecular signaling, *Phys. Rev. Lett.* **103**, 148102 (2009).

36. P. C. Bressloff, Stochastic switching in biology: From genotype to phenotype *J. Phys. A.* **50**, 133001 (2017).

37. H. C. Berg and E. M. Purcell, Physics of chemoreception, *Biophys. J.* **20**, 193–239 (1977).

38. D. Shoup, G. Lipari, and A. Szabo, Diffusion-controlled bimolecular reaction rates. The effect of rotational diffusion and orientation constraints, *Biophys. J.* **36**, 697–714 (1981).

39. R. Zwanzig and A. Szabo, Time dependent rate of diffusion-influenced ligand binding to receptors on cell surfaces, *Biophys. J.* **60**, 671–678 (1991).

40. D. S. Grebenkov, Searching for partially reactive sites: Analytical results for spherical targets, *J. Chem. Phys.* **132**, 034104 (2010).

41. J. G. Powles, M. J. D. Mallett, G. Rickayzen, and W. A. B. Evans, Exact analytic solutions for diffusion impeded by an infinite array of partially permeable barriers, *Proc. R. Soc. London A.* **436**, 391–403 (1992).

42. B. Sapoval, M. Filoche, and E. Weibel, Smaller is better — but not too small: A physical scale for the design of the mammalian pulmonary acinus, *Proc. Nat. Acad. Sci. U S A* **99**, 10411–10416 (2002).

43. D. S. Grebenkov, M. Filoche, B. Sapoval, and M. Felici, Diffusion-reaction in branched structures: Theory and application to the lung acinus, *Phys. Rev. Lett.* **94**, 050602 (2005).

44. M. Filoche and B. Sapoval, Can one hear the shape of an electrode? II. Theoretical Study of the Laplacian Transfer, *Eur. Phys. J. B.* **9**, 755–763 (1999).

45. D. S. Grebenkov, M. Filoche, and B. Sapoval, Mathematical basis for a general theory of Laplacian transport towards irregular Interfaces, *Phys. Rev. E.* **73**, 021103 (2006).

46. K. R. Brownstein and C. E. Tarr, Importance of classical diffusion in NMR studies of water in biological cells, *Phys. Rev. A.* **19**, 2446–2453 (1979).

47. D. S. Grebenkov, NMR survey of reflected Brownian motion, *Rev. Mod. Phys.* **79**, 1077–1137 (2007).

48. D. S. Grebenkov, From the microstructure to diffusion NMR, and back, in: *Diffusion NMR of Confined Systems*, R. Valiullin (ed.), RSC Publishing, Cambridge, (2016).

49. S. Condamin, O. Bénichou, V. Tejedor, R. Voituriez, and J. Klafter, First-passage time in complex scale-invariant media, *Nature* **450**, 77–80 (2007).

50. O. Bénichou, C. Chevalier, J. Klafter, B. Meyer, and R. Voituriez, Geometry-controlled kinetics, *Nature Chem.* **2**, 472–477 (2010).

51. O. Bénichou and R. Voituriez, From first-passage times of random walks in confinement to geometry-controlled kinetics, *Phys. Rep.* **539**, 225–284 (2014).

52. M. Freidlin, *Functional Integration and Partial Differential Equations, Annals of Mathematics Studies*, Princeton University Press, Princeton, New Jersey (1985).

53. Z. Schuss, *Theory and Applications of Stochastic Differential Equations.* Wiley, New York (1980).

54. D. W. Stroock and S. R. S. Varadhan, Diffusion processes with boundary conditions, *Commun. Pure Appl. Math.* **24**, 147–225 (1971).

55. Z. Ma and R. Song, Probabilistic methods in Schrodinger equations, in Seminar on Stochastic Processes 1989, pp. 135–164, *Progr. Probab.* **18**, Birkhauser, Boston (1990).

56. V. G. Papanicolaou, The probabilistic solution of the third boundary value problem for second order elliptic equations, *Probab. Th. Rel. Fields* **87**, 27–77 (1990).

57. G. N. Milshtein, The solving of boundary value problems by numerical integration of stochastic equations, *Math. Comp. Sim.* **38**, 77–85 (1995).
58. R. F. Bass, K. Burdzy, and Z.-Q. Chen, On the Robin problem in Fractal Domains, *Proc. London Math. Soc.* **96**, 273–311 (2008).
59. Z. Schuss, *Brownian Dynamics at Boundaries and Interfaces in Physics, Chemistry and Biology.* Springer, New York (2013).
60. D. S. Grebenkov, M. Filoche, and B. Sapoval, Spectral properties of the Brownian self-transport operator, *Eur. Phys. J. B.* **36**, 221–231 (2003).
61. D. S. Grebenkov, Partially reflected Brownian motion: A stochastic approach to transport phenomena, in: Focus on Probability Theory, L. R. Velle (ed.), Nova Science Publishers, Hauppauge (2006), pp. 135–169.
62. D. S. Grebenkov, Residence times and other functionals of reflected Brownian motion, *Phys. Rev. E.* **76**, 041139 (2007).
63. A. Singer, Z. Schuss, A. Osipov, and D. Holcman, Partially reflected diffusion, *SIAM J. Appl. Math.* **68**, 844–868 (2008).
64. D. S. Grebenkov, Scaling properties of the spread harmonic measures, *Fractals* **14**, 231–243 (2006).
65. D. S. Grebenkov, Analytical representations of the spread harmonic measure, *Phys. Rev. E.* **91**, 052108 (2015).
66. J. B. Garnett and D. E. Marshall, *Harmonic Measure.* Cambridge University Press, Cambridge, England (2005).
67. D. S. Grebenkov, A. A. Lebedev, M. Filoche, and B. Sapoval, Multifractal properties of the harmonic measure on Koch boundaries in two and three dimensions, *Phys. Rev. E.* **71**, 056121 (2005).
68. D. S. Grebenkov, What makes a boundary less accessible, *Phys. Rev. Lett.* **95**, 200602 (2005).
69. K. D. Cole, J. V. Beck, A. Haji-Sheikh, and B. Litkouhi, *Heat Conduction Using Green's Functions,* Taylor & Francis (2nd edn.), Boca Raton (2010).
70. B. Sapoval, J. S. Andrade Jr, A. Baldassari, A. Desolneux, F. Devreux, M. Filoche, D. S. Grebenkov, and S. Russ, New simple properties of a few irregular systems, *Physica A* **357**, 1–17 (2005).
71. B. B. Mandelbrot, *The Fractal Geometry of Nature,* W. H. Freeman, New York (1982).
72. G. M. Viswanathan, V. Afanasyev, S. V. Buldyrev, E. J. Murphy, P. A. Prince, and H. E. Stanley, Levy flight search patterns of wandering albatrosses, *Nature* **381**, 413–415 (1996).
73. P. Levitz, Random flights in confining interfacial systems, *J. Phys. Cond. Matt.* **17**, S4059–S4074 (2005).
74. P. Levitz, D. S. Grebenkov, M. Zinsmeister, K. Kolwankar, and B. Sapoval, Brownian flights over a fractal nest and first passage statistics on irregular surfaces, *Phys. Rev. Lett.* **96**, 180601 (2006).
75. P. Levitz, P. A. Bonnaud, P.-A. Cazade, R. J.-M. Pellenq, and B. Coasne, Molecular intermittent dynamics of interfacial water: Probing adsorption and bulk confinement, *Soft Matter* **9**, 8654–8663 (2013).

76. M. Filoche and D. S. Grebenkov, The toposcopy, a new tool to probe the geometry of an irregular interface by measuring its transfer impedance, *Eur. Phys. Lett.* **81**, 40008 (2008).

77. D. S. Grebenkov, Subdiffusion in a bounded domain with a partially absorbing-reflecting boundary, *Phys. Rev. E.* **81**, 021128 (2010).

78. A. Singer, Z. Schuss, D. Holcman, and R. S. Eisenberg, Narrow Escape, Part I, *J. Stat. Phys.* **122**, 437–463 (2006).

79. D. S. Grebenkov, R. Metzler, and G. Oshanin, Towards a full quantitative description of single-molecule reaction kinetics in biological cells, *Phys. Chem. Chem. Phys.* **20**, 16393–16401 (2018).

80. T. Prüstel and M. Tachiya, Reversible diffusion-influenced reactions of an isolated pair on some two dimensional surfaces, *J. Chem. Phys.* **139**, 194103 (2013).

81. D. S. Grebenkov, First passage times for multiple particles with reversible target-binding kinetics, *J. Chem. Phys.* **147**, 134112 (2017).

82. A. R. Kansal and S. Torquato, Prediction of trapping rates in mixtures of partially absorbing spheres, *J. Chem. Phys.* **116**, 10589–10597 (2002).

83. D. S. Grebenkov, *Efficient Monte Carlo methods for simulating diffusion-reaction processes in complex systems*, in: *First-Passage Phenomena and Their Applications*, R. Metzler, G. Oshanin, and S. Redner (eds.), World Scientific Press, Singapore (2014).

84. M. Galanti, D. Fanelli, S. D. Traytak, and F. Piazza, Theory of diffusion-influenced reactions in complex geometries, *Phys. Chem. Chem. Phys.* **18**, 15950–15954 (2016).

85. D. S. Grebenkov and S. D. Traytak, Semi-analytical computation of Laplacian Green functions in three-dimensional domains with disconnected spherical boundaries, *J. Comput. Phys.* **379**, 91–117 (2019).

Chapter 9

Survival, Absorption, and Escape of Interacting Diffusing Particles

Tal Agranov* and Baruch Meerson[†]

Racah Institute of Physics, Hebrew University of Jerusalem, Jerusalem 91904, Israel

tal.agranov@mail.huji.ac.il
[†]meerson@mail.huji.ac.il

At finite concentrations of reacting molecules, kinetics of diffusion-controlled reactions is affected by intrareactant interactions. As a result, multiparticle reaction statistics cannot be deduced from single-particle results. Here, we briefly review a recent progress in overcoming this fundamental difficulty. We show that the fluctuating hydrodynamics and macroscopic fluctuation theory provide a simple, general, and versatile framework for studying a whole class of problems of survival, absorption, and escape of interacting diffusing particles.

1. Introduction

Kinetics of many diffusion-controlled reactions is affected by intra-reactant interactions. This happens when the density of the reacting molecules is not too small. Although the importance of interactions may have been recognized for a long time, there has been very little progress in their account in theory. Here we will briefly review, and slightly generalize, one promising approach toward solving this long-standing problem.[1–6] We will consider several prototypical gas settings. The first group of settings — interior settings — deals with interacting diffusing molecules inside a domain (think about a living cell). The second group — exterior settings — deals with molecules surrounding a domain. In both cases, the domain boundary, or part of it, absorbs the molecules upon impact, signaling that a reaction occurred.

The interior settings give simplified descriptions of inter-cellular transport in the living cell, where molecules search for a correct location within a cell membrane. The efficacy of the inter-cellular transport is determined by the absorption rate of the molecules.[7] The interior settings include the *narrow escape problem*, see the right panel of Fig. 1, which is well studied in the case of non-interacting diffusing molecules trying to escape from a closed domain via a small hole in its boundary.[7–12]

The exterior problems are different but closely related. The case when the boundary of the domain is fully absorbing is known as the target search, or target survival problem.[13–15] This describes the situation where the molecules of one reactant — a minority — can be viewed as big and immobile, whereas the molecules of another reactant — a majority — are small and mobile. A different scenario happens when molecules get absorbed only through some absorbing patches — receptors — distributed on the otherwise reflecting domain boundary,[16,17] see the left panel of Fig. 1.

If the diffusing molecules are treated as non-interacting random walkers (RWs), the calculation of the effective reaction rates and its

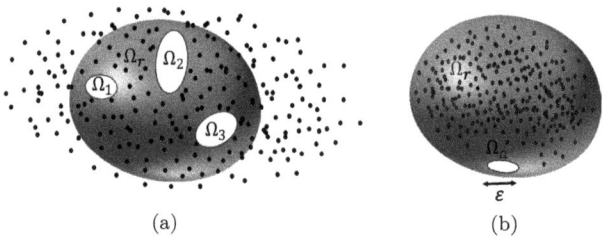

(a)   (b)

Fig. 1.   (a) An exterior problem with multiple absorbing patches shown in white. The rest of the boundary is reflecting. (b) Narrow escape of multiple particles through a small hole in the boundary, the rest of the boundary being reflecting.

fluctuation statistics boils down to calculating a single-particle probability. Interactions invalidate the single-particle picture and make the problem very difficult. Fortunately, a new simplification emerges if there are sufficiently many interacting diffusing particles in the relevant region of space. In this case one can use the fluctuating hydrodynamics, which goes back to Landau and Lifshitz,[18] and a large deviation theory for it. As a convenient and well-controlled first-principle model, Refs. 1–5 adopted diffusive lattice gases, where the fluctuating hydrodynamics is well established.[19] The corresponding large-deviation theory has recently become available under the name of macroscopic fluctuation theory (MFT).[20]

2. Fluctuating Hydrodynamics and the MFT

Fluctuating hydrodynamics is a coarse-grained description of a gas of particles in terms of the particle number density $\rho(\mathbf{x}, t)$.[19,21] The average particle density of a lattice gas obeys a diffusion equation $\partial_t \rho = \nabla \cdot [D(\rho)\nabla\rho]$, whereas macroscopic fluctuations are described by the conservative Langevin equation

$$\partial_t \rho = -\nabla \cdot \mathbf{J}, \quad \mathbf{J} = -D(\rho)\nabla\rho - \sqrt{\sigma(\rho)}\boldsymbol{\eta}(\mathbf{x}, t), \tag{1}$$

where $\boldsymbol{\eta}(\mathbf{x}, t)$ is a zero-mean Gaussian noise, delta-correlated in space and in time.[19,21,22] The diffusivity $D(\rho) \geq 0$ and the mobility $\sigma(\rho) \geq 0$ are to be obtained, for each lattice gas, from the microscopic model. The simplest case, whose coarse-grained behavior coincides with that of non-interacting Brownian particles is the gas of RWs,[23] where one has $D(\rho) = D_0$ and $\sigma(\rho) = 2D_0\rho$.[19] A model with interactions, which we will focus on, is the symmetric simple exclusion process (SSEP)[19] which accounts, in a simple way, for excluded volume interactions. The SSEP's average behavior coincides with that of the RW's, as they share the same density-independent diffusivity D_0. Their fluctuations, however, are different as the SSEP's mobility $\sigma(\rho) = 2D_0\rho(1 - \rho a^3)$ is a nonlinear function of ρ.[19,21] Here a is the lattice constant which we set to unity, so that $0 \leq \rho \leq 1$.

Like many other lattice gases, the SSEP behaves in its dilute limit as non-interacting RWs.

To develop a large-deviation theory for Eq. (1), one starts from a path integral for the probability of observing a joint density and flux histories $\rho(\mathbf{x}, t), \mathbf{J}(\mathbf{x}, t)$, constrained by the continuity equation (1)

$$\mathcal{P} = \int \mathcal{D}\rho \mathcal{D}\mathbf{J} \prod_{\mathbf{x}, t} \delta(\partial_t \rho + \nabla \cdot \mathbf{J}) \exp(-\mathcal{S}),$$

$$\mathcal{S}\left[\rho(\mathbf{x}, t), \mathbf{J}(\mathbf{x}, t)\right] = \int_0^T dt \int d^3\mathbf{x} \frac{[\mathbf{J} + D(\rho)\nabla\rho]^2}{2\sigma(\rho)}. \tag{2}$$

In the interior and exterior settings, presented above, we condition the process on a specified (zero or non-zero) particle absorption current by a given time T. Therefore, we need to evaluate the path integral over only those density and flux histories which led to the specified current. Assuming that all characteristic length scales involve large numbers of particles, the dominant contribution for $\mathcal{P}$ comes from the *optimal fluctuation*: The most probable history $\rho(\mathbf{x}, t)$, $\mathbf{J}(\mathbf{x}, t)$.[24] The ensuing minimization procedure yields the Euler–Lagrange equation which can be cast into a Hamiltonian form, known as the MFT equations.[20] The minimization procedure also generates problem-specific boundary conditions. Evaluating the minimum action S over the solutions to the minimization problem yields the desired probability $\mathcal{P}$ up to a preexponential factor,

$$-\ln \mathcal{P} \simeq S \equiv \min_{\rho, \mathbf{J}} \mathcal{S}\left[\rho(\mathbf{x}, t), \mathbf{J}(\mathbf{x}, t)\right]. \tag{3}$$

In general, the minimization problem is not solvable analytically. Considerable simplifications arise in the limits of very long and very short times compared to a characteristic diffusion time of the problem, see below. We will first address the long-time limit, where the optimal gas density and flux become *stationary*, and devote Section 5 to the non-stationary regime relevant for short times. Following Refs. 1–5, we will consider two types of initial conditions. The first is a random (or annealed) initial condition, where particles are randomly distributed in space with an average density ρ_0. It describes the situation where the gas has enough time to equilibrate before the process starts (for example, before the receptor becomes available). The other is deterministic (or quenched) initial condition with a uniform density ρ_0. When considering long times, the details of the initial condition become irrelevant. In contrast, the short-time statistics strongly depends on the initial condition.

3. The Exterior Problem

Suppose a gas of diffusing particles fills the whole space outside of a simply connected $3d$ domain of a linear size L. The domain boundary Ω is composed of a reflecting part Ω_r and a complementary absorbing part Ω_a. Whenever a particle hits Ω_a, it is immediately absorbed, which sets

$$\rho(\mathbf{x} \in \Omega_a, t) = 0. \tag{4}$$

Whenever a particle hits Ω_r, it is reflected, which sets a zero-flux boundary condition

$$\mathbf{J}(\mathbf{x} \in \Omega_r, t) \cdot \hat{n} = 0, \tag{5}$$

where $\hat{n}$ denotes a local unit vector normal to the domain boundary and directed into the domain. For a fully absorbing domain, there is no reflecting part. The simpler latter setting is known as "the target search problem"[13-15]; it captures the essence of many diffusion-controlled chemical reactions. A more involved setting is a domain whose boundary has several disjoint absorbing patches (receptors).[16,17] For both random and deterministic initial conditions, the boundary condition at infinity is

$$\rho(|\mathbf{x}| \to \infty, t) = \rho_0. \tag{6}$$

The quantity of interest is the probability $\mathcal{P}(n, T, \rho_0)$ that N gas particles were absorbed during the time interval $0 < t < T$, where $n \equiv N/T$ is the absorption current. For multiple absorbing patches, one is interested in the corresponding multivariate probability.

At times T much longer than the diffusion time $L^2/D(\rho_0)$, the system reaches a non-equilibrium steady state, where the average gas density $\bar{\rho}(\mathbf{x})$ is independent of time. In its turn, the average number of absorbed particles, $\bar{N}$, is proportional to time, so that the average absorption current $\bar{n} = \bar{N}/T$ is independent of time. Similarly, for a whole class of lattice gases the *optimal* density and flux, conditioned on a specified current $n \neq \bar{n}$, also become stationary.[a] As a result, $\mathcal{P}(n, T, \rho_0)$ exponentially decays with time T. A similar situation occurs in the context of stationary fluctuations of current in diffusive lattice gases, driven by density reservoirs at the boundaries. There the stationarity of the optimal gas density and flux is known under the name of the "additivity principle",[25] and we use this term here as well.

3.1. *Target survival*

The authors of Ref. 1 considered a fully absorbing domain ($\Omega_a = \Omega$) and studied the probability $\mathcal{P}(n = 0, T, \rho_0)$ that not a single particle gets absorbed by time T. This probability is often called the survival probability; it is a key quantity in determining the distribution of absorption times of the first particle. The latter is given by $\mathcal{P}_{\text{first}}(T, \rho_0) = -\partial_T \mathcal{P}(n = 0, T, \rho_0)$. As a result, the mean absorption time of the first particle, which determines the average reaction rate, is $\langle T \rangle = \int_0^T dt \mathcal{P}(n = 0, T, \rho_0)$.

Previously, the target survival has been extensively studied, by exploiting single particle results, in the case when the particles are non-interacting RWs. The probability that the target survives until a long time T decays exponentially in time,

$$- \ln \mathcal{P}(n = 0, T, \rho_0) \simeq T s(\rho_0) \tag{7}$$

with the decay rate[26–34]

$$s(\rho_0) = 4\pi C D_0 \rho_0, \tag{8}$$

---

[a]The full time-dependent solution of the problem develops two narrow boundary layers in time, at $t = 0$ and $t = T$. They only give a subleading contribution to the action (3), see e.g., Refs. 1,2.

where C is the electrical capacitance of a conductor whose shape is Ω. For a sphere of radius R one has $C = R$. As shown in Ref. 1, the long-time expression (7) holds for interacting lattice gases as well, and the steady-state MFT calculations yield model-specific $s(\rho_0)$. Here is a scheme of the calculations. As one can show, the stationary particle flux, optimal for survival, vanishes everywhere.[1,2] In other words, the fluctuating contribution to the optimal flux exactly counterbalances the deterministic contribution, thus preventing the particles from being absorbed. One is left with finding the optimal density profile. Upon the ansatz $\mathbf{J} = 0$ and $\rho = \rho(\mathbf{x})$ in Eq. (2), the action $\mathcal{S}$ becomes proportional to T, and the problem reduces to minimizing the *action rate* functional

$$\mathfrak{s}\left[\rho\left(\mathbf{x}\right)\right] = \int d^3\mathbf{x} \frac{[D(\rho)\nabla\rho]^2}{2\sigma(\rho)}, \tag{9}$$

subject to the boundary conditions (4) and (6). It is convenient to make the transformation $u(\mathbf{x}) = f\left[\rho\left(\mathbf{x}\right)\right]$, where[1,b]

$$f(\rho) = \int_0^\rho dz \frac{D(z)}{\sqrt{\sigma(z)}}. \tag{10}$$

We denote the inverse function, f^{-1}, by F. The transformation (10) reduces the minimization problem to solving the Laplace's equation

$$\nabla^2 u = 0. \tag{11}$$

Returning to the original variables, the solution is given in terms of the effective electrostatic potential around a conductor with boundary Ω kept at unit voltage $\phi(\mathbf{x})$. In simple cases (e.g., when Ω is a disk, a sphere or a spheroid), $\phi(\mathbf{x})$ can be found explicitly.[35] The stationary density profile, optimal for the particle survival, is a function

---

[b]Convergence of the integral (10) puts some limitations on the behavior of $D(\rho)$ and $\sigma(\rho)$ at small densities. As an example, let $D(\rho \to 0) \sim \rho^\alpha$ and $\sigma(\rho \to 0) \sim \rho^\beta$. Then the integral converges at $\rho \to 0$ if and only if $2\alpha - \beta + 2 > 0$. This condition holds in the examples we consider here.

of this potential alone

$$\rho(\mathbf{x}) = F\{f(\rho_0)[1 - \phi(\mathbf{x})]\}. \tag{12}$$

The action rate (9), evaluated over the solution (12), yields the decay rate $s(\rho_0)$ entering Eq. (7). It is given by the electrostatic energy created by a conductor Ω held at voltage $f(\rho_0)$

$$s(\rho_0) = 2\pi C f^2(\rho_0). \tag{13}$$

Remarkably, the entire effect of interactions is encoded in the density dependence $f(\rho_0)$, coming from the nonlinear transformation (10). The geometry dependence is universal for all gases of this class and is given by the capacitance C. When specialized to the RWs, Eq. (13) reduces to Eq. (8), as to be expected.

For the SSEP Eqs. (10) and (13) yield

$$s(\rho_0) = 4\pi C D_0 \arcsin^2(\sqrt{\rho_0}). \tag{14}$$

This decay rate is larger than that of the Rws (8), as to be expected because of the effective mutual repulsion of the particles, see the Fig. 2(a). Earlier works[36–40] on the target survival for the SSEP only established some bounds on $\mathcal{P}(n = 0, T, \rho_0)$.

3.2.   *Full statistics of absorption*

When conditioning on arbitrary $n > 0$, we impose the constraint $\oint_{\mathbf{x} \in \Omega} \mathbf{J} \cdot \hat{n} = n$. The stationary version of Eq. (1) is $\nabla \cdot \mathbf{J} = 0$. This fact, alongside with the boundary conditions and the additional assumption that the field $\mathbf{J}$ is irrotational, uniquely defines $\mathbf{J}$.[c,d] It is given in terms of the average steady-state flux field

$$\bar{\mathbf{J}}(\mathbf{x}) = -D(\bar{\rho})\nabla\bar{\rho}, \tag{15}$$

---

[c]One also needs to use the fact that the component of the flux, transverse to the domains boundary, vanishes. This property holds for the optimal irrotational flux field.
[d]Within linearized MFT equations, the vorticity of the flux field vanishes.[4] It would be very interesting to find out whether a nonzero vorticity can emerge, in nontrivial geometries, beyond small fluctuations.

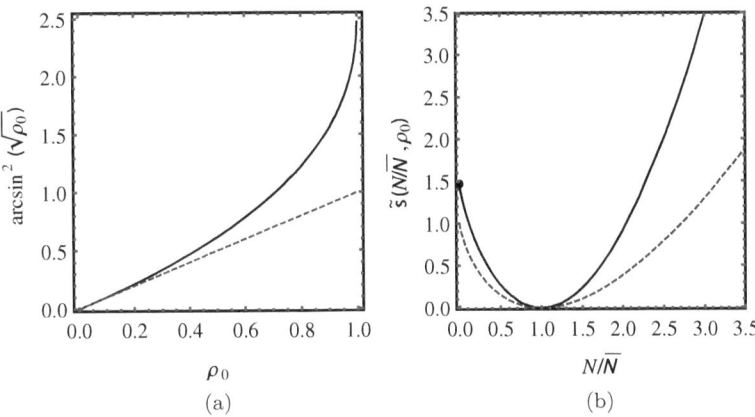

Fig. 2. (a) The function $\arcsin^2 \sqrt{\rho_0}$ which describes the density dependence of the decay rate, Eq. (14), of the target survival probability for the SSEP. The solid line is the same quantity for the RWs, Eq. (8). (b) The large-deviation function $\tilde{s}(N/\bar{N}, \rho_0)$ from Eq. (21), which describes the full statistics of absorption in the exterior problem for the SSEP with $\rho_0 = 0.75$. The dashed line is the same quantity for the RWs, see Eq. (22). The fat point at $N = 0$ shows the survival probability (14).

which is also solenoidal and irrotational but obeys $\oint_{\mathbf{x}\in\Omega} \bar{\mathbf{J}} \cdot \hat{n} = \bar{n}$. Then it follows that the optimal absorption flux is simply

$$\mathbf{J} = \frac{n}{\bar{n}}\bar{\mathbf{J}}, \tag{16}$$

see also Refs. 3, 41. The average flux (15) is given by the effective electrostatic potential $\phi(\mathbf{x})$, defined in the previous section:[4,16]

$$\bar{\mathbf{J}} = V(\rho_0)\nabla\phi(\mathbf{x}), \quad V(\rho_0) \equiv \int_0^{\rho_0} D(z)dz. \tag{17}$$

This potential plays the role of the natural "spatial coordinate" of the problem. The optimal density field ρ is a function of the potential alone, and the problem is effectively 1D with respect to this coordinate[3,4,41]

$$\rho(\mathbf{x}) = \rho_1[\phi(\mathbf{x})]. \tag{18}$$

The function $\rho_1(u)$ is defined on the segment $u \in [0,1]$ and obeys the boundary conditions $\rho_1(0) = \rho_0$ and $\rho_1(1) = 0$. It is to be determined by a 1D variational problem which involves the minimization of the action (3). Upon the ansatz $\mathbf{J} = (n/\bar{n})\bar{\mathbf{J}}$ and $\rho(\mathbf{x}) = \rho_1[\phi(\mathbf{x})]$ in Eq. (2), the action becomes proportional to T, so one needs to minimize an action rate functional. After some algebra the problem is reduced to minimizing a 1D functional in terms of $\rho_1(u)$[3,4,41]

$$\mathfrak{s}[\rho_1(u)] = 4\pi C \times \int_0^1 \frac{\left\{\frac{n}{4\pi C} + D[\rho_1(u)]\rho_1'(u)\right\}^2}{2\sigma[\rho_1(u)]} du, \qquad (19)$$

where, as in the previous section, C is the capacitance of the absorbing domain. The same functional appears in the context of the long-time statistics of the current in a lattice gas on a segment, driven by two reservoirs with different gas densities at the segment's ends.[25] In the latter setting, the probability of having the current n decays exponentially in time, $-\ln \mathcal{P}_{1d}(n, T, \rho_0) \simeq T s_{1d}(n, \rho_0)$. The action rate $s_{1d}(n, \rho_0)$ is simply related to the action rate, obtained by minimization of Eq. (19) $s(n, \rho_0)$

$$s(n, \rho_0) = 4\pi C s_{1d}\left(\frac{n}{4\pi C}, \rho_0\right). \qquad (20)$$

This sets a universal relation between the different problems.[3,4,41] The geometry enters only through the capacitance of the domain.

The 1D problem is exactly solvable in quadratures.[3,25] For the SSEP, the result can be written as[3]

$$-\ln \mathcal{P}(N, T, \rho_0) \simeq \bar{N}\tilde{s}\left(\frac{N}{\bar{N}}, \rho_0\right), \qquad (21)$$

where $\bar{N} = 4\pi C D_0 \rho_0 T$. The function $\tilde{s}(N/\bar{N}, \rho_0)$ is shown in the right panel of Fig. 2. In the limit of $\rho_0 \to 0$ the function $\tilde{s}(N/\bar{N}, \rho_0)$ describes the RWs and corresponds to the $N \gg 1$ limit of the Poisson distribution with mean $\bar{N}$:

$$\tilde{s}\left(\frac{N}{\bar{N}}, \rho_0 \to 0\right) = \frac{N}{\bar{N}} \ln \frac{N}{\bar{N}} - \frac{N}{\bar{N}} + 1. \qquad (22)$$

3.3. *Multiple absorbing patches*

Reference 4 considered particle absorption by *multiple* patches Ω_i, $i = 1, 2, \ldots, s$, distributed on an otherwise reflecting boundary, see the left panel of Fig. 1. The results brought some surprises. To start with, the optimal particle flux field, conditioned on a specified joint absorption statistics $\{n_i\}_{i=1}^{s}$, exhibits a large-scale *vorticity* $\boldsymbol{\omega} = \nabla \times \mathbf{J} \neq 0$.[4] The vorticity emerges even when the particles are non-interacting RWs, and for any geometry, as long as there are more than one absorbing patch. This makes the problem more involved as one should consider a joint variational problem for the flux and the density given by Eq. (3). A simplification comes when considering the statistics of typical, small fluctuations, $\delta n_i = n_i - \bar{n}_i \ll \bar{n}_i$, of the absorption currents around their mean values. Here one can linearize the MFT equations around the mean values $\bar{\rho}$ and $\bar{\mathbf{J}}$. The resulting solution[4] describes a multivariate Gaussian distribution

$$\mathcal{P} \simeq \frac{T^{s/2}}{(2\pi)^{s/2} \left[\det \boldsymbol{C}\right]^{1/2}} \exp\left(-\frac{T}{2} \sum_{i,j=1}^{s} \delta n_i C_{ij}^{-1} \delta n_j \right). \qquad (23)$$

Here $\boldsymbol{C}$ is an $s \times s$ positive-definite symmetric matrix which depends on ρ_0 and on the geometry of the problem, but is independent of time. Equation (23) suffices for the evaluation of the variance of the joint probability distribution.[4,42] Each diagonal element of $\boldsymbol{C}$ describes the variance of the current into the corresponding patch

$$\overline{\delta n_i^2} = \frac{C_{ii}}{T}, \qquad (24)$$

where the overline denotes averaging with respect to the Gaussian distribution (23). The off-diagonal elements of $\boldsymbol{C}$ describe cross-correlations between the currents into different patches

$$\overline{\delta n_i \delta n_j} = \frac{C_{ij}}{T}. \qquad (25)$$

The optimal density field can again be presented via an electrostatic analogue which involves s characteristic potentials $\phi_i(\mathbf{x})$. Each of the potentials appears when the corresponding conducting patch Ω_i is held at unit voltage, the rest of the conducting patches are grounded, and the Neumann boundary condition is specified at the reflecting part of the boundary. The potentials ϕ_i-s can be found explicitly in simple cases.[4,35] The covariance matrix C is given in terms of the volume integrals involving the characteristic potentials

$$C_{ij} = \int d\mathbf{x}\, \sigma(\bar{\rho}) \nabla \phi_i \cdot \nabla \phi_j. \tag{26}$$

Remarkably, general properties of the cross-correlations turn out to be independent of the system's geometry, and are determined solely by the functions $D(\rho)$ and $\sigma(\rho)$.[4] Of course, there are no cross-correlations if the particles do not interact. What is the sign of cross-correlations for an interacting gas? As Ref. 4 showed, if

$$D(\bar{\rho})\sigma''(\bar{\rho}) < D'(\bar{\rho})\sigma'(\bar{\rho}), \tag{27}$$

for any value of $\bar{\rho} \in [0, \rho_0]$, then the currents into different patches $i \neq j$ are all anti-correlated, $\overline{\delta n_i \delta n_j} < 0$, regardless of the system's geometry. In particular, this is always true for the SSEP.

Interestingly, the same condition (27) guarantees the validity of the additivity principle (that is, stationarity of the optimal density profile in the long-time limit) for an arbitrary value of current,[43,44] and also determines the sign of the two-point *density* correlation function,[45,46] in single-current systems.

4. The Interior Problem

For many non-interacting particles the theory is based on the well-established single-particle results[7-11,47,48] As in the exterior problem, the long-time survival probability in this case decays exponentially in time, $-\ln \mathcal{P}(T, \rho_0) \simeq T s(\rho_0)$. The geometry dependence of $s(\rho_0)$ is, however, different[7-11,47,48]

$$s(\rho_0) = D_0 \mu_0^2 \rho_0 V, \tag{28}$$

where V is the domain's volume, and μ_0^2 is the principal eigenvalue of the eigenvalue problem $\nabla^2 \Psi + \mu^2 \Psi = 0$ inside the domain with the mixed boundary conditions $\Psi(\mathbf{x} \in \Omega_a, t) = \nabla \Psi(\mathbf{x} \in \Omega_r, t) \cdot \hat{n} = 0$.

What happens for interacting particles? As in Subsection 3.1, the optimal flux field, conditioned on the survival of all particles, vanishes identically. As a result, we can determine the optimal density profile for survival by minimizing the same action rate functional (9), but now the integration is carried over the space inside the domain. One distinct feature of the interior survival problem is conservation of the total number of particles, which enters the variational problem as a constraint,

$$\int d^3\mathbf{x}\,\rho(\mathbf{x}) = \rho_0 V \tag{29}$$

and calls for a Lagrange multiplier Λ. The transformation of variables (10) proves useful in the interior case as well. The resulting Euler–Lagrange equation for u has the form of a nonlinear Poisson equation,[2]

$$\nabla^2 u + \Lambda \frac{dF(u)}{du} = 0 \tag{30}$$

with the mixed boundary conditions[e],

$$u(\mathbf{x} \in \Omega_a) = 0, \quad \nabla u(\mathbf{x} \in \Omega_r) \cdot \hat{n} = 0. \tag{31}$$

For the RWs Eq. (10) yields $F(u) = u^2/2D_0$, and Eq. (30) becomes the Helmholtz equation

$$\nabla^2 u + \mu^2\, u = 0, \tag{32}$$

with $\mu^2 \equiv \Lambda/D_0$ playing the role of the eigenvalue. The minimum action is achieved for the fundamental mode, and the resulting expression for $\mathfrak{s}[u(\mathbf{x})] = s(\rho_0)$ reproduces the result quoted in Eq. (28).[2,5] For the SSEP, upon rescaling $U = \sqrt{2/D_0}\,u$ and $C = \Lambda/D_0$, Eq. (30) becomes the stationary sine-Gordon equation

$$\nabla^2 U + C \sin U = 0. \tag{33}$$

---

[e]The condition $u(\mathbf{x} \in \Omega_a) = 0$ is inherited from $\rho(\mathbf{x} \in \Omega_a) = 0$ due to the definition (10). The condition $\nabla u(\mathbf{x} \in \Omega_r) \cdot \hat{n} = 0$ results from a boundary term that appears when minimizing the action (9).

4.1. *Particle survival inside a fully absorbing domain*

Equation (33) can be solved exactly in some simple geometries. Among them are a 1D segment (where the problem is exactly solvable for any gas model) and a rectangle.[2] For a sphere of radius R one can solve Eq. (33) numerically, and also explore analytically the low- and high-density limits. In the dilute limit $\rho_0 \ll 1$ one reproduces the RWs result (28) which becomes

$$s_{\text{RWs}}(\rho_0) = \frac{4\pi^3}{3} R D_0 \rho_0, \qquad (34)$$

At the other extreme, $\rho_0 \to 1$, the stationary optimal density profile ρ stays very close to 1 across most of the domain, and drops to 0 in a narrow boundary layer of characteristic width $\delta = 1 - \rho_0$ along the domain boundary. As a result, the problem becomes effectively 1D in the direction normal to the domain boundary. The solution for this 1D problem can be found exactly, and the action rate, Eq. (9), mostly comes from the boundary layer. The final result, for a general domain shape, is

$$s(\rho_0) \simeq \frac{D_0 A^2}{V(1 - \rho_0)}, \qquad (35)$$

where A is the surface area of the boundary. For a sphere of radius R one obtains

$$s(\rho_0) \simeq \frac{12\pi D_0 R}{1 - \rho_0}. \qquad (36)$$

Figure 3 shows the numerically found $s(\rho_0)/D_0 R$, alongside with the asymptotics (34) and (36).[2]

4.2. *Narrow escape of interacting particles*

In the narrow escape problem, particles can escape only through a small escape hole Ω_a, of size $\epsilon \ll L$, see the right panel of Fig. 1. The mean escape time (MET) of the first particle in this setting determines the rates of important processes in molecular and cellular biology.[7,49–51]

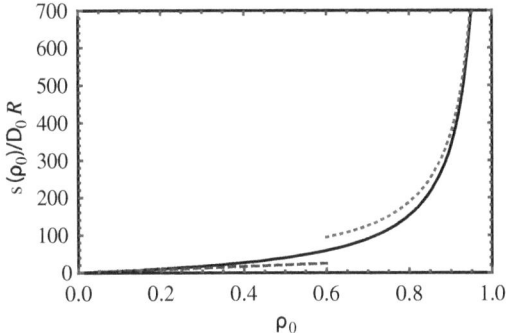

Fig. 3. Solid line: The rescaled action rate $s(\rho_0)/D_0 R$ for a sphere, versus ρ_0, obtained by numerically solving Eq. (33) and using Eq. (9). Also shown are the low-density asymptotic (34) (dashed line) and the high-density asymptotics (36) (dotted line).

For the non-interacting RWs, one can evaluate μ_0^2 in Eq. (28) perturbatively with respect to the small parameter ϵ/L. In the leading order μ_0^2 can be expressed through the electrical capacitance C_ϵ of the conducting patch Ω_a in an otherwise empty space: $\mu_0^2 \simeq 2\pi C_\epsilon/V$.[52] The capacitance C_ϵ scales as ϵ. When Ω_a is a disk of radius ϵ, one has $C_\epsilon = 2\epsilon/\pi$.[35] The resulting survival probability decay rate (28) is[5,47]

$$s(n_0, \epsilon) \simeq 2\pi C_\epsilon D_0 \rho_0. \tag{37}$$

For interacting particles one can exploit the small parameter ϵ/L in a similar way.[5] The leading-order contribution to the action rate (9) comes from only a vicinity of the escape hole. That is, to leading order in ϵ/L, the solution for a finite domain coincides with the one for a gas of particles occupying the infinite half-space on one side of an infinite reflecting plane with the hole Ω_a on it. This reduces the problem to the *unconstrained* minimization procedure of the exterior survival problem of Subsection 3.1. The solution is therefore given by Eq. (12) where $\phi(\mathbf{x})$ is the electrostatic potential of a conducting patch Ω_a kept at unit voltage on an otherwise insulating infinite plane. If the escape hole is a circle, $\phi(\mathbf{x})$ can be found explicitly.[35]

Then Eq. (9) yields the decay rate of the non-escape probability to order ϵ/L

$$s\left(\rho_0, \epsilon\right) \simeq \pi C_\epsilon f^2(\rho_0). \tag{38}$$

As in the exterior survival problem, the gas-specific interactions are encoded in the density dependence $f(\rho_0)$, whereas the geometry dependence C_ϵ is universal. To leading order it only depends on the shape of the escape hole and is independent of the domain shape. For the SSEP inside a domain with a small circular escape hole of radius ϵ one obtains[5]

$$s(\rho_0, \epsilon) \simeq 4D_0\epsilon \arcsin^2(\sqrt{\rho_0}). \tag{39}$$

The density dependence of Eq. (39) is the same as in Eq. (14), see Fig. 2. As argued in Ref. 5, for the SSEP with random initial condition, the exponential decay of $\mathcal{P}$ with time T holds as soon as T is much longer than the diffusion time across the escape hole.[5,9,47] For sufficiently low gas densities, $\rho_0\epsilon^3 \ll 1$, the MET of the first particle, $\langle T \rangle$ is also much longer than this diffusion time, and is thus given by $\langle T \rangle \simeq 1/s(\rho_0, \epsilon)$.[5,47]

5. Short-Time Statistics: Non-Stationary Fluctuations

At short times, $T \ll L^2/D\left(\rho_0\right)$, the particle absorption statistics, in both exterior and interior settings, strongly depend on the initial condition, whereas the optimal density profile explicitly depends on time.[1] Here we must return to the full time-dependent MFT formulation given by Eqs. (2) and (3). A universal simplification comes from the fact that, for very short times, the domain size is irrelevant. As a result, the process is effectively 1D in the direction normal to the absorbing part of the boundary,[1,5,47] and the absorption statistics can be expressed through that of a gas on the infinite half-line $x > 0$, with absorbing boundary conditions at $x = 0$.[29,31,33,34] The particle *survival* probability on the half-line is well studied.[53–57] For the

RWs with random initial conditions one obtains $-\ln \mathcal{P}_{1d}^{(\text{rand})} \simeq (2/\sqrt{\pi})\rho_0\sqrt{D_0 T}$. The corresponding result for the deterministic (or quenched) setting differs by a numerical factor.[1,54] Remarkably, for the SSEP one obtains the same stretched-exponential decay with time as for the RWs: $-\ln \mathcal{P}_{1d} \simeq s_{1d}(\rho_0)\sqrt{D_0 T}$, but the density dependence $s_{1d}(\rho_0)$ is now different for the different initial conditions. The low-density expansion of $s_{1d}(\rho_0)$ was recently calculated: $s_{1d}^{(\text{rand})}(\rho_0) = (2/\sqrt{\pi})[\rho_0 + (\sqrt{2}-1)\rho_0^2 + \cdots]$.[1,58] For larger ρ_0, $s_{1d}(\rho_0)$ can be computed numerically.[1] To evaluate the survival probability $\mathcal{P}(T,\rho_0)$, one should multiply the action $s_{1d}(\rho_0)$ by the surface area A of the absorbing part Ω_a[1,5,47]

$$-\ln \mathcal{P}(T,\rho_0) \simeq A s_{1d}(\rho_0)\sqrt{D(\rho_0)T}. \tag{40}$$

At sufficiently high densities, $\rho_0 L^3 \gg 1$, the short-time expression (40) suffices for the evaluation of the MET. Indeed, in this regime the MET is much shorter then the diffusion time across the domain, L^2/D_0, and we obtain $\langle T \rangle \simeq 2[A^2 D(\rho_0) s_{1d}^2(\rho_0)]^{-1}$.[5,47] For the narrow escape problem the relevant diffusion time scale is ϵ^2/D_0.[5,47] As an example, consider a circular absorbing patch of radius ϵ. In this case we have for the RWs $\langle T_{\text{RWs}} \rangle^{(\text{rand})} \simeq (2\pi D_0 \rho_0^2 \epsilon^4)^{-1}$.[5,47] For the SSEP the MET is shorter because of the effective particle repulsion: $\langle T \rangle^{(\text{rand})} \simeq \langle T_{\text{RWs}} \rangle^{(\text{rand})} \left[1 - 2(\sqrt{2}-1)\rho_0 + \cdots\right]$.

6. Summary

The fluctuating hydrodynamics and MFT provide a simple, general, and versatile framework for the study of kinetics of diffusion-controlled reactions in multiparticle systems where intrareactant interactions are important. We demonstrated the versatility of these approaches in several exterior and interior settings of particle survival, absorption and escape. More complicated settings and geometries can be also considered. The approach can be extended in different directions. For example, it can accommodate simple reactions among, and a finite lifetime of, the particles.

Acknowledgments

This research was supported by the United States-Israel Binational Science Foundation (BSF) (Grant No. 2012145) and by the Israel Science Foundation (Grant No. 807/16).

References

1. B. Meerson, A. Vilenkin, and P. L. Krapivsky, *Phys. Rev. E.* **90**, 022120 (2014).
2. T. Agranov, B. Meerson, and A. Vilenkin, *Phys. Rev. E.* **93**, 012136 (2016).
3. B. Meerson, *J. Stat. Mech.* P05004 (2015).
4. T. Agranov and B. Meerson, *Phys. Rev. E.* **95**, 062124 (2017).
5. T. Agranov and B. Meerson, *Phys. Rev. Lett.* **120**, 120601 (2018).
6. T. Agranov, P. L. Krapivsky, and B. Meerson, arXiv:1901.00153 (2019).
7. P. C. Bressloff and J. M. Newby, *Rev. Mod. Phys.* **85**, 135 (2013).
8. M. J. Ward and J. B. Keller, *SIAM J. Appl. Math.* **53**, 770 (1993).
9. I. V. Grigoriev, Y. A. Makhnovskii, A. M. Berezhkovskii, and V. Y. Zitserman, *J. Chem. Phys.* **116**, 9574 (2002).
10. O. Bénichou and R. Voituriez, *Phys. Rep.* **539**, 225 (2014).
11. D. Holcman and Z. Schuss, *Stochastic Narrow Escape in Molecular and Cellular Biology*. Springer, New York (2015).
12. T. Chou and M. R. D'Orsogna, in: *First-Passage Phenomena and Their Applications*, R. Metzler, G. Oshanin, and S. Redner (eds.), World Scientific, Singapore (2013).
13. S. A. Rice, *Comprehensive Chemical Kinetics*. Elsevier, Amsterdam (1985).
14. C. Mejia-Monasterio, G. Oshanin, and G. Schehr, *J. Stat. Mech.* P06022 (2011).
15. B. Meerson and S. Redner, *Phys. Rev. Lett.* **114**, 198101 (2015).
16. H. C. Berg and E. M. Purcell, *Biophys. J.* **20**, 193 (1977).
17. H. C. Berg, *Random Walks in Biology*. Princeton University Press, Princeton, USA (1993).
18. L. D. Landau and E. M. Lifshitz, *Statistical Physics*. Pergamon Press, London (1958).
19. H. Spohn, *Large-Scale Dynamics of Interacting Particles*. Springer-Verlag, New York (1991).
20. L. Bertini, A. De Sole, D. Gabrielli, G. Jona Lasinio, and C. Landim. *Rev. Mod. Phys.* **87**, 593 (2015).
21. C. Kipnis and C. Landim, *Scaling Limits of Interacting Particle Systems*. Springer, New York (1999).
22. T. M. Liggett, *Stochastic Interacting Systems: Contact, Voter, and Exclusion Processes*. Springer, New York (1999).

23. P. L. Krapivsky, S. Redner, and E. Ben-Naim, *A Kinetic View of Statistical Physics.* Cambridge University Press, Cambridge (2010).

24. P. C. Martin, E. D. Siggia, and H. A. Rose, *Phys. Rev. A.* **8**, 423 (1973).

25. T. Bodineau and B. Derrida, *Phys. Rev. Lett.* **92**, 180601 (2004).

26. G. Zumofen, J. Klafter, and A. Blumen, *J. Chem. Phys.* **79**, 5131 (1983).

27. M. Tachiya, *Radiat. Phys. Chem.* **21**, 167 (1983).

28. S. Redner and K. Kang, *J. Phys. A.* **17**, L451 (1984).

29. A. Blumen, G. Zumofen, and J. Klafter, *Phys. Rev. B.* **30**, 5379(R) (1984).

30. A. Blumen, J. Klafter, and G. Zumofen, in: *Optical Spectroscopy of Glasses*, I. Zchokke (ed.), Reidel, Dordrecht (1986), p. 199.

31. S. F. Burlatsky and A. A. Ovchinnikov, *Sov. Phys. JETP.* **65**, 908 (1987).

32. G. Oshanin, O. Benichou, M. Coppey, and M. Moreau, *Phys. Rev. E.* **66**, 060101(R) (2002).

33. R. A. Blythe and A. J. Bray, *Phys. Rev. E.* **67**, 041101 (2003).

34. A. J. Bray, S. N. Majumdar, and G. Schehr, *Adv. Phys.* **62**, 225 (2013).

35. J. D. Jackson, *Classical Electrodynamics.* Wiley, New York (1999).

36. V. Kuzovkov and E. Kotomin, *Phys. Rev. Lett.* **72**, 2105 (1994).

37. S. F. Burlatsky, M. Moreau, G. Oshanin, and A. Blumen, *Phys. Rev. Lett.* **75**, 585 (1995).

38. D. P. Bhatia, M. A. Prasad, and D. Arora, *Phys. Rev. Lett.* **75**, 586 (1995).

39. K. Seki and M. Tachiya, *Phys. Rev. E.* **80**, 041120 (2009).

40. K. Seki, M. Wojcik, and M. Tachiya, *J. Chem. Phys.* **134**, 094506 (2011).

41. E. Akkermans, T. Bodineau, B. Derrida and O. Shpielberg, *EPL* **103**, 20001 (2013).

42. P. L. Krapivsky and B. Meerson, *Phys. Rev. E.* **86**, 031106 (2012).

43. L. Bertini, A. De Sole, D. Gabrielli, Jona-Lasinio and C. Landim, *J. Stat. Phys.* **123**, 237 (2006).

44. O. Shpielberg and E. Akkermans, *Phys. Rev. Lett.* **116**, 240603 (2016).

45. L. Bertini, A. De Sole, D. Gabrielli, G. Jona-Lasinio, and C. Landim, *J. Stat. Phys.* **135**, 857 (2009).

46. T. Sadhu and B. Derrida, *J. Stat. Mech.* (2016) 113202.

47. S. Ro and Y. W. Kim, *Phys. Rev. E.* **96**, 012143 (2017).

48. K. Basnayake, C. Guerrier, Z. Schuss, and D. Holcman, arXiv:1711.01330.

49. D. Coombs, R. Straube, and M. Ward, SIAM. *J. Appl. Math.* **70**, 302 (2009).

50. S. A. Gorski, M. Dundr, and T. Misteli, *Curr. Opin. Cell Biol.* **18**, 284 (2006).

51. D. Holcman, Z. Schuss, and E. Korkotian, *Bio. J.* **87**, 81 (2004).

52. J. W. S. Baron Rayleigh, *The Theory of Sound.* Dover, New York, (2nd edn.), Vol. 2. (1945).

53. M. Tachiya, *Radiat. Phys. Chem.* **21**, 167 (1983).

54. G. Zumofen, J. Klafter, and A. Blumen, *J. Chem. Phys.* **79**, 5131 (1983).

55. S. Redner and K. Kang, *J. Phys. A Math. Gen.* **17**, L451 (1984).

56. J. Franke and S. N. Majumdar, *J. Stat. Mech.* P05024 (2012).

57. S. Redner and B. Meerson, *J. Stat. Mech.* P06019 (2014).
58. J. E. Santos and G. M. Schütz, *Phys. Rev. E.* **64**, 036107 (2001).
59. V. Elgart and A. Kamenev, *Phys. Rev. E.* **70**, 041106 (2004).
60. T. Bodineau and M. Lagouge, *J. Stat. Phys.* **139**, 201 (2010).
61. B. Meerson and P. V. Sasorov, *Phys. Rev. E.* **83**, 011129 (2011).
62. P. I. Hurtado, A. Lasanta, and A. Prados, *Phys. Rev. E.* **88**, 022110 (2013).
63. B. Meerson, *J. Stat. Mech.* P05004 (2015).

Chapter 10

Polymer Reaction Kinetics Beyond Markov Hypothesis

O. Bénichou[*], T. Guérin[†], and R. Voituriez[*,‡,§]

[*]*Laboratoire de Physique Théorique de la Matière Condensée,*
CNRS UMR 7600, Case Courrier 121, Université Paris 6,
4 Place Jussieu, 75255 Paris Cedex, France
[†]*Laboratoire Ondes et Matière d'Aquitaine, University of Bordeaux,*
Unité Mixte de Recherche 5798,
CNRS, F-33400 Talence, France
[‡]*Laboratoire Jean Perrin, CNRS UMR 7600, Case Courrier 121,*
Université Paris 6, 4 Place Jussieu,
75255 Paris Cedex, France
[§]*voiturie@lptmc.jussieu.fr*

We review recent theoretical works that enable the accurate computation of the mean first passage time (MFPT) to a target site for non-Markovian random walkers in confinement on the example of a tagged monomer of a polymer chain looking for a reactive target. We show that the MFPT of this non-Markovian process can be calculated accurately by computing the distribution of the positions of all the monomers in the chain at the instant of reaction. Such a theory goes beyond Markovian approximations and can be used to derive asymptotic relations that generalize the scaling dependence with the volume and the initial distance to the target derived for Markovian walks. We finally briefly discuss in conclusion results obtained recently on more general non-Markovian processes, highlighting the importance of non-Markovian effects.

1. Introduction

How long does it take for a diffusing particle to reach a reactive target site? This time is known in the random walk literature as a first passage time (FPT), and it has generated a considerable amount of work for many years in the context of diffusion limited reactions and beyond.[1-3] The importance of FPT relies on the fact that many physical properties, such as fluorescence quenching,[4] neuron dynamics,[5] or resonant activation[6] to name a few, are controlled by first passage events. More recently, FPTs have regained interest in the context of random search strategies.[7-11]

The basic formulation of the problem is as follows. Consider a random walker of position $\mathbf{r}(t)$ at time t in a confined domain, and ask for the FPT statistics of $\mathbf{r}(t)$ to one (or several) targets. A target can represent a critical level to reach before another event is triggered, or a reactive site. First passage properties depend on (i) the properties of the stochastic process, and (ii) the geometrical parameters of the problem, such as the initial distance to the target, its size, or the size and the shape of the confining domain.

The determination of FPT statistics for random walks in confined geometries is not only a theoretical challenge. It is a general question that arises as soon as molecules diffuse in confined media such as, for example, biomolecule-diffusing cells and undergoing a series of reactions at specific sites. An estimate of the time needed to go from one point to another is then an essential step in the understanding of the kinetics of the whole process. Particular attention has been paid to the mean FPT, denoted MFPT $\langle \mathbf{T} \rangle$. The MFPT is a complex quantity, which results from averaging trajectories characterized by a broad range of time scales: some trajectories reach the target almost directly and contribute to very short time scales, whereas other trajectories first hit the domain boundary before reaching the target, and can be shown to contribute to much longer time scales.[12,13] We will focus in this chapter on the MFPT.

So far, explicit determinations of MFPTs in confined geometries have been most of the time limited to Markov processes.[14-24] An important aspect of the problem is however the potential existence of *memory* effects for some random walks, for which the future trajectories are not determined uniquely by the position of the walker at initial time, but depends also on the past trajectory. Such non-Markovian dynamics are found in many examples of particles moving stochastically in complex fluids,[25-27] or molecules bound to fluctuating extended structures, as in the model of a tagged monomer of a polymer chain,[28] which seems to be observed for example in the motion of chromatin fibers.[29-31] The key question then arises of how the memory properties of the random walk influence the MFPT, and of how to take them into account quantitatively.

In this chapter, we present recent results in the analysis of FPT statistics of non-Markovian processes, on the example of a reactive monomer of a Rouse polymer chain that looks for a target in a confined domain. We synthesize recent works presented more extensively in Ref. 32. Section 2 presents the Rouse model and its basic properties, and introduces a first Markovian approximation of the problem. In Section 3, we introduce the full non-Markovian analysis and show that the calculation of the MFPT requires the determination of the distribution of the conformation of the full chain at the instant of encounter with the target. This analysis is first presented in 1 space dimension, and extended to 3D spaces in Section 4. We finally briefly review in the conclusion results obtained recently on more general non-Markovian processes, highlighting the importance of non-Markovian effects.

Let us finally stress that we will consider in this chapter only MFPTs of random walks in confinement. For a recent review of first-passage processes in the absence of confinement, and in particular persistence properties of non-equilibrium processes, the reader is referred to Ref. 33.

2. MFPT for Non-Markovian Processes and the Markovian Approximation

Most of available theoretical approaches to determine mean first passage times (MFPTs) of random walks make use of Markovian hypotheses, which are only approximations for processes with memory. For example, it was shown in Ref. 16 that the MFPT of Markovian random walks can be expressed as time integrals of the propagator of the process, which is ill defined for generic processes with memory. Determining the MFPT of non-Markovian processes requires specific methods that we present here. We consider a stochastic process $\mathbf{X}(t)$ in continuous time t, characterized by stationary increments. The non-Markovian character of a process often arises from the interaction of the variable of interest, $\mathbf{X}(t)$, with other degrees of freedom whose dynamics are characterized by time scales comparable to that of motion.

A typical example of non-Markovian walker is given by a tagged monomer of a polymer chain,[34] or of a colloidal particle in a complex fluid.[25–27,35] Here we consider the overdamped dynamics of N particles of positions $\mathbf{x}_1, \ldots, \mathbf{x}_N$ in dimension d, which interact through harmonic interactions. The corresponding probability density function (pdf) then satisfies the following Fokker–Planck equation:

$$\frac{\partial P}{\partial t} = \sum_{i,j=1}^{N} \nabla_i A_{ij} \mathbf{x}_j P + D \sum_{i=1}^{N} \nabla_i^2 P, \tag{1}$$

where the connectivity matrix A is symmetric, and D is the individual diffusion coefficient of each particle. The case of a flexible (Rouse) chain,[36] where a polymer is simply described by a chain of beads with friction coefficient ζ linked by springs of stiffness k, is obtained when A takes the tridiagonal form:

$$A = \frac{k}{\zeta} \begin{pmatrix} 1 & -1 & 0 & .. & .. & .. \\ -1 & 2 & -1 & 0 & .. & .. \\ 0 & -1 & 2 & -1 & .. & .. \\ .. & .. & .. & .. & .. & .. \\ .. & .. & 0 & -1 & 2 & -1 \\ .. & .. & .. & 0 & -1 & 1 \end{pmatrix}. \tag{2}$$

We denote by $\mathbf{X}(t)$ the position of one monomer, called below reactive monomer, located at position q in the chain, $\mathbf{X}(t) = \mathbf{x}_q(t)$. We study the MFPT of $\mathbf{X}(t)$ to a fixed target in a confining domain of

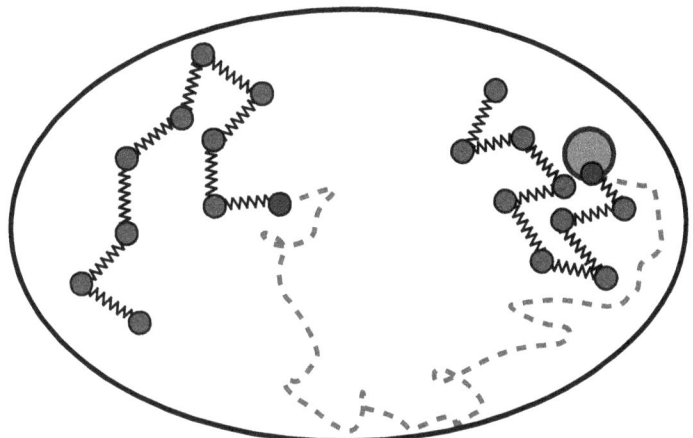

Fig. 1. Schematic of the MFPT problem in the non-Markovian case: A monomer of a Rouse chain is searching for a target in a confined domain.

volume V (see Fig. 1). We assume that V is much larger than the gyration radius of the polymer, and that both the initial position $\mathbf{X}(t = 0)$ and the target position are far from the confining boundaries. We focus on the simple case of the Rouse model. However, depending on the choice of matrix A, models of the type Eq. (1) can be used to describe more complicated structures, such as Gaussian semiflexible chains[37] (in which successive bonds tend to be aligned), chains with dihedral interactions,[38] the dynamics of interfaces,[39] etc, and the theory described here does not depend on the precise shape of the matrix A. In the following, we describe the essential steps to obtain a precise estimate of the MFPT in this non-Markovian problem.

The Eq. (1) is generally simplified by using normal mode analysis. We define the orthogonal matrix Q such that $A = Q\mathrm{diag}(\lambda_1, \ldots, \lambda_N)Q^{-1}$, with $\mathrm{diag}(\lambda_1, \ldots, \lambda_N)$ a diagonal matrix of eigenvalues λ_i, with $\lambda_1 = 0 < \lambda_2 < \cdots < \lambda_N$. The modes $\mathbf{a}_i$ are defined by

$$\mathbf{a}_i(t) = \sum_{j=1}^{N} Q_{ij}\mathbf{x}_j(t), \qquad \mathbf{x}_i(t) = \sum_{j=1}^{N} Q_{ji}\mathbf{a}_j(t). \qquad (3)$$

The evolution of the probability density $P(\{\mathbf{a}\}, t)$ to observe the system with a configuration $\{\mathbf{a}\} = \{\mathbf{a}_1, \ldots, \mathbf{a}_N\}$ at time t in the absence of absorption now satisfies the simplified Fokker–Planck equation

$$\frac{\partial P}{\partial t} = \sum_{i=1}^{N} \frac{\partial}{\partial \mathbf{a}_i} \left(\lambda_i \, \mathbf{a}_i P + D \frac{\partial}{\partial \mathbf{a}_i} P \right), \tag{4}$$

where each mode evolves independently of the others. In terms of the modes, the variable $\mathbf{X}$ is expressed as the linear combination

$$\mathbf{X}(t) = \sum_{i=1}^{N} b_i \mathbf{a}_i(t), \qquad b_i \equiv Q_{qi}. \tag{5}$$

We study the problem of finding the MFPT of $\mathbf{X}(t)$ to a spherical target of radius a located around the position $\mathbf{X} = \mathbf{0}$. For simplicity, we restrict our analysis to initial configurations such that $\mathbf{X} = \mathbf{X}_0$ and such that the rest of the polymer is at equilibrium

$$P_{\text{ini}}(\{\mathbf{a}\})$$
$$= |b_1|^d \left[\frac{\lambda_2 \ldots \lambda_N}{(2\pi D)^{N-1}} \right]^{d/2} \exp\left(- \sum_{i=2}^{N} \frac{\lambda_i \mathbf{a}_i^2}{2D} \right) \delta\left(\sum_{i=1}^{N} b_i \mathbf{a}_i - \mathbf{X}_0 \right). \tag{6}$$

The above distribution is Gaussian, with mean vector $\langle \mathbf{a}_i \rangle_{\text{ini}} = \delta_{i,1} \mathbf{X}_0 / b_1$, and covariance matrix $\text{Cov}(a_{i\alpha}, a_{j\beta}) = \sigma_{ij}^{\text{eq},*} \delta_{\alpha\beta}$ with α, β the spatial coordinates $\{x, y, z\}$, and $\sigma_{ij}^{\text{eq},*}$ is defined by

$$\sigma_{ij}^{\text{eq},*} = D \begin{cases} \delta_{ij}/\lambda_i & \text{if } i, j \geq 2, \\ -b_j/(b_1 \lambda_j) & \text{if } j \geq 2, i = 1, \\ \sum_{q=2}^{N} b_q^2/(\lambda_q b_1^2) & \text{if } i = j = 1. \end{cases} \tag{7}$$

Note that $\sum_j \sigma_{ij}^{\text{eq},*} b_j = 0$, which imposes $\mathbf{X} = \mathbf{X}_0$ at $t = 0$, as seen by the presence of the delta function in Eq. (6). Interestingly, $\sigma_{ij}^{\text{eq},*}$ does not depend on the value of $\mathbf{X}_0$. Now, let $\delta_{\alpha\beta}\gamma_{ij}(t)$ be the covariance of $a_{i\alpha}$ and $a_{j\beta}$ at a time t after the initial time, in unbounded space and in the absence of target. It is known[40] that the Fokker–Planck equation (4) admits Gaussian solutions and that the evolution of the covariance matrix γ_{ij} satisfies the following equation

$$\partial_t\,\gamma_{ij} = -(\lambda_i + \lambda_j)\gamma_{ij} + 2D\delta_{ij}. \tag{8}$$

The value of $\gamma_{ij}(t)$ is found by solving Eq. (8) with the initial condition $\gamma_{ij}(0) = \sigma_{ij}^{\text{eq},*}$ given by Eq. (7)

$$\gamma_{ij}(t) = \delta_{ij}D\left(1 - e^{-2\lambda_i t}\right)/\lambda_i + e^{-\lambda_i t}e^{-\lambda_j t}\sigma_{ij}^{\text{eq},*}. \tag{9}$$

We define the mean square displacement function $\psi(t)$ as the variance of X_α at t, given that the initial distribution of the monomers was $P_{\text{ini}}(\{\mathbf{a}\})$ given by (6). Because $\mathbf{X} = \sum b_i a_i$, we have $\psi(t) = \sum_{i,j=1}^{N}b_i b_j \gamma_{ij}(t)$, and therefore

$$\psi(t) \equiv \frac{1}{d}\langle[\mathbf{X}(t) - \mathbf{X}_0]^2\rangle = 2Db_1^2 t + 2D\sum_{j\geq 2}\frac{b_j^2}{\lambda_j}(1 - e^{-\lambda_j t}), \tag{10}$$

where d is the space dimension. Then, we can define an "effective propagator" (or "reduced Green function") in free space

$$P(\mathbf{X}, t|\{\mathbf{X}_0, \text{eq}\}, 0) = \frac{1}{[2\pi\psi(t)]^{d/2}}\exp\left\{-\frac{(\mathbf{X} - \mathbf{X}_0)^2}{2\psi(t)}\right\}. \tag{11}$$

Note that this represents the probability of finding the reactive monomer at position $\mathbf{X}$ at time t, given that it was observed at position $\mathbf{X}_0$ at the initial time $t = 0$, *and that the rest of the chain was at equilibrium at this initial time.* We stress that the effective propagator (11) does not satisfy the Chapman–Kolmogorov equation

$$P(\mathbf{X}, t|\{\mathbf{X}_0, \text{eq}\}, 0) \neq \int d\mathbf{X}' P(\mathbf{X}, t|\{\mathbf{X}', \text{eq}\}, t')P(\mathbf{X}', t'|\{\mathbf{X}_0, \text{eq}\}, 0), \tag{12}$$

reflecting the fact that the process $\mathbf{X}(t)$ is not Markovian. It is however tempting to use (11) as an effective propagator in the expression
of the MFPT derived for Markovian processes[16,24]

$$\lim_{V \to \infty} \frac{\langle \mathbf{T} \rangle_{\text{Markovian}}}{V} = \int_0^\infty \frac{dt}{[2\pi\psi(t)]^{d/2}}$$

$$\times \left\{ \exp\left(-\frac{a^2}{2\psi(t)}\right) - \exp\left(-\frac{|\mathbf{X}_0|^2}{2\psi(t)}\right) \right\}. \quad (13)$$

We call this assumption the Markovian assumption, which can be
shown to amount to the classical Wilemski–Fixman approximation.[41,42] This formula suggests that as in the Markovian case, the
MFPT is proportional to the volume, and that its dependance with
the initial distance is contained in the infinite space effective propagator, which depends on $\psi(t)$. More precisely, in the case of the Rouse
chain, it can be shown by calculating explicitly the eigenvalues λ_i
and coefficients b_i that

$$\psi(t) \simeq \begin{cases} 2Dt & \text{if } t \ll \tau_0, \\ \kappa t^{1/2} & \text{if } \tau_0 \ll t \ll N^2\tau_0 , \\ 2D_{\text{cm}}t & \text{if } t \gg \tau_0 N^2, \end{cases} \quad (14)$$

where $D_{\text{cm}} = Db_1^2 = D/N$ is the diffusion coefficient of the center-
of-mass, $\tau_0 = \zeta/k$ is the relaxation time of a single bond, and
$\kappa = 4l_0^2/\sqrt{\pi\tau_0}$ for an end-monomer, with $l_0 = \sqrt{k_BT/k}$ the equilibrium bond length. Equation (14) states that, at very short times
scales, a monomer behaves as if it were disconnected from the rest
of the chain and freely diffused. At intermediate time scales, all the
internal time scales contribute to the motion and it is known that the
monomer motion becomes subdiffusive. Non-Markovian subdiffusion
is the hallmark of polymer dynamics, with exponents that depend on
the model considered.[34] At larger time scales, the tagged monomer
recovers a diffusive motion, due to the motion of the polymer center-
of-mass.

Using the properties (14) of the function $\psi(t)$, we can infer those
the the MFPT in the Markovian approximation by using the formula

(13) if $d = 1$ (with a target size $a = 0$)[43]

$$\langle \mathbf{T} \rangle_{\text{Markovian}} \simeq V \times \begin{cases} 2X_0/D & \text{if } X_0 \ll l_0, \\ 0.316X_0^2 N^{3/2}/(Dl_0) & \text{if } l_0 \ll X_0 \ll l_0 N^{1/2}, \\ 2X_0/D_{\text{cm}} & \text{if } X_0 \gg l_0 N^{1/2}. \end{cases}$$

$$(15)$$

Interestingly, in the intermediate scale regime, the MFPT is thus found to scale as X_0^2, in contradiction with the results $T \simeq X_0^3$ obtained from a naive application of the Markovian theory of[16,24] with a walk dimension $d_w = 4$ and a space dimension $d = 1$. In fact, we will see that in this regime non-Markovian corrections lead to the scaling $T \simeq X_0^3$ (see Eq. (27)).

In 3D, the Markovian approximation of the MFPT reads, in the limit of large initial distance between the target and the polymer ($|X_0| \to \infty$)

$$\langle \mathbf{T} \rangle_{\text{Markovian}} \simeq \begin{cases} V/(4\pi Da) & \text{if } a \ll l_0/N^{1/2}, \\ V/(4\pi D_{\text{cm}} a_{\text{eff}}), & \text{if } l_0/N^{1/2} \ll a \ll l_0 N^{1/2}, \\ a_{\text{eff}} \simeq 0.87 l_0 N^{1/2} \\ V/(4\pi D_{\text{cm}} a) & \text{if } a \gg l_0 N^{1/2}, \end{cases}$$

$$(16)$$

where a_{eff} is an effective reactive radius. Comparing with the expression for a Markovian diffusive walker,[16,24] we observe that, at very small target sizes, the MFPT is the same as that of an isolated reactive monomer and is essentially controlled by the microscopic diffusion coefficient of the individual monomer. The most interesting regime is the intermediate regime, where the MFPT is that of an effective particle of diffusion coefficient D_{cm}, that reacts with a target whose size a_{eff} is larger than the "true" target size a. This size a does not enter in the expression of the MFPT, reflecting the compact feature of the walk at length scales between l_0 and $l_0 N^{1/2}$. Finally, for target sizes much larger than the gyration radius, the polymer behaves as a single particle of diffusion coefficient $D_{\text{cm}} = D/N$.

3. The Renewal Equation: Non-Markovian Case

We now adapt the Renewal equation method that was used in the case of Markovian processes to this non-Markovian problem, following the works of the authors of Refs. 43 and 44. We focus here on the 1D problem for simplicity; generalization to 3D will be described in Section 4. The target is assumed to be located at position $X = 0$. While the dynamics of the position of the reactive monomer $X(t)$ is non-Markovian, we can use the fact that the problem with N degrees of freedom is Markovian to write the following Renewal equation, valid for any configuration $\{a\}$ such that $X = \sum_i b_i a_i = 0$

$$P(\{a\}, t | \{\text{ini}\}, 0) = \int_0^t dt' \int d\{a'\} \, f(\{a'\}, t') P(\{a\}, t - t' | \{a'\}, 0).$$
(17)

Here, $f(\{a'\}, t')$ is the probability density to reach the target for the first time at time t', with a configuration $\{a'\} = (a'_1, \ldots, a'_N)$. In turn, $P(\{a'\}, t | \{a\}, 0)$ is the propagator of the chain in the presence of confinement, but in the absence of absorption at $X = 0$, and $P(\{a'\}, t | \{\text{ini}\}, 0)$ is the probability to observe the configuration $\{a'\}$ at t, starting from the initial distribution $P_{\text{ini}}(\{a\})$ (Eq. (6)). We introduce the splitting probability distribution $\pi(\{a\})$ that represents the probability density of observing a configuration $|a\rangle$ when the reaction takes place

$$\pi(\{a\}) \equiv \int_0^\infty dt \, f(\{a\}, t).$$
(18)

Taking the Laplace transform of the renewal Eq. (17) and developing for small values of the Laplace variable, we obtain that the distribution π is normalized, and that

$$\langle \mathbf{T} \rangle P_{\text{eq}}(\{a\}) = \int_0^\infty dt [P(\{a\}, t | \pi, 0) - P(\{a\}, t | \{\text{ini}\}, 0)],$$
(19)

which is valid for all the conformations $\{a\}$ such that $\sum_i b_i a_i = 0$. Here, we have introduced $P_{\text{eq}}(\{a\})$, that represents the probability of observing a given configuration in the equilibrium state in confinement. If the confining volume is sufficiently large, the stationary distribution is equal to

$$P_{eq}(\{a\}) \simeq \frac{|b_1|}{V} \left[\frac{\lambda_2 \dots \lambda_N}{(2\pi D)^{N-1}} \right]^{1/2} \exp\left[-\sum_{i=2}^{N} \frac{\lambda_i a_i^2}{2D} \right], \qquad (20)$$

which means that the center-of-mass is uniformly distributed in the volume V, while all the other internal degrees of freedom of the chain are at equilibrium and do not feel the confining volume.

In (19), the quantity $P(\{a\}, t|\pi, 0)$ is the probability of a configuration $\{a\}$ at t given that the configuration at $t = 0$ is taken from the splitting probability π. It is given by the convolution equation

$$P(\{a\}, t|\pi, 0) = \int d\{a'\}\pi(\{a'\})P(\{a\}, t \mid \{a'\}, 0). \qquad (21)$$

The Eqs. (19,21) together with the normalization of π form an integral equation that completely defines the splitting probability π and the mean first reaction time T. Its use is however limited, since it requires to solve for an unknown function of N variables, with N potentially large.

To go further, two additional steps can be followed:

(i) *Large volume limit*: In this limit, all the terms appearing in the right-hand side of Eq. (19) can be replaced by their expression in unbounded space, as previously done in the case of Markovian walkers, see Refs. 16 and 24.

(ii) *Gaussian approximation of the splitting distribution*: We assume that the splitting probability distribution $\pi(\{a\})$ can be accurately described by a multivariate Gaussian distribution, with mean vector m_i^π and covariance matrix σ_{ij}^π that are determined by solving a set of self-consistent equations. This Gaussian approximation is well supported by numerical simulations (Fig. 2).

With these assumptions, the evolution of the average value $\mu_i^\pi(t)$ of the mode a_i at a time t after the first passage to the target satisfies

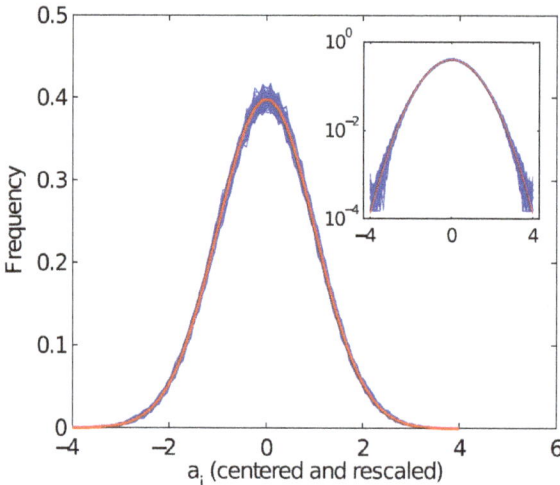

Fig. 2. Histograms of mode positions a_i, after they have been centered and rescaled to have a zero mean and variance 1, obtained from numerical (Langevin) simulations of the evolution of a Rouse chain of $N = 20$ monomers, whose first monomer is seeking for a target located at $X = 0$. The figure is a superposition of 200 histograms $\pi(a_i; X_0)$, for all initial distances X_0 appearing in Fig. 3, and all values of i ($1 \leq i \leq N$). Red curve: normalized Gaussian distribution. Inset: same plot, in semi-log scale.

$\partial_t \mu_i^\pi = -\lambda_i \mu_i^\pi$, leading to

$$\mu_i^\pi(t) = m_i^\pi \, e^{-\lambda_i t}. \tag{22}$$

We next define $X_\pi(t)$ as the average of X at a time t after the first passage to the target. Because $X = \sum_j b_j a_j$, it is simply given by

$$X_\pi(t) \equiv \sum_{i=1}^{N} b_i \mu_i^\pi(t) = -\sum_{i=2}^{N} b_i m_i^\pi (1 - e^{-\lambda_i t}), \tag{23}$$

where we have used the fact that $X_\pi(0) = 0$. For the sake of simplicity, we assume that the covariance matrix σ_{ij}^π can be approximated by the covariance of equilibrium configurations constrained by $X = 0$: $\sigma_{ij}^\pi \simeq \sigma_{ij}^{\text{eq},*}$ given by Eq. (7). We call this additional approximation the "stationary covariance approximation". With these hypotheses, the distribution $P(\{a\}, t | \pi, 0)$ of configurations at a time t after the first passage is also a Gaussian, with covariance matrix $\gamma_{ij}(t)$ (Eq. (9)), and mean vector $\mu_i^\pi(t)$.

The value of the modes at the instant of first passage, m_i^π, will be deduced from a set of self-consistent equations. The derivation of these equations requires the calculation of the following general Gaussian integrals

$$
\int d\{a\} a_i \, \delta \left(\sum_{j=1}^N b_j a_j \right) \exp \left\{ -\frac{1}{2} \sum_{j,k=1}^N (a_j - \mu_j)(\gamma^{-1})_{jk}(a_k - \mu_k) \right\}
$$
$$
= \left[\mu_i - \frac{\langle e_i | \gamma | b \rangle}{\langle b | \gamma | b \rangle} \langle b | \mu \rangle \right] \left[\frac{\det(2\pi\gamma)}{2\pi \langle b | \gamma | b \rangle} \right]^{1/2} \exp \left[-\frac{\langle b | \mu \rangle^2}{2 \langle b | \gamma | b \rangle} \right], \quad (24)
$$

where we adopted the notation $\langle v | M | u \rangle = \sum_{i,j} v_i M_{ij} u_j$ for any symmetric matrix M and vectors $\{u\}, \{v\}$, and $|e_i\rangle$ is the vector whose elements are all zero, except for the ith which takes the value 1. Using the formula (24), the result of the multiplication of the integral equation (19) by $a_i \delta(\sum_j b_j a_j)$ gives the following set of equations, for any i between 2 and N

$$
0 = \int_0^\infty \frac{dt}{\psi(t)^{1/2}} \left\{ \exp \left(-\frac{[X_\pi(t)]^2}{2\psi(t)} \right) \left[m_i^\pi e^{-\lambda_i t} - \frac{b_i(1 - e^{-\lambda_i t}) X_\pi(t)}{\lambda_i \, \psi(t)} \right] \right.
$$
$$
\left. + \exp \left(-\frac{X_0^2}{2\psi(t)} \right) \frac{b_i(1 - e^{-\lambda_i t}) X_0}{\lambda_i \, \psi(t)} \right\}. \quad (25)
$$

The Eqs. (25), together with the expression of $X_\pi(t)$ given by (23), form a set of N self-consistent nonlinear equations that completely define the moments m_i^π. An expression of the MFPT is obtained from (19), after multiplication by $\delta(\sum_i b_i a_i)$ and integration over all configurations

$$
\frac{\langle \mathbf{T} \rangle}{V} = \int_0^\infty \frac{dt}{[2\pi\psi(t)]^{1/2}} \left\{ \exp \left(-\frac{[X_\pi(t)]^2}{2\psi(t)} \right) - \exp \left(-\frac{X_0^2}{2\psi(t)} \right) \right\}.
$$
$$
(26)
$$

This expression finally gives the MFPT, and fully takes into account non-Markovian effects.

Some remarks can be done at this stage:

(i) *Structure of the equations and scaling with the volume.* Comparing Eq. (26) with the expression of the MFPT obtained in the Markovian approximation (13), as well as with those obtained for Markovian walkers,[16,24] we see that the MFPT is still proportional to the volume. An key ingredient of this approach is the quantity $X_\pi(t)$, which represents the fictive averaged trajectory followed by the walker in the future of the first passage event. This directly depends on the averaged configuration of the chain at the first passage event, described by the m_i^π. Note that the Markovian approximation is recovered by setting $m_i^\pi = 0$.

(ii) *Validity of the non-Markovian theory.* In Fig. 3, we compare the predictions of the MFPT for different theories to the results of simulations for the Rouse chain. If the Markovian approximation

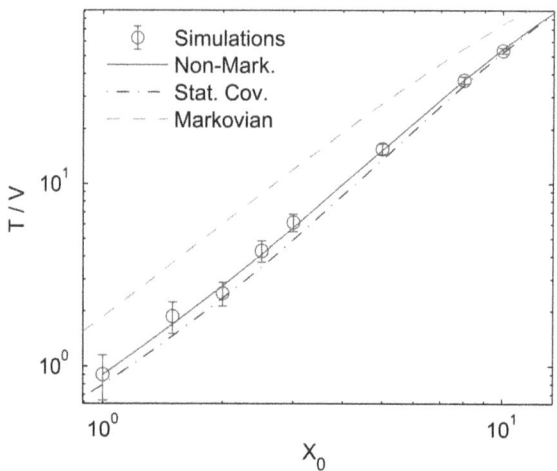

Fig. 3. MFPT of the first monomer of a Rouse chain to reach a target located at $X = 0$, as a function of initial distance X_0 for $N = 20$ monomers. Symbols: mid-circles: stochastic simulations; dashed upper line: Markovian approximation (13); lower dashed line: non-Markovian prediction obtained through stationary covariance approximation by numerically solving (25); middle solid line: non-Markovian theory without doing the stationary covariance approximation, described in Ref. 43. The polymer evolves inside a confining volume of radius $R = 40.25l_0$. Lengths and times are in units of l_0 and τ_0, respectively.

clearly overestimates the MFPT, the non-Markovian theory is in quantitative agreement with the simulations. Note that the "stationary covariance approximation" is accurate, as shown in Fig. 3.

(iii) *Meaning of the non-Markovian theory.* It is clear in Fig. 3 that the non-Markovian theory predicts MFPTs that are much lower than in the Markovian approximation. This can be understood by considering the average position of the monomers at the instant of the reaction, $\langle x_i \rangle_\pi = \sum_{j=1}^N Q_{ji} m_j^\pi$ (see (3)), represented in Fig. 4. It is seen on this plot that the positions of the non-reactive monomers of the chain are shifted *on average* at the instant of first passage. Therefore, the non-Markovian theory takes into account the events, due to fluctuations, which bring the first monomer to the target, while the rest of the monomers, and the polymer center of mass, are still far from the target. This leads to a smaller estimate of the MFPT than in the Markovian approximation, which does not account for these events correctly, since it assumes that the reactive conformations of the chain are equilibrium conformations.

(iv) *Strongly non-Markovian regime.* A careful analysis[43] of the asymptotic properties of Eqs. (25) and (26) reveals that

$$\langle \mathbf{T} \rangle \simeq V \times \begin{cases} 2X_0/D & \text{if } X_0 \ll l_0, \\ \tilde{\kappa}\, X_0^3 & \text{if } l_0 \ll X_0 \ll l_0 N^{1/2}, \\ 2NX_0/D & \text{if } l_0 N^{1/2} \ll X_0 \end{cases} \qquad (27)$$

with $\tilde{\kappa}$ a constant independent of N and X_0. Comparing with (15), we see that the regime at intermediate length scales is different from that predicted by the Markovian approximation. Here one finds $T \sim V X_0^{d_w - d_f}$ with $d_w = 4$ and $d_f = 1$, as could be expected from a naive application of Refs. 16 and 24, originally derived for scale invariant Markovian processes. In addition, the scalings for small X_0 and large X_0 are the same as predicted for Markovian processes (see Eq. (15)). At very short length scales, the process reduces to the free diffusion of an

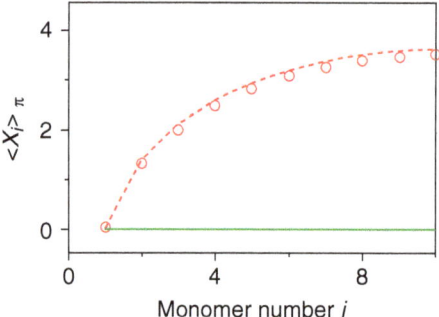

Fig. 4.   Average of monomer positions at the instant of the first passage of the first monomer to the target, for $N = 10$ and X_0 large in 1D. Symbols: circles: Numerical (Langevin) simulations; upper dashed line: Non-Markovian theory; lower solid line: Markovian theory. The unit length is l_0.

isolated reactive monomer, while at very large length scales it is controlled by the diffusion of the center of mass.

4.   Non-Markovian Theory in 3D

We now consider the extension of the theory to 3D. The problem consists in determining the MFPT of a reactive monomer to a target of radius $a > 0$. The main difference with the 1D case is that the extension a of the target must now be taken into account. We denote by $\hat{\mathbf{u}}$ the "arrival direction", defined as the unit vector normal to the target sphere at the position of the reactive monomer at its first passage to the target. Anticipating that, when the confinement volume is large, the MFPT will depend only on the initial distance between the polymer and the target, we assume isotropic initial conditions. Then, the arrival direction vector $\hat{\mathbf{u}}$ is isotropically distributed. The main hypothesis of the non-Markovian theory is that the distribution of conformations of the chain at the instant of first passage, given that the arrival direction is $\hat{\mathbf{u}}$, is a multivariate Gaussian, with an average vector $m_i^\pi \hat{\mathbf{u}}$. In the stationary covariance approximation, we also assume that the covariance matrix is isotropic and given by Eq. (7). We demote by $X_\pi(t)$ the average position of the reactive monomer at a time t after the first passage in the direction $\hat{u}$ by

$$\mathbf{X}_\pi(t) = X_\pi(t)\hat{u} = \left[a - \sum_{i=2}^{N} b_i m_i^\pi (1 - e^{-\lambda_i t}) \right] \hat{u}. \qquad (28)$$

By construction, $X_\pi(0) = a$. Then, following the same steps as in 1D (see Ref. 43), we obtain the set of nonlinear self-consistent equations that define m_i^π

$$\int_0^\infty dt \left\{ \left[\frac{X_\pi(t) m_i^\pi e^{-\lambda_i t}}{3} + \left(1 - \frac{[X_\pi(t)]^2}{3\psi(t)} \right) \frac{b_i(1 - e^{-\lambda_i t})}{\lambda_i} \right] e^{-[X_\pi(t)]^2/[2\psi(t)]} \right.$$

$$\left. - \left(1 - \frac{|\mathbf{X}_0|^2}{3\psi} \right) \frac{b_i(1 - e^{-\lambda_i t})}{\lambda_i} e^{-|\mathbf{X}_0|^2/[2\psi(t)]} \right\} \frac{1}{\psi(t)^{5/2}} = 0. \qquad (29)$$

Once the moments m_i^π are determined, the MFPT is calculated from

$$\frac{\langle \mathbf{T} \rangle}{V} = \int_0^\infty \frac{dt}{[2\pi\psi(t)]^{\frac{3}{2}}} \left[\exp\left(-\frac{[X_\pi(t)]^2}{2\psi(t)} \right) - \exp\left(-\frac{|\mathbf{X}_0|^2}{2\psi(t)} \right) \right]. \qquad (30)$$

Comparison with the expression (13) in the Markovian approximation shows that the key ingredient of the non-Markovian theory is to take into account the mean reactive trajectory $X_\pi(t)$ followed by the polymer after the reaction. The MFPT is represented as a function of the initial distance between the reactants in Fig. 5, where one observes that the use of the non-Markovian theory improves significantly the accuracy of the MFPT predictions.

A careful asymptotic analysis for large V of the Eqs. (28–30) shows that

$$\langle \mathbf{T} \rangle \simeq \begin{cases} V/(4\pi D a) & \text{if } a \ll l_0/\sqrt{N}, \\ V/(4\pi D_{\mathrm{cm}} a_{\mathrm{eff}}), \quad a_{\mathrm{eff}} \simeq 0.71 l_0 \sqrt{N} & \text{if } l_0/\sqrt{N} \ll a \ll l_0\sqrt{N}, \\ V/(4\pi D_{\mathrm{cm}} a) & \text{if } a \gg l_0\sqrt{N}. \end{cases}$$
$$(31)$$

Comparing with the asymptotic expression (16) of the MFPT in the Markovian approximation, it appears that non-Markovian effects vanish in the regimes where the dynamics of the reactive monomer is purely diffusive (very large and very small target sizes in (31)). In the

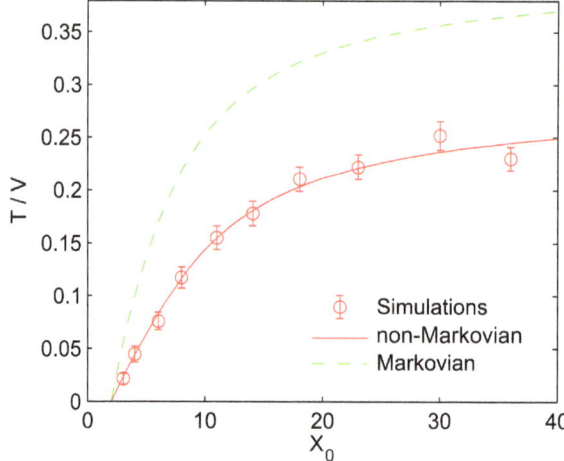

Fig. 5. MFPT in 3D for a Rouse chain with $N = 20$ monomers. The reactive monomer is the first monomer of the chain, the target is centered around 0 and its radius is $a = 2l_0$. Symbols: mid-circles: Simulations; continuous red line: Non-Markovian theory (29, 30); upper dashed line: MFPT in the Markovian approximation (13).

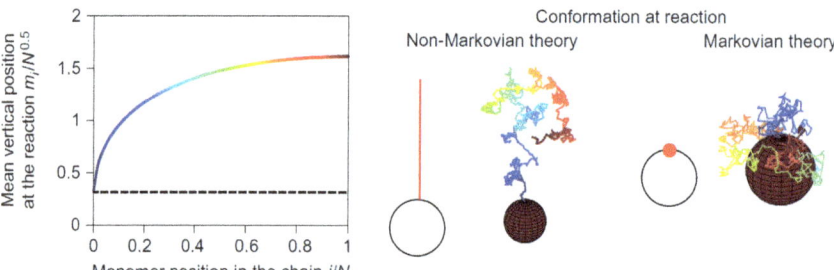

Fig. 6. Predicted polymer conformations at reaction. Left: Average vertical position (the reaction is assumed to take place along the vertical axis, $\hat{\mathbf{u}} = \mathbf{e}_z$) of the monomers when the reaction takes place for the reaction between the first monomer and a target (Continuous line: Prediction of non-Markovian theory, dashed line: Markovian approximation). We also plot the sketch of the polymer shape when the reaction takes place, and an example of conformation drawn from the splitting probability distribution (left), which is in marked contrast with the stationary distribution (right). The position of a monomer in the chain is represented by a color code. Parameters: $N = 800$ and $a = 0.32l_0 N^{1/2}$.

intermediate regime however, which involves a subdiffusive scaling, the non-Markovian effects are quantitatively important, since the effective reaction radius a_{eff} in (31) is significantly larger than in the Markovian theory. The configuration of the polymer at the instant of reaction is represented in Fig. 6. Similarly to the 1D case, the non-Markovian theory predicts that the chain is elongated in the direction of reaction, thereby taking into account the events where the first monomer hits the target due to fluctuations, while the center-of-mass is still far from it.

5. Concluding Remarks and Perspectives

To conclude let us summarize the main content of this Chapter. We have presented a theoretical framework which enables the accurate determination of the mean FPT (MFPT) for non-Markovian random walks in bounded domains, on the example of the Rouse chain (note that the simpler example of persistent random walks was analyzed in Ref. 45). Taking the Rouse model as a paradigm of polymer dynamics, we developed a theory of polymer reaction kinetics that takes into account the non-Markovian features that characterize the dynamics of polymers. This approach shows that non-Markovian effects can be quantitatively important, and that the reaction kinetics crucially depend on the non-equilibrium statistics of polymer conformations at the instant of the reaction. We show quantitatively that the typical reactive conformation of the polymer is more extended than the equilibrium conformation, which leads to reaction times that are significantly shorter than those predicted by existing Markovian theories. These result apply to both reactions with fixed targets, and cyclization reactions. Together, our results provide a better understanding of the complex kinetics of polymer reactions involved, for example, in the formation of loops of RNA or polypeptides chains.

The results presented in this chapter have opened several promising perspectives. First, the analysis of the MFPT in confinement for non-Markovian processes was performed on the example of the Rouse chain. Similar methods can be developed to deal with more general Gaussian non-Markovian processes,[46,47] or even non-Gaussian processes.[48] This includes the very important case of Fractional Brownian Motion, which is a Gaussian model for anomalous diffusion presenting strongly non-Markovian features. There is a growing number of experimental observations and theoretical works indicating that this kind of stochastic motion is involved in a number of different fields, such as in the motion of telomeres in the nucleus,[49] polymer translocation,[50,51] single file diffusion[52] or the motion of particles in complex fluids.[25–27,53,54] Determining the first-passage properties for this kind of random processes could therefore find applications in very different fields. In the context of polymer physics, the case of more realistic polymer dynamics models, such as semi-flexible chains,[55] models involving dihedral interactions,[38] the Zimm model[56] that incorporates hydrodynamics interactions, or models of branched polymers[57] also deserve a particular interest.

Second, it should be noted that in this chapter, we focused only on the first moment of the FPT distribution. In general, the distribution of the FPT can be expected to be quite broad, since trajectories hitting the target range from very short ones when the target is found immediately, to very long ones when the walker explores the whole volume before reaching the target position. Hence, fluctuations can be important in FPT problems,[58,59] and the analysis of the FPT statistics at all time scales requires the determination of the full FPT distribution. This in addition can give access to further first-passage observables, such as participation ratios discussed in Refs. 58 and 59. In the case of Markovian walkers, it was demonstrated that the FPT distribution falls into well defined universality classes after a rescaling by the MPFT,[12,13] which is therefore a very important determinant of the distribution. Recently the full distribution of the FPT in confinement for non-Markovian processes was analysed, showing that similar properties still hold.

Acknowledgment

Support from European Research Council starting Grant FPTOpt-277998 is acknowledged.

References

1. N. G. Van Kampen, *Stochastic Processes in Physics and Chemistry.* North-Holland, Amsterdam (1992).
2. S. Redner, *A guide to First- Passage Processes.* Cambridge University Press, Cambridge, England (2001).
3. R. Metzler, G. Oshanin, and S. Redner, *First Passage Problems: Recent Advances.* World Scientific, Singapore (2014).
4. S. A. Rice, *Diffusion-Limited Reactions.* Elsevier, Amsterdam (1985).
5. H. C. Tuckwell, *Introduction to Theoretical Neurobiology.* Cambridge University Press, Cambridge (1988).
6. C. R. Doering and J. C. Gadoua, *Phys. Rev. Lett.* **69**, 2318 (1992).
7. O. Bénichou, M. Coppey, M. Moreau, P.-H. Suet, and R. Voituriez, Optimal search strategies for hidden targets. *Phys. Rev. Lett.* **94**(19), 198101 (2005).
8. O. Bénichou, M. Coppey, M. Moreau, P. H. Suet, and R. Voituriez, A stochastic model for intermittent search strategies. *J. Phys. Cond. Matt.* **17**(49), S4275–S4286 (2005).
9. O. Bénichou, C. Loverdo, M. Moreau, and R. Voituriez, Optimizing intermittent reaction paths. *Phys. Chem. Chem. Phys.* **10**(47), 7059–7072 (2008).
10. O. Bénichou, C. Loverdo, M. Moreau, and R. Voituriez, Intermittent search strategies. *Rev. Mod. Phys.* **83**(1), 03 (2011).
11. M. Sheinman, O. Bénichou, Y. Kafri, and R. Voituriez, Classes of fast and specific search mechanisms for proteins on dna. *Rep. Prog. Phys.* **75**(2), 026601 (2012).
12. O. Bénichou, C. Chevalier, J. Klafter, B. Meyer, and R. Voituriez, Geometry-controlled kinetics. *Nat Chem.* **2**(6), 472–477 (2010).
13. B. Meyer, C. Chevalier, R. Voituriez, and O. Benichou, Universality classes of first-passage-time distribution in confined media. *Phys. Rev. E.* **83**(5), 051116 (2011).
14. Jae Dong Noh and Heiko Rieger, Random walks on complex networks. *Phys. Rev. Lett.* **92**(11), 118701–118704 (2004).
15. S. Condamin, O. Bénichou, and M. Moreau, First-passage times for random walks in bounded domains. *Phys. Rev. Lett.* **95**(26), 260601 (2005).
16. S. Condamin, O. Bénichou, V. Tejedor, R. Voituriez, and J. Klafter, First-passage times in complex scale-invariant media. *Nature* **450**(7166), 77–80 (2007).

17. S. Condamin, O. Bénichou, and M. Moreau, Random walks and Brownian motion: A method of computation for first-passage times and related quantities in confined geometries. *Phys. Rev. E. Stat. Nonlin. Soft. Matter. Phys.* **75**(2 Pt 1), 021111 (2007).

18. Z. Schuss, A. Singer, and D. Holcman, The narrow escape problem for diffusion in cellular microdomains. *Proc. Natl. Acad. Sci. USA*, **104**(41), 16098–16103 (2007).

19. S. Condamin, V. Tejedor, R. Voituriez, O. Bénichou, and J. Klafter, Probing microscopic origins of confined subdiffusion by first-passage observables. *Proc. Natl. Acad. Sci. USA*, **105**(15), 5675–5680 (2008).

20. C. P. Haynes and A. P. Roberts, Global first-passage times of fractal lattices. *Phys. Rev. E.* **78**(4), 041111–041119 (2008).

21. M. J. Ward and J. B. Keller, Strong localized perturbations of eigenvalue problems. *SIAM J. Appl. Math.* **53**(3), 770–798 (1993).

22. V. Tejedor, O. Bénichou, and R. Voituriez, Global mean first-passage times of random walks on complex networks. *Phys. Rev. E.* **80**(6), 065104 (2009).

23. C. Chevalier, O. Bénichou, B. Meyer, and R. Voituriez, First-passage quantities of brownian motion in a bounded domain with multiple targets: A unified approach. *J. Phys. A. Math. Theor.* **44**, 025002 (2011).

24. O. Bénichou and R. Voituriez, From first-passage times of random walks in confinement to geometry-controlled kinetics. *Phys. Rep.* **539**(4), 225–284 (2014).

25. M. Weiss, Single-particle tracking data reveal anticorrelated fractional brownian motion in crowded fluids. *Phys. Rev. E.* **88**(1), 010101 (2013).

26. D. Ernst, M. Hellmann, J. Köhler, and M. Weiss, Fractional brownian motion in crowded fluids. *Soft Matter* **8**(18), 4886–4889 (2012).

27. T. Turiv, I. Lazo, A. Brodin, B. I. Lev, V. Reiffenrath, V. G. Nazarenko, and O. D. Lavrentovich, Effect of collective molecular reorientations on brownian motion of colloids in nematic liquid crystal. *Science* **342**(6164), 1351–1354 (2013).

28. D. Panja, Generalized langevin equation formulation for anomalous polymer dynamics. *J. Stat. Mech. Theor. Exp.* (2010).

29. S. C. Weber, A. J. Spakowitz, and J. A. Theriot, Bacterial chromosomal loci move subdiffusively through a viscoelastic cytoplasm. *Phys. Rev. Lett.* **104**(23), 238102 (2010).

30. K. Burnecki, E. Kepten, J. Janczura, I. Bronshtein, Y. Garini, and A. Weron, Universal algorithm for identification of fractional brownian motion. A case of telomere subdiffusion. *Biophys. J.* **103**(9), 1839–1847 (2012).

31. H. Hajjoul, J. Mathon, H. Ranchon, I. Goiffon, J. Mozziconacci, B. Albert, P. Carrivain, J.-M. Victor, O. Gadal, K. Bystricky, and A. Bancaud, High-throughput chromatin motion tracking in living yeast reveals the flexibility of the fiber throughout the genome. *Genome Res.* **23**(11), 1829–1838 (2013).

32. O. Bénichou, T. Guérin, and R. Voituriez, Mean first-passage times in confined media: From markovian to non-markovian processes. *J. Phys. A: Math. Theor.* **48**(16), 163001 (2015).

33. A. J. Bray, S. N. Majumdar, and G. Schehr, Persistence and first-passage properties in nonequilibrium systems. *Adv. Phys.* **62**(3), 225–361 (2013).
34. D. Panja, Anomalous polymer dynamics is non-markovian: memory effects and the generalized langevin equation formulation. *J. Stat. Mech. Theor. Exp.* **2010**(06), P06011 (2010).
35. V. Démery, H. Jacquin, and O. Bénichou, Generalized langevin equations for a driven tracer in dense soft colloids: Construction and applications. *arXiv preprint arXiv:1401.5515*, 2014.
36. M. Doi and S. F. Edwards, *The Theory of Polymer Dynamics*. Clarendon Press, Oxford University Press, New York (1988).
37. M. Bixon and R. Zwanzig, Optimized rouse-zimm theory for stiff polymers. *J. Chem. Phys.* **68**, 1896 (1978).
38. M. Dolgushev and A. Blumen, Dynamics of discrete semiflexible chains under dihedral constraints: Analytic results. *J. Chem. Phys.* **138**(20), 204902 (2013).
39. J. Krug, H. Kallabis, S. N. Majumdar, S. J. Cornell, A. J. Bray, and C. Sire, Persistence exponents for fluctuating interfaces. *Phys. Rev. E.* **56**(3), 2702–2712 (1997).
40. N. G. Van Kampen, *Stochastic Processes in Physics and Chemistry*. North-Holland Personnal Library, 3rd edn. Amsterdam (1992).
41. G. Wilemski and M. Fixman, Diffusion-controlled intrachain reactions of polymers. 1. Theory. *J. Chem. Phys.* **60**(3), 866–877 (1974).
42. G. Wilemski and M. Fixman, Diffusion-controlled intrachain reactions of polymers. 2. Results for a pair of terminal reactive groups. *J. Chem. Phys.* **60**(3), 878–890 (1974).
43. T. Guérin, O. Bénichou, and R. Voituriez, Reactive conformations and non-Markovian kinetics of a rouse polymer searching for a target in confinement. *Phys. Rev. E.* **87**, 032601 (2013).
44. T. Guérin, O. Bénichou, and R. Voituriez, Non-Markovian polymer reaction kinetics. *Nat. Chem.* **4**, 568–573 (2012).
45. V. Tejedor, R. Voituriez, and O. Bénichou, Optimizing persistent random searches. *Phys. Rev. Lett.* **108**(8), 088103 (2012).
46. N. Levernier, O. Bénichou, and R. Voituriez, Mean first-passage time of an anisotropic diffusive searcher. *J. Phys. A.* **50**(2), 024001 (2017).
47. T. Guérin, N. Levernier, O. Bénichou, and R. Voituriez, Mean first-passage times of non-markovian random walkers in confinement. *Nature.* **534**(7607), 356–359 (2016).
48. N. Levernier, O. Bénichou, T. Guérin, and R. Voituriez, Universal first-passage statistics in aging media. *Phys. Rev. E.* **98**(2), 022125 (2018).
49. I. Bronstein, Y. Israel, E. Kepten, S. Mai, Y. Shav-Tal, E. Barkai, and Y. Garini, Transient anomalous diffusion of telomeres in the nucleus of mammalian cells. *Phys. Rev. Lett.* **103**(1), 018102 (2009).
50. V. V. Palyulin, T. Ala-Nissila, and R. Metzler, Polymer translocation: The first two decades and the recent diversification. *Soft Matter* **10**(45), 9016–9037 (2014).

51. D. Panja, G. T. Barkema, and A. B. Kolomeisky, Through the eye of the needle: Recent advances in understanding biopolymer translocation, *J. Phys.: Cond. Matt.* **25**(41), 413101 (2013).
52. L. Lizana, T. Ambjörnsson, A. Taloni, E. Barkai, and M. A. Lomholt, Foundation of fractional langevin equation: Harmonization of a many-body problem, *Phys. Rev. E.* **81**(5), 051118 (2010).
53. Jae-Hyung Jeon, V. Tejedor, S. Burov, E. Barkai, C. Selhuber-Unkel, K. Berg-Sørensen, L. Oddershede, and R. Metzler, In vivo anomalous diffusion and weak ergodicity breaking of lipid granules. *Phys. Rev. Lett.* **106**(4), 048103 (2011).
54. Jae-Hyung Jeon, N. Leijnse, L. B. Oddershede, and R. Metzler, Anomalous diffusion and power-law relaxation of the time averaged mean squared displacement in worm-like micellar solutions. *N. J. Phys.* **15**(4), 045011 (2013).
55. T. Guérin, M. Dolgushev, O. Bénichou, R. Voituriez, and A. Blumen, Cyclization kinetics of gaussian semiflexible polymer chains. *Phys. Rev. E.* **90**(5), 052601 (2014).
56. N. Levernier, M. Dolgushev, O. Benichou, A. Blumen, T. Guerin, and R. Voituriez, Non-markovian closure kinetics of flexible polymers with hydrodynamic interactions. *J. Chem. Phys.* **143**(20), 204108 (2015).
57. M. Dolgushev, G. Berezovska, and A. Blumen, Branched semiflexible polymers: Theoretical and simulation aspects. *Macromol. Theor. Simul.* **20**(8), 621–644 (2011).
58. C. Mejía-Monasterio, G. Oshanin, and G. Schehr, First passages for a search by a swarm of independent random searchers. *J. Stat. Mech. Theor. Exp.* **2011**(06), P06022 (2011).
59. T. G. Mattos, C. Mejía-Monasterio, R. Metzler, and G. Oshanin, First passages in bounded domains: When is the mean first passage time meaningful? *Phys. Rev. E.* **86**(3), 031143 (2012).

Chapter 11

Reaction Kinetics
in the Few-Encounter Limit

David Hartich and Aljaž Godec*

Mathematical Biophysics Group,
Max-Planck-Institute for Biophysical Chemistry,
37077 Gottingen, Germany
*agodec@mpibpc.mpg.de

The classical theory of chemical reactions can be understood in terms of diffusive barrier crossing, where the rate of a reaction is determined by the inverse of the mean first passage time (FPT) to cross a free energy barrier. Whenever a few reaction events serve to trigger a response or the energy barriers are not high, the mean first passage time alone does not suffice to characterize the kinetics, i.e., the kinetics do not occur on a single timescale. Instead, the full statistics of the FPT are required. We present a spectral representation of the FPT statistics that allows us to understand and accurately determine FPT distributions over several orders of magnitudes in time. A canonical narrowing of the first-passage density is shown to emerge whenever several molecules are searching for the same target, which is termed the *few-encounter limit*. The few-encounter limit is essential in all situations in which already the first encounter triggers a response, such as misfolding-triggered aggregation of proteins or protein transcription regulation.

1. Introduction

Since Smoluchowski's[1] and Kramers'[2] seminal contributions, first passage time (FPT) theory has been a paradigm for studying chemical kinetics;[3-7] see also Refs. 8–12 for extensive reviews. Extensions of these original ideas led to theories of diffusion-controlled reaction kinetics in fractal[13,14] and heterogeneous media,[15-18] surface-mediated reactions,[19,20] and search processes involving swarms of agents,[21] to name but a few. Notably, in contrast to extensively studied nearest-neighbor random walks (see e.g., Ref. 12), the FPT statistics in multiply connected Markov-state dynamics, aside from a few studies on simple enzyme models[22-24] and recent numerical approximation schemes based on Bayesian inference,[25,26] are barely explored.[27]

The importance of understanding the full FPT statistics is meanwhile well established.[12,21,28-30] For example, it was proposed to be essential for explaining the so-called proximity effect in gene regulation, according to which direct reactive trajectories boost the speed and precision of gene regulation.[31,32] The full FPT statistics were also shown to be required for a quantitative description of misfolding-triggered protein aggregation[33] and various nucleation-limited phenomena.[34,35] Underlying the kinetics in these systems is the FPT problem of n-independent simultaneous trajectories,[30,33] which we refer to as kinetics in the *few-encounter limit* and which will be the focus of this chapter.

We will limit the discussion to FPT phenomena of reversible Markovian dynamics in bounded domains or confining potentials, which renders all FPT moments finite and probability densities asymptotically exponential.[12,21,28-30,33] We discuss effectively 1D diffusion processes in arbitrary potentials $U(x)$ and jump processes with arbitrary transition matrices. Hyperspherically symmetric diffusion processes in d dimensions will be treated via a mapping onto radial diffusion with a repulsive potential $U(x) = -(d-1)\ln(x)$ in units of thermal energy,[9,12,28-30] i.e., $k_B T \equiv 1$.

The chapter is organized as follows. Generic single-molecule FPT concepts are introduced in Subsection 2.1. Subsection 2.2 outlines a spectral expansion of the FPT density. Subsection 2.3 relates the single-molecule FPT problem to the corresponding many-particle problem, while two examples of FPT statistics in discrete- and continuous state-space dynamics are presented in Subsection 2.4. In working out these examples, we utilize a recently proven duality between first passage and relaxation processes — an algorithmic tool that allows determining the full FPT distribution analytically from a simpler relaxation process (Appendix). We conclude with an outlook in Section 3.

2. First Passage Time Statistics

2.1. *The single-particle setting*

Let x_t denote the dynamics of a reaction coordinate, e.g., the position of a particle in a potential $U(x)$ (see Fig. 1(a)) or in a circular domain with a central target (see Fig. 1(b) and 1(c)). Suppose that x_t obeys a Markovian equation of motion. The reaction kinetics are then characterized by the FPT — the first instance x_t reaches a given threshold

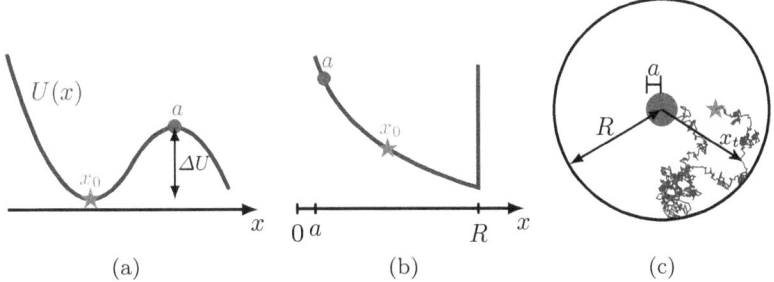

(a)   (b)   (c)

Fig. 1.   Crossing of a free energy barrier. (a) Energy landscape $U(x)$ from a local potential minimum a over a barrier a with barrier height ΔU. (b) Geometry-induced potential $U(x) = -(d-1)\ln x$ for a diffusive search in a d-dimensional domain with radius R as illustrated in (c).

a, defined formally as

$$t_a(x_0) = \min\{t|x_t = a\}. \tag{1}$$

The stochasticity of x_t renders the FPT, $t_a(x_0)$, a stochastic variable. The statistics of $t_a(x_0)$ is fully characterized by the survival probability

$$\mathcal{P}_a(t|x_0) \equiv \mathrm{Prob}[t_a(x_0) \geq t], \tag{2}$$

which quantifies the probability that the reaction did not occur before t. $\mathcal{P}_a(t|x_0)$ decays monotonically from $\mathcal{P}_a(t|x_0) = 1$ to $\mathcal{P}_a(t = \infty|x_0) = 0$ with a slope that is nothing but the FPT density

$$\wp_a(t|x_0) \equiv -\frac{\partial}{\partial t}\mathcal{P}_a(t|x_0). \tag{3}$$

The kth moment of $t_a(x_0)$ can be determined via

$$\langle t_a(x_0)^k \rangle = \int_0^\infty t^k \wp_a(t|x_0)\mathrm{d}t = k \int_0^\infty t^{k-1}\mathcal{P}_a(t|x_0)\mathrm{d}t, \tag{4}$$

where the last equality follows from Eq. (3) by partial integration. Notably, $\langle t_a(x_0)^k \rangle$ are typically dominated by the long-time behavior of $\wp_a(t|x_0)$.[30,33] While the full FPT density is generally hard to determine, simple integral formulas exist for the moments of the FPT under diffusive dynamics.[36]

However, as we show later, the moments of the FPT in the single-particle setting in fact provide very little information about the kinetics in many-particle systems. Namely, few-encounter and nucleation kinetics for example, are typically governed by short[34,35] or intermediate timescales.[33]

2.2. Spectral expansion of first-passage distributions

For reversible Markovian dynamics, the FPT density allows the expansion

$$\wp_a(t|x_0) = \sum_{k>0} w_k(x_0)\mu_k e^{-\mu_k t}, \tag{5}$$

where μ_k^{-1} denotes the kth FPT-scale such that μ_k is a rate, and $w_k(x_0)$ is the corresponding weight of the kth mode. In contrast to μ_k, $w_k(x_0)$ depends on the starting position x_0. For convenience, we drop the functional dependence of both w_k and μ_k on a. The weights satisfy the normalization condition $\sum_k w_k(x_0) = 1$, and the positivity of $\wp_a(t|x_0)$ implies $w_1 > 0$. Specifically, if energetic or kinetic barriers are high enough, a separation of timescales emerges ($\mu_2 \gg \mu_1$), such that the FPT distribution becomes approximately $\wp_a(t|x_0) \simeq w_1(x_0)\mu_1 e^{-\mu_1 t}$ with $w_1(x_0) \simeq 1$ if x_0 is located *before* the highest energy barrier.[30] Note that for finite discrete-state systems, the sum in Eq. (5) is finite. The survival probability analogously becomes

$$P_a(t|x_0) = \int_t^\infty \wp_a(t'|x_0)\mathrm{d}t' = \sum_{k>0} w_k(x_0)e^{-\mu_k t}. \qquad (6)$$

Using the spectral expansion, the Laplace transform of $\wp_a(t|x_0)$ reads

$$\tilde{\wp}_a(s|x_0) = \int_0^\infty e^{-st}\wp_a(t|x_0)\mathrm{d}t = \sum_{i>0} \frac{w_k(x_0)\mu_k}{s+\mu_k} \qquad (7)$$

and the kth moment of the single-particle FPT is given by

$$\langle t_a(x_0)^k \rangle = k! \sum_{i>0} w_i(x_0)\mu_i^{-k}. \qquad (8)$$

When $\mu_2 \gg \mu_1$, $\langle t_a(x_0)^k \rangle$ is typically dominated by the slowest timescale, i.e., $\langle t_a(x_0)^k \rangle \simeq k! w_1(x_0)/\mu_1^k$, which is usually quite accurate in problems such as the one used in Fig. 1 (see also Ref. 33).

In general, it can be difficult to determine both first passage eigenvalues $\{\mu_k(x)\}$ and their corresponding weights $\{w_k(x)\}$. However, we have recently derived an analytical theory that allows us to determine the spectral representation of $\wp_a(t|x_0)$ from the corresponding dual relaxation spectrum,[33,37] which is summarized in the Appendix.

2.3. The many-particle setting and kinetics in the few-encounter limit

Suppose that now n particles starting from the same position x_0 at time $t = 0$ are searching independently for the same target. Once the first molecule hits the target, a "catastrophic" response is triggered (e.g., aggregation of misfolded proteins, induction/inhibition of gene transcription, etc.), or the target disappears such as in foraging problems. In order to understand such "nucleation-type phenomena", details about the FPT distribution become relevant.[12,28,30,38] The n-particle survival probability is simply the product of the single-particle survival probabilities[21]

$$\mathcal{P}_a^{(n)}(t|x_0) \equiv \mathcal{P}_a(t|x_0)^n. \tag{9}$$

The probability density that the first of the n particles hits a for the first time at time t then becomes, using Eqs. (3) and (9),[21,33,37]

$$\wp_a^{(n)}(t|x_0) \equiv n\wp_a(t|x_0)\mathcal{P}_a(t|x_0)^{n-1}$$

$$= n\wp_a^{(n)}(t|x_0) \left[\int_t^\infty \wp_a(\tau|x_0)\mathrm{d}\tau \right]^{n-1}. \tag{10}$$

We will henceforth omit the superscript (1) in denoting the single-particle scenario, i.e., $\wp_a \equiv \wp_a^{(1)}$ and $\langle \cdots \rangle \equiv \langle \cdots \rangle^{(1)}$. Analogously to Eq. (4), the moments of the FPT in the n-particle case read

$$\langle t_a(x_0)^k \rangle^{(n)} \equiv \int_0^\infty t^k \wp_a^{(n)}(t|x_0)\mathrm{d}t, \tag{11}$$

which according to Eq. (10) can be determined solely from $\wp_a(t|x_0)$

$$\langle t_a(x_0)^k \rangle^{(n)} = n\langle t_a(x_0)^k \mathcal{P}_a[t_a(x_0)|x_0]^{n-1} \rangle$$

$$= n \int_0^\infty t^k \mathcal{P}_a(t|x_0)^{n-1} \wp_a(t|x_0)\mathrm{d}t. \tag{12}$$

Due to the term $\mathcal{P}_a(t|x_0)^{n-1}$ in Eq. (12) one needs, for any finite value of $k \geq 1$, formally an infinite number of single-particle moments, to determine $\langle t_a(x_0)^k \rangle^{(n)}$. Hence, many-particle nucleation-type kinetics cannot be understood in terms of single-particle mean FPT.[33–35]

Even if $\langle t_a(x_0)^k \rangle$ is accurately characterized by long-time asymptotics, the latter do not provide accurate results for $\langle t_a(x_0)^k \rangle^{(n)}$, which can be orders of magnitude off.[33] The severe insufficiency of single-particle moments arises from the sharp sigmoidal shape of $\mathcal{P}_a(t|x_0)^{n-1}$ within the many-particle average (12) (e.g., see lower panel of Fig. 2 for an illustration). We note that utilizing long-time

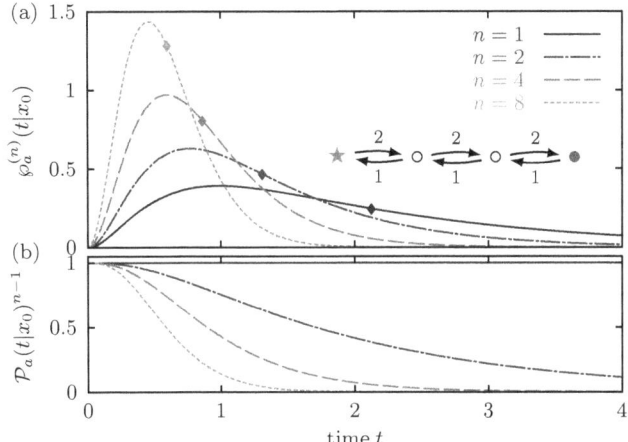

Fig. 2. n-particle density $\wp_a^{(n)}(t|x_0)$ (a) and the second term in the product of Eq. (10), $\mathcal{P}_a(t|x_0)^{n-1}$, (b) for a four-state random walk. Diamonds depict the respective mean FPTs. More details about $\wp_a$ and $\mathcal{P}_a$ are given in Subsection 2.5.

asymptotics can lead in general to both an overestimation or an underestimation of $\langle t_a(x_0)^k \rangle^{(n)}$, depending on the initial conditions.[33] For $n \to \infty$, short-time asymptotics sets in, for which it has been found that $\langle t_a(x_0) \rangle^{(n)} \propto 1/\ln(n)$ for overdamped diffusive first passage problems[34,35] (see also Ref. 21).

Two generic phenomena emerge as the particle number n increases: (i) $\langle t_a(x_0) \rangle^{(n)}$ reduces and (ii) the width of $\wp_a^{(n)}(t|x_0)$ concurrently decreases (see Fig. 2). Both are independent of the details of dynamics and directly follow from a progressively sigmoidal shape of $\mathcal{P}_a(t|x_0)^{n-1}$. These features are particularly important for explaining the so-called proximity effect — the spatial proximity of co-regulated genes — in transcription regulation.[30] The generic origin of these effects provides an explanation of the robustness of the proximity effect (see Ref. 30 for more details on the biological aspect).

2.4. Determining first passage time statistics from relaxation spectra

Having established that the full FPT statistics are required for a correct physical description of reaction kinetics in the few-encounter limit, we now present, with two illustrative examples, a canonical method to determine $\wp_a(t|x_0)$ from the corresponding relaxation spectrum.

We consider two classes of processes, diffusion in effectively 1D potentials and reversible Markovian jump-processes, in more detail (see e.g., Fig. 3). We call x_t a relaxation process if, in contrast to the first passage problem, the dynamics does not terminate upon reaching a threshold. More precisely, for a diffusion process (see Fig. 3(a)) the probability density $P(x,t|x_0)$ to find a particle starting from x_0 at position x at time t satisfies the Fokker–Planck equation

$$\frac{\partial}{\partial t}P(x,t|x_0) = \hat{L}_{\mathrm{FP}}P(x,t|x_0) \equiv \frac{\partial}{\partial x}D(x)\left[U'(x) + \frac{\partial}{\partial x}\right]P(x,t|x_0),$$
(13)

where $U(x)$ is the potential $(U' \equiv \partial_x U)$ and $D(x)$ the diffusion landscape. Relaxation dynamics conserves probability, i.e., $\int P(x,t|x_0)\mathrm{d}x = 1$ for all t, which is obtained either with natural boundary condition or a "reflecting barrier", which would in turn imply $[U(x) - \partial_x]P(x,t|x_0)|_{x=a} = 0$. For jump-processes (see Fig. 3(b)) the probability to find the system at state $x_t = x$ if it started at x_0 obeys a master equation

$$\frac{\partial}{\partial t}P(x,t|x_0) = \hat{L}_{\mathrm{ME}}P(x,t|x_0) \equiv \sum_{x'} L_{xx'}P(x',t|x_0),$$
(14)

where $L_{xx'}$ is the transition rate from state x' to x if $x \neq x'$ and $L_{xx} = -\sum_{x'\neq x}L_{x'x}$ is the negative rate of leaving state x', guaranteeing conservation of probability $\sum_x L_{xx'} = 0$, i.e., $\sum_x P(x,t|x') = 1$

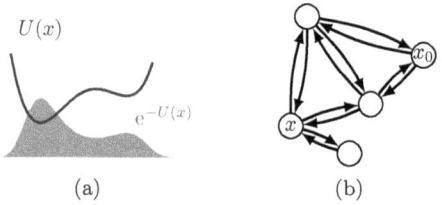

(a)   (b)

Fig. 3.   Schematic of (a) diffusive dynamics, (b) Markovian jump-process.

for all t and x'. Moreover, reversibility requires the rates to obey detailed balance $\ln(L_{xx'}/L_{x'x}) = U(x') - U(x)$.[39] Both classes of reversible stochastic dynamics allow an expansion of the operator $\hat{L} = \hat{L}_{\text{FP}}, \hat{L}_{\text{ME}}$ in a real bi-orthogonal eigenbasis, such that

$$P(x,t|x_0) = \sum_{k \geq 0} \psi_k^{\text{R}}(x)\psi_k^{\text{L}}(x_0)e^{-\lambda_k t}, \tag{15}$$

where λ_k is the kth eigenvalue of operator $\hat{L}$ (with $0 = \lambda_0 < \lambda_1 \leq \lambda_2 \leq \cdots$), and ψ_k^{R} (ψ_k^{L}) are the corresponding right (left) eigenvectors satisfying $\hat{L}\psi_k^{\text{R}} = -\lambda_k \psi_k^{\text{R}}$ and $\psi_k^{\text{L}}(x) = \mathcal{N}_k^{-1}e^{U(x)}\psi_k^{\text{R}}(x)$. The normalization for $\hat{L} = \hat{L}_{\text{FP}}$ reads $\mathcal{N}_k = \int e^{U(x)}[\psi_k^{\text{R}}(x)]^2 \mathrm{d}x$, whereas for $\hat{L} = \hat{L}_{\text{ME}}$ the integral in x becomes a sum. Note that the zeroth eigenvector ($k = 0$) is given by $\psi_0^{\text{R}}(x) = e^{-U(x)}$, such that $\psi_0^{\text{R}}(x)\psi_0^{\text{L}}(x_0) = P^{\text{eq}}(x)$ is the Boltzmann distribution.

The terms $k > 0$ in the sum of Eq. (15) relax to zero exponentially fast with rates λ_k, and the corresponding eigenfunctions $\psi_k^{\text{R}}(x)\psi_k^{\text{L}}(x_0)$ quantify the redistribution of the probability mass. For potential landscapes with n energy basins (e.g., $n = 2$ in left panel of Fig. 3) we generally expect at least one (or the last) gap at $\lambda_{n-1} \ll \lambda_n$ in the relaxation spectrum.

For any stationary Markov process, the renewal theorem[40]

$$P(a,t|x_0) = \int_0^t P(a,t-\tau|a)\wp_a(\tau|x_0)\mathrm{d}\tau \tag{16}$$

connects the propagator of relaxation dynamics to the FPT density. It has the following intuitive interpretation: if a particle starting from x_0 is found at position $x_t = a$ at time t, then it must have reached it for the first time before that time $\tau \leq t$ and then returned to (or stayed at) a in the remaining time interval $t-\tau$. Laplace transforming Eq. (16), where a convolution in the time domain becomes a product, translates Eq. (16) to $\tilde{P}(a,s|x_0) = \tilde{P}(a,s|a)\tilde{\wp}_a(s|x_0)$, i.e.,

$$\tilde{\wp}_a(s|x_0) = \frac{\tilde{P}(a,s|x_0)}{\tilde{P}(a,s|a)}. \tag{17}$$

Comparing Eq. (17) with the first passage density (7), one can easily verify that poles of the FPT distribution $\tilde{\wp}_a(s|x_0)$, which are located at the first passage rates $\mu_k = -s$, are zeros of the diagonal of the propagator $\tilde{P}(a, s|a)$.[41]

In the Appendix, we present an explicit and exact duality relation that allows for an explicit inversion of Eq. (17) to the time domain. Briefly, $\wp_a(t|x_0)$ is obtained in three steps: (i) the first step is to realize that the first passage and relaxation timescales interlace, $\lambda_{k-1} \leq \mu_k \leq \lambda_k$, which is then utilized in (ii) the second step is to express all first passage rates $\{\mu_k\}$ in terms of series of determinants of almost triangular matrices (A.2). (iii) The third an final step involves the Cauchy residue theorem to determine the first passage weights $\{w_k\}$ from Eq. (A.7), leading to

$$w_k(x_0) = \frac{\sum_{l \geq 0} (1 - \lambda_l/\mu_k)^{-1} \psi_l^R(a) \psi_l^L(x_0)}{\sum_{l \geq 0} (1 - \lambda_l/\mu_k)^{-2} \psi_l^R(a) \psi_l^L(a)}. \tag{18}$$

For the full details, we refer the reader to the Appendix or Refs. 33 and 37. In the following, we apply the duality to determine FPT densities of a simple four-state Markov process and a diffusion in a rugged potential.

2.5. *Four-state Markov jump process*

For illustratory purposes, we consider a simple four-state biased random walk as shown in the inset of Fig. 2 with a transition matrix

$$\mathbf{L} = \begin{pmatrix} -2 & 1 & 0 & 0 \\ 2 & -3 & 1 & 0 \\ 0 & 2 & -3 & 1 \\ 0 & 0 & 2 & -1 \end{pmatrix}, \tag{19}$$

whose eigenvalues are $\{\lambda_0, \lambda_1, \lambda_2, \lambda_3\} = \{0, 1, 3, 5\}$ and the corresponding eigenvectors can be obtained in a straightforward manner. We fix the initial and target state to $x_0 = 1$ and $a = 4$, respectively. The diagonal and off-diagonal relaxation propagators then have the simple forms

$$P(a,t|a) = \frac{8}{15} + \frac{e^{-t}}{4} + \frac{e^{-3t}}{6} + \frac{e^{-5t}}{20},$$

$$P(a,t|x_0) = \frac{8}{15} + e^{-t} + \frac{2e^{-3t}}{3} + \frac{e^{-5t}}{5},$$

(20)

whereas a similarly compact analytical formula for $\wp_a(t|x_0)$ cannot be found. In the Appendix (Section A.2), we use the duality between relaxation and first passage processes to determine $\{\mu_1, \mu_2, \mu_3\} \simeq \{0.657, 2.529, 4.814\}$ and $\{w_1(x_0), w_2(x_0), w_3(x_0)\} \simeq \{1.565, -0.740, 0.175\}$. The resulting single-particle FPT probability density is depicted with the solid line in the upper panel of Fig. 2. The dash-dotted line (here $n = 2$) in the lower panel depicts the corresponding single-particle survival probability $P_a(t|x_0)$. We note that the short-time limit yields $\wp_a(t|x_0) = 4t^2 + \mathcal{O}(t)^3$, which arises from the two intermediate states between x_0 and a (see model scheme from Fig. 2). The vanishing first passage density $\wp_a \to 0$ (short-time limit) causes a strong narrowing of the many-particle first passage density $\wp_a^{(n)}(t|x_0) = n\wp_a(t|x_0)P_a(t|x_0)^{n-1}$, since the survival probability "pushes" the probability mass to short times for increasing values of n (see Fig. 2).

2.6. *Diffusive exploration of a rugged energy landscape*

As a second example, we analyze $\wp_a(t|x_0)$ for a diffusive barrier crossing in a rugged multiwell potential, which is particularly relevant for protein folding and misfolding kinetics[42-45] and biochemical association reactions.[46]

We generate a single rugged potential landscape as a sum of a harmonic potential and a truncated Karhunen–Loève expansion of a Wiener process

$$U(x) = \frac{x^2}{4} + \sum_{k=1}^{N} z_k \frac{\sin[(2k-1)x]}{(2k-1)}.$$

(21)

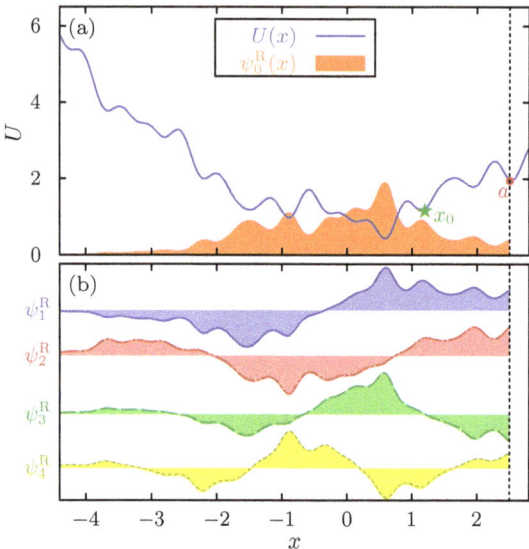

Fig. 4. (a) Rugged potential landscape $U(x)$ from Eq. (21) and the corresponding Boltzmann measure with $(z_1, \ldots, z_7) = (-0.14, -1.04, 0.77, -1.32, -0.61, -1.66, -2.67)$. (b) The first four excited right eigenfunctions corresponding to $U(x)$ with $D(x) = 1$.

We truncate the expansion at $N = 7$ and sample z_k from a normal distribution. Once $\{z_k\}$ are determined, they are kept fixed. In Fig. 4(a), we depict $U(x)$ and its corresponding equilibrium probability density $\psi_0^{\mathrm{R}}(x) \propto \mathrm{e}^{-U(x)}$. We numerically determine the first 45 eigenvalues $\{\lambda_k\}$ and eigenfunctions $\{\psi_k^{\mathrm{R}}\}$ of the Fokker–Planck operator (13), from which the first four excited relaxation eigenmodes are illustrated in Fig. 4(b). The relaxation eigenfunctions determine the redistribution of probability during the approach to equilibrium. Using Newton's series of almost triangular matrices from Eq. (A.2), we determine the FPT-scales μ_k^{-1}, which interlace with the relaxation timescales[33,37] as illustrated in the upper panel of Fig. 5. Specifically, between any two consecutive FPT-scales (blue circles) we find exactly one relaxation timescale (red triangles) and vice versa. In particular, the slowest FPT-scale occurs on a longer timescale than the slowest relaxation timescale. This can be explained by the fact that the slowest first passage mode requires *all* trajectories to reach the target, whereas the slowest relaxation mode

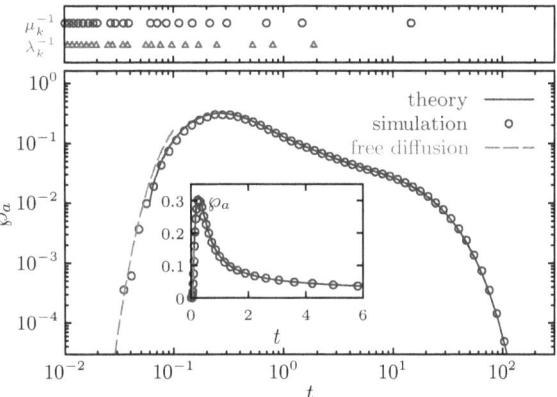

Fig. 5. The FPT density for particle starting from $x_0 = 0.6$ and to $a = 2.5$ within potential from Fig. 4. The inset shows the FPT density on a linear scale. The upper panel superimposes the FPT scales μ_k^{-1} (open circles) and the relaxation timescales λ_k^{-1} (open triangles). The symbols are obtained using the theory outlined in the Appendix, and the symbols denote results of Brownian dynamics simulations of 10^6 trajectories.

only reflects that most trajectories have reached the equilibrium distribution.

In the lower panel of Fig. 5, we present results for the full FPT density using the analytical theory from the Appendix (blue solid line), together with results of extensive Brownian dynamics simulations of 10^6 trajectories, which perfectly agree with the theory. The inset depicts $\wp_a(t|x_0)$ on a linear scale. The short-time limit for a freely diffusing particle in form of a Lévi–Smirnov density, also know as Sparre Anderson result,[30,47] is shown as dashed green line (see also Refs. 21, 34 and 35 for further discussions on the short-time limit). Intuitively, diffusion is faster than advection on short-timescales ($\propto \sqrt{t}$ versus $\propto t$ behavior), rendering the actual potential shape less relevant for $t \to 0$.

3. Concluding Perspectives

The mean and higher moments of the FPT in a single-particle setting were shown to be inherently insufficient for characterizing many-particle FPT kinetics within the few-encounter limit. To correctly describe few-encounter kinetics, one has to go beyond a description limited to FPT moments and determine the full FPT distribution. It was shown how to achieve this utilizing a duality relation between relaxation and first passage process[33,37] outlined in the Appendix. The method is applicable to a broad class of reversible Markov dynamics that includes discrete Markovian jump-processes in any dimension and Markovian diffusion in effectively 1D potential landscapes.

The duality relation can in fact be considered as an analytical algorithmic tool for determining FPT distributions, which was demonstrated with a simple four-state model in full detail. The analysis of the n-particle FPT distribution revealed a reduced mean FPT and a canonical narrowing of the FPT distribution in the few-encounter limit as the number of particles increases. This narrowing arises due to a combination of the short-time cutoff in the FPT density ($\wp_a \to 0$ for $t \to 0$) and an inherent many-particle speed-up, which together render the n-particle kinetics deterministic in the limit $n \to \infty$. In the case of a diffusive exploration of (rugged) energy landscapes, the short-time behavior is dominated by free diffusion, rendering the shape of the potential essentially irrelevant.[34,35]

It will be interesting to extend the applications of the theory outlined in the Appendix and to explore the physical consequences of few-encounter kinetics also in narrow escape problems[48–53] and diffusion on higher-dimensional graphs. Extending the work to irreversible dynamics will be challenging, whereas long-time asymptotics are still accessible.[30,37]

Appendix. Duality Between Relaxation and First-Passage Processes

A.1. General Case

In this appendix, we review the duality relation from Refs. 33 and 37 that allows us to determine analytically the spectral representation of the FPT density in Eq. (5) from the propagator in Eq. (15) in three steps.

The first step is to realize that the relaxation timescales $\{\lambda_k^{-1}\}$ and FPT-scales μ_k^{-1} interlace[33,37]

$$\lambda_{k-1} \leq \mu_k \leq \lambda_k. \tag{A.1}$$

We note that this interlacing of timescales can be related to Cauchy's interlacing theorem for real symmetric matrices.[54] The interlacing has also been demonstrated for simple 1D processes.[41]

The second step is based on an explicit Newton iteration that allows, after some rather involved algebra,[33,37] to exactly express the first passage rates μ_k as a series of determinants of almost triangular matrices $\boldsymbol{A}_n(k)$

$$\mu_k = \bar{\mu}_k + \sum_{n=1}^{\infty} f_0(k)^n f_1(k)^{1-2n} \det \boldsymbol{A}_n(k), \tag{A.2}$$

where $\bar{\mu}_k \equiv (\lambda_k + \lambda_{k-1})/2$, $f_n(k) = \partial_s^n F(k^*, s)|_{s=-\bar{\mu}_k}$ with

$$F(k, s) = (s + \lambda_k)\tilde{P}(a, s|a) \tag{A.3}$$

and the index function

$$k^* \equiv k^*(k) = \begin{cases} k & \text{if } F(k, -\bar{\mu}_k) < 0 \\ k - 1 & \text{else} \end{cases} \tag{A.4}$$

that guarantees $f_0(k)$ to be negative, and we used the almost triangular $(n-1) \times (n-1)$ matrices with elements

$$\mathcal{A}_n^{i,j}(k) = \frac{f_{i-j+2}(k)\Theta(i-j+1)}{(i-j+2)!} \begin{cases} i+j-1 & \text{if } j=1, \\ n(i-j+1)+j-1 & \text{if } j>1, \end{cases} \tag{A.5}$$

where Θ is the discrete Heaviside step function ($\Theta(l) = 1$ if $l \geq 0$) and $\det \mathbfcal{A}_1(k) = 1$. Moreover, we have explicitly[33,37]

$$f_0(k) = \psi_{k^*}^{\mathrm{L}}(a)\psi_{k^*}^{\mathrm{R}}(a) + \sum_{l|l\neq k^*} \psi_l^{\mathrm{L}}(a)\psi_l^{\mathrm{R}}(a)\frac{(\bar{\mu}_k - \lambda_{k^*})}{(\bar{\mu}_k - \lambda_l)},$$

$$f_{n\geq1}(k) = n! \sum_{l|l\neq k^*} \psi_l^{\mathrm{L}}(a)\psi_l^{\mathrm{R}}(a)\frac{(\lambda_l - \lambda_{k^*})}{(\bar{\mu}_k - \lambda_l)^{n+1}}.$$

(A.6)

The third step is a straightforward application of the residue theorem, delivering the first passage weights

$$w_k(x_0) = \frac{\tilde{P}(a, s|x_0)}{\mu_k \partial_s \tilde{P}(a, s|a)}\bigg|_{s=-\mu_k} = \frac{\sum_{l\geq0}(1 - \lambda_l/\mu_k)^{-1}\psi_l^{\mathrm{R}}(a)\psi_l^{\mathrm{L}}(x_0)}{\sum_{l\geq0}(1 - \lambda_l/\mu_k)^{-2}\psi_l^{\mathrm{R}}(a)\psi_l^{\mathrm{L}}(a)},$$

(A.7)

where $\tilde{P}(a, s|x_0)$ is the Laplace transform of Eq. (20). We note that Eq. (A.2) and (A.7) are exact relations that fully characterize the first passage kinetics.

A.2. Four-State Model

We now evaluate $\wp_a(t|x_0)$ for the model from Section 2.5 step-by-step. First, the Laplace transform of the first line of Eq. (20), $\tilde{P}(a, s|a)$, is inserted into Eqs. (A.3) and (A.4), giving $k^* = k$ for $k = 1, 2, 3$. Second, Eq. (A.6) yields

$$\begin{pmatrix} f_0(1) \\ f_0(2) \\ f_0(3) \end{pmatrix} = \begin{pmatrix} -\frac{11}{45} \\ -\frac{1}{3} \\ -\frac{1}{3} \end{pmatrix},$$

$$\begin{pmatrix} \frac{f_n(1)}{n!} \\ \frac{f_n(2)}{n!} \\ \frac{f_n(3)}{n!} \end{pmatrix} = \begin{pmatrix} \frac{1}{3}(-\frac{2}{5})^{n+1} + \frac{1}{5}(-\frac{2}{9})^{n+1} - \frac{2^{n+4}}{15} \\ \frac{1}{10}[(-3)^{-n-1} - 2^{3-n} - 5] \\ \frac{1}{3}[-2^{1-2n} - 3^{-n} - 1] \end{pmatrix}, \qquad \text{(A.8)}$$

with $n > 0$. Note that k^* is chosen to guarantee the negativity of $f_0(k)$. Third, inserting the $f_n(k)/n!$ into the almost triangular matrix Eq. (A.5) and evaluating Newton's series (A.2) yields the exact FPT-scales, which numerically are given by $\{\mu_1, \mu_2, \mu_3\} \simeq \{0.657, 2.529, 4.814\}$. Finally, the weights from Eq. (A.7) yield $\{w_1(x_0), w_2(x_0), w_3(x_0)\} \simeq \{1.565, -0.740, 0.175\}$, which fully determines $\wp_a(t|x_0)$.

Acknowledgment

The financial support from the German Research Foundation (DFG) through the Emmy Noether Program "GO 2762/1-1" (to AG) is gratefully acknowledged.

References

1. M. von Smoluchowski, *Phys. Z.* **17**, 557–585 (1916).
2. H. Kramers, *Physica* **7**, 284–304 (1940).
3. A. Szabo, K. Schulten, and Z. Schulten, *J. Chem. Phys.* **72**, 4350–4357 (1980).
4. E. Ben-Naim, S. Redner, and F. Leyvraz, *Phys. Rev. Lett.* **70**, 1890–1893 (1993).
5. G. Oshanin, A. Stemmer, S. Luding, and A. Blumen, *Phys. Rev. E.* **52**, 5800–5805 (1995).
6. T. Guérin, N. Levernier, O. Bénichou, and R. Voituriez, *Nature* **534**, 356–359 (2016).
7. Y. Li, D. Debnath, P. K. Ghosh, and F. Marchesoni, *J. Chem. Phys.* **146**, 084104 (2017).
8. P. Hänggi, P. Talkner, and M. Borkovec, *Rev. Mod. Phys.* **62**, 251–341 (1990).
9. S. Redner, *A Guide to First-passage Processes*. Cambridge University Press, Cambridge (2001).
10. A. J. Bray, S. N. Majumdar, and G. Schehr, *Adv. Phys.* **62**, 225–361 (2013).
11. R. Metzler, G. Oshanin, and S. Redner, (eds.), *First-Passage Phenomena and Their Applications*. World Scientific Publishing, Singapore (2014).
12. O. Bénichou and R. Voituriez, *Phys. Rep.* **539**, 225–284 (2014).
13. R. Kopelman, *Science* **241**, 1620–1626 (1988).
14. D. ben Avraham and S. Havlin, *Diffusion and Reactions in Fractals and Disordered Systems*. Cambridge University Press, Cambridge (2000).
15. P. C. Bressloff and J. M. Newby, *Rev. Mod. Phys.* **85**, 135–196 (2013).
16. A. Godec and R. Metzler, *Phys. Rev. E.* **91**, 052134 (2015).
17. G. Vaccario, C. Antoine, and J. Talbot, *Phys. Rev. Lett.* **115**, 240601 (2015).
18. A. Godec and R. Metzler, *Sci. Rep.* **6**, 20349 (2016).
19. J.-F. Rupprecht, O. Bénichou, D. S. Grebenkov, and R. Voituriez, *J. Stat. Phys.* **158**, 192–230 (2015).
20. D. S. Grebenkov, *Phys. Rev. Lett.* **117**, 260201 (2016).

21. C. Mejía-Monasterio, G. Oshanin, and G. Schehr, *J. Stat. Mech.* P06022 (2011).
22. B. Munsky, I. Nemenman, and G. Bel, *J. Chem. Phys.* **131**, 235103 (2009).
23. G. Bel, B. Munsky, and I. Nemenman, *Phys. Biol.* **7**, 016003 (2010).
24. R. Grima and A. Leier, *J. Phys. Chem. B.* **121**, 13–23 (2017).
25. D. Schnoerr, B. Cseke, R. Grima, and G. Sanguinetti, *Phys. Rev. Lett.* **119**, 210601 (2017).
26. M. F. Weber and E. Frey, *Rep. Prog. Phys.* **80**, 046601 (2017).
27. D. Schnoerr, G. Sanguinetti, and R. Grima, *J. Phys. A.: Math. Theor.* **50**, 093001 (2017).
28. O. Bénichou, C. Chevalier, J. Klafter, B. Meyer, and R. Voituriez, *Nat. Chem.* **2**, 472–477 (2010).
29. B. Meyer, C. Chevalier, R. Voituriez, and O. Bénichou, *Phys. Rev. E.* **83**, 051116 (2011).
30. A. Godec and R. Metzler, *Phys. Rev. X.* **6**, 041037 (2016).
31. G. Kolesov, Z. Wunderlich, O. N. Laikova, M. S. Gelfand, and L. A. Mirny, *Proc. Natl. Acad. Sci. USA*, **104**, 13948–13953 (2007).
32. P. Fraser and W. Bickmore, *Nature* **447**, 413–417 (2007).
33. D. Hartich and A. Godec, *New J. Phys.* **20**, 112002 (2018).
34. N. G. van Kampen, *J. Stat. Phys.* **70**, 15–23 (1993).
35. H. van Beijeren, *J. Stat. Phys.* **110**, 1397–1410 (2003).
36. C. W. Gardiner, *Handbook of Stochastic Methods*, 3rd edn. Springer, Berlin (2004).
37. D. Hartich and A. Godec, *J. Stat. Mech.* **2019**, 024002 (2019).
38. D. Grebenkov, R. Metzler, and G. Oshanin, *Phys. Chem. Chem. Phys.* **20**, 16393–16401 (2018).
39. N. G. van Kampen, *Stochastic Processes in Physics and Chemistry*, North-Holland Personal Library, Elsevier, Amsterdam (3rd edn.) (2007).
40. A. J. F. Siegert, *Phys. Rev.* **81**, 617–623 (1951).
41. J. Keilson, *J. Appl. Prob.* **1**, 247–266 (1964).
42. F. Noé, S. Doose, I. Daidone, M. Löllmann, M. Sauer, J. D. Chodera, and J. C. Smith, *Proc. Natl. Acad. Sci. USA.* **108**, 4822–4827 (2011).
43. H. Yu, D. R. Dee, X. Liu, A. M. Brigley, I. Sosova, and M. T. Woodside, *Proc. Natl. Acad. Sci. USA.* **112**, 8308–8313 (2015).
44. K. Neupane, A. P. Manuel, and M. T. Woodside, *Nat. Phys.* **12**, 700 (2016).
45. D. R. Dee and M. T. Woodside, *Prion* **10**, 207–220 (2016).
46. K. Schulten, Z. Schulten, and A. Szabo, *J. Chem. Phys.* **74**, 4426–4432 (1981).
47. E. Sparre Andersen, *Math. Scand.* **1**, 263–285 (1953); *Math. Scand.* **2**, 194–222 (1954).
48. A. Singer, Z. Schuss, and D. Holcman, *J. Stat. Phys.* **122**, 465–489 (2006).
49. Z. Schuss, A. Singer, and D. Holcman, *Proc. Natl. Acad. Sci. USA.* **104**, 16098–16103 (2007).
50. J. Reingruber and D. Holcman, *Phys. Rev. Lett.* **103**, 148102 (2009).
51. S. Pillay, M. J. Ward, A. Peirce, and T. Kolokolnikov, *Multiscale Model. Simul.* **8**, 803–835 (2010).

52. S. A. Isaacson, A. J. Mauro, and J. Newby, *Phys. Rev. E.* **94**, 042414 (2016).
53. D. S. Grebenkov and G. Oshanin, *Phys. Chem. Chem. Phys.* **19**, 2723–2739 (2017).
54. R. Grone, K. H. Hoffmann, and P. Salamon, *J. Phys. A.: Math. Theor.* **41**, 212002 (2008).

Chapter 12

Spatially Inhomogeneous Search Strategies

Anne Hafner and Heiko Rieger*

Department of Theoretical Physics and Center for Biophysics, Saarland University, 66123 Saarbrücken, Germany

*h.rieger@mx.uni-saarland.de

The efficiency of intracellular reactions which are driven by motor-assisted transport strongly depends on the spatial organization of the cytoskeleton. The cytoskeleton is a highly complex filament network which is generally neither homogeneous nor isotropic. In cells with a centrosome, microtubules emanate radially from the center, whereas actin filaments populate the cortex underneath the plasma membrane in a random manner. While intermittent search strategies with stochastic transitions between a slow reactive phase and a fast non-reactive phase have been shown to be advantageous in homogeneous, isotropic environments, the effect of a realistic global cytoskeleton topology has only very recently gained scientific interest. In this chapter, we review the progress in analyzing the efficiency of spatially inhomogeneous search strategies.

1. Introduction

Random search processes are ubiquitous in nature and are fundamental to chemical kinetics: Two reaction partners performing a random motion in space first have to find each other before they eventually can bind to each other. In particular, at low concentrations

of the reaction partner the random search process becomes the
rate-limiting factor. In conventional reaction-diffusion systems, the
reaction-partners are subject to thermal Brownian motion, and
the efficiency of the search process depends solely on the diffusion
constant D, the initial distance of the reaction partners R, the reaction range a, and the size of the search domain V. For instance, in
three space dimensions the mean first passage time (MFPT) for a
purely diffusive search process is $T^{3d}_{\text{MFPT}} = V/(4\pi D)(a^{-1} - R^{-1})$ and
in two dimensions, $T^{3d}_{\text{MFPT}} = A/(2\pi D) \ln R/a$.[1]

In biological systems, in particular in living cells, the reaction
kinetics are frequently enhanced by various mechanisms: One example is reactions involved in genomic transcription, where facilitated
diffusion enhances the efficiency of the search of DNA-binding proteins for their specific binding site on a DNA molecule by alternating
between linear 1D diffusion along the DNA molecule and 3D volume
excursion events between successive dissociation from and rebinding
to DNA.[2,3] Another prominent example is active intracellular transport of proteins and also larger objects like vesicles, endosomes, and
mitochondria, which are equipped with molecular motors that can
randomly bind to and unbind from the actin or microtubule filaments
of the cell's cytoskeleton.[4] The resulting motion of these particles
alternates stochastically between two modes: a diffusive mode and a
ballistic mode along the direction of the filament when a molecular
motor is bound. Both examples are also representative of intermittent
search, which means that during the fast motility mode the searcher
cannot find the target (i.e., particles cannot bind).[5]

It has been shown that switching between the two motility modes
with certain rates can dramatically increase the search efficiency
defined by the MFPT to find the target,[6,7] implying an enhanced
reaction kinetics for molecular motor-assisted search processes in
cells.[8] The specific values of the transition rates between the two
motility modes is commonly denoted as a "search strategy" to remind
one of the fact that parameters could be varied to optimize the search
efficiency. If these parameters are constant in space, we denote this as
a "spatially homogeneous" strategy, and when they can vary in space,

we denote this as a "spatially inhomogeneous" strategy. In this sense, the cellular cytoskeleton represents, with respect to motor-assisted random search, a spatially inhomogeneous search strategy, since cytoskeletal filaments are not homogeneously distributed in space but have a characteristic spatial organization as detailed below.

In this chapter, we will review the recent progress in analyzing the efficiency of spatially inhomogeneous search strategies. Their advantage is obvious whenever additional information about the target location is available, like a preferential location in a particular subvolume of the search domain or at its boundary. But we will see that inhomogeneous search strategies can even be superior to homogeneous strategies in cases when the random target location is homogeneously distributed. In the following, we will specifically address three standard search problems encountered in cellular chemical kinetics, as illustrated in Fig. 1: The narrow escape problem (the target is a specific area on the domain boundary), the reaction kinetics problem (the target is randomly distributed in the search domain), and the reaction-escape problem, which is a combination of the first two. Then we will discuss the generality of the results and give an outlook to future applications.

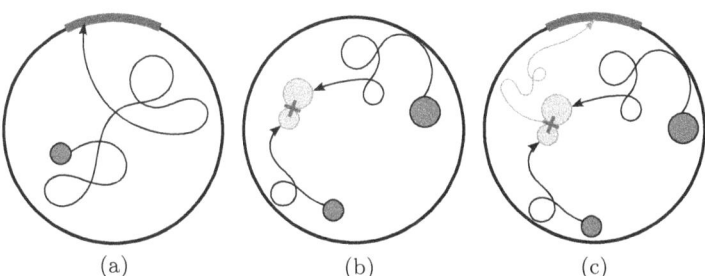

(a) (b) (c)

Fig. 1. Three standard search problems are encountered in cellular chemical kinetics: (a) The narrow escape problem which emerges during transport to a specific region on the domain boundary. (b) The reaction kinetics problem which considers the arrival of a searcher at a motile or immotile target in the bulk of the domain. (c) The combined reaction-escape problem which occurs when cargo must be delivered to a narrow area on the boundary only after tethering to another particle beforehand.

2. The Cytoskeleton — A Specific Spatially Inhomogeneous Search Strategy

The efficiency of motor-assisted transport strongly depends on the spatial organization of the cytoskeleton. The cytoskeleton is a highly complex filament network which is generally neither homogeneous nor isotropic. In cells with a centrosome, microtubules emanate radially from the center, whereas actin filaments are accumulated with random orientations in the cortex underneath the plasma membrane.[4]

While intermittent search strategies with stochastic transitions between a slow reactive phase and a fast non-reactive phase have been shown to be advantageous in homogeneous, isotropic environments,[5–11] the effect of a realistic global cytoskeleton topology has only very recently gained scientific interest.

A spatially inhomogeneous diffusion constant is included in models of surface-mediated diffusion.[12–19] In Ref. 12, Bénichou et al. investigated the narrow escape problem in 2D and 3D spheres S of radius R. A particle performs alternating phases of bulk diffusion with diffusion constant D_2 and surface-mediated diffusion along the surface of the sphere ∂S with diffusion constant D_1. When detaching from the surface after an exponentially distributed time scale with rate λ, the particle is radially delocated at a distance $a \ll R$ from the surface into the bulk, where it exhibits bulk diffusion until it reaches the surface again and eventually the small target on ∂S is detected. Remarkably, the MFPT of a searcher, which is initially uniformly distributed on ∂S, can be minimized as a function of the desorption rate λ in dependence of D_1/D_2. In Ref. 17, Calandre et al. further showed that the MFPT to a target in the bulk, reminiscent of the reaction problem, can be minimized as a function of the desorption rate λ, if the surface diffusion constant D_1 is sufficiently large (in particular, larger than the bulk diffusion constant D_2). Consequently, a spatial inhomogeneity of the diffusion constant can substantially increase the search efficiency.

Cherstvy et al. investigated the transport of particles from the center to the surface of a circular disk.[20] The particle performs

Brownian motion with a diffusion constant $D(r) = D_0 \frac{A}{A+r^2}$, $A > 0$, which is a function of the radial position r. The diffusion constant is thus the highest close to the center and gradually decreases with increasing distance. For $r \gg A$, the diffusion constant scales like a power-law $D(r) \sim 1/r^2$, whereas for $r \ll A$, diffusion is almost Brownian. With the aid of computer simulations, Cherstvy *et al.* found that the time scale $t_{1/2}$, at which the fastest half of the population arrives at the membrane, is defined by two asymptotes. Namely, the one with the slowest diffusivity $D(r = R)$ and the one with average diffusivity $\langle D \rangle = \int_a^R D(r) \mathrm{d}r / (2(R^2 - a^2))$, such that $t_{1/2}$ scales like R^4 in the leading order.[20]

Ando *et al.* investigated the influence of the topology of the cytoskeleton on the transport efficiency of particles which travel from the nucleus to an arbitrary position alongside the membrane.[21] Their model system is a 2D circular disk of radius $R = 10$ μm, which possesses a nucleus of radius R_n. Tracer particles are initially positioned on the surface of the nucleus. In the cytoplasm, they perform Brownian motion with diffusion constant $D = 0.011$ μm^2/s. The cytoskeleton is modeled as a shell of width w whose inner radius is positioned at R_a. In order to account for active transport, the diffusion constant is increased within this shell to $D_a = 100D$. They found that the MFPT can be minimized for shells positioned close to the nucleus if $R_n \gtrsim R/4$. Ando *et al.* further explicitly simulated filaments with fixed length which are randomly distributed in the cytoplasm. Tracer particles experience alternations of ballistic motion alongside the filaments and diffusion in the bulk. They found that the transport efficiency from the nucleus to the membrane is increased if the filament polarities collectively point toward the membrane.

3. Spatially Inhomogeneous Cytoskeleton Enhances Intracellular Reaction Kinetics

In essence, the specific spatial organization of the cytoskeleton represents, in conjunction with motor-assisted transport, a search strategy in spatially inhomogeneous environments, which is intermittent if the searcher cannot find the target (i.e., bind to the reaction partner) in the ballistic mode. In order to study the efficiency of spatially inhomogeneous search strategies, a random walk model with two alternating motility modes was formulated in Refs. 22–25: (i) a ballistic motion state at velocity v, which is associated to directed transport by molecular motors between binding and unbinding events, and (ii) a diffusive state with diffusivity D in which motors are unbound. Note that the case $D = 0$ corresponds to a model with arrest states studied in Refs. 24 and 25. The limit of a vanishing diffusion constant is biologically relevant for intracellular cargo, such as vesicles, mitochondria, or macromolecules, which experience size-dependent subdiffusion in the crowded cytoplasm and thus undergo effectively stationary states.[26–29] But more importantly, since a single cargo is typically attached to several motor proteins concurrently, a full dissociation of the filament is rather unlikely.[30] Instead, arrest states at filament crossings are observed.[31–37] The speed v is assumed to be constant. Transitions between the motility modes are determined by a constant attachment rate k and detachment rate k'. Generally, the rates can also be space-dependent.

The cytoskeleton structure in a spherical cell of radius R is idealized by the probability density $\rho_\Omega(\boldsymbol{r})$ to choose a direction Ω conditionally on the switch from the diffusive to the ballistic mode at position $\boldsymbol{r}$ and can, for simplicity, be parameterized as follows:

$$\rho_\Omega(\boldsymbol{r}) = \begin{cases} p\,\delta(\Omega - \Omega'(\boldsymbol{r})) + (1-p)\,\delta(\Omega - \Omega'(-\boldsymbol{r})) & \text{for } 0 < r < R - \delta, \\ 1/2\pi \text{ (in 2D)}; 1/4\pi \text{ (in 3D)} & \text{for } R - \delta < r < R, \end{cases}$$

(1)

where $\Omega'(\boldsymbol{r})$ denotes the direction defined by the position vector $\boldsymbol{r}$ and p denotes the probability to move radially outward. The probability p represents the contribution of kinesins and dyneins on the apparent motion of cargo. For $p = 1$, the transport is solely managed by kinesins in the cell interior, and for $p = 0$ only dyneins are active. In contrast,

$p = 0.5$ is associated to an equal distribution of active kinesins and dyneins on the cargo. Instead of isotropic distributions in the periphery, the framework allows to study arbitrary distributions of actin filaments in the cortex, see Refs. 24 and 25. The exact distribution of orientations is objective of ongoing research, but it is reported that actin filaments align to microtubules[38] and are tangentially oriented to the membrane in cellular blebs.[39]

The distribution $\rho_\Omega(\boldsymbol{r})$ together with the state transition rates k and k' defines a search strategy which is generally inhomogeneous and anisotropic. However, $\delta = R_\mathrm{m}$ leads to a spatially homogeneous and isotropic search strategy.

The proposed model allows the study of diverse search tasks. Here, the focus is on the three different, biologically relevant search problems mentioned in the introduction: The narrow escape, the reaction and the reaction-escape problem. At time $t = 0$, the particle starts at position $\boldsymbol{r_0}$, which may either be the cell center or a uniformly distributed position within the cell. Apart from the stochastic detachment events with rate k', a ballistically moving particle switches automatically to the diffusive mode at the MTOC ($r = 0$), at the inner border of the actin cortex ($r = R - \delta$), and at the cell membrane ($r = R$). The particle is propagated until termination of the respective search problem. For the narrow escape problem, the search is terminated when the particle hits the plasma membrane at the exit zone of opening angle α_exit, as illustrated in Fig. 2. In the case of the reaction problem, the search is terminated by encounter of searcher and target particle $|\boldsymbol{r}^\mathrm{S} - \boldsymbol{r}^\mathrm{T}| \leq R_\mathrm{d}$, see Fig. 2, if both particles are in the diffusive state. In the reaction-escape problem, searcher and target particle first have to react before the product particle can be transported to a specific zone on the membrane of the cell. The efficiency of a search strategy, defined by a specific cytoskeleton organization and transition rates, is measured in terms of the MFPT to target detection with the aid of an event-driven Monte Carlo algorithm.[40] In the following dimensionless spatial and temporal coordinates $\boldsymbol{r} \mapsto \boldsymbol{r}/R$, $t \mapsto vt/R$ and parameters $D \mapsto D/vR$, $k \mapsto Rk/v$, $k' \mapsto Rk'/v$, $\delta \mapsto \delta/R_\mathrm{m}$ are used.

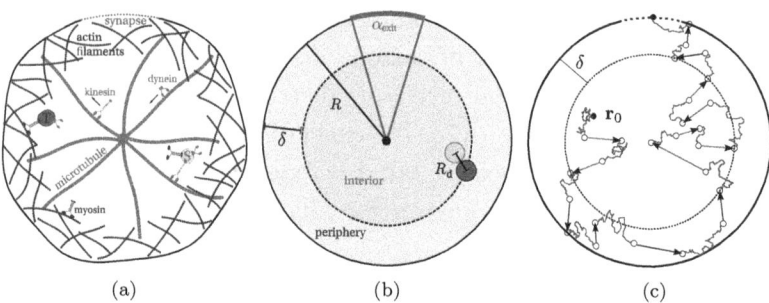

Fig. 2. (a) Targeted intracellular transport by molecular motors is a stochastic process which strongly depends on the spatially inhomogeneous organization of the cytoskeleton. (b) The cytoskeleton structure in a spherical cell of radius R is idealized by introducing a well-defined actin cortex of width δ. An exit zone on the cell boundary is characterized by the angle α_{exit} and the detection distance of two particles is determined by R_{d}. (c) In the narrow escape problem, particles which are initially located at r_0 perform stochastic motion with alternating phases of ballistic and diffusive motion until they reach the exit. (Parts (a) and (b) are reproduced from Ref. 25 with the permission of Elsevier; Part (c) is reproduced from Ref. 22 with the permission of American Physical Society.)

3.1. *Narrow escape problem*

First, the search for a specific area on the domain boundary is considered, which is the so-called narrow escape problem.[41,42] A prominent example is directed secretion by immune cells which requires active transport of toxic vesicles toward the immunological synapse in order to kill tumorigenic or virus infected cells.[43-46]

In order to demonstrate the gain in search efficiency by a spatially inhomogeneous search strategy, corresponding to $0 < \delta < 1$, a homogeneous cytoskeleton with $\delta = 1$ is considered to determine the optimal transition rates $k_{\text{opt}}(D)$ and $k'_{\text{opt}}(D)$. It turns out that the optimal detachment rate $k'_{\text{opt}} = 0$ is zero for all D, whereas the optimal attachment rate $k_{\text{opt}}(D)$ increases with decreasing diffusivity D, such that $k_{\text{opt}} = \infty$ for $D = 0$. Consequently, a motion pattern without directional changes in the bulk of the cell constitutes an optimal search strategy for a homogeneous cytoskeleton.

In Refs. 22–25, it is investigated whether an inhomogeneous filament structure ($\delta < 1$) has the potential to solve the narrow escape problem more efficiently than its homogeneous counterpart.

To answer this question, the influence of the actin cortex width δ and the probability for radially outward transport p on the MFPT are evaluated for the optimal parameters $k_{opt}(D)$ and k'_{opt}. Remarkably, for large probabilities of anterograde transport p the MFPT exhibits a minimum at small widths of the actin cortex $0 < \delta < 1$. The minimum is most pronounced for arrest states, i.e., $D = 0$, but it is also found for $D > 0$, as shown in Fig. 3. Moreover, exclusive radial outward transport ($p = 1$) in the ballistic mode represents the best strategy for the escape problem, which is plausible since the target location is on the boundary. This phenomenon is largely robust against changes in the transition rates k and k'. A small cortex width δ significantly reduces the MFPT, which emphasizes the

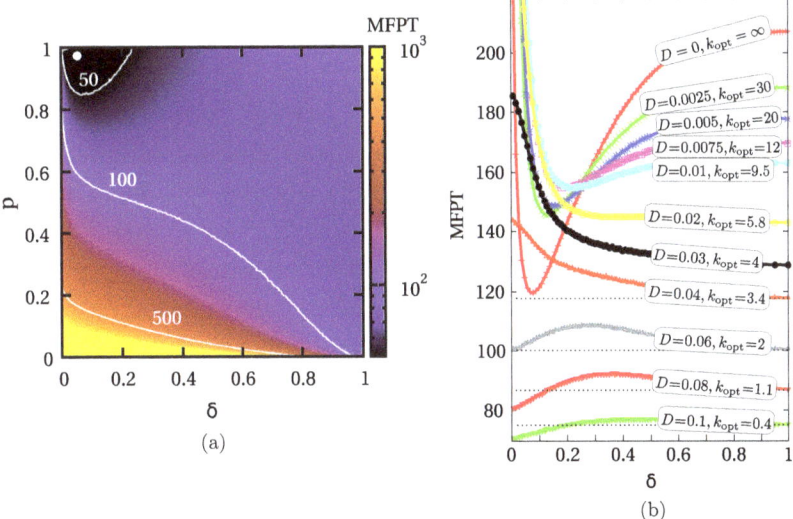

(a)

(b)

Fig. 3. Narrow escape problem for inhomogeneous spatial organizations of the cytoskeleton. (a) The MFPT versus δ and p for $D = 0$, $k'_{opt} = 0$, and $k_{opt} = \infty$, which constitutes the optimal choice of transition rates for a spatially homogeneous cytoskeleton, i.e., $\delta = 1$, in the case of $\alpha_{exit} = 0.1$ and 2D domains. (b) MFPT as a function of δ for different diffusivities D and $p = 1$ (i.e., exclusively radial outward transport in the ballistic mode) using the optimal rates $k'_{opt} = 0$ and $k_{opt}(D)$ for $\delta = 1$ in the case of $\alpha_{exit} \approx 0.1433$ and 3D domains. (Part (a) is reproduced from Ref. 25 with the permission of Elsevier; Part (b) is reproduced from Ref. 22 with the permission of American Physical Society.)

general enhancement of the search efficiency by a spatially inhomogeneous filament structure. Note that for the narrow escape problem, it is irrelevant whether the search is intermittent or not since the searching particle switches always to the diffusive mode when reaching the boundary, in particular when reaching the escape area.

3.2. Reaction problem

Next, the efficiency of spatially inhomogeneous search strategies for reaction with an immobile particle is addressed. When the searcher is in the diffusive mode and its position r comes closer to the target than $|r^S - r^T| \leq R_d$, the search is successfully finished and for the moment it is assumed that the search is intermittent, i.e., the searcher cannot find the target in the ballistic mode. The target position r^T is either homogeneously distributed within the search domain or it is preferentially located in a specific subvolume close to the center $r^T \leq 0.5$ with probability w.

In contrast to the narrow escape problem, non-trivial optimal transition rates arise for the homogeneous reaction problem. The optimal attachment rate k_{opt} and detachment rate k'_{opt} depend on the diffusivity D and the detection radius R_d. But for $D = 0$, $k_{opt} = \infty$ (i.e., the absence of arrests) is optimal for all values of R_d.

In order to investigate the impact of inhomogeneous cytoskeleton organizations, the MFPT to an immobile bulk target is evaluated in Refs. 22–25 as shown in Fig. 4 in dependence of p_{antero} and δ, where the optimal transition rates for the homogeneous counterparts are applied. For all widths δ of the actin cortex, an optimal strategy to detect an immobile target within the cell is defined by $p = 0.5$ even if the target is preferentially located close to the center. For a fixed cortex width $\delta = 0.1$, the MFPT is minimized for p close to 0.5 even for large w and also for fixed non-optimal rates k, k', as shown in Fig. 4(c). While for $D = 0$ a homogeneous isotropic cytoskeletal network with $\delta = 1$ is most efficient, as shown in Fig. 4(a), for $D \neq 0$ again a thin cortex $\delta \ll 1$ may yield a much smaller search time for $p = 0.5$, as displayed in Fig. 4(b).

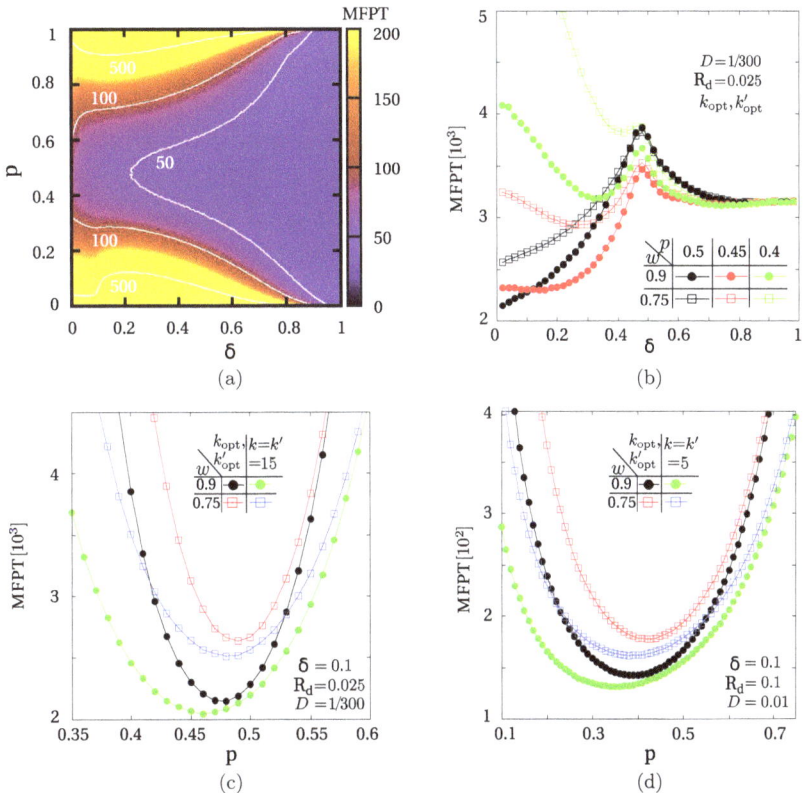

Fig. 4. Reaction problem for inhomogeneous spatial organizations of the cytoskeleton. (a) The MFPT versus δ and p for $D = 0$ and $R_d = 0.1$ in 2D spheres. The optimal transition rates for the homogeneous counterpart are applied, i.e., $k'_{opt} = 7$, $k_{opt} = \infty$. (b) MFPT as a function of δ for different values of p and w in a 3D search process with intermittent diffusion with $D = 1/300$ and $R_d = 0.025$. The optimal rates $k_{opt}(D, R_d)$ from the homogeneous case $\delta = 1$ are applied. (c) MFPT as in (b) but as a function of p for a fixed cortex width $\delta = 0.1$ and different rates k and k' and probabilities w. (d) The same as in (c) but for $D = 0.01$ and $R_d = 0.1$. (Part (a) is reproduced from Ref. 25 with the permission of Elsevier; Parts (b)–(d) are reproduced from Ref. 22 with the permission of American Physical Society.)

Note that similar results are obtained when the searcher can find the target also during the ballistic motion phase (i.e., non-intermittent search) and an optimal strategy for motile targets is given by $p = 0$ and $\delta = 0$, as studied in Ref. 25 for $D = 0$.

3.3. *Reaction-escape problem*

Finally, the efficiency of inhomogeneous search strategies for the combination of reaction and escape problem is discussed. Cargo first has to bind to a reaction partner before the product can be delivered to a specific area on the cell boundary. A prominent example is the docking of lytic granules at the immunological synapse of cytotoxic T-lymphocytes that requires the pairing with CD3 endosome beforehand.[47] The total MFPT of the reaction-escape problem is composed of the $\text{MFPT}_{\text{react}}$ for the reaction problem and the $\text{MFPT}_{\text{escape}}$ for the following escape problem of the product particle to the exit zone on the membrane, i.e., $\text{MFPT} = \text{MFPT}_{\text{reac}} + \text{MFPT}_{\text{esc}}$.

In order to explore the influence of the spatial inhomogeneity of the cytoskeleton, in Refs. 22–25 the MFPT for the reaction-escape problem is measured as a function of the cortex width δ for the transition rates k_{opt} and k'_{opt} which are optimal in the homogeneous case

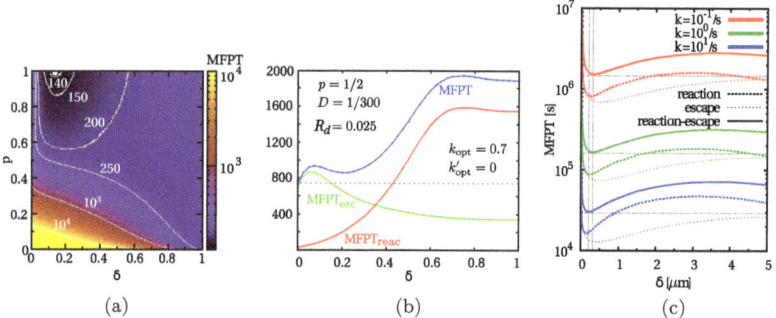

(a)   (b)   (c)

Fig. 5.   Reaction-escape problem for inhomogeneous spatial organizations of the cytoskeleton. (a) The MFPT in dependence of the cortex width δ and the probability for radially outward transport p for 2D search processes with $D = 0$. Transition rates $k'_{\text{opt}} = 5$, $k_{\text{opt}} = 22$, and parameters $R_{\text{d}} = 0.1$, $\alpha_{\text{exit}} = 0.1$ are applied. (b) MFPT for the reaction, the escape, and the combined reaction-escape problem as a function of δ in a 3D search process with intermittent diffusion with $D = 1/300$. The optimal rates k_{opt}, k'_{opt}, and parameters $p = 0.5$, $R_{\text{d}} = 0.025$, $\alpha_{\text{exit}} \approx 0.1433$ are applied. (c) The MFPT of the escape, the reaction and the reaction-escape problem in dependence of δ for a 3D cell of radius $R = 5\,\mu\text{m}$ with $\alpha_{\text{exit}} = 0.2$, $R_{\text{d}} = 0.1\,\mu\text{m}$, $v = 1\,\mu\text{m}/s$, $p_{\text{antero}} = 1$, $k' = 10/s$, $k \in \{10^{-1}/\text{s}; 10^{0}/\text{s}; 10^{1}/\text{s}\}$, and $D = 0$. (Parts (a) and (c) are reproduced from Ref. 25 with the permission of Elsevier; Part (b) is reproduced from Ref. 22 with the permission of American Physical Society.)

$\delta = 1$. Figure 5(a) shows that in the case of $D = 0$ a superior search strategy for the reaction-escape problem is defined by a high probability p of radially outward motion and a thin actin cortex δ. But a small cortex δ also reduces the total MFPT in comparison to the homogeneous strategy $\delta = 1$ for intermittent diffusion with $D \neq 0$. In general, inhomogeneous search strategies with $0 < \delta < 1$ which are more efficient than the homogeneous counterpart also exist for non-optimal transition rates k and k'. Remarkably, Fig. 5(c) indicates that the optimal width of the actin cortex $\delta^{\mathrm{opt}} = 0.3\,\mu$m, predicted by our model under biologically reasonable conditions, is in good agreement with experimental data.[48,49]

4. Discussion and Outlook

We reviewed the efficiency of spatially inhomogeneous intermittent search strategies and pointed out the importance of the spatial organization of the cytoskeleton for targeted intracellular transport, which occurs when cargo particles have to find reaction partners or specific target areas inside a cell. Remarkably, the confinement of randomly oriented filaments to a thin cortex is not a handicap for the cell, but can substantially increase the efficiency of diverse transport tasks. The best strategy for the narrow escape problem is to allow only radially outward transport from the center toward a thin cortex underneath the boundary, where multidirectional transport is possible. This thin cortex allows an accelerated random motion along the boundary to find the escape region. A similar result holds for the reaction problem, in which the target is located in the bulk of the domain: here again, superior inhomogeneous strategies exist, but the optimal probability for radially outward transport is now around $p = 0.5$. The reaction-escape problem combines both scenarios and the optimal forward/backward radial transport probability p depends on the size ratio of target and escape region. The basic mechanism underlying a higher search efficiency is actually reminiscent of an acceleration of purely diffusive search kinetics by following boundaries with an increased diffusivity.[12–19] For intracellular reaction, kinetics cells are able to economically realize efficient search strategies by intermittent transport on a cytoskeleton with specific spatial structure. Instead of supporting a resource demanding isotropic homogeneous filament network it is sufficient, and often even more efficient, to establish just a thin actin cortex underneath the cell membrane.

It is shown that first passage times of reactions in biological cells actually are broadly distributed, such that the most likely value may deviate significantly from the mean.[51–53] Consequently, it is worth studying the full distribution of first passage times for spatially inhomogeneous search strategies in more detail. And, in particular, considering multiple searcher and target particles within a cell is promising and opens a new range of questions. Spatially inhomogeneous search strategies potentially also reduce the cover time to several targets.[50] The study of extreme statistics (i.e., when does the first x particle arrive at a given target) is certainly also relevant for various biochemical reactions.

Acknowledgments

This work was financially supported by the German Research Foundation (DFG) within the Collaborative Research Center SFB 1027. We thank Karsten Schwarz for fruitful collaboration and Olivier Bénichou, Gleb Oshanin, Ralf Metzler, and Raphaël Voituriez for stimulating discussions.

References

1. S. Condamin, O. Bénichou, and M. Moreau, First-passage times for random walks in bounded domains, *Phys. Rev. Lett.* **95**, 260601 (2005).
2. O. G. Berg, R. B. Winter, and P. H. von Hippel, Diffusion-driven mechanisms of protein translocation on nucleic acids. 1. Models and theory, *Biochemistry* **20**, 6929 (1981); P. H. von Hippel, From Simple DNA–Protein interactions to the macromolecular machines of gene expression, *Ann. Rev. Biophys. Biomol. Struc.* **36**, 79 (2007).
3. J. Gorman and E. C. Greene, Visualizing one-dimensional diffusion of proteins along DNA, *Nature Struct. Mol. Biol.* **15**, 768 (2008).
4. B. Alberts, *et al.*, *Molecular Biology of the Cell.* Garland, New York, 6th edn. (2014).

5. O. Bénichou, C. Loverdo, M. Moreau, and R. Voituriez, Intermittent search strategies, *Rev. Mod. Phys.* **83**, 81 (2011).
6. O. Bénichou, M. Coppey, M. Moreau, P. H. Suet, and R. Voituriez, Optimal search strategies for hidden targets, *Phys. Rev. Lett.* **94**, 198101 (2005).
7. C. Loverdo, O. Bénichou, M. Moreau, and R. Voituriez, Robustness of optimal intermittent search strategies in one, two, and three dimensions, *Phys. Rev. E.* **80**, 031146 (2009).
8. C. Loverdo, O. Benichou, M. Moreau, and R. Voituriez, Enhanced reaction kinetics in biological cells, *Nat. Phys.* **4**, 134–137 (2008).
9. C. Loverdo, O. Bénichou, M. Moreau, and R. Voituriez, Reaction kinetics in active media, *J. Stat. Mech.* **9**, P02045 (2009).
10. O. Bénichou, M. Coppey, M. Moreau, and R. Voituriez, Intermittent search strategies: When losing time becomes efficient, *Europhys. Lett.* **75**, 349–354 (2006).
11. O. Bénichou, C. Loverdo, M. Moreau, and R. Voituriez, Two-dimensional intermittent search processes: An alternative to Lévy flight strategies, *Phys. Rev. E.* **74**, 020102 (2006).
12. O. Bénichou, D. Grebenkov, P. Levitz, C. Loverdo, and R. Voituriez, Optimal reaction time for surface-mediated diffusion, *Phys. Rev. Lett.* **105**, 150606 (2010).
13. O. Bénichou, D. S. Grebenkov, P. E. Levitz, C. Loverdo, and R. Voituriez, Mean first-passage time of surface-mediated diffusion in spherical domains, *J. Stat. Phys.* **142**, 657–685 (2011).
14. T. Calandre, O. Bénichou, D. S. Grebenkov, and R. Voituriez, Interfacial territory covered by surface-mediated diffusion, *Phys. Rev. E.* **85**, 051111 (2012).
15. J.-F. Rupprecht, O. Bénichou, D. S. Grebenkov, and R. Voituriez, Kinetics of active surface-mediated diffusion in spherically symmetric domains, *J. Stat. Phys.* **147**, 891–918 (2012).
16. J.-F. Rupprecht, O. Bénichou, D. S. Grebenkov, and R. Voituriez, Exact mean exit time for surface-mediated diffusion, *Phys. Rev. E.* **86**, 041135 (2012).
17. T. Calandre, O. Bénichou, and R. Voituriez, Accelerating search kinetics by following boundaries, *Phys. Rev. Lett.* **112**, 230601 (2014).
18. T. Calandre, O. Bénichou, D. S. Grebenkov, and R. Voituriez, Splitting probabilities and interfacial territory covered by two-dimensional and three-dimensional surface-mediated diffusion, *Phys. Rev. E.* **89**, 012149 (2014).
19. O. Bénichou, D. Grebenkov, L. Hillairet, L. Phun, R. Voituriez, and M. Zinsmeister, Mean exit time for surface-mediated diffusion: Spectral analysis and asymptotic behavior, *Anal. Math. Phys.* **5**, 321–362 (2015).
20. A. G. Cherstvy, A. V. Chechkin, and R. Metzler, Particle invasion, survival, and non-ergodicity in 2d diffusion processes with space-dependent diffusivity, *Soft Matter* **10**, 1591–1601 (2014).

21. D. Ando, N. Korabel, K. C. Huang, and A. Gopinathan, Cytoskeletal network morphology regulates intracellular transport dynamics, *Biophys. J.* **109**, 1574–1582 (2015).

22. K. Schwarz, Y. Schröder, B. Qu, M. Hoth, and H. Rieger, Optimality of spatially inhomogeneous search strategies, *Phys. Rev. Lett.* **117**, 068101 (2016).

23. K. Schwarz, Y. Schröder, and H. Rieger, Numerical analysis of homogeneous and inhomogeneous intermittent search strategies, *Phys. Rev. E.* **94**, 042133 (2016).

24. A. E. Hafner and H. Rieger, Spatial organization of the cytoskeleton enhances cargo delivery to specific target areas on the plasma membrane of spherical cells, *Phys. Biol.* **13**, 066003 (2016).

25. A. E. Hafner and H. Rieger. Spatial cytoskeleton organization supports targeted intracellular transport. *Biophys. J.* **114**, 1420–1432 (2018).

26. K. Luby-Phelps, P. E. Castle, D. L. Taylor, and F. Lanni, Hindered diffusion of inert tracer particles in the cytoplasm of mouse 3t3 cells, *P. Natl. Acad. Sci. USA.* **84**, 4910–4913 (1987).

27. O. Seksek, J. Biwersi, and A. S. Verkman, Translational diffusion of macromolecule-sized solutes in cytoplasm and nucleus. *J. Cell Biol.* **138**, 131–142 (1997).

28. M. Arrio-Dupont, G. Foucault, M. Vacher, P. F. Devaux, and S. Cribier, Translational diffusion of globular proteins in the cytoplasm of cultured muscle cells. *Biophys. J.* **78**, 901–907 (2000).

29. M. Weiss, M. Elsner, F. Kartberg, and T. Nilsson, Anomalous subdiffusion is a measure for cytoplasmic crowding in living cells. *Biophys. J.* **87**, 3518–3524 (2004).

30. C. Appert-Rolland, M. Ebbinghaus, and L. Santen, Intracellular transport driven by cytoskeletal motors: General mechanisms and defects. *Phys. Rep.* **593**, 1–59 (2015).

31. M. Y. Ali, E. B. Krementsova, G. G. Kennedy, R. Mahaffy, T. D. Pollard, K. M. Trybus, and D. M. Warshaw, Myosin va maneuvers through actin intersections and diffuses along microtubules. *P. Natl. Acad. Sci. USA.* **104**, 4332–4336 (2007).

32. J. L. Ross, M. Y. Ali, and D. M. Warshaw, Cargo transport: molecular motors navigate a complex cytoskeleton. *Curr. Opin. Cell Biol.* **20**, 41–47 (2008).

33. J. L. Ross, H. Shuman, E. L. F. Holzbaur, and Y. E. Goldman, Kinesin and dynein-dynactin at intersecting microtubules: Motor density affects dynein function. *Biophys. J.* **94**, 3115–3125 (2008).

34. H. W. Schröder, C. Mitchell, H. Shuman, E. L. F. Holzbaur, and Y. E. Goldman, Motor number controls cargo switching at actin-microtubule intersections in vitro. *Curr. Biol.* **20**, 687–696 (2010).

35. Š. Bálint, I. Verdeny-Vilanova, Á. S. Álvarez, and M. Lakadamyali, Correlative live-cell and superresolution microscopy reveals cargo transport dynamics at microtubule intersections. *P. Natl. Acad. Sci. USA.* **110**, 3375–3380 (2013).

36. M. Lakadamyali, Navigating the cell: How motors overcome roadblocks and traffic jams to efficiently transport cargo. *Phys. Chem. Chem. Phys.* **16**, 5907 (2014).

37. I. Verdeny-Vilanova, F. Wehnekamp, N. Mohan, Á. S. Álvarez, J. S. Borbely, J. J. Otterstrom, D. C. Lamb, and M. Lakadamyali, 3d motion of vesicles along microtubules helps them to circumvent obstacles in cells. *J. Cell Sci.* **130**, 1904–1916 (2017).

38. M. P. López, F. Huber, I. Grigoriev, M. O. Steinmetz, A. Akhmanova, G. H. Koenderink, and M. Dogterom, Actin-microtubule coordination at growing microtubule ends, *Nat. Commun.* **5**, 4778 (2014).

39. M. Bovellan, Y. Romeo, M. Biro, A. Boden, C. Priyamvada, A. Yonis, M. Vaghela, M. Fritzsche, D. Moulding, R. Thorogate, A. Jégou, A. J. Thrasher, G. Romet-Lemonne, P. P. Roux, E. K. Paluch, and G. Charras, Cellular control of cortical actin nucleation, *Curr. Biol.* **24**, 1628–1635 (2014).

40. K. Schwarz and H. Rieger, Efficient kinetic Monte Carlo method for reaction-diffusion problems with spatially varying annihilation rates, *J. Comput. Phys.* **237**, 396–410 (2013).

41. Z. Schuss, A. Singer, and D. Holcman, The narrow escape problem for diffusion in cellular microdomains, *Proc. Nat. Acad. Sci. USA.* **104**, 16098 (2007).

42. Z. Schuss, The narrow escape problem — a short review of recent results, *J. Sci. Comput.* **53**, 194 (2012).

43. A. Grakoui, S. K. Bromley, C. Sumen, M. M. Davis, A. S. Shaw, P. M. Allen, and M. L. Dustin, The immunological synapse: A molecular machine controlling t cell activation, *Science* **285**, 221 (1999).

44. S. K. Bromley, W. R. Burack, K. G. Johnson, K. Somersalo, T. N. Sims, C. Sumen, M. M. Davis, A. S. Shaw, P. M. Allen, and M. L. Dustin, The immunological synapse, *Ann. Rev. Immun.* **19**, 375 (2001).

45. K. L. Angus and G. M. Griffiths, Cell polarization and the immunological synapse, *Curr. Opinion Cell Biol.* **25**, 85 (2013).

46. A. T. Ritter, K. L. Angus, and G. M. Griffiths, The role of the cytoskeleton at the immunological synapse, *Immunol. Rev.* **256**, 107 (2013).

47. B. Qu, V. Pattu, C. Junker, E. C. Schwarz, S. S. Bhat, C. Kummerow, M. Marshall, U. Matti, F. Neumann, M. Pfreundschuh, U. Becherer, H. Rieger, J. Rettig, and M. Hoth, Docking of lytic granules at the immunological synapse in human ctl requires vti1b-dependent pairing with cd3 endosomes, *J. Immunol.* **186**, 6894 (2011).

48. G. Salbreux, G. Charras, and E. Paluch, Actin cortex mechanics and cellular morphogenesis, *Trends Cell Biol.* **22**, 536–545 (2012).

49. A. G. Clark, K. Dierkes, and E. Paluch, Monitoring actin cortex thickness in live cells, *Biophys. J.* **105**, 570–580 (2013).

50. M. Chupeau, O. Bénichou, and R. Voituriez, Cover times of random searches, *Nat. Phys.* **11**, 844–848 (2015).

51. C. Mejía-Monasterio, G. Oshanin, and G. Schehr, First passages for a search by a swarm of independent random searchers, *J. Stat. Mech.* P06022 (2011).

52. T. G. Mattos, C. Mejía-Monasterio, R. Metzler, and G. Oshanin, First passages in bounded domains: When is the mean first passage time meaningful? *Phys. Rev. E.* **86**, 031143 (2012).

53. D. S. Grebenkov, R. Metzler, and G. Oshanin, Towards a full quantitative description of single-molecule reaction kinetics in biological cells, *Phys. Chem. Chem. Phys.* (2018), DOI: 10.1039/C8CP02043D.

Chapter 13

Markov and Non-Markov Transport Processes within Bulk-Mediated Surface Diffusion Schemes

Horacio S. Wio[*,‡] and Jorge A. Revelli[†]

Instituto de Física Interdisciplinary Sistemas Complejos,
Universitat de les Illes Balears-CSIC, Palma de Mallorca, Spain
†*Instituto de Física E. Gaviola,*
Universidad Nacional de Córdoba-CONICET,
FaMAF-UNC, Córdoba, Argentina
‡*horacio@ifisc.uib-csic.es*

Bulk-mediated surface diffusion is a process where a particle randomly alternates between surface and bulk diffusions. Such a topic is present in several areas, for instance, various chemical and biochemical processes such as reactions in porous media or trafficking in living cells. Recently, several theoretical studies were carried out, including both non-interactive and interactive approaches for particle diffusion on the surfaces. Some exact remarkable results were obtained for the particle's mean and variance functions on the boundary, before its exit in the case of a 2D domain. Here, we review part of our studies on bulk-mediated surface diffusion approach, based on master equation frameworks. Both, Markovian and non-markovian dynamics were treated. The models took into account the motion over finite and infinite spaces. Analytical expressions were obtained for the probability distributions, and the models were numerically validated, providing in this way a first insight of the framework. We also discuss possible extensions and applications of this phenomenon.[a]

1. Introduction

Interfaces and dynamics of particles adsorbed on surfaces are seminal problems of physics that have attracted high interest both from experimentalists and theorists. We are dealing with phenomena that can be described by the movement of particles on a plane surrounded by a bulk in such a way that allows for the particle dynamics jumping between the surface-bound and bulk states. As a result, particles intermittently diffuse on the surface and in the bulk. This phenomenon is widely known as the bulk-mediated surface diffusion (BMSD). Such a dynamic shows up on small scales in biology as well as in technology applications. For instance, the dynamics of gene regulation is a paradigmatic example of this phenomenon. Here, the dimensionality reduction mechanism[1] was introduced to simulate the association of a protein with its operator site on a DNA chain, which was shown to happen significantly faster than expected for a diffusion-controlled reaction. The reaction rate can be explained[2] by considering a process in which non-specifically bound proteins not only diffuse in one direction along the DNA but also intermittently unbind to undergo bulk diffusion.

Also, proton motion across a cell shows a behavior that can be explained by BMSD dynamics. The proton concentration gradient between two cellular compartments is an important component of the cellular energy system. Together with the transmembrane electrical potential difference, it generates a proton field that is necessary to synthesize ATP.[3] Because measurements of the bulk-to-bulk proton field correlate poorly with measured ATP yields, a simple chemiosmotic theory is inadequate. To explain this field, a stationary model has been introduced[4,5] in which protons are thought to be continuously pumped across the membrane. These protons are then

---

[a]Dedicated to the memory of our friend Carlos E. Budde.

retained at the surface by a barrier that separates the surface volume from the bulk, so that a sufficiently high proton force arises between the two faces of the membrane. Experiments[6,7] and calculations[8,9] show that the proton concentration is indeed markedly higher on the membrane surface than in the bulk.

From the theoretical point of view, many works on BMSD have been made. Such studies were performed in terms of scaling arguments, master equations, and simulations.[10–22] Recently, Berezhkovskii *et al.* proposed a formalism that significantly simplifies the analysis of BMSD.[23] The formalism is extended to bulk-mediated surface transport in the presence of bias, i.e., when the particle has arbitrary drift velocities on the surface and in the bulk. Bénichou *et al.* developed a spectral approach to the escape problem in which the mean exit time is explicitly expressed through the eigenvalues of the related self-adjoint operator.[24] This representation is particularly well suited to investigate the asymptotic behavior of the mean exit time in the limit of large desorption rate. Based on a random-walk approach, Chechkin *et al.*[25] derived the diffusion equations for surface and bulk diffusion including the surface-bulk coupling. From these exact dynamic equations, they analytically obtained the propagator of the effective surface motion, showing a superdiffusive, Cauchy-type behavior on the surface.

All examples mentioned above share the fact that particles can move across all the spaces, that is not only on the surface but on the bulk as well. However, for different reasons it is important to determine a (effective) diffusion coefficient just on the bounding surface where particles move. In this chapter, we propose to revisit the approach to the problem which exploits the fact that the particle propagation over the surface is determined by the cumulative times spent by the particle on the surface and in the bulk. These cumulative times are random variables, the sum of which is equal to the total observation time. It is important that these times can be analyzed by considering a relatively simple two-state problem that describes transitions of the particle between the surface and the bulk. The developed formalism begins by describing diffusion in multilayer media, which deals with diffusion without any constraints on the particle position at time t. Furthermore, this formalism is modified to study the particle propagation under the constraint that the particle is restricted to move in a finite bulk. We also deal with Markovian and non-Markovian mechanisms. One of the advantages of the proposed approach is that it treats the cases of normal and anomalous diffusion (finite and infinite bulk layer thicknesses, respectively). The chapter is organized in the following way. In Section 2, the theoretical framework is developed, introducing the typical set of equations that we use to describe the BMSD, briefly explaining the techniques used to solve them. In Section 3, some results are depicted. In Section 4, some consequences of the obtained results are discussed and projections of the BMSD research are remarked.

2. Theoretical Framework

2.1. *The unbound bulk case*

Let a particle be making a random walk in a semi-infinite discrete space. The position of the particle is defined by the integer numbers $n, m, l \geq 1$, each of them denotes the x, y, and z directions, respectively. We define the plane $z = 1$ as the surface where particles make their surface diffusion. This plane is unbounded while in the z-direction the particle can move from $l = 1$ to ∞. We calculate $P(n, m, l; t)$, the conditional probability that particle will be at n, m, l at time t given that it was at n_0, m_0, l_0 at time t_0 establishing the following master equation system:

$$\partial_t P(n, m, l; t) = \gamma P(n, m, 2; t) - \delta P(n, m, 1; t),$$

$$\partial_t P(n, m, 2; t) = \alpha[P(n + 1, m, 2; t) + P(n - 1, m, 2; t)$$
$$- 2P(n, m, 2; t)] + \beta[P(n, m + 1, 2; t)$$
$$+ P(n, m - 1, 2; t) - 2P(n, m, 2; t)] + \gamma P(n, m, 3; t)$$
$$- \delta P(n, m, 1; t) + 2\gamma P(n, m, 2; t),$$

$$\partial_t P(n, m, l; t) = \alpha[P(n + 1, m, l; t) + P(n - 1, m, l; t)$$
$$- 2P(n, m, l; t)] + \beta[P(n, m + 1, l; t)$$
$$+ P(n, m - 1, l; t) - 2P(n, m, l; t)]$$
$$+ \gamma P(n, m, l + 1; t) - \delta P(n, m, l - 1; t)$$
$$+ 2\gamma P(n, m, l; t), \tag{1}$$

where the first line corresponds to l, and the third to $l \geq 3$. The parameters α, β, and γ are the transition rates per unit time in the x, y, and z directions, and δ represents the desorption transition rate. After some analytical calculations (see Refs. 15–22 for details), it is possible to get $P_{\text{Plane}}(t)$, the probability that the particle is on plane $z = 1$ at time t, in the Laplace space as

$$P_{\text{Plane}}(s) = \frac{N(s)}{D(s)}, \tag{2}$$

where

$$N(s) = \gamma[2\gamma + s - \sqrt{s(4\gamma + s)}]$$

and

$$D(s) = \gamma s[2\gamma + s - \sqrt{s(4\gamma + s)}] + \delta[\sqrt{s(4\gamma + s)} - 3s]$$
$$+ s[\sqrt{s(4\gamma + s)} - s].$$

A useful experimental quantity is the mean square position of the particle on the plane, which is related to the variance distribution over the plane $z = 1$ at time t,

$$\langle r(t)^2 \rangle = \sum P(m, n, l = 1, t | 0, 0, l_0 = 1, t_0)[m^2 + n^2]. \tag{3}$$

In the Laplace space, this quantity can be expressed again as a quotient,

$$\langle r(s)^2 \rangle = \frac{N(s)}{D(s)}, \tag{4}$$

where now

$$N(s) = 4\delta\gamma(\alpha + \beta)[s^2(\sqrt{s(4\gamma + s)} - 3) + 3\gamma^2(\sqrt{s(4\gamma + s)} - 3) - 2\gamma^2],$$

and

$$D(s) = \sqrt{s(4\gamma + s)}[\gamma s(2\gamma + s - \sqrt{s(4\gamma + s)} + \gamma)$$
$$+ \delta((\sqrt{s(4\gamma + s)} - 3) + s(\sqrt{s(4\gamma + s)} - s))]^2.$$

2.2.   The finite bulk case

The framework developed in the preceding subsection is now exploited in order to study the effects of the finite bulk size. In this case, Eq. (1) becomes a finite set of equations since there is a top plane defined as $z = L$ (L representing the top layer number). In contrast to what happens in the infinite bulk system where there is an infinite number of equations (each equation represents the dynamics on a specific layer), now there is a finite number of equations. The conditional probability in the $L - t$ plane is represented by

$$\partial_t P(n, m, L; t) = \gamma P(n, m, L - 1; t) - \delta P(n, m, L; t)$$
$$+ \alpha[P(n + 1, m, L; t) + P(n - 1, m, L; t)$$
$$- 2P(n, m, 2; t)] + \beta[P(n, m + 1, 2; t)$$
$$+ P(n, m - 1, 2; t) - 2P(n, m, 2; t)]. \tag{5}$$

Transforming this set of equations into Fourier and Laplace spaces (see again Refs. 15–22), it is possible to reduce it into the matrix form

$$[s\tilde{I} - \tilde{H}]\tilde{G} = \tilde{I}, \tag{6}$$

where

$$G(k_x, k_y, l, s) = G(k_x, k_y, l, s | 0, 0, l_0, t = 0)$$
$$= \int e^{-st} \sum e^{k_x n + k_y m} P(n, m, l, t) dt. \tag{7}$$

Finally, similarly as in the infinite case, the mean squared displacement on the $z = 1$ plane is calculated by

$$\langle r(s)^2 \rangle = -[\partial_{k_x}^2 + \partial_{k_y}^2][\tilde{G}]_{k_x = k_y = 0}. \tag{8}$$

2.3. *The non-Markovian desorption dynamics case*

So far, we have discussed the case where the dynamics of the particles both, on the surface and in the bulk, were Markovian. Now, we are going to address the case where the desorption is governed by a non-Markovian process. The set of equations is similar to Eqs. (1), i.e., we have studied the infinite case, however, in the present case the first two equations are modified by including a memory kernel in the following way:

$$\partial_t P(n, m, l; t) = \gamma P(n, m, 2; t) - \int dt' K(t') P(n, m, 1; t - t')$$

$$+ \partial_t P(n, m, 2; t) = \int dt' K(t') P(n, m, 1; t - t')$$

$$+ \gamma P(n, m, 3; t) - \gamma P(n, m, 2; t)$$

$$+ \alpha [P(n + 1, m, 2; t) + P(n - 1, m, 2; t)$$

$$- 2P(n, m, 2; t)] + \beta [P(n, m + 1, 2; t)$$

$$+ P(n, m - 1, 2; t) - 2P(n, m, 2; t)],$$

$$\partial_t P(n, m, l \geq 3; t) = \gamma P(n, m, l + 1; t) - \gamma P(n, m, l; t)$$

$$+ \alpha [P(n + 1, m, l; t) + P(n - 1, m, l; t)$$

$$- 2P(n, m, l; t)] + \beta [P(n, m + 1, l; t)$$

$$+ P(n, m - 1, l; t) - 2P(n, m, l; t)]. \tag{9}$$

Here, $K(t)$ represents the memory Kernel acting in all sites of the first two layers ($l = 1$ and $l = 2$). Following a procedure similar to the one used in previous cases (again, see Refs. 15–22), it is possible to get the mean squared displacement in the Fourier–Laplace spaces as a quotient

$$L\{\langle r(t)^2 \rangle\} = \langle r(s)^2 \rangle = \frac{N(s)}{D(s)}, \tag{10}$$

where now

$$N(s) = 4K(s)\gamma(\alpha + \beta)[2\gamma^3 + s^2(\sqrt{s(4\gamma + s)} - s)$$

$$- 3\gamma^2(\sqrt{s(4\gamma + s)} - 3s) - 2\gamma^2 s(\sqrt{s(4\gamma + s)} - 3s)],$$

and

$$D(s) = s\sqrt{s(4\gamma + s)}[\gamma s(2\gamma + s - \sqrt{s(4\gamma + s)} + \gamma)$$
$$+ K(s)\gamma((\sqrt{s(4\gamma + s)} - 3s) + s(\sqrt{s(4\gamma + s)} - s))]^2.$$

Operating in an analogous form to the infinite Markovian case, the probability of finding the particle on the $z = 1$ plane can also be expressed as a quotient

$$P(z = 1, s) = \frac{N(s)}{D(s)}, \tag{11}$$

with

$$N(s) = [\gamma(2\gamma + s - \sqrt{s(4\gamma + s)})],$$

and

$$D(s) = s\gamma s(2\gamma + s - \sqrt{s(4\gamma + s)}$$
$$+ K(s)\gamma((\sqrt{s(4\gamma + s)} - 3s) + s(\sqrt{s(4\gamma + s)} - s)).$$

It is worth remarking here that, as is well known, within the CTRW scheme, there exists a relation between $\psi(s)$, the waiting time density for the desorption dynamics, and the memory kernel (Refs. 15–22) given by

$$K(s) = \frac{s\psi(s)}{1 - \gamma s}.$$

3. Results

3.1. *The unbounded bulk case*

Clearly, Eqs. (2) and (4) represent theoretical results for the probability of finding a particle on plane $z = 1$ and the mean square displacement for particles that undergo a surface motion characterized by a composition of both, a 2D random walk over the studied surface, and a 3D random walk on an unbounded bulk that surrounds the surface.

Both equations were obtained in the Laplace space. In order to find the temporal functions, it is necessary to calculate the inverse Laplace transforms. In general, it is a difficult task. However, the asymptotic behavior for large t can be obtained by means of the Tauberian theorems[26] as

$$P(t) \Rightarrow \frac{\sqrt{\gamma}}{s} \frac{(\alpha + \beta)}{\Gamma(1/2)} t^{-1/2},$$

$$\langle r(t)^2 \rangle \Rightarrow \frac{\sqrt{\gamma}}{s} \frac{(\alpha + \beta)}{\Gamma(3/2)} t^{1/2}. \tag{12}$$

The last equation shows that, for large t, a $t^{1/2}$ dependence for the mean squared displacement arises. This fact indicates that in such a regime the system tends toward a typical diffusion behavior.

Besides, simulations were performed to validate the theoretical results. Figure 1 shows both, on the left-hand side (LHS) the temporal evolution of the probability that the particle is at plane $z = 1$, while the right-hand side (RHS) depicts the mean square displacement as a function of t for three different desorption values.

In both cases, we depict the comparison of the numerical inverse Laplace transforms and the simulation of the mentioned quantities, using δ as a parameter. It is important to remark the excellent agreement that exists between numerical results and the corresponding simulations. It is also worth remarking here that even though the large t behavior is diffusive, as indicated before, there exists a local time interval (or transient time) where it is possible to find a super-diffusive-like behavior.

3.2. *The bilayer case*

In this subsection, we show some results concerning the problem of a bilayer system, as an analytical example of application of the general BMSD theory on finite systems. A bilayer consists of two layers where one of them is the surface were the particle motion is studied, while the other one represents "the bulk".

From Eq. (7), it is possible to obtain the probability to find the particle at plane $z = 1$ by considering the matrix element G_{11} and

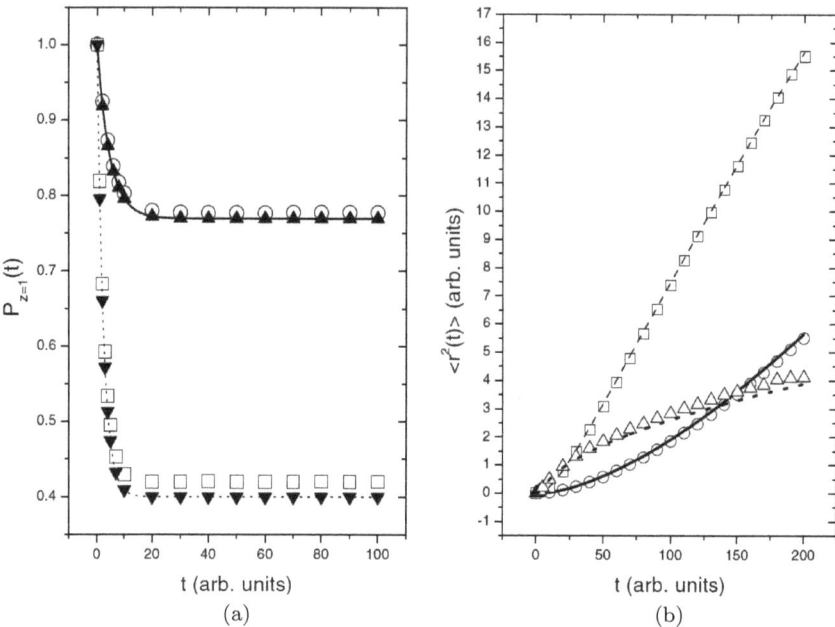

Fig. 1.   (a) Probability $P(z = 1; t)$ versus time, for three different values of δ. Circles: simulations with $\delta = 0.01$, solid line is the theoretical curve. Squares: simulations with $\delta = 0.1$, and the dashed line is the theoretical curve. Triangles: simulations with $\delta = 1.0$, and the dotted line is the theoretical curve. (b) $\langle r^2(t) \rangle$ versus time, for three different values of δ as on the right. In all cases, we have used $\alpha = \beta = \gamma = 1.0$, and the number of realizations is 106. (Reproduced from Ref. 14 with the permission of Springer Nature.)

taking the particular case with $L = 2$

$$P(z = 1, t) = \frac{\gamma + \delta e^{-(\gamma+\delta)t}}{\gamma + \delta}. \tag{13}$$

The variance can be considered as composed of two different contributions, associated to two different motions, as indicated in the following equations:

$$\langle r(t)^2 \rangle = \langle r(t)^2 \rangle_p + \langle r(t)^2 \rangle_v, \tag{14}$$

where the first term corresponds to the motion on the $z = 1$ plane, and is given by

$$\langle r(t)^2 \rangle_p = s(\gamma + \delta) \left[\frac{-2\gamma\delta}{(\gamma + \delta)^3} [e^{-(\gamma + \delta)t} - 1] \right.$$
$$\left. + \frac{(2\gamma\delta + \delta^2)}{(\gamma + \delta)^2} t e^{-(\gamma + \delta)t} + \frac{\gamma^2 t}{(\gamma + \delta)^2} \right],$$

while the second term, that corresponds to the bulk, is given by

$$\langle r(t)^2 \rangle_v = s(\gamma + \delta) \left[\frac{-4\gamma\delta}{(\gamma + \delta)^3} [e^{-(\gamma + \delta)t} - 1] \right.$$
$$\left. + \frac{2\gamma\delta}{(\gamma + \delta)^2} t e^{-(\gamma + \delta)t} + \frac{2\gamma\delta t}{(\gamma + \delta)^2} \right].$$

In the following figures, we show theoretical numerical results and some simulations for the bilayer case. In Fig. 2, on the LHS we show the probability to be at plane $z = 1$, while on the RHS we depict results for the mean square displacement. The results for two desorption coefficients (δ) are shown. As it is apparent from the figures, there is an excellent agreement between the theoretical and simulation results.

3.3. *Infinite non-Markovian case*

In order to apply non-Markovian concepts developed in the previous section, the following function's family was exploited:

$$\psi(t) = \frac{\theta \alpha (\theta \alpha t)^{\alpha - 1}}{\Gamma(\alpha)} e^{-\theta \alpha t}, \tag{15}$$

where α is a positive integer and $\Gamma[\alpha]$ is the factorial function. The parameter α indicates the non-Markovian character of the process. When $\alpha = 1$, the process is Markovian, while for $\alpha \neq 1$ the process is non-Markovian. The parameter θ represents the average desorption rate.

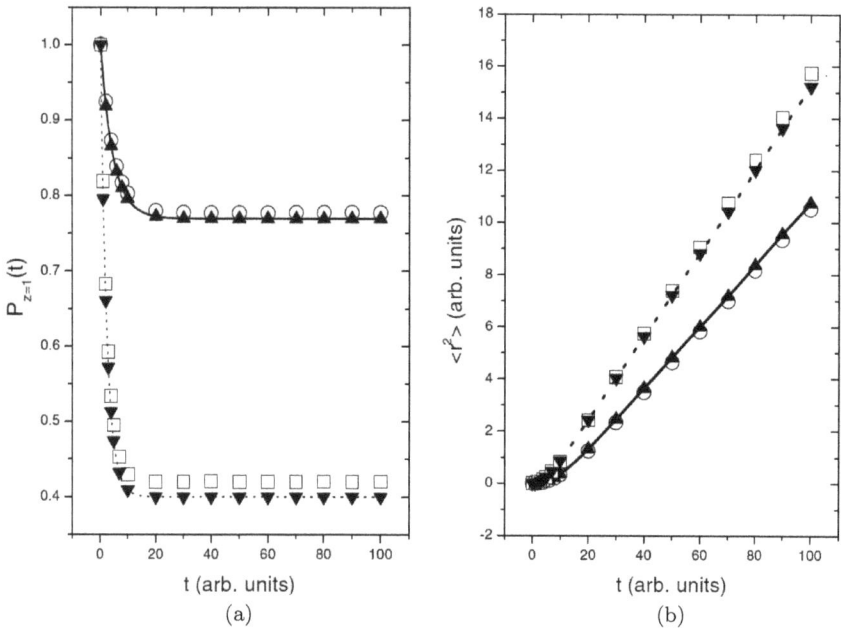

Fig. 2. (a) Temporal evolution of the $P(z = 1, t)$. We have shown two cases: (i) Upside triangles represent theoretical points, the continuous line indicates the theoretical-numerical results and circles are the simulations data for $\delta = 0.1$. (ii) Downside triangles correspond to theoretical points, the dashed line represents theoretical numerical results and squares the simulation data for $\delta = 0.5$. (b) time evolution of $\langle r^2(t) \rangle$ with parameters as in the left case. (Reproduced from Ref. 15 with the permission of Springer Nature.)

The temporal asymptotic limit for the probability of being on the surface and the mean square displacement can be obtained analyzing Eqs. (10) and (11) in the limit for $s \ll 1$. The obtained results are

$$P(z = 1, t) \sim \sqrt{\gamma} \langle t \rangle s^{-1/2},$$

$$\langle r(t)^2 \rangle_p \sim \frac{\sqrt{\gamma} \langle t \rangle (\alpha + \beta)}{2} s^{-3/2}. \tag{16}$$

Finally, by means of using tauberian theorems once again, it is possible to obtain

$$\langle r(t)^2 \rangle_p \sim \frac{\sqrt{\gamma} \langle t \rangle (\alpha + \beta)}{2\Gamma(3/2)} t^{1/2}.$$

These equations represent the asymptotic time behavior for any non-Markovian desorption process. It is important to remark that the bulk non-Markovian dynamic has not been treated here.

Figure 3 depicts numerical and Monte Carlo simulations, on the LHS, for the probability of finding a particle on plane $z = 1$, while on the RHS the variance for the non-Markovian dynamics is indicated in this section.

4. Discussions and Conclusions

In this work, we have reviewed the general framework of the relevant phenomenon of the BMSD dynamics. This is just a brief contribution of previous results,[14–22] shown in order to take into account some possible extensions to the cited problem.

A theoretical model based on a set of Master Equations which describe the motion of particles within a simple cubic lattice was introduced. This model can be considered general in the sense that both kinds of particle movement were included: on the surface and on the bulk. This problem is solved by using techniques of the Laplace and Fourier transformation and exploiting Dyson's formula.

The equations for the probability of finding particles on the observed surface and the mean square displacements were obtained. It was possible to get analytical results for these quantities in the Laplace – Fourier spaces. In some particular cases it was possible to invert these quantities.[15–22]

Although models were derived under the assumption of discrete lattice space and in the limit of diluted system, we illustrate the practical uses of the theoretical approach and the properties of the BMSD by considering in detail several important examples, for instance, diffusion in unbounded and finite bulk as well. Besides, it was considered that non-Markovian desorption dynamics are a first step to include processes with memory. The developed approach conforms the theoretical ground for a systematic study of surface-mediated processes which are relevant for chemical and biochemical reactions in porous catalysts and living cells. For instance, studies on the effect

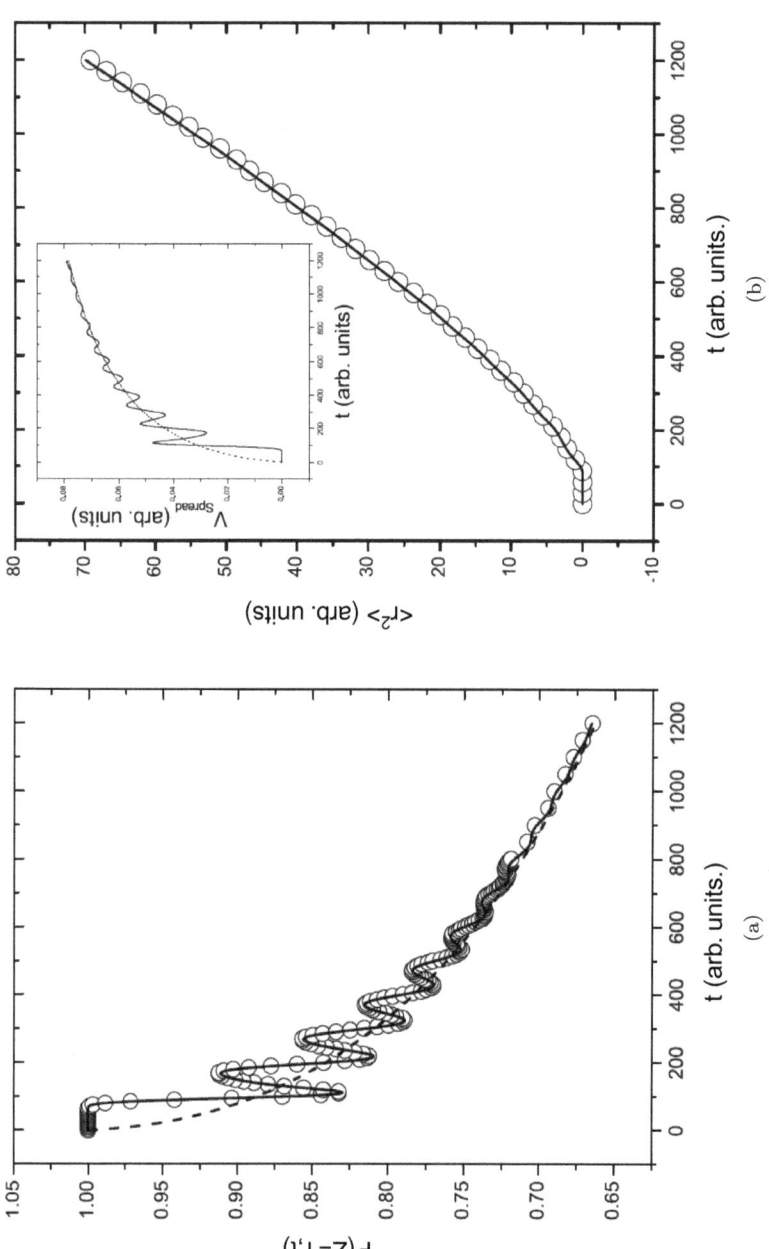

Fig. 3. (a) Temporal evolution of the $P(z = 1, t)$. It represented the case for $\theta = 0.01$. Dot line depicts Markovian evolution, meanwhile, continuous line and open circles (which correspond to Monte Carlo simulations) represent the non-Markovian case for $a = 2$. (b) $\langle r^2(t) \rangle$ for $\theta = 0.01$. The solid line and open circles (which correspond to Monte Carlo simulations) depict the non-Markovian case with $a = 75$. In the insert, the spread velocity versus t for this case (solid line) and the Markovian evolution (dot line) can be seen. (Reproduced from Ref. 17 with the permission of IOP science.)

of the bulk-mediated excursions of reactive species on catalytic reactions and implications for the cellular energy machinery[25] were carried out and explained by considering BMSD phenomena. From the mathematical point of view, the remarkable accuracy of the solution remains striking and have been the trigger of new and complementary model approaches, Refs. 25, 27–33 are deepening the theoretical structures that involve new geometric shapes and non-Markovian dynamics.

Summarizing, BMSD is still an open and actual research physics topic with important applications, among others, in biology and chemical sciences. Such applications require both adequate experimental set ups and new theoretical frameworks which are under study nowadays.

Acknowledgments

The authors acknowledge fruitful discussions and collaborations with D. Prato, F. Rojo, and C. E. Budde Jr.

References

1. G. Adam and M. Delbruck, *Structural Chemistry and Molecular Biology*, A. Rich and N. Davidson (eds.), Freeman, San Francisco (1968), pp. 198–215.
2. H. Richter and M. Eigen, Diffusion controlled reaction rates in spheroidal geometry applications to repression–operator association and membrane bound enzymes, *Biophys. Chem.* **2**, 255 (1974).
3. R. A. Capaldi and R. Aggeler, Mechanism of the F1F0-type ATP synthase, a biological rotary motor. *Trends Biochem. Sci.* **27**, 154–160 (2002).
4. A. Y. Mulkidjanian, J. Heberle, and D. A. Cherepanov, Protons at interfaces: implications for biological energy conversion, *Biochim. Biophys. Acta Bioenerg.* **1757**, 913–930 (2006).
5. D. A. Cherepanov, B. A. Feniouk, and A. Y. Mulkidjanian, Low dielectric permittivity of water at the membrane interface: Effect on the energy coupling mechanism in biological membranes, *Biophys. J.* **85**, 1307–1316 (2003).
6. M. Brändén, T. Sandén, and P. Brzezinski, Localized proton microcircuits at the biological membrane-water interface, *Proc. Natl. Acad. Sci. USA.* **103**, 19766–19770 (2006).
7. T. Sandén, L. Salomonsson, P. Brzezinski, and J. Widengren, Surface-coupled proton exchange of a membrane-bound proton acceptor, *Proc. Natl. Acad. Sci. USA.* **107**, 4129–4134 (2010).

8. A. M. Smondyrev and G. A. Voth, Molecular dynamics simulation of proton transport near the surface of a phospholipid membrane, *Biophys. J.* **82**, 1460–1468 (2002).

9. T. Yamashita and G. A. Voth, Properties of hydrated excess protons near phospholipid bilayers, *J. Phys. Chem. B.* **114**, 592–603 (2010).

10. O. V. Bychuk and B. O'Shaughnessy, Anomalous diffusion at liquid surfaces, *Phys. Rev. Lett.* **74**, 1795 (1995).

11. O. V. Bychuk and B. O'Shaugnessy, Anomalous surface diffusion: A numerical study, *J. Chem. Phys.* **101**, 772 (1994).

12. O. V. Bychuk and B. O'Shaugnessy, Role of bulk-surface exchange in diffusion at liquid surfaces: Non-fickian relaxation kinetics, *Langmuir* **10**, 3260 (1994).

13. R. Valiullin, R. Kimmich, and N. Fatkullin, Lèvy walks of strong adsorbates on surfaces: Computer simulation and spin-lattice relaxation, *Phys. Rev. E.* **56**, 4371 (1997).

14. J. A. Revelli, C. E. Budde, D. Prato, and H. S.Wio, Bulk mediated surface diffusion: The infinite system case, *Eur. Phys. J. B.* **36**, 245–251 (2003).

15. J. A. Revelli, C. E. Budde, D. Prato, and H. S.Wio, Bulk mediated surface diffusion: The finite bulk case, *Eur. Phys. J. B.* **37**, 205–212 (2004).

16. J. A. Revelli, C. E. Budde, D. Prato, and H. S. Wio, Bulk mediated surface diffusion: Non Markovian desorption with finite first moment, *Eur. Phys. J. B.* **43**, 65–71 (2005).

17. J. A. Revelli, C. E. Budde, D. Prato, and H. S. Wio, Bulk mediated surface diffusion: Non Markovian dynamics for the desorption process, *New J. Phys.* **7**, 16 (2005).

18. J. A. Revelli, C. E. Budde, D. Prato, and H. S. Wio, Bulk mediated surface diffusion: Infinite biased dynamics, *J. Phys. Cond. Matt.* **17**, S4175 (2005).

19. J. A. Revelli, C. E. Budde, and H. S.Wio, Bulk mediated surface diffusion: Non Markovian and biased behavior, *COMPLEXUS MUNDI: Emergent Patterns in Nature, Proc. of FRACTAL 2006*, Vienna, Austria, M. Novak (ed.), World Scientific, Singapore, (2005).

20. J. A. Revelli, C. E. Budde, and H. S. Wio, Bulk mediated surface diffusion: Return probability in an infinite system, *J. Phys. Cond. Mat.* **19**, 065127 (2007).

21. F. Rojo, H. S. Wio, and C. Budde, Narrow-escape-time problem: The imperfect trapping case, *Phys. Rev. E.* **86**, 031105 (2012).

22. F. Rojo, C.E. Budde Jr., H. S. Wio, and C. E. Budde, Enhanced transport through desorption-mediated diffusion, *Phys. Rev. E.* **87**, 012115 (2013).

23. A. M. Berezhkovskii, L. Dagdug, and S. M. Bezrukov, A new approach to the problem of bulk-mediated surface diffusion, *J. Chem. Phys.* **143**, 084103 (2015).

24. O. Bénichou, D. S. Grebenkov, L. Hillairet, L. Phun, R. Voituriez, and M. Zinsmeister, Mean exit time for surface-mediated diffusion: Spectral analysis and asymptotic behavior, *Anal. Math. Phys.* **5**(4), 321362 (2015).

25. A. V. Chechkin, I. M. Zaid, M. A. Lomholt, I. M. Sokolov, and R. Metzler Bulk-mediated diffusion on a planar surface: Full solution, *Phys. Rev. E* **86**, 041101 (2012).

26. E. W. Montroll and B. J. West, in: *Fluctuation Phenomena*, E. W. Montroll and J. L. Lebowitz (eds.), North Holland, Amsterdam (1979).

27. M. Coppey, O. Bénichou, J. Klafter, M. Moreau, and G. Oshanin, Catalytic reactions with bulk-mediated excursions: Mixing fails to restore chemical equilibrium, *Phys. Rev. E.* **69**, 036115 (2004).

28. T. Calandre, O. Bénichou, D. S. Grebenkov, and R. Voituriez, Interfacial territory covered by surface-mediated diffusion, *Phys. Rev. E.* **85**, 051111 (2012).

29. M. G. Wolf, H. Grubmller, and G. Groenhof, Anomalous surface diffusion of protons on lipid membranes, *Biophys. J.* **107**, 76–87 (2014).

30. J.-F. Rupprecht, O. Bénichou, D. S. Grebenkov, and R. Voituriez, Kinetics of active surface-mediated diffusion in spherically symmetric domains, *J. Stat. Phys.* **147**, (5) 891918 (2012).

31. A. V. Chechkin, I. M. Zaid, M. A. Lomholt, I. M. Sokolov, and R. Metzler, Bulk-mediated surface diffusion along a cylinder: Propagators and crossovers. *Phys. Rev. E.* **79**, 04015 (2009).

32. M. Levesque, O. Bénichou, and B. Rotenberg, Molecular diffusion between walls with adsorption and desorption, *J. Chem. Phys.* **138**, 034107 (2013).

33. V. V. Ignatyuk, Kinetic equation approach to the description of quantum surface diffusion: Non-Markovian effects versus jump dynamics, *Phys. Rev. E.* **80**, 041133 (2009).

Chapter 14

Diffusion to Capture and the Concept of Diffusive Interactions

Marta Galanti*, Duccio Fanelli[†], Sergey D. Traytak[‡],
and Francesco Piazza[§],[¶]

*Department of Environmental Health Sciences,
Mailman School of Public Health, Columbia University,
722 West 168th Street, New York, NY 10032, USA
[†]Dipartimento di Fisica e Astronomia, Università di Firenze,
INFN and CSDC, Via Sansone 1,
50019 Sesto Fiorentino, Firenze, Italy
[‡]Semenov Institute of Chemical Physics Russian Academy of Sciences,
4 Kosygina St., 117977 Moscow, Russia
[§]Université d'Orléans, Centre de Biophysique Moléculaire (CBM),
CNRS UPR4301, Rue C. Sadron, 45071, France
[¶]Francesco.Piazza@cnrs-orleans.fr

Diffusion to capture is an ubiquitous phenomenon in many fields of biology and physical chemistry, with implications as diverse as ligand–receptor binding on eukaryotic and bacterial cells, nutrient uptake by colonies of unicellular organisms, and the functioning of complex core–shell nanoreactors. Whenever many boundaries compete for the same diffusing molecules, they inevitably shield a variable part of the molecular flux from each other. This gives rise to the so-called *diffusive interactions* (*DI*), which can substantially reduce the influx to a collection of reactive boundaries depending chiefly on their geometrical configuration.

In this chapter, we provide a pedagogical discussion of the main mathematical aspects underlying a rigorous account of DIs. Starting from a striking and deep result on the mean-field description of ligand binding to a receptor-covered cell, we develop step-by-step mathematical description of DIs in the stationary case through the use of translational addition theorems for spherical harmonics. We provide several enlightening illustrations of this powerful mathematical theory, including diffusion to capture ensembles of reactive boundaries within a spherical cavity.

1. Introduction

It is common knowledge that the first step of any chemical or biochemical reaction proceeding in an inert fluid phase is the mutual diffusive encounter of reactants[1]

$$A + B \underset{k_{off}}{\overset{\kappa_{on}(t)}{\rightleftharpoons}} A \cdot B \rightleftharpoons \ldots \rightleftharpoons P. \tag{1}$$

In the above general scheme, the encounter complex $A \cdot B$ is transformed reversibly into a (series of) products (collectively indicated by P), through a variable number of intermediate steps that depend on the specific reaction considered. Irrespective of whether the reaction (1) is considered under thermodynamical equilibrium or non-equilibrium conditions, the second-order rate constant κ_{on} is proportional to the effective relative diffusion coefficient of the species A and B and describes the formation of the encounter complex. For this reason, these kinds of reactions are also known in the physical chemistry community as *diffusion-influenced reactions*. The time dependence of κ_{on} refers to the possibility of investigating transient kinetics effects as opposed to steady-state (equilibrium or non-equilibrium) kinetics.

Several mathematical difficulties emerge when one tries to develop a general kinetic theory of bulk irreversible diffusion-influenced reactions. One way around this problem is to treat *trapping* and *target* models,[2] which are simpler than the original one. In the first model, a particle, say, B diffuses toward many sinks A that are often assumed to be static. Conversely, the target problem describes a situation where many sinks A diffuse to a static particle B.[3] The two settings, even if equivalent under certain conditions (see Ref. 2), describe in general different problems at high density of reactants. In the present chapter, we shall concentrate on the trapping model for reactions of the kind denoted in (1).

Diffusion in domains with one connected smooth boundary is a fairly well-understood and well-characterized phenomenon from the general standpoint of mathematical physics.[4] However, real-life situations are often very complex and make mathematical descriptions challenging. For example, this is the case of biochemical reactions taking place in living media, such as the cell interior or the extracellular matrix (e.g., paracrine delivery),[5] where confinement, crowding (excluded-volume) effects,[6,7] and non-specific interactions among diffusing species and with all sorts of cellular structures[8] make it very difficult to elaborate quantitative models for the calculation of diffusive encounter rates.[9–13]

Among all the possible effects on diffusion-influenced encounters and reactions arising in non-ideal conditions, in this short chapter we shall concentrate on the so-called *diffusive interactions* (*DI*). As we shall see in the following, these describe a fundamental mechanism of competition among different reactive boundaries, competing for the same diffusive molecular flux. This effect was discussed for the first time as far back as in 1953 by Frisch and Collins in terms of a *competition* phenomenon.[14] A decade later, Reck and Prager referred to the same phenomenon simply as an *interaction*.[15] Although the same problem has been investigated later in different contexts, it seems that there is invariably a reference to a generic *competition* mechanism,[16,17] until the term DI was introduced by Traytak[18] in analogy with hydrodynamical interactions for Stokes flow in many-body systems.

Picture for example a concentration field of small ligand molecules (e.g., signaling hormones or growth factors) diffusing in the extracellular matrix or in the bacterial periplasm looking for an available receptor on a cell membrane to form a complex. For the sake of the argument, let us imagine that far away from the receptor-covered surface the ligand bulk concentration is constant and that the ligand–receptor affinity is large enough to consider receptors as *sinks*, i.e., perfectly absorbing units. H. Berg has famously termed this general scheme of diffusive problems *diffusion to capture*.[19] If the surface density of receptors is low, then the diffusion problem is additive and the overall capture rate for the whole ensemble of receptors (e.g., number of ligands diffusing to a receptor per unit time) is well estimated by the sum of the individual rates. In this case, a many-body problem can be solved as many identical two-body problems. However, receptors on cell membranes are typically very densely packed in clusters.[20–26] In this case, any two identical receptors sitting at close separation will *screen* a portion of the diffusive ligand flux to each other. Overall, a complex pattern of many-body screening effects will arise, reflecting the many-body geometrical arrangement of receptors, with the consequence of reducing the overall capture rate corresponding to the array of receptors. A similar scenario is relevant for the case of multivalent molecules, i.e., molecules carrying more than one binding site.[27–31] By the same token, DI among the different active sites will give rise to a similar negative cooperativity, which will reduce the overall capture rate with respect to an equivalent number of isolated sites.

Mathematically, if many-body effects are relatively well characterized for unbounded systems of distributed sinks,[2,32–37] the case of DI for sinks or partially absorbing boundaries located in a finite domain is more challenging from a mathematical standpoint. This scenario is relevant in many fields, ranging from catalysis in composite nanostructures[38–40] to nutrient uptake by dense colonies of microorganisms.[41]

A full treatment of time-dependent DI is extremely hard to perform and *de facto* limited to simple cases.[42,43] Conversely, several methods have been used to tackle this problem for different geometries in the stationary state, where more theoretical approaches are available, such as renormalization group[18,44] and the *method of irreducible Cartesian tensors*[18] and the *generalized method of separation of variables* (GMSV),[45,46] based on addition theorems (AT) for solid harmonics.[45–48] It is also possible to combine such methods with methods based on dual-series relations[49] to deal with the case of DI among inhomogeneous reactive boundaries with active and reflecting patches.[50–52]

In this short chapter, we will provide a concise account of DI arising in many-body systems consisting of spherical fully absorbing and partially absorbing boundaries within a finite domain. In Section 2, we will lay out the basic ideas of the mathematical method employed to compute the diffusive encounter rate for two isolated spherical molecules. In Section 3, these ideas will be used to estimate many-body effects in the mean-field approximation and illustrate the main physical features of DI. In Section 4, we will show how using multipole expansion methods coupled to translational AT for solid harmonics allow one to solve the problem exactly to any desired level of accuracy for arbitrary geometries. This method constitutes a powerful tool that can be employed to tackle a wide host of important problems in physical chemistry and biology.

2. Bimolecular Diffusive Encounters as Two-Body Boundary Problems

The polish physicist Marian Ritter von Smolan Smoluchowski (1872–1917), besides being a skilled watercolor painter and an exquisite pianist, during the first two decades of the 20th century laid the bases of the mathematical theory of diffusion processes.[53,54] The calculation of diffusive encounter rates in ideal conditions follows directly from his ideas. Let us imagine a solution containing two spherical molecules A and B, with radii R_A and R_B, diffusion coefficients D_A and D_B, and concentrations (number densities) c_A and c_B. Our goal is to determine the rate of A–B encounters dictated by their relative diffusive motion as a function of the concentrations. The full many-body problem is exceedingly hard to treat analytically. However, as it was first recognized by Smoluchowski, one can reduce it to the effective two-body problem of relative diffusion of a single A–B pair under certain hypotheses. However, as already noted by Szabo,[55] the commonly accepted hypothesis of high dilution of both species is not enough. The first step toward the equivalent two-body problem is that one species be much more diluted than the other. Yet, not even this is enough. Let us imagine that the particles of kind A are sufficiently diluted, i.e., $c_A \ll c_B$, so that one can concentrate on a single A particle surrounded by many B particles, say N of them. It is not difficult to show that the $(N + 1)$-body Smoluchowski equation describing the diffusion of a single A molecule within a sea of B particles contains cross-terms that make it non-separable if $D_A \neq 0$.[2] Therefore, bimolecular encounters between A and B molecules can be modeled as an equivalent two-body problem provided that

(i) Both species should be highly diluted, so that mutual interactions can be safely neglected.

(ii) One species (A) must be much more diluted than the other, so that the full problem can be reduced to study the fate of a single particle surrounded by many particles of the other species.

(iii) The diffusion coefficient of the highly diluted species should be much smaller than that of the other species (from N-body to two-body). A consequence of this is that the relative diffusion coefficient essentially coincides with the one of the mobile species, i.e., $D = D_A + D_B \simeq D_B$.

Under these conditions, the rate of encounters can be computed by solving the following stationary diffusion problem (i.e., Laplace equation) for the local concentration field $c(\boldsymbol{r})$ of B molecules.

$$\nabla^2 c = 0, \tag{2a}$$

$$c|_{\partial\Omega} = 0, \tag{2b}$$

$$\lim_{r \to \infty} c = c_B. \tag{2c}$$

This describes the non-equilibrium steady state arising as a constant bulk concentration c_B is maintained far from the reactive boundary $\partial\Omega$, which acts as a perfectly absorbing sink. This is the contact surface of the two molecules, which is the sphere $\mathcal{S}_R$ of radius $R = R_A + R_B$, since both A and B molecules are spherical by assumption. Physically, this describes the pseudo-first order (annihilation) reaction

$$A + B \overset{\kappa_S}{\rightarrow} A,$$

whose rate $k_S = \kappa_S c_B$ coincides with the overall flux into the reactive surface $\mathcal{S}_R$, i.e., the number of B molecules crossing $\mathcal{S}_R$ per unit time. Here, we shall use the Greek letter κ to denote rate constants (dimensions of inverse concentration times inverse time) and the Latin letter k to denote rates (dimensions of inverse time). The rate can be computed straightforwardly as the incoming flux across $\mathcal{S}_R$, that is

$$k_S = - \int_{\mathcal{S}_R} \boldsymbol{J} \cdot \hat{n} \, dS, \tag{3}$$

where $\boldsymbol{J} = -D\nabla c$ is the relative diffusion current (Fick's first law).

The solution to the spherically symmetric boundary problem (2) can be computed straightforwardly, yielding

$$c(r) = c_B \left(1 - \frac{R}{r} \right), \tag{4}$$

which, using Eq. (3), immediately gives the so-called Smoluchowski rate constant κ_S

$$\kappa_S = 4\pi D R. \tag{5}$$

The rate of encounter is thus $k_S = 4\pi D R c_B$ molecules per unit time disappearing across the absorbing boundary $\mathcal{S}_R$. Note that this is proportional to the *linear* size of the latter, a distinctive signature of the diffusive dynamics.

2.1. *Finite reaction probability and radiation boundary conditions*

Scheme (2) describes the reactive boundary as a perfect sink, which amounts to the consideration that B particles are annihilated the moment they reach contact distance. This is meant to describe a (long-lived) binding event. However, in reality the two reacting partners first approach diffusively to contact distance forming the so-called *encounter complex*. Subsequently, this can either dissociate (this is the case if for example the two partners were not mutually oriented in a favorable manner) or proceed to form a stable contact. This more realistic situation can be accommodated for within the above mathematical formalism thanks to an intuition put forward by Collins and Kimball in 1949.[56] The idea is to replace the perfectly absorbing boundary condition (2b) with a *radiation* boundary condition (in more mathematical terms Robin boundary condition) that interpolates between perfectly absorbing (Dirichlet type) and perfectly reflecting (von Neumann type) boundary conditions. Namely, the boundary value problem (2) is replaced by

$$\nabla^2 c = 0, \tag{6a}$$

$$\left(4\pi D R^2 \frac{\partial c}{\partial r} - \kappa^* c \right)\Big|_{\partial\Omega} = 0, \tag{6b}$$

$$\lim_{r\to\infty} c = c_B. \tag{6c}$$

The boundary condition (6b) stipulates that the particle flux across the reactive contact surface is proportional to the local concentration of ligands B. The proportionality constant, the *intrinsic* rate constant κ^*, can be considered as describing the physical mechanism underlying the chemical fixation of the encounter complex. In the limit $\kappa^* \to 0$, the surface becomes perfectly reflecting, i.e., no reaction can occur. In the opposite limit, $\kappa^* \to \infty$ (mathematically, it is necessary to divide Eq. (6b) by κ^* before taking the limit) one recovers the perfectly absorbing boundary, which is thus seen as corresponding to infinitely fast chemical fixation step. As we shall see, the latter case is known as the *diffusion-limited* regime, where the diffusive encounter is the rate-limiting step of the reaction. Diffusion-limited regime is *as-fast-as-one-can-go*, other situations corresponding to a finite intrinsic reaction rate necessarily proceeding slower than that.

The solution to the problem (6) can be computed as straightforwardly as before, yielding

$$c(r) = c_B \left[1 - \left(\frac{h}{1+h} \right) \frac{R}{r} \right], \tag{7}$$

where $h = \kappa^*/\kappa_S$, which, using Eq. (3), gives

$$\kappa = \kappa_S \left(\frac{h}{1+h} \right). \tag{8}$$

In general, $\kappa < \kappa_S$. This case in the physical chemistry community is often indicated with the specific term *diffusion-influenced* regime (or reaction), as opposed to the diffusion-limited regime, $h \to \infty$, where $\kappa = \kappa_S$.

3. Approximate Evaluation of Diffusive Interactions: A Surprising Lesson in Biology

The concept of diffusive interactions is best introduced through a simple, yet astonishing classical result. Let us consider a cell, which we model as a spherical surface of radius R, uniformly covered with M receptors, which we model as small absorbing circular patches of radius a. The rest of the cell surface is supposed to be reflecting. The problem of computing the rate of absorption of this partially absorbing cell was first famously considered by Berg and Purcell in 1977.[57] Here, we shall follow the appealing re-derivation by Shoup and Szabo[58] of the same result. The main idea is to treat the receptor-covered cell as a partially absorbing sphere, in the sense of radiation boundary conditions. According to Shoup and Szabo's argument, the corresponding intrinsic reaction rate constant can be computed as the ratio between the rate constant of M isolated circular disks on an otherwise reflecting surface and the Smoluchowski rate constant of the entire cell, namely,

$$\kappa^* = 4Da \times M, \tag{9}$$

where we used the classical result $k_a = 4Da$ for the rate constant of a small absorbing disk on an infinite reflecting plane.[59] Using Eq. (8), the rate constant corresponding to the partially absorbing sphere is easily found, namely,

$$\kappa = \kappa_S \left(\frac{Ma}{\pi R + Ma} \right). \tag{10}$$

A surprising finding emerges if we plug realistic figures in Eq. (10). The typical size of a cell is around 10 μm, while the typical size of a receptor is of the order of 1.5 nm. If we calculate how many receptors are needed to reduce the rate constant to only one half that of the fully covered cell, i.e., κ_S, we find $M \simeq 10^4$, which is the correct order of magnitude for the average number of receptors of a given family present on a cell's surface at any given time.[60] Is this a large number? A quick calculation shows that the fraction of cell surface covered by as many receptors is $\simeq 10^{-4}$! To summarize, an active surface fraction as low as 10^{-4} only yields a factor of reduction of 2 in the rate of capture. The surprising finding is that an extremely sparse uniform distribution of receptors is as effective an absorber as a fully covered cell.

We can now ask a deep and intriguing question. What would be the rate if all the M receptors were clustered in one single active patch covering the same surface fraction? The answer to this question is the result of the classical calculation of the rate to an active spherical cap on an otherwise reflecting sphere,[52] $\kappa = f_c(\theta_0)\kappa_S$, where $f_c(\theta_0) \leq 1$ is a steric factor describing the diffusion to the active cap of aperture θ_0. In the monopole approximation (MOA), one has

$$f_c(\theta_0) = \frac{\sin\theta_0 + \theta_0}{2\pi - (\sin\theta_0 + \theta_0)}.$$ (11)

Incidentally, according to the general physics of diffusion, f_c can be approximated as the *square root* of the surface fraction covered by the cap,[52] namely,

$$f_c(\theta_0) \approx \sqrt{\frac{\Delta S_{\mathrm{cap}}(\theta_0)}{4\pi R^2}} = \sqrt{M}\left(\frac{a}{2R}\right).$$ (12)

Combining Eq. (10) and Eq. (12), we can compute the ratio between the steric factor f_u corresponding to a sparse uniform configuration of the M receptors and that of the cluster configuration, f_c, namely,

$$\frac{f_u}{f_c} \simeq \frac{2\sqrt{M}}{\pi} + \mathcal{O}(a/R).$$ (13)

For a number of receptors M of the order of $10^4 \div 10^5$, one finds $f_u/f_c \simeq 10^2$. This is a first, striking manifestation of the anti-cooperative effects caused by DI. Summarizing, (i) the rate of capture for a ligand diffusing to a cell uniformly and very sparsely covered with receptors is essentially as large as that of a fully covered cell and (ii) about 100 times larger than in the case where all the receptors would be clustered in a single active patch. It is intriguing to observe that in many cases receptors are indeed densely clustered on the cell surface. Famously, this is the case of chemotaxis receptors in bacteria such as *E. coli*, forming extended patches at the cell poles.[20,23,24] One might argue that there should be other biochemical or structural constraints that offset such strong reduction to the rate of capture.

4. The Generalized Method of Separation of Variables Allows One to Solve the Problem Semi-Analytically

A precursor idea of the GMSV, first discussed in 1944 by S. K. Mitra[61] relating to Laplace equation with two disconnected spherical boundaries, goes back to the well-known paper by Lord Rayleigh on the conductivity of heat and electricity in a medium with regularly arranged obstacles.[62]

In the theory of partial differential equations (PDE), a 3D (bounded or unbounded) domain $\Omega \subset \mathbb{R}^3$ is called a *canonical domain* for a given PDE if the classical solution to this equation may be expanded in an absolutely and uniformly convergent series with respect to corresponding basis solutions in the Hilbert space $L_2(\partial\Omega)$. Remarkably, the GMSV allows one to find semi-analytical solutions of various boundary value problems for Laplace equation in all known 3D canonical domains and their combinations thereof.[63] The GMSV can be thought of comprising five separate logical steps:

(a) reduction of the boundary value problem to its non-dimensional standard form,

(b) determination of the basis solutions to the equation in a given canonical domain,

(c) application of the linear superposition principle,

(d) application of the reexpansion (addition) theorems in order to impose the boundary conditions,

(e) reduction of the problem to an infinite system of linear algebraic equations and its solution.

In principle, one would like to solve the problem of diffusion to ensembles of absorbing or partially absorbing boundaries exactly. Although the GMSV can be used to deal with (general) canonical domains, we will limit ourselves here to only spherical boundaries.

4.1. Diffusive interactions between two spheres

The general power of the GMSV and its main features can be most clearly appreciated by discussing a simple problem, namely, that of diffusion to a pair of spherical sinks of radii a_1 and a_2 located at the origin (Ω_1) and along the z axis at $z = \ell$, (Ω_2). The diffusion of ligands (particles B) should be described in the 3D smooth oriented manifold $\Omega^- = \mathbb{R}^3 \backslash \overline{\Omega}_1 \cup \overline{\Omega}_2$, which can be referred to as the *concentration manifold*.[a] With reference to the logical sequence of the GMSV, we proceed as follows.

(a) To find the standard non-dimensional form of the problem, we consider the reduced concentration field of B particles that is regular at infinity, that is,

$$u(\boldsymbol{r}) = 1 - u(\boldsymbol{r})/c_B,$$

and non-dimensional radial coordinates $\xi_i = r_i/a_i$. The non-dimensional standard form of the original boundary value problem reads

$$\nabla^2 u = 0 \quad \text{in} \quad \Omega^-, \tag{14a}$$

_____

[a]It is expedient to introduce also the *partial domains* $\mathcal{D}_i = \mathbb{R}^3 \backslash \overline{\Omega}_i$, so that $\Omega^- = \mathcal{D}_1 \cap \mathcal{D}_2$.

$$u|_{\xi_i=1} = 1, \tag{14b}$$

$$u|_{\xi_i \to \infty} \to 0. \tag{14c}$$

(b) The appropriate basis functions for this problem are scalar axially symmetric *regular* and *irregular solid spherical harmonics* with respect to the two spherical coordinate systems for Ω_i (see cartoon in Fig. 1)

$$\psi_n^+(r_i, \theta_i) = r_i^n \, P_n(\mu_i), \qquad \psi_n^-(r_i, \theta_i) = r_i^{-n-1} P_n(\mu_i), \tag{15}$$

where $P_n(\mu_i)$ is a Legendre polynomial of degree n, with $\mu_i = \cos \theta_i$. Solid spherical harmonics form a *canonical basis*, $\{\psi_n^+(r_i, \theta_i)\}_{n=0}^{\infty}$ and $\{\psi_n^-(r_i, \theta_i)\}_{n=0}^{\infty}$, for harmonic functions in Ω_i and $\mathcal{D}_i$, respectively.

(c) For $N > 2$, it is impossible to introduce a global coordinate system (e.g., bispherical coordinates for $N = 2$, such as in Ref. 64). Hence, in general one should introduce appropriate local coordinates in Ω^-. The solution to the problem (14) can be expressed as

$$u(\boldsymbol{r}) = u_1(\boldsymbol{r}_1) + u_2(\boldsymbol{r}_2) \quad \text{for} \quad \boldsymbol{r}_i \in \mathcal{D}_i, \tag{16}$$

where

$$u_i(\boldsymbol{r}_i) = \sum_{n=0}^{\infty} A_n^i \psi_n^-(\xi_i, \theta_i) \quad \text{in} \quad \mathcal{D}_i, \tag{17}$$

are absolutely and uniformly convergent series expansions of irregular spherical harmonics. The unknown coefficients A_n^i should be determined by imposing the boundary conditions (14b) for $i = 1, 2$. In order to do so, we have to express the function $u_1(\boldsymbol{r}_1)$ in the local coordinates of Ω_2 and vice versa.

(d) This can be accomplished through AT.[47] For the present axially symmetric problem, one has

$$\xi_1^{-(k+1)} P_k(\mu_1) = \sum_{n=0}^{\infty} U_{nk}^{21} \, \xi_2^n P_n(\mu_2) \quad \text{for } \xi_2 < \ell, \tag{18}$$

$$\xi_2^{-(k+1)} P_k(\mu_2) = \sum_{n=0}^{\infty} U_{nk}^{12} \, \xi_1^n P_n(\mu_1) \quad \text{for } \xi_1 < \ell, \tag{19}$$

where $\epsilon_i = a_i/\ell < 1$ and the so-called *mixed-basis matrices elements* read

$$U^{12}_{nk} = (-1)^k \binom{n+k}{n} \epsilon_1^n \epsilon_2^{k+1}, \tag{20a}$$

$$U^{21}_{nk} = (-1)^n \binom{n+k}{n} \epsilon_1^{k+1} \epsilon_2^n. \tag{20b}$$

Equation (19) needs to be used when imposing that $u(\boldsymbol{r})$ satisfy Eq. (14b) for $i = 1$, and Eq. (18) needs to be used for $i = 2$.

(e) This procedure leads to the following infinite system of linear algebraic equations of the II kind (ISLAE), comprising in general as many equations as there are boundaries,

$$\begin{cases} A^1_n + \sum_{k=0}^{\infty} U^{12}_{nk} A^2_k = \delta_{n0}, \\ \sum_{k=0}^{\infty} U^{21}_{nk} A^1_k + A^2_n = \delta_{n0}. \end{cases} \tag{21}$$

It may be proved that the system (21) can be truncated to obtain a solution to any desired accuracy through the so-called *reduction method*.[65]

The overall rate of capture k, i.e., the total flux into the two-sphere system, is given by

$$k = -2\pi D a_1 \int_{-1}^{1} \frac{\partial u}{\partial \xi_1}\bigg|_{\xi_1=1} d\mu_1 - 2\pi D a_2 \int_{-1}^{1} \frac{\partial u}{\partial \xi_2}\bigg|_{\xi_2=1}$$

$$d\mu_2 = k_{S_1} A^1_0 + k_{S_2} A^2_0, \tag{22}$$

where we have used the general property of Legendre polynomials $\int_{-1}^{1} P_n(\mu)\, d\mu = 2\delta_{n0}$ and introduced the two Smoluchowski rates, $k_{S_i} = 4\pi D a_i c_B$.

The simplest analytical approximation of the exact solution is the MOA, which consists in keeping only the $n = 0, k = 0$ terms in the system (21). It is not difficult to see that this yields

$$k = k_{S_1} \left(\frac{1 - \epsilon_2}{1 - \epsilon_1\epsilon_2} \right) + k_{S_2} \left(\frac{1 - \epsilon_1}{1 - \epsilon_1\epsilon_2} \right) \leq k_{S_1} + k_{S_2}. \tag{23}$$

The case of two equal sinks provides some immediate insight into the anti-cooperativity of DI. If $a_1 = a_2$, Eq. (23) reduces to the well-known result[17]

$$k = \frac{2k_S}{1 + \epsilon} = 2k_S \frac{\ell}{a + \ell}. \tag{24}$$

It can be appreciated that $k \to 2k_S$ in the limit of infinite separation, $\ell \to \infty$. The MOA predicts a maximum reduction $k/2k_S = 2/3$ of the rate of capture (i.e., maximum strength of DIs) at contact distance, $\ell = 2a$. This has to be compared with the exact value,[64] $k/2k_S = \log 2 \approx 0.693$. It is interesting to observe that DIs are long range, that is $1 - k/2k_S \simeq a/\ell$ for separations larger than a few radii: DIs are entropic forces that decay with distance like Coulomb and gravitational interactions.

One may wonder how good an approximation is the MOA. It turns out that for assemblies of perfectly absorbing sinks, it is indeed an extremely good approximation, as is apparent from Fig. 1. For $\ell = 3a$, the relative error is less than 1 %. It can be shown that the relative error decreases rapidly, $\propto \ell^{-4}$, until $\ell \simeq 10a$ (approximately 0.01 %), and then decreases more slowly, $\propto \ell^{-1}$. The reasons why the MOA is so good an approximation have been investigated in Ref. 18.

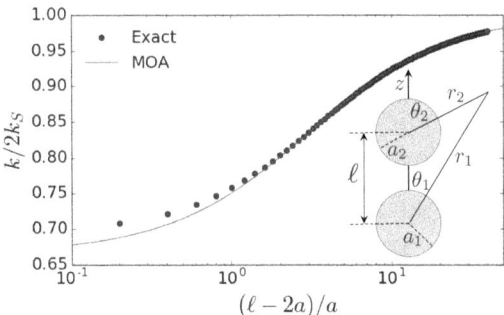

Fig. 1. Rate of capture for two sinks of radius $a_1 = a_2 = a$ separated by a distance ℓ. Comparison of the exact result and the MOA approximation.

4.2. *Diffusive interactions are weaker among multiple partially reactive boundaries*

A system of partially reactive boundaries experiences weaker DI. This can be easily seen and quantified by repeating the above calculations for two spheres endowed with intrinsic rate constants κ_1 and κ_2. This entails replacing boundary conditions (14b) with radiation conditions, namely,

$$\left[\frac{\partial u}{\partial \xi_i} - h_i(u - 1)\right]_{\xi_i=1} = 0, \tag{25}$$

where $h_i = \kappa_i^*/\kappa_{S_i}$, $i = 1, 2$. In this case, it is not difficult to take the same steps as in the above derivation and compute the new matrices U^{12}, U^{21}. The MOA in this case gives

$$k = k_{S_1}\left[\frac{q_1(1 - q_2\epsilon_2)}{1 - q_1 q_2 \epsilon_1 \epsilon_2}\right] + k_{S_2}\left[\frac{q_2(1 - q_1\epsilon_1)}{1 - q_1 q_2 \epsilon_1 \epsilon_2}\right], \tag{26}$$

where $q_i = h_i/(1 + h_i)$. The case of two identical, partially absorbing spheres gives immediately a reduction with respect to additivity

$$\frac{k}{2k_S q} = \frac{1 + h}{1 + h(1 + \epsilon)} \geq \frac{1}{1 + \epsilon}. \tag{27}$$

DI are therefore less prominent for partially absorbing boundaries. It is easy to check that the maximum strength of DIs (i.e., at contact distance) is reduced by an intrinsic reaction rate κ^* by a factor $(3 + 3h)/(2 + 3h) > 1$ in the MOA. However, it should be emphasized that the MOA performs increasingly worse the lower the value of h, and more multipoles should be considered beyond the $n = 0$ term to achieve the same accuracy as in the limit $h \to \infty$.

5. Many Spherical Boundaries Arranged Arbitrarily in Space

The trick of using AI to express multipole expansions in local reference frames centered on two different disconnected spherical boundaries can be extended with no conceptual difficulties to the case of many spheres of arbitrary size, intrinsic reaction rate constant, and position in 3D space. Let us consider the finite spherical domain $\Omega = \Omega_0 \backslash \bigcup_{\alpha=1}^{N} \overline{\Omega}_\alpha$, represented in Fig. 2, filled with N spherical reactive boundaries. Let us introduce the non-dimensional normalized ligand density $u(\mathbf{r}) = c(\mathbf{r})/c_B$ and the variables $\xi_\alpha = r_\alpha/R_\alpha$, $\xi_0 = r_0/R_0$, normalized to the radii of the respective reactive boundaries. We need to solve the following boundary problem:

$$\nabla^2 u = 0, \tag{28a}$$

$$\left(\frac{\partial u}{\partial \xi_\alpha} - h_\alpha u \right)\bigg|_{\partial \Omega_\alpha} = 0 \qquad \forall\, \alpha = 1, 2, \ldots, N, \tag{28b}$$

$$\left(\frac{\partial u}{\partial \xi_0} + h_0 (u - 1) \right)\bigg|_{\partial \Omega_0} = 0. \tag{28c}$$

Again, we have introduced the parameters $h_\alpha = \kappa_\alpha^*/4\pi D R_\alpha$ that determine the reactivity of the αth sphere. The boundary condition (28c) on the inner surface of the *container* sphere Ω_0 is a radiation-type boundary condition and has the following meaning. One should imagine that the ligand concentration is c_B outside Ω_0 (even if formally the problem is not defined there) and that there is a membrane separating the inner compartment Ω from the exterior, whose non-dimensional permeability is proportional to h_0. In the limit $R_0 \to \infty$, one recovers the open-boundary problem with the boundary condition $\lim_{R_0 \to \infty} c = c_B$. Furthermore, it is not difficult to show that if one considers the problem (28) for a single sink at the center of Ω_0 and an equivalent problem (single sink) in the open domain but with $D = \{D_{\text{in}}$ for $r \leq R_0|\ D_{\text{out}}$ for $r > R_0\}$, then the two problems are equivalent provided $h_0 = D_{\text{out}}/D_{\text{in}}$. Hence, one may think of the problem (28) as describing diffusion of ligand to a set of spheres within a spherical container such that the ligand concentration outside the container is fixed (c_B), as well as the ratio h_0 between the ligand diffusion coefficient outside the container (bulk) and in the interior, the latter parameter playing the role of the non-dimensional permeability of the (imaginary) membrane at $\partial\Omega_0$.

By virtue of the superposition principle for the Laplace equation, the problem (28) admits a solution in Ω as a sum of linear combinations of regular (inside Ω_0) and irregular harmonics (outside each Ω_α), namely,

$$
u = u_0^+ + \sum_{\alpha=1}^{N} u_\alpha^-
$$

$$
= \sum_{n=0}^{\infty} \sum_{m=-n}^{n} A_{mn}\xi_0^n Y_{mn}(\boldsymbol{r}_0) + \sum_{\alpha=1}^{N} \sum_{n=0}^{\infty} \sum_{m=-n}^{n} B_{mn}^\alpha \xi_\alpha^{-n-1} Y_{mn}(\boldsymbol{r}_\alpha),
$$

(29)

where $Y_{mn}(\boldsymbol{r}_\alpha) = P_n^m(\cos\theta_\alpha)e^{im\phi_\alpha}$ are solid harmonics referring to the local reference frame centered on the αth boundary (see Fig. 2). The coefficients A_{mn} and B_{mn}^α should be determined by imposing the boundary conditions. In the neighborhood of each boundary, one has to express all the bases as a function of the local coordinates. More precisely, in the neighborhood of each $\partial\Omega_\alpha$, u_0^+ and u_β^- ($\beta \neq \alpha$) have to be expressed as a function of the $\boldsymbol{r}_\alpha$ coordinates, and similarly, in the neighborhood of $\partial\Omega_0$, every u_α^- has to be written as a

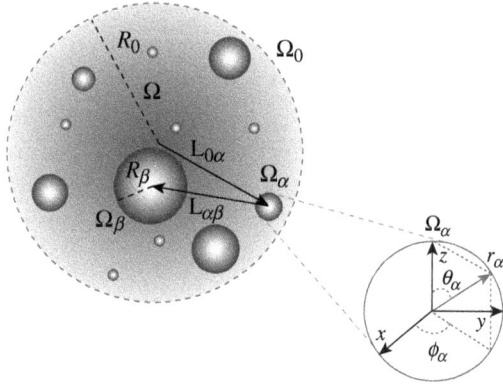

Fig. 2.   Schematic representation of the domain Ω with the relevant local coordinate systems, radii, and position vectors.

function of r_0. For this purpose, one can make use of the translational AT for solid harmonics.[47] This operation requires some care, as one out of three possible ATs must be selected for each pair of boundaries depending on the geometry. These rules are summarized in the appendix.

5.1. *Many spheres inside a spherical cavity*

Diffusion-influenced reactions inside a spherical cavity are of great importance in various applications; however, the simple Smoluchowski rate is often incorrectly used to describe the kinetics of these reactions.[66,67]

The rate of capture of a sink of radius R_1 at the center of a spherical cavity of radius R_0 outside which there is a constant bulk ligand density c_B is given by the solution of the following problem:

$$\nabla^2 u = 0, \tag{30a}$$

$$u|_{r=R_1} = 0, \tag{30b}$$

$$\left(R_0 \frac{\partial u}{\partial r} + h_0(u - 1) \right)\Big|_{r=R_0} = 0, \tag{30c}$$

where $u(r) = c(r)/c_B$ and $h_0 = D_{out}/D_{in}$ is a parameter gauging the permeability of the internal boundary of the spherical cavity. Here, we assume that D_{out}, D_{in} are the ligand diffusion coefficients outside and inside the cavity, respectively. The solution to the problem (30) is straightforward, and the capture rate by (total flux into) the sink yields

$$\frac{k}{k_{S_1}} = \frac{h_0}{\epsilon + h_0(1 - \epsilon)}, \tag{31}$$

where $\epsilon = R_1/R_0$. We see that $k \to k_{S_1} = 4\pi D_{in} c_B$ in the limit of infinite cavity $R_0 \to \infty$. For a finite cavity with a fixed ligand concentration outside, Eq. (31) has a simple interpretation: the rate of capture is enhanced for $h_0 > 1$, that is, when $D_{out} > D_{in}$. In the limit of infinitely absorbing boundary (or conversely, infinitely viscous interior), the rate of capture is enhanced by a factor $1/(1-\epsilon)$.

This becomes very large as the sink approaches the inner surface of the cavity.

This simple result may have interesting implications for the diffusion of ligands within the bacterial periplasm. This region, comprised between the outer cell membrane and an inner (cytoplasmic) membrane, can be as wide as 40% of the total volume in gram-negative bacteria and is typically a very shallow layer in gram-positive bacteria. The periplasm is filled with a thick gel-like, highly crowded matrix[68] and is lined up with many arrays of receptors on the inner cytoplasmic membrane, facing the outer membrane (interior of the cavity). Many ligands, such as those related to chemotaxis, diffuse to receptors within the inner membrane (at $r = R_1$). Since typically $(R_0 - R_1)/R_0 \ll 1$ and the periplasm is very crowded,[68] one has $D_{\text{in}} \ll D_{\text{out}}$ and $1 - \epsilon \ll 1$, which would thence boost the rate of capture.

Using the general AT for solid harmonics (see details reported in the appendix), we are now in a position to answer many interesting questions related to such problems.

5.1.1. Two spheres inside a spherical cavity

It is interesting to investigate diffusion interactions between two sinks in a finite domain. Let us consider the simple case of two identical perfect sinks arranged symmetrically along a diameter of a spherical cavity with respect to the center. Let us denote ℓ as the center–center distance, R_1 as the size of the sinks, and R_0 as the size of the cavity, whose internal surface is made perfectly absorbing. Problem (28) can be solved as described in the appendix. The results are summarized in Fig. 3. In agreement with what was discussed in the previous section, one can appreciate that the rate of the two confined sinks is larger than in the absence of cavity. In particular, the rate increases abruptly as the sinks approach the inner boundary of the cavity. This is a direct consequence of the assumption that the ligand density at the cavity interface is equal to the bulk density. Another non-trivial observation is that the normalized rate now depends on the size of the sink: large sinks have more capture power with respect to the

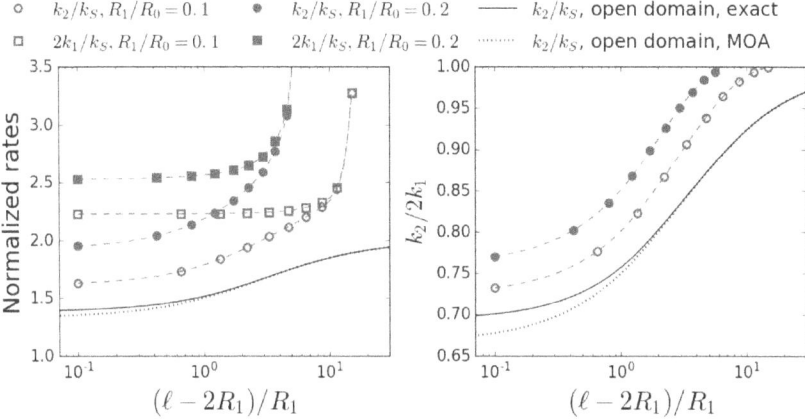

Fig. 3. Rate of capture by two identical spheres of radius R_1 inside a spherical cavity of radius R_0 with absorbing inner boundary ($h_0 \to \infty$). With reference to a frame with the origin at the center of Ω_0, the two spheres are placed symmetrically along the z axis at a center–center separation $\ell \in [2R_1, 2(R_0 - R_1)]$. Left: normalized rates of the two-sphere system compared with twice the rate of one isolated sink at the same position vs rescaled center–center distance. The rates are normalized to the rate of one isolated sink in the open domain, $k_S = 4\pi D R_1 c_B$. Right: measure of DI. Symbols and lines in the right panel refer to the same cases as in the left panel.

open-domain, non-confined setting than small ones. Concerning the rate of capture of single confined sinks, one remarks that prediction (31) in the limit $h_0 \to \infty$ for a sink at the center of the cavity is still accurate when the sink is displaced up to a distance of $\simeq 7 \div 8 R_1$ from the center (constant curves with squares in Fig. 3, left panel).

The rates increase in a cavity and DI decrease. This effect is illustrated in the right panel of Fig. 3. The larger the embedded sinks, the greater the overall rate and correspondingly the weaker the DI. For example, for $R_1/R_0 = 0.2$, the DI are practically gone ($k_2 = 2k_1$) already for $\ell \simeq 5 \div 6 R_1$, i.e., when the outer surface of the sinks is at a distance of about $0.2R_0$ from the inner surface of the cavity.

5.1.2. *Many sinks on a spherical inner layer inside a spherical cavity*

It is instructive to use the method described above to investigate the rate of capture of many equivalent sinks arranged randomly at a given distance from the center on a spherical layer. In the open domain, the rate of such ensembles of sinks is strongly reduced due to DI. For example, the average capture rate of random configurations of $N = 50$ non-overlapping sinks of size R_1 at a distance $d = 8.85R_1$ from the center is $k_{50} = (8.33 \pm 0.01)k_S$, with $k_S = 4\pi DR_1c_B$ (average over 100 independent configurations). DI reduce by a staggering 85% of the overall capture rate of the ensemble with respect to as many isolated sinks. We have learned that confining sinks within a cavity helps sustain the capture rate due to the proximity (exterior of the cavity) of the bulk concentration (effectively reducing the ligand depletion region). In fact, the same ensembles of sinks within a cavity of radius $R_0 = 10R_1$, i.e., close to the inner surface of the cavity, display a rate of capture $k_{50}^c = (54.2 \pm 0.2)k_S$. This corresponds to a situation of even *positive* cooperativity. This situation is found, for example, in the bacterial periplasm. It is reasonable to assume that ligands, whose concentration is constant outside the cell, diffuse very slowly in the periplasm as compared to the bulk, which justifies the assumption $h_0 \gg 1$. It is fascinating to think that such a complex, double-membrane architecture could be an evolutive answer to the requirement of maximizing the diffusive flux of (possibly low-concentration) ligands to a set of membrane-bound receptors.

Figure 4 reveals what happens to the individual capture rates for a large set of equivalent configurations of receptors on the inner membrane of an imaginary periplasmatic layer. Each receptor-sink is seen to capture on average the same amount of flux it would capture if it was isolated at the center of the cavity (see Eq. (31) for $h_0 \to \infty$), i.e., about $1/(1 - R_1/R_0) \approx 1.11$ in units of k_S. This somewhat surprising fact is due to the close proximity of the sinks to the inner surface of the cavity (see also again Fig. 3). If the cavity disappears, this figure drops down to about $0.15\,k_S$ (left histogram in Fig. 4). This is another manifestation of the virtual suppression of DI for sinks close to the absorbing inner surface of a cavity. Furthermore, it can be observed that the intrinsic variability of the capture rate around the ensemble average is reduced when DI are strong (width of the left histogram in Fig. 4). This means that when DIs are weaker,

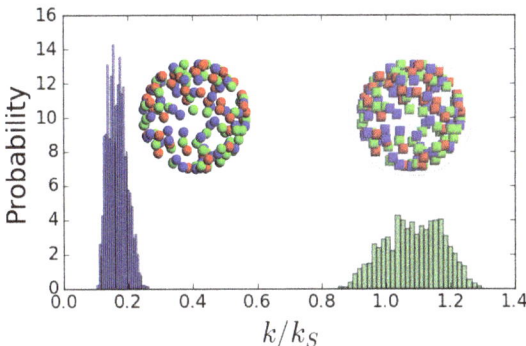

Fig. 4. Histograms of the individual capture rate of $N = 50$ identical sinks of radius $R_1 = R_0/10$ arranged at a fixed distance $d = 0.95 \times (R_0 - R_1)$ from the center of a spherical cavity with absorbing inner wall (right) and in the infinite domain (left). The histograms are normalized to the capture rate of an isolated sink in the open domain, $k_S = 4\pi D R_1 c_B$. The histograms refer to a population of 100 independent random configurations of the $N = 50$ sinks, 3 of which are pictured explicitly to illustrate the geometry of the problem.

not only does the ensemble recover a large rate of capture on average, but some of the receptor-sinks can individually attain large peaks of capture rate.

6.   Summary

In this short, mostly pedagogical chapter, we have described the phenomenon of diffusion to capture, which has important implications in a wide range of fields in biology and physical chemistry. We have shown how, under certain circumstances, the problem of bimolecular encounters and reactions can be solved as a two-body stationary diffusion boundary problem. This theoretical framework immediately leads to some surprising conclusions. One of the most striking findings concerns the rate of ligand capture by a receptor-covered cell. The classic mean-field solution of this problem shows that a fraction of surface coverage as low as 10^{-4} (approximately 10^4 receptors of 1.5 nm size on the surface of a cell of size 10 μm) ensures that the overall rate of capture is of the same order (reduced by a factor of 2) as for a fully covered surface. Moreover, we have shown that if the same number of active receptors are all moved into an active cluster covering the same surface fraction, the overall rate of capture drops by a factor of up to 10^2. This is a first manifestation of DI, which describe the interference among diffusive fluxes to neighboring reactive boundaries.

In order to provide a rigorous mathematical description of DI, we have considered in detail the classic problem of diffusion to two neighboring sinks at a center-to-center distance ℓ in the open domain. Although this problem can be solved by using bispherical coordinates,[64] we have followed another, more general, approach based on translational AT for spherical harmonics.[47] The exact solution of the problem, expressed in the form of an infinite series of multipoles, can be surprisingly well approximated for perfectly absorbing spheres by the monopole term alone, showing that DI are long range, i.e., decrease as ℓ^{-1}.

The mathematical strategy based on AT can be easily extended to compute the rate of capture of an ensemble of spheres of arbitrary size, intrinsic reactivity (κ^*), and arranged in arbitrary configurations in 3D, both in the open domain and within a spherical cavity. This theory, developed in Ref. 46, is described in detail in the appendix. Although the applications of such theoretical framework are countless, we have examined here two simple examples. First, we have studied the case of two sinks within a cavity, whose solution shows that DI are generally reduced in a finite domain with an absorbing inner surface, concomitantly with the enhancement of the rate of capture. This phenomenon is due to the fact that, in this modeling strategy, the density of ligands reaches its constant bulk value outside the cavity (whose surface is modeled as a permeable membrane), which enhances the rate of capture of a given boundary with respect to the open domain. Interestingly, now the relative position of the boundary within the cavity obviously makes a difference, the rate of capture increasing massively as the boundary approaches the inner surface of the cavity. At the same time, if many sinks are present, the DI among them are virtually suppressed for many-body configurations close to the inner surface of the cavity. The second and final example studied, namely, many independent configurations of sinks close to the inner surface of the cavity, shows this clearly. Finally, we have argued that this problem, while interesting on purely theoretical grounds, might also have important implications in ligand–receptor interactions in biology. Notably, ligand diffusion to receptors on the cytoplasmatic membrane in the periplasmatic space in bacteria provides an example of this problem. In this specific case, ligand diffusion in the crowded, gel-like periplasm is likely to be strongly reduced with respect to the mobility in the bulk outside, which justifies modeling the inner surface of the outer (cell) membrane as an absorbing boundary. The fascinating speculation that follows from these results is that such a complex architecture might have been designed by evolution to maximize the ligand–receptor binding rate. This would make sense, as such receptors are mostly chemotaxis receptors, used by bacteria to sense gradients of nutrients (small molecules).

Acknowledgments

This work was partially supported within the framework of the state task program of the FASO Russia (Theme 0082-2014-008, No AAAA-A17-117040310008-5).

Appendix. Rules for Selecting the Appropriate Addition Theorem

The AT for spherical harmonics allows one to express a combination of spherical harmonics, written in multiple coordinate systems, as a function of any one of them. Depending on the type of spherical harmonics that one needs to reexpand (regular or irregular) and on the geometry of the domain, one among three ATs has to be chosen in each specific case. Let us suppose to have spherical harmonics $u^+(\boldsymbol{r}_\beta)$ and $u^-(\boldsymbol{r}_\beta)$ written in a spherical coordinate system centered on S_β, that we want to express at a given point P as a function of the S_α-coordinate system (see Fig. 2). The relation $\boldsymbol{r}_\beta = \mathbf{L}_{\beta\alpha} + \mathbf{r}_\alpha$ holds.

The regular harmonics $u^+(\mathbf{r}_\beta)$ are always expressed as a function of the regular harmonics $u^+(\mathbf{r}_\alpha)$, namely,

$$r_\beta^n Y_{mn}(\mathbf{r}_\beta) = \sum_{q=0}^{n} \sum_{g=-q}^{q} \frac{(n+m)!}{(n-q+m-g)!(q+g)!} L_{\beta\alpha}^{n-q} Y_{m-g,n-q}$$

$$\times (\mathbf{L}_{\beta\alpha}) r_\alpha^q Y_{gq}(\mathbf{r}_\alpha). \tag{A.1}$$

If one has to reexpand an irregular harmonics $u^-(\mathbf{r}_\beta)$, two cases are possible, depending on the ratio between the distance $L_{\beta\alpha}$ between the centers of the old and new reference frames, and the norm of the vector $\mathbf{r}_\alpha$ expressing the position of P in the new frame S_α. More precisely, if $|\mathbf{r}_\alpha| < |\mathbf{L}_{\beta\alpha}|$, then one has to write the irregular harmonics as a function of the regular harmonics centered on S_α, namely,

$$r_\beta^{-n-1} Y_{mn}(\mathbf{r}_\beta) = \sum_{q=0}^{\infty} \sum_{g=-q}^{q} (-1)^{q+g} \frac{(n-m+q+g)!}{(n-m)!(q+g)!}$$

$$\times L_{\beta\alpha}^{-(n+q)-1} Y_{m-g,n+q}(\mathbf{L}_{\beta\alpha}) r_\alpha^q Y_{gq}(\mathbf{r}_\alpha). \tag{A.2}$$

Conversely, if $|\mathbf{r}_\alpha| > |\mathbf{L}_{\beta\alpha}|$, then one has to write the irregular harmonics as a function of the irregular harmonics centered in S_α

$$r_\beta^{-n-1} Y_{mn}(\mathbf{r}_\beta) = \sum_{l=0}^{\infty} \sum_{s=-n}^{n} \frac{(-1)^{l+s}(n+l-m+s)!}{(n-m)!(l+s)!}$$

$$\times L_{\alpha\beta}^{l} Y_{sl}(-\mathbf{L}_{\alpha\beta}) r_\alpha^{-(n+l)-1} Y_{m-s,n+l}(\mathbf{r}_\alpha). \tag{A.3}$$

To summarize, one can use the following scheme to change variables from system S_β to S_α (see also Fig. A.1)

- $u^+(\mathbf{r}_\beta) = f(u^+(\mathbf{r}_\alpha))$,

- $u^-(\mathbf{r}_\beta) = \begin{cases} f(u^+(\mathbf{r}_\alpha)) & \text{if } |\mathbf{r}_\alpha| < |\mathbf{L}_{\beta\alpha}|, \\ f(u^-(\mathbf{r}_\alpha)) & \text{if } |\mathbf{r}_\alpha| > |\mathbf{L}_{\beta\alpha}|. \end{cases}$

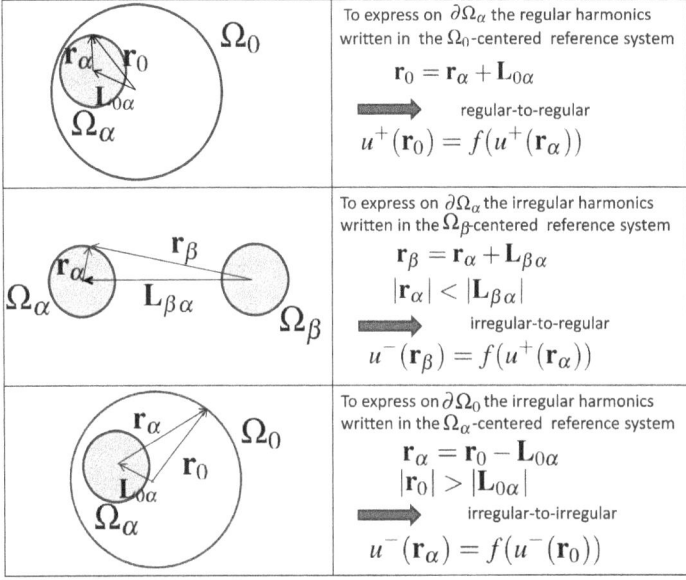

Fig. A.1. Scheme for the the application of the AT (A.1), (A.2), and (A.3) to express the boundary conditions in the local coordinates on $\partial\Omega_0$ and on each $\partial\Omega_\alpha$. The choice of the appropriate AT depends on the ratio between the distance between the centers of each pair of reference systems and the norm of the position vector in the new reference system.

A.1. The solution to the problem

By using the above AT, the solution to the problem (28) can be cast in the form of the following infinite-dimensional linear system:

$$
\begin{cases}
-B_{gq}^\alpha + \sum_{n=0}^\infty \sum_{m=-n}^n \left(A_{mn} H_{m,n}^{(\alpha,g,q)} \mathbf{1}_{q \leq n} + \sum_{\beta=1,\beta \neq \alpha}^N B_{mn}^\beta W_{m,n}^{(\alpha,\beta,g,q)} \right) = 0, \\
A_{gq} + \sum_{\alpha=1}^N \sum_{n=0}^q \sum_{m=-n}^n B_{mn}^\alpha V_{g,q}^{\alpha,m,n} \mathbf{1}_{\{g-(q-n)\leq m \leq g+(q-n)\}} = \delta_{(g,q)=(0,0)},
\end{cases}
\tag{A.4}
$$

where

$$
\begin{aligned}
V_{g,q}^{\alpha,m,n} &= -\frac{h_0 + q + 1}{(q - h_0)} \frac{(-1)^{q-n+m-g}(q - g)!}{(n - m)!(q - n + m - g)!} \\
&\quad \times \eta_{0\alpha}^{q-n} \chi_\alpha^{n+1} Y_{m-g,q-n}(-\mathbf{L}_{0\alpha}),
\end{aligned}
\tag{A.5}
$$

$$H_{m,n}^{(\alpha,g,q)} = \frac{(q-h_\alpha)}{(h_\alpha+q+1)}\binom{n+m}{q+g}\chi_\alpha^q \eta_{0\alpha}^{n-q} Y_{m-g,n-q}(\boldsymbol{L_{0\alpha}}), \quad (A.6)$$

$$W_{m,n}^{(\alpha,\beta,g,q)} = \frac{(q-h_\alpha)}{(h_\alpha+q+1)}(-1)^{q+g}\frac{(n-m+q+g)!}{(n-m)!(q+g)!}$$

$$\times \eta_{\beta\alpha}^{-(n+q)-1}\chi_\alpha^q \chi_\beta^{n+1} Y_{m-g,n+q}(\boldsymbol{L_{\beta\alpha}}), \quad (A.7)$$

with $\chi_\alpha := \frac{R_\alpha}{R_0}$ and $\eta_{\alpha\beta} := \frac{L_{\alpha\beta}}{R_0}$ with $\eta_{\alpha\beta} = \eta_{\beta\alpha}$. The system (A.4) can be cast in matrix form as follows:

$$
\begin{bmatrix}
\mathbb{1} & V^1 & V^2 & \cdots & V^N \\
H^1 & -\mathbb{1} & W^{1,2} & \cdots & W^{1,N} \\
H^2 & W^{2,1} & -\mathbb{1} & \cdots & W^{2,N} \\
\vdots & \vdots & \vdots & \ddots & \vdots \\
H^N & W^{N,1} & W^{N,2} & \cdots & -\mathbb{1}
\end{bmatrix}
\times
\begin{bmatrix}
A_{00} \\
\vdots \\
A_{N_M N_M} \\
B_{00}^1 \\
\vdots \\
B_{N_M N_M}^1 \\
\vdots \\
B_{00}^N \\
\vdots \\
B_{N_M N_M}^N
\end{bmatrix}
=
\begin{bmatrix}
1 \\
\vdots \\
0 \\
0 \\
\vdots \\
0 \\
\vdots \\
0 \\
\vdots \\
0
\end{bmatrix}.
$$

A.2. The rate

The capture rate for a ligand with diffusion coefficient D by a selected boundary Ω_α can be computed easily as the total incoming flux, namely,

$$k_\alpha = -\int_{\partial\Omega_\alpha} \boldsymbol{J}_\alpha \cdot \hat{n}\, dS, \quad (A.8)$$

where $\boldsymbol{J}_\alpha = -D\nabla_\alpha c$ is the current to the αth boundary. It is not difficult to see from the general form of the solution (29) and general properties of the Legendre polynomials $P_n^m(\mu_\alpha)$ that Eq. (A.8) gives

$$\frac{k_\alpha}{k_{S_\alpha}} = -B_{00}^\alpha, \tag{A.9}$$

where $k_{S_\alpha} = 4\pi D R_\alpha c_B$ is the Smoluchowski rate of capture corresponding to an isolated sink of radius R_α in the infinite domain.

References

1. S. Rice, *Diffusion-Limited Reactions*. Elsevier, Amsterdam (1985).
2. F. Piazza, G. Foffi, and C. De Michele, Irreversible bimolecular reactions with inertia: From the trapping to the target setting at finite densities, *J. Phys. Cond. Matt.* **25**(24), 245101 (2013).
3. S. Torquato, *Random Heterogenous Materials*. Interdisciplinary Applied Mathematics, Springer-Verlag, New York (2002).
4. J. Crank, *The Mathematics of Diffusion*. Oxford Science Publications (1975).
5. M. Labowsky and T. M. Fahmy, Diffusive transfer between two intensely interacting cells with limited surface kinetics, *Chem. Eng. Sci.* **74**, 114–123 (2012).
6. H. X. Zhou, G. Rivas, and A. P. Minton, Macromolecular crowding and confinement: Biochemical, biophysical, and potential physiological consequences, *Ann. Rev. Biophys.* **37**, 375–397 (2008).
7. A. S. Verkman, Solute and macromolecule diffusion in cellular aqueous compartments, *Trends Biochem. Sci.* **27**(1), 27–33 (Jan., 2002).
8. K. Luby-Phelps, Cytoarchitecture and physical properties of cytoplasm: volume, viscosity, diffusion, intracellular surface area, *Int. Rev. Cytol.* **192**, 189–221 (2000).
9. A. M. Berezhkovskii and A. Szabo, Theory of Crowding Effects on Bimolecular Reaction Rates, *J. Phys. Chem. B.* **120**(26), 5998–6002 (2016).
10. S. D. Traytak, Ligand binding in a spherical region randomly crowded by receptors, *Phys. Biol.* **10**(4), 045009 (2013).
11. F. Piazza, N. Dorsaz, C. De Michele, P. De Los Rios, and G. Foffi, Diffusion-limited reactions in crowded environments: A local density approximation, *J. Phys. Cond. Mat.* **25**(37), 375104 (2013).
12. N. Dorsaz, C. De Michele, F. Piazza, P. De Los Rios, and G. Foffi, Diffusion-limited reactions in crowded environments, *Phys. Rev. Lett.* **105**(12), 120601 (2010).
13. C. Echevería, K. Tucci, and R. Kapral, Diffusion and reaction in crowded environments, *J. Phys. Cond. Matt.* **19**(6), 065146 (2007).
14. H. L. Frisch and F. C. Collins, Diffusional processes in the growth of aerosol particles. II, *J. Chem. Phys.* **21**(12), 2158–2165 (1953).
15. R. A. Reck and S. Prager, Diffusion-controlled quenching at higher quencher concentrations, *J. Chem. Phys.* **42**(9), 3027–3032 (1965).

16. V. A. Borzilov and A. S. Stepanov, On the derivation of the equation for condensation of an array of drops, *Izv. Atmos. Ocean. Phys.* **7**, 164–172 (1971).

17. J. M. Deutch, B. U. Felderhof, and M. J. Saxton, Competitive effects in diffusion-controlled reactions, *J. Chem. Phys.* **64**, 4559 (1976).

18. S. Traytak, The diffusive interaction in diffusion-limited reactions: The steady-state case, *Chem. Phys. Lett.* **197**, 247–254 (9, 1992).

19. H. C. Berg, *Random Walks in Biology*. Princeton University Press (1993).

20. V. Sourjik, Receptor clustering and signal processing in *E. coli* chemotaxis. *Trends Microbiol.* **12**(12), 569–76 (2004).

21. J.-D. Lelièvre, F. Petit, L. Perrin, F. Mammano, D. Arnoult, J.-C. Ameisen, J. Corbeil, A. Gervaix, and J. Estaquier, The density of coreceptors at the surface of CD4+ T cells contributes to the extent of human immunodeficiency virus Type 1 viral replication-mediated T cell death, *AIDS Res. Hum. Retrov.* **20**(11), 1230–1243 (2004).

22. C. DeLisi and F. W. Wiegel, Effect of nonspecific forces and finite receptor number on rate constants of ligand-cell bound-receptor interactions, *Proc. Nat. Acad. Sci. USA* **78**(9), 5569–5572 (1981).

23. A. Briegel, D. R. Ortega, E. I. Tocheva, K. Wuichet, Z. Li, S. Chen, A. Müller, C. V. Iancu, G. E. Murphy, M. J. Dobro, I. B. Zhulin, and G. J. Jensen, Universal architecture of bacterial chemoreceptor arrays, *Proc. Nat. Acad. Sci. USA.* **106**(40), 17181–17186 (2009).

24. Š. Bálint, M. L. Dustin, A. Bruckbauer, F. Batista, S. Banjade, J. Okrut, D. King, J. Taunton, M. Rosen, R. Vale, Z. Guo, R. Vishwakarma, M. Rao, S. Mayor, D. Klenerman, A. Aricescu, and S. Davis, Localizing order to boost signaling, *eLife.* **6**, 1055–1068 (2017).

25. S. Angioletti-Uberti, Exploiting receptor competition to enhance nanoparticle binding selectivity, *Phys. Rev. Lett.* **118**(6), 68001 (2017).

26. W. S. Hlavacek, R. G. Posner, and A. S. Perelson, Steriuc effects on multivalent ligand-receptor binding: Exclusion of ligand sites by bound cell surface receptors, *Biophys. J.* **76**, 3031–3043 (1999).

27. G. Vauquelin and S. J. Charlton, Exploring avidity: Understanding the potential gains in functional affinity and target residence time of bivalent and heterobivalent ligands, *Brit. J. Pharmacol.* **168**(8), 1771–1785 (2013).

28. A. Todorovska, R. C. Roovers, O. Dolezal, A. A. Kortt, H. R. Hoogenboom, and P. J. Hudson, Design and application of diabodies, triabodies and tetrabodies for cancer targeting, *J. Immunol. Methods* **248**(1–2), 47–66 (2001).

29. N. Nuñez-Prado, M. Compte, S. Harwood, A. Álvarez-Méndez, S. Lykkemark, L. Sanz, and L. Álvarez-Vallina, The coming of age of engineered multivalent antibodies, *Drug Discov. Today* **20**(5), 588–594 (2015).

30. C. Fasting, C. A. Schalley, M. Weber, O. Seitz, S. Hecht, B. Koksch, J. Dernedde, C. Graf, E.-W. Knapp, and R. Haag, Multivalency as a chemical organization and action principle, *Angew. Chem. Inter. Edit.* **51**(42), 10472–10498 (2012).

31. M. Mammen, S.-K. Choi, and G. M. Whitesides, Polyvalent interactions in biological systems: Implications for design and use of multivalent ligands and inhibitors, *Angew. Chem. Internat. Edit.* **37**(20), 2754–2794 (1998).

32. I. V. Gopich, A. M. Berezhkovskii, and A. Szabo, Concentration dependence of the diffusion controlled steady-state rate constant, *J. Chem. Phys.* **117**, 2987–2988 (2002).
33. A. M. Berezhkovskii, D. J. Bicout, and G. H. Weiss, Target and trapping problems: From the ballistic to the diffusive regime, *J. Chem. Phys.* **110**, 1112 (1999).
34. S. Yuste, G. Oshanin, K. Lindenberg, O. Bénichou, and J. Klafter, Survival probability of a particle in a sea of mobile traps: A tale of tails, *Phys. Rev. E.* **78**(2), 021105 (2008).
35. S. D. Traytak and M. Tachiya, Concentration dependence of fluorescence quenching by ionic reactants, *J. Phys. Cond. Matt.* **19**(6), 065111 (2007).
36. J. Keizer, Diffusion effects on rapid bimolecular chemical reactions, *Chem. Rev.* **87**(1), 167–180 (1987).
37. B. U. Felderhof and J. M. Deutch, Concentration dependence of the rate of diffusion-controlled reactions, *J. Chem. Phys.* **64**(11), 4551–4558 (1976).
38. M. Galanti, D. Fanelli, S. Angioletti-Uberti, M. Ballauff, J. Dzubiella, and F. Piazza, Reaction rate of a composite core-shell nanoreactor with multiple nanocatalysts, *Phys. Chem. Chem. Phys.* **18**(30), 20758–20767 (2016).
39. F. Piazza and S. D. Traytak, Diffusion-influenced reactions in a hollow nanoreactor with a circular hole, *Phys. Chem. Chem. Phys.* **17**(16), 10417–10425 (2015).
40. Y. Lu and M. Ballauff, Thermosensitive core–shell microgels: From colloidal model systems to nanoreactors, *Prog. Polym. Sci.* **36**(6), 767–792 (2011).
41. A. Sozza, F. Piazza, M. Cencini, F. De Lillo, and G. Boffetta, Point-particle method to compute diffusion-limited cellular uptake, *Phys. Rev. E.* **97**(2), 23301 (2018).
42. S. D. Traytak, The diffusive interaction in diffusion-limited reactions: the time-dependent case, *Chem. Phys.* **193**, 351–366 (1995).
43. S. D. Traytak, On the time-dependent diffusive interaction between stationary sinks, *Chem. Phys. Lett.* **453**, 212–216 (2008).
44. S. D. Traytak, Competition effects in steady-state diffusion-limited reactions: Renormalization group approach, *J. Chem. Phys.* **105**(24), 10860–10867 (1996).
45. E. Gordelyi, S. Crouch, and S. Mogilevskaya, Transient heat conduction in a medium with multiple spherical cavities, *Int. J. Num. Meth. Eng.* **77**, 751–775 (2009).
46. M. Galanti, D. Fanelli, S. D. Traytak, and F. Piazza, Theory of diffusion-influenced reactions in complex geometries, *Phys. Chem. Chem. Phys.* **18**(23), 15950–15954 (2016).
47. P. M. Morse and H. Feshbach, *Methods of Theoretical Physics.* Vol. 2, McGraw-Hill Science/Engineering/Math (1953).
48. M. J. Caola, Solid harmonics and their addition theorems, *J. Phys. A. Math. Gen.* **11**(2), L23 (1978).
49. I. N. Sneddon, *Mixed Boundary Value Problems in Potential Theory.* North-Holland Pub. Co., Amstardem (1966).

50. S. D. Traytak and A. V. Barzykin, Diffusion-controlled reaction on a sink with two active sites, *J. Chem. Phys.* **127**(21), 215103 (2007).

51. F. Piazza, P. D. L. Rios, D. Fanelli, L. Bongini, and U. Skoglund, Anticooperativity in diffusion-controlled reactions with pairs of anisotropic domains: A model for the antigen–antibody encounter, *Eur. Biophys. J.* **34**(7), 899–911 (2005).

52. S. D. Traytak, Diffusion-controlled reaction rate to an active site, *Chem. Phys.* **192**, 1–7 (1995).

53. M. von Smoluchowski, Drei vortrage ubër diffusion brownsche molekular bewegung und koagulation von kolloidteichen, *Physik Z.* **17**, 557–571 (1916).

54. M. von Smoluchowski, Versuch einer matematischen theorie der koagulationskinetic kolloider lösungen, *Z Phys. Chem.* **92**, 129–168 (1917).

55. A. Szabo, Theory of diffusion-influenced fluorescence quenching, *J. Phys. Chem.* **93**(19), 6929–6939 (1989).

56. F. C. Collins and G. E. Kimball, Diffusion-controlled reaction rates, *J. Coll. Sci.* **4**(4), 425–437 (1949).

57. H. C. Berg and E. M. Purcell, Physics of chemoreception, *Biophys. J.* **20**(2), 193–219 (1977).

58. D. Shoup and A. Szabo, Role of diffusion in ligand binding to macromolecules and cell-bound receptors, *Biophys. J.* **40**(1), 33–39 (1982).

59. T. L. Hill, Effect of Rotation on the Diffusion-Controlled Rate of Ligand-Protein Association, *Proc. Nat. Acad. Sci. USA.* **72**(12), 4918–4922 (1975).

60. D. A. Lauffenburger and J. J. Linderman, *Receptors: Models for Binding, Trafficking, and Signaling.* Oxford University Press, New York (1993).

61. S. K. Mitra, A new method of solution of the boundary value problem of laplace's equation relating to two spheres, *Bull. Calcutta Math. Soc.* **36**, 31–39 (1944).

62. L. Rayleigh, On the influence of obstacles arranged in rectangular order upon the properties of a medium, *Philos. Mag. Ser. 5.* **34**(211), 481–502 (1892).

63. S. D. Traytak and D. S. Grebenkov, Diffusion-influenced reaction rates for active "sphere-prolate spheroid" pairs and Janus dimers, *J. Chem. Phys.* **148**, 024107(11) (2018).

64. R. Samson and J. M. Deutch, Exact solution for the diffusion controlled rate into a pair of reacting sinks, *J. Chem. Phys.* **67**(2), 847–847 (1977).

65. L. V. Kantorovich and G. P. Akilov, *Functional analysis.* Pergamon Press, Oxford (1982).

66. K. Sneppen and G. Zocchi, *Physics Molecular Biology.* Cambridge UP (2005).

67. A. Vazquez, Optimal cytoplasmatic density and flux balance model under macromolecular crowding effects, *J. Theor. Biol.* **264**(2), 356–359 (2010).

68. B. C. McNulty, G. B. Young, and G. J. Pielak, Macromolecular Crowding in the Escherichia coli Periplasm Maintains α-Synuclein Disorder, *J. Mol. Biol.* **355**(5), 893–897 (2006).

Chemical Reactions for Molecular and Cellular Biology

O. Shukron, U. Dobramysl, and D. Holcman*

*Group of Applied Mathematics and Computational Biology,
Ecole Normale Supérieure, Paris, France*

*david.holcman@ens.fr

This chapter is dedicated to modeling, analysis, and some simulations of chemical reactions associated with molecular and cellular processes. What is the meaning of chemical reactions today? We propose to discuss this question in the context of cellular biology. With the advance of experimental microscopy, new areas have emerged focusing on quantifying molecular dynamics inside cellular microcompartments. Some key questions are: How long does it take to activate a molecular pathway that has physiological consequence. How many molecules are participating in the chemical reaction? What defines the molecular noise and how it alters the signal transduction? Chemical activation can also occur through polymer loops in the cell nucleus.

We introduce here several concepts originating from stochastic processes related to the activation and search times used to studied *in vivo* rate constants. Examples are chemical reaction at neuronal synapses, signal transduction in photoreceptors, looping DNA to activate a gene promoter, positive feedback loop transcription factor for cellular patterning in the developing tissue, and many more.

"This chapter is dedicated to the late Pr. Zeev Schuss, exceptional applied mathematician of broad interest who has impacted several times the theory of applications of stochastic processes, asymptotic of PDEs, simulations of chemical reactions and many more. Zeev was a passionate scientist, a talent which allowed him to have a second career after retirement. We had a wonderful time discussing science nonstop all day long, being driven by no duties, nobody to please, no other constraints or demand than satisfying our own curiosity and imagination, but always with the goal of understanding how a piece of biology was working with the most accurate description that we could. Learning a field because it was necessary was never a limitation for Zeev. Those were the values that drove us and kept us moving nonstop. Our vision was also summarized in Ref. 1," D. Holcman.

1. Introduction

The theory of chemical reaction traditionally deals with computing the rate of production of a species from another. The production rates have been computed for decades when species are homogenously distributed in containers.[2] In parallel, the pharmaceutical and oil industries have developed optimized procedures for oil cracking using control theory.[3] More than 20 years ago, to reduce the exploration of the very large dimensional space of chemical compound, molecular dynamics simulations were developed, which had a large impact in the field of drug design. A new industry was born based on developing the theory of relatively short simulations,[4,5] generating billions of dollars in revenue and leading to innovative start-up companies to design optimal drugs.

With the advance of experimental microscopy, genetic engineering, the refinement of fluorescence probes, and continuous improvement of electrophysiology, our understanding of cell function has been based on chemical reaction pathways, revealing that a few molecules (after being amplified) could play a key role (see review of chemical reactions in phototransduction in Ref. 6).

Developing a theory for interpreting and/or quantifying data at a single molecule level remains challenging, because measurements are often indirect. For example, the signal of slow voltage dyes needed to be deconvolved to recover the spatio-temporal voltage map in electrical cells.[7] Another example is the effort of the past decades to measure and quantify the molecular response to a single photon. If the biochemical players are identified for most of them, it is still very difficult in rod and cone photoreceptors to reconstruct the refined molecular machinery involved in light modulation and adaptation, especially how the geometry of the outer segment of the cell can change during aging or certain degenerative disease, such as retinitis pigmentosa, thus affecting the organization of the chemical reactions. Connecting chemical reactions with local geometry at a nanometer level remains challenging. Other examples are calcium dynamics and G-protein cascade that occupy a central role in many signal transduction.[8]

Another type of chemical reaction occurring at a molecular level is the process of gene activation and regulation. Indeed, a promoter site is a key binding site that promotes protein expression following activation. For that purpose, a gene can be activated by transcription factors (TFs) either after direct binding to the site or after DNA has generated a loop to bring the TFs to the promoter site. But this process depends on the 3D chromatin organization, the regulation of which remains poorly understood. TFs search can be amplified by a positive feedback loop that could lead to full gradual expression of specific proteins as predicted by FPE approach, which is missed in the mass-action law approximation of the equation, reviewed in Ref. 9. Indeed, bridging the continuum to the discreet description remains a delicate question; the reduction to the mass-action equation remains especially inappropriate at intermediate concentration and does not capture that molecules that could remain at low levels in some cells and at high level in others.[10]

The renewal of interest in modeling chemical reactions for molecular and cellular biology has been motivated by the exploding computational power and design of accurate simulations to characterize chemical reactions inside a cell, accounting also for the subcellular geometrical organization of the micro-compartment. In particular, when the number of produced molecules is indirectly measured, such as in transduction, the chemical reaction theory is used not only to retrieve the unperturbed behavior and estimate rate constant but also to discover molecular behavior. Finally, when the number of molecules is small, chemical reactions should be modeled at a stochastic level.[9,11] In that case, it is necessary to employ the two key rate constants — backward and forward rates — very carefully: if the first one is the reciprocal time for two molecules to dissociate, the second one is the reciprocal mean time for the two molecules to meet, which is largely influenced by the local geometry and the distribution of the substrate. We will dedicate a section to modeling of stochastic chemical reactions and will specify the conditions for using Brownian simulations when Gillespie's algorithm fails. We will also discuss the concept of the mean time to a threshold, which is the time when a small number of binding sites are activated stochastically. Another possibility of activation is to wait for a threshold to be reached, driven by the arrival of the fastest particles. The associated time scale falls into the extreme statistics description and provides a new conceptual framework for cellular activation, especially when there is no equilibrium state.

Mixing of stochastic and deterministic chemical reactions plays a key role in understanding how sensory cells convert an external signal into a cellular response. For example, retinal photoreceptors can convert even a single photon into a cellular hyperpolarization. Most of the chemical reactions have been identified, but to account for the entire photoresponse, the geometry of rod and cone outer segments have to be taken into account.[12] In particular, an interesting finding was that some chemical rates seem to have evolved across species to accommodate for the large or small cell geometry, leading to a slow or fast photon response, at the expense of the robustness, reviewed in Refs. 6 and 12. The chemical rates in general are finely tuned, because they serve to generate a fast photoresponse, but also lead to inherent background noise. Cellular geometry and chemical rate constants seem to be correlated in physiology. This novel paradigm is currently investigated in most of the transduction cells, such as olfactory or sensory cells. We will not discuss these questions much and instead refer the reader to Ref. 12 and associated references.

This chapter is organized as follows: we will first recall the Markov approach used for stochastic chemical reactions and the associated mean times for a compound to reach a given threshold. We will then discuss the rate constant associated with the first particle to arrive to a target site that defines the chemical reaction in many physiological amplification problem. We then move to chemical reactions in the context of neuronal growth and migration motivated by two questions: how cells know where they are located and how to simulate Brownian particles generated at a source. The reviewed modeling is associated with a source triangulation: recovery of the source location from fluxes. We end this chapter with the chemical reaction associated with loop formation in the context of polymer model for modeling chromatin dynamics: Indeed by forming a DNA loop, a TF can activate the expression of a gene. We review the physical laws associated with DNA looping for different polymer models.

2. Stochastic Chemical Reaction, MTT Theory, and Extreme Statistics

Chemical kinetics is based on mass-action laws formulated in differential equations when species are well mixed. When a small number of substrate and reactant molecules are involved in microdomains, reaction-diffusion equations do not provide a complete description of stochastic chemical reactions. The small populations of reactant and substrate give rise to large fluctuations and rare binding events in the number of bound substrate sites, which can have drastic consequences such as activating a physiological pathway.

An alternative and fruitful approach is to coarse-grain the binding and unbinding reactions to the time scale of the mean first passage time of the diffusing reactant into and out of a small binding site (which is also the narrow escape time[13]). This approach leads to a Markovian jump process description of the stochastic dynamics of the binding and unbinding of molecules. The goal of this section is to present such an approach, which was developed over the years by the authors of Refs. 11, 14 and 15.

2.1. Introduction to stochastic chemical reactions

The simplest example of a chemical reaction is between two definite species: the mobile reactant M, which diffuses freely in a bounded domain Ω, and the stationary substrate S (e.g., a protein), which binds M. The boundary $\partial\Omega$ of a domain Ω is partitioned into an absorbing part $\partial\Omega_a$ (e.g., protein channels, pumps, exchangers, another substrate that forms permanent bonds with M, and so on) and a reflecting part $\partial\Omega_r$ (e.g., a cell membrane). In this model, the size of species M is neglected. In terms of chemical kinetics the binding of M to S follows the chemical equation

$$M + S_{\text{free}} \underset{k_{-1}}{\overset{k_1}{\rightleftharpoons}} MS, \tag{1}$$

where k_1 is defined as the forward binding rate constant, k_{-1} is the backward binding rate constant, and S_{free} is the unbound substrate. The reaction occurs where M molecules diffuse in Ω independently of each other and when bound are released independently of each other at exponential waiting times with rate k_{-1}. The time for the binding of a single M molecule is the first passage time to diffuse to a small portion of the boundary, $\partial\Omega_a$, which is absorbing and represents the active surface of the free substrate (receptor), whereas the remaining part of $\partial\Omega$ is reflecting. Thus, it is the narrow escape time (NET)[16] to $\partial\Omega_a$. Due to the small target and the deep binding potential well of the binding site, the binding and unbinding of M to S are rare events on the time scale of diffusion.[17,18] This implies that the probability distribution of binding times is approximately exponential[19] with rate $\lambda_1 = 1/\bar{\tau}_1$, where $\bar{\tau}_1$ is the MFPT to $\partial\Omega_a$. When there are S binding sites, $k(t)$ of which are unbound, there are $N = [M - S + k]^+ = \max\{0, M - S + k\}$ free diffusing molecules in Ω. The arrival time of a molecule to the next unbound site is well approximated by an exponential law with state-dependent instantaneous rate (see discussion in Ref. 11)

$$\lambda_k = \frac{Nk}{\bar{\tau}_1} = \frac{k[M - S + k]^+}{\bar{\tau}_1}.$$

The number $k = k(t)$ of unbound receptors at time t is a Markovian birth–death process with states $0, 1, 2, \ldots, \max\{M - S, 0\}$ and transition rates $\lambda_{k \to k+1} = \lambda_k$, $\lambda_{k \to k-1} = \mu = k_{-1}$. The boundary conditions are $\lambda_{S \to S+1} = 0$ and $\lambda_{0 \to -1} = 0$.

Setting $P_k(t) = \Pr\{k(t) = k\}$, the Kolmogorov equations for the transition probabilities take the form

$$\dot{P}_k(t) = -\left[\lambda_k + k_{-1}(S - k)\right] P_k(t)$$
$$+ \lambda_{k+1} P_{k+1}(t) + k_{-1}(S - k + 1) P_{k-1}(t), \qquad (2)$$

for $k = [S - M] + 1, \ldots, S - 1$, with the initial and boundary equations

$$P_{k,q}(0) = \delta_{k,S} \delta_{q,0},$$
$$\dot{P}_{(S-M)^+}(t) = -k_{-1} S P_{[S-M]^+}(t) + \lambda_1 P_{(S-M)^+ +1}(t),$$
$$\dot{P}_S(t) = -\lambda_S P_S(t) + k_{-1} P_{S-1}(t). \qquad (3)$$

Most statistical properties about the chemical reactions can be extracted from solving system 2, when possible. We calculate here the average number of unbound (or bound) sites $\bar{k}(t)$, which for $t \to \infty$ is given by $\bar{k}_\infty = \sum_{j=(S-M)^+}^{S} j P_j$, where $P_j = \lim_{t \to \infty} P_j(t)$. Similarly, the stationary variance of the number of unbound sites is

$$\sigma^2(M, S) = \bar{k}_\infty^2 - (\bar{k}_\infty)^2, \qquad (4)$$

where $\bar{k}_\infty^2 = \sum_{j=(S-M)^+}^{S} j^2 P_j$. The results of the Markovian model (2)–(3), using a direct induction from the steady-state equations, are

$$P_S = \cfrac{1}{1 + \sum_{k=1}^{S-(S-M)^+} \cfrac{\prod_{i=S-k+1}^{S} i(M - S + i)^+}{k!(\bar{\tau}_1 k_{-1})^k}}, \qquad (5)$$

$$\langle k_\infty \rangle = P_S \sum_{k=S-1}^{(S-M)^+} (S-k)^+ \frac{\prod_{i=S-k+1}^{S} i(M-S+i)^+}{k!(\bar{\tau}_1 k_{-1})^k},$$

$$\langle k_\infty^2 \rangle = P_S \sum_{k=S-1}^{(S-M)^+} [(S-k)^+]^2 \frac{\prod_{i=S-k+1}^{S} i(M-S+i)^+}{k!(\bar{\tau}_1 k_{-1})^k},$$

$$\sigma_S^2(M) = \langle k_\infty^2 \rangle - \langle k_\infty \rangle^2, \tag{6}$$

see Ref. 11 for further details.

2.2.   *The mean time the number of bound molecules reaches a threshold (MTT)*

Another application of the Markovian model is the calculation of the mean time the number of bound molecules reaches a threshold (MTT). In a cellular context, the MTT can be used to characterize the stability of chemical processes, especially when they underlie a biological function. The above Markovian description leads to an estimate of the MTT in terms of fundamental parameters, such as the number of molecules, ligands, and the forward and backward binding rates. As shown next, the MTT depends nonlinearly on the threshold T.

We recall that the concept of the number of molecules reaching a threshold is ubiquitous in biology. For example, in the patterning process, occurring in embryo development, cell differentiation is controlled by a gradient concentration of morphogens, and interestingly the cell fate can change by a small difference in that concentration (the concentration of the decapentaplegic gene (DPP) in insects can activate different genes at different thresholds[20–23]). Another example concerns the first step of cellular division, where chromosomes need to be attached before being separated. Another example is the case of neuronal synapses, where synaptic plasticity, a long lasting process underlying learning and memory,[24] can be induced when the concentration of calcium reaches a certain threshold.[25] Another example concerns the cellular response to a double-stranded DNA (dsDNA) break: The cell can "sense" the number of breaks and may decide to undergo apoptosis or not. Interestingly, a single dsDNA break can be detected, and this event is sufficient to activate a global cellular response.

The MMT was initially formulated as follows[14]: Consider M Brownian particles (e.g., molecules) that can bind to immobile targets S in a microdomain. The number $[MS](t)$ of MS bound particles at time t is modeled generically by equation (1). The first time the number $[MS](t)$ reaches the threshold T is defined as

$$\tau_T = \inf\{t > 0 : [MS](t) = T\}, \tag{7}$$

and its expected value is $\mathbb{E}\left(\tau_T\right) = \bar{\tau}_T$. To compute $\mathbb{E}\left(\tau_T\right)$, we first consider an ensemble of initially free targets distributed on the surface of a closed microdomain, assuming a vanishing backward rate $k_{-1} = 0$ and then assuming $k_{-1} > 0$. The dynamical system for the transition probabilities of the Markov process $MS(t)$ is similar to that in (1), except for the absorbing boundary condition at the threshold T, which gives (2) (see Ref. 14). When the binding is irreversible ($k_{-1} = 0$), $\bar{\tau}_T$ is the sum of the forward rates

$$\tau_T^{\text{irrev}} = \frac{1}{\lambda_0} + \frac{1}{\lambda_1} + \cdots + \frac{1}{\lambda_{T-1}} = \frac{1}{\lambda} \sum_{k=0}^{T-1} \frac{1}{(M_0 - k)(S_0 - k)}. \tag{8}$$

In particular, when $M_0 = S_0$ and $M_0 \gg 1$, (8) becomes asymptotically $\tau_T^{\text{irrev}} \approx T/\lambda M_0(M_0 - T)$. In addition, when the diffusing molecules largely exceed the number of targets ($M_0 \gg S_0, T$), (8) gives the asymptotic formulas

$$\tau_T^{\text{irrev}} \approx \begin{cases} \frac{1}{\lambda M_0} \log \frac{S_0}{S_0-T} & \text{for } M_0 \gg S_0, T, \\ \frac{1}{\lambda S_0} \log \frac{M_0}{M_0-T} & \text{for } S_0 \gg M_0, T, \\ \frac{T}{\lambda M_0 S_0} & \text{for } M_0, S_0 \gg T. \end{cases} \tag{9}$$

Figure 1 shows the plot of τ_T^{irrev} for several values of the threshold T, compared to Brownian simulations in a circular disk $\Omega = D(R)$ with reflecting boundary, except at the targets. When $k_{-1} > 0$, the

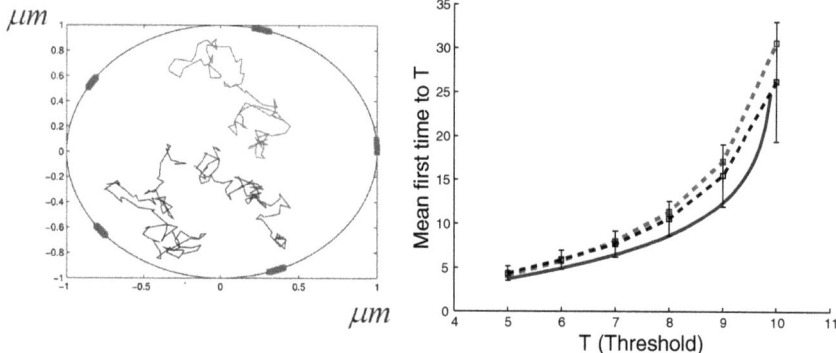

Fig. 1. Left: trajectories of diffusing molecules in a circular disk containing five binding sites on the boundary. Right: MFPT τ_T^{irrev} to threshold T as a function of T in the irreversible case $(k_{-1} = 0)$. Brownian simulations (upper dashed line), formula (8) (middle black dashed line), and its approximation (9) (lower solid line). The other parameters are $S_0 = 15$, $M_0 = 10$, $\varepsilon = 0.05$, $D = 0.1\ \mu m^2 s^{-1}$ and the radius of the disk is $R = 1\ \mu m$ (200 runs).

asymptotic formulas are obtained by solving the Markov equations (A.2) with a different boundary condition (see the Appendix and Ref. 14) and are given by

$$\bar{\tau}_T \approx \begin{cases} \tau_T^{\mathrm{irrev}} + \frac{k_{-1}}{(\lambda M_0)^2}[\frac{T}{S_0-T} - \log(1 + \frac{T}{S_0-T})] & \text{for } M_0 \gg S_0, T, \\[2mm] \tau_T^{\mathrm{irrev}} + \frac{k_{-1}}{(\lambda S_0)^2}[\frac{T}{M_0-T} - \log(1 + \frac{T}{M_0-T})] & \text{for } S_0 \gg M_0, T, \\[2mm] \tau_T^{\mathrm{irrev}} + \frac{k_{-1}}{2\lambda^2}(\frac{T}{M_0^2})^3 & \text{for } S_0 = M_0 \gg T. \end{cases}$$

Thus, $\bar{\tau}_T$ varies quadratically with the NET $\bar{\tau} = 1/\lambda$, and is a nonlinear increasing function of T. These computations are quite general and can be applied to describe the mean time to reach a threshold for any chemical reaction. Changing the threshold modulates the threshold time in an efficient way. To conclude, the time to threshold T τ_T varies quadratically with the Narrow Escape Time $\bar{\tau}$ $\left(\text{as } \bar{\tau} = \frac{1}{\lambda}\right)$; however, it increases nonlinearly with T. These computations are quite general and can be applied to describe the mean time to a given number of bound molecules for any chemical reactions.

In particular, by changing the threshold, chemical pathways can be modulated.[14,15]

2.3.   *MTT with killing*

When diffusing molecules can be killed before hitting the binding sites, the threshold may never be reached. The probability to reach the threshold and the conditional mean time MTT can be computed using a 2D Markov chain for the joint probability density function (PDF) $p_{k,m}(t)$ that at time t, there are k bound molecules and $m-k$ free diffusing molecules.

$$p_{k,m}(t) = P\{|MS|(t) = k, w(t) = m, |MS|(0) = 0, w(0) = M_0\},$$

where $w(t) = |M|(t) + |MS|(t)$. In the irreversible case, the transitions to the state (k, m) between time t and $t + \Delta t$ can only occur from the states $(k - 1, m), (k, m)$, and $(k, m + 1)$. To compute the probability that the threshold T is achieved, we study the chain (k, m) of having k bound molecules and the remaining $m - k$ free molecules. By imposing a boundary condition at (T, m) and $(k, T - 1)$, we will obtain the probability that the threshold is reached before the molecules are degraded by summing over the state (T, m) for $m = T \dots M_0$. We represent the transition diagram between states in Fig. 2, and the master equations are

$$\dot{p}_{0,M_0} = -(\lambda M_0 S_0 + \mu M_0)p_{0,M_0},$$

$$\dot{p}_{0,m} = \mu(m+1)p_{0,m+1} - (\lambda S_0 m + \mu m)p_{0,m} \text{ for } T-1 < m < M_0,$$

$$\dot{p}_{k,M_0} = \lambda(S_0 - k + 1)(M_0 - k + 1)p_{k-1,M_0}$$
$$\quad - (\lambda(S_0 - k) + \mu)(M_0 - k)p_{k,M_0} \text{ for } 0 < k < T,$$

$$\dot{p}_{T,m} = \lambda(S_0 - T + 1)(m - T + 1)p_{T-1,m} \text{ for } T \leq m \leq M_0,$$

$$\dot{p}_{k,T-1} = \mu(T - k)p_{k,T} \text{ for } 0 \leq k \leq T - 1,$$

$$\dot{p}_{k,m} = \lambda(S_0 - k + 1)(m - k + 1)p_{k-1,m} + \mu(m + 1 - k)p_{k,m+1}$$
$$\quad - (\lambda(S_0 - k) + \mu)(m - k)p_{k,m}$$
$$\text{for } T - 1 < m < M_0, 0 < k < T,$$

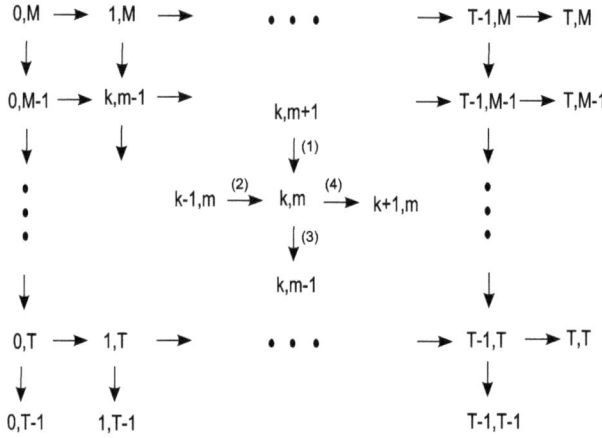

Fig. 2. Diagram of transition between states. Transition rates (1): $\mu(m - k + 1)$ (2): $\lambda(S - k + 1)(m - k + 1)$ (3): $\lambda(S - k)(m - k)$ (4): $\mu(m - k)$ where λ and μ are, respectively, the forward binding rate and the killing rate.

where λ and μ are, respectively, the forward binding rate and the killing rate.

The initial condition reads $p_{0,M_0}(0) = 1$ and the normalization condition is

$$\sum_{k=0,m=0}^{M_0,N_0} p_{k,m} = 1. \tag{10}$$

To derive the steady-state probabilities $p_{T,m}(\infty)$ and $p_{k,T-1}(\infty)$, we shall now consider that the only absorbing states (T, m) and $(k, T - 1)$ are given for $T \leq m \leq M_0$ and $0 \leq k \leq T - 1$ (see Fig. 2). Thus, by integrating system (10) over time, we obtain

$$q_{0,M_0} = \frac{1}{M_0(\lambda S_0 + \mu)}, \tag{11}$$

$$q_{0,m} = \frac{1}{m(\lambda S_0 + \mu)} \left(\frac{\mu}{\lambda S_0 + \mu}\right)^{M_0 - m} \quad \text{for } T - 1 < m < M_0,$$

$$q_{k,M_0} = \frac{\lambda^k S_0!}{(S_0 - k)!(M_0 - k) \prod_{j=0}^{k}(\mu + \lambda(S_0 - j))} \quad \text{for } 0 < k < T,$$

$$p_{T,m}(\infty) = \lambda(S_0 - T + 1)(m - T + 1)q_{T-1,m} \text{ for } T \leq m \leq M_0,$$

$$p_{k,T-1}(\infty) = \mu(T - k)q_{k,T} \text{ for } 0 \leq k \leq T - 1,$$

$$q_{k,m} = \frac{(\mu q_{k,m+1} + \lambda(S_0 - k + 1)(m - k + 1)q_{k-1,m})}{(m - k)(\mu + \lambda(S_0 - k))}$$

$$\times (m + 1 - k) \text{ for } T - 1 < m < M_0, 0 < k < T, \quad (12)$$

where for $0 \leq k \leq T - 1$, $T \leq m \leq M_0$,

$$q_{k,m} = \int_0^\infty p_{k,m}(t)dt. \tag{13}$$

The reader can see Ref. 26 for the computation for probability and the conditional MTT.

2.4. *The probability to reach the threshold*

The probability P_T to reach the threshold is equal to the probability to reach any of the states (T, m) for $m = T, \ldots, M_0$, that is

$$P_T = \sum_{m=T}^{M_0} p_{T,m}(\infty). \tag{14}$$

To estimate P_T, we study the system of Eqs. (11–12), and for $0 < k < T$ and $T \leq m < M$, we have derived in the Appendix A

$$q_{k,m} = \frac{\left(\prod_{i=0}^{k-1} \frac{\lambda(S_0-i)}{\lambda(S_0-i)+\mu}\right) \sum_{0\leq i_1,\ldots,i_{M_0}-m\leq k} \prod_{ij} \frac{\mu}{\lambda(S_0-i_j)+\mu}}{(\lambda(S_0 - k) + \mu)(m - k)}. \tag{15}$$

We proceed with the computation of the probability P_T to reach the threshold. We get from (12), for $T \leq m < M_0$, that

$$p_{T,m}(\infty) = \lambda(S_0 - T + 1)(m - T + 1)q_{T-1,m} \tag{16}$$

$$= \left(\prod_{i=0}^{T-1} \frac{\lambda(S_0 - i)}{\lambda(S_0 - i) + \mu}\right) \sum_{0\leq i_1,\ldots,i_{M_0}-m\leq T-1} \prod_{i_k} \frac{\mu}{\lambda(S_0 - i_k) + \mu}.$$

$$\tag{17}$$

Using Eq. (12),

$$q_{0,m} = \frac{1}{m(\lambda S_0 + \mu)} \left(\frac{\mu}{\lambda S_0 + \mu} \right)^{M_0 - m} \quad \text{for } T - 1 < m < M_0, \quad (18)$$

we also have for $m = M_0$,

$$p_{T,M_0}(\infty) = \left(\prod_{i=0}^{T-1} \frac{\lambda(S-i)}{\lambda(S-i) + \mu} \right). \quad (19)$$

Finally, considering that the probability is given by

$$P_T = \sum_{m=T}^{M_0} p_{T,m}(\infty), \quad (20)$$

we finally obtain, introducing the variable $\frac{\mu}{\lambda}$ and T,

$$p\left(\frac{\mu}{\lambda}, T \right) = P_T = \left(\prod_{i=0}^{T-1} \frac{\lambda(S-i)}{\lambda(S-i) + \mu} \right),$$

$$\left(1 + \sum_{0 \le i_1 \le T-1} \frac{\mu}{\lambda(S-i_1) + \mu} + \cdots \right.$$

$$\left. + \sum_{0 \le i_1 \le \cdots \le i_{M_0-T} \le T-1} \prod_{i_k} \frac{\mu}{\lambda(S-i_k) + \mu} \right). \quad (21)$$

We show in Fig. 3 the graph of P_T as a function of the threshold T for various values of the ratio $r = \frac{\mu}{\lambda}$. When $\frac{T}{S} \ll 1$, we obtain the approximation

$$p\left(\frac{\mu}{\lambda}, T \right) \approx \left(\frac{\lambda S}{\lambda S + \mu} \right)^T \left(1 + \frac{\mu}{\lambda S + \mu} |G_1| + \cdots + \right.$$

$$\left(\frac{\mu}{\lambda S + \mu} \right)^i |G_i| + \cdots$$

$$\left. + \left(\frac{\mu}{\lambda S + \mu} \right)^{M-T} |G_{M_0-T}| \right), \quad (22)$$

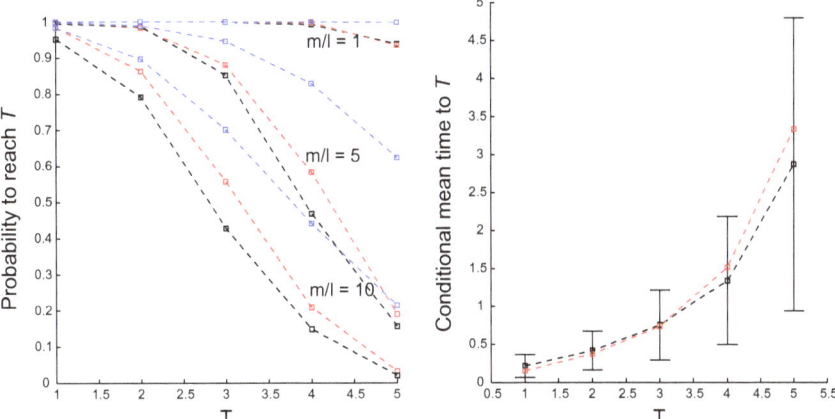

Fig. 3. Probability and MTT in a microdomain where ligands can be killed before binding. Left: we plot the probability P_T to reach the threshold as a function of T, for different values of the ratio $\frac{\mu}{\lambda}$ (1,5,10). The exact formula (21) (dotted red line) is compared with the approximation (24)(dotted blue line). We also compare with Brownian simulations of M molecules (dotted black line), with diffusion coefficient D moving inside a circular disk of radius R. The binding sites are of size ϵ. Right: we plot the conditional MTT as function of T for $\epsilon = 0.05$. The mean first passage time (the initial position is at the center) to a binding site of size ϵ is approximated by $\frac{1}{\lambda} = \tau = \left(\frac{R^2}{D}\right)\left(\log\left(\frac{4\pi}{\epsilon}\right) + \frac{1}{4}\right)$ (see Ref. 16). The other parameters are $R = 1$ μm, $D = 1$ $\mu m s^{-2}$, $S = 5$, $M = 10$. Number of Brownian simulations = 250.

where $|G_k|$ is the number of k-tuples $(i_1, \ldots, i_k)$ where $0 \leq i_1 \leq \cdots \leq i_k \leq T - 1$. Using $|G_k| = \begin{pmatrix} T - 1 + k \\ k \end{pmatrix}$, we obtain

$$p\left(\frac{\mu}{\lambda}, T\right) \approx \left(\frac{\lambda S}{\lambda S + \mu}\right)^T \sum_{k=0}^{M-T} \begin{pmatrix} T - 1 + k \\ k \end{pmatrix} \left(\frac{\mu}{\lambda S + \mu}\right)^k, \quad (23)$$

which can written as (see Appendix):

$$p\left(\frac{\mu}{\lambda}, T\right) \approx \sum_{k=T}^{M} \begin{pmatrix} M \\ k \end{pmatrix} \left(\frac{\lambda S}{\lambda S + \mu}\right)^k \left(\frac{\mu}{\lambda S + \mu}\right)^{M-k}. \quad (24)$$

This formula has the following interpretation: when there are many binding sites compared to the number of diffusing molecules, the binding events become independent. Consequently, the probability to bind can be approximated by a Bernoulli distribution of parameter $\frac{\lambda S}{\lambda S + \mu}$, and the probability of the binding number is a binomial distribution of parameters $(M, \frac{\lambda S}{\lambda S + \mu})$. Finally, the probability to reach the threshold is equivalent to having at least T bounds, and thus we obtain formula (24). Interestingly, as shown in Fig. 3, already for $r = 10$, the exact formula cannot be well approximated by the approximation (22) and the analytical solution should be used. The probability is a decreasing, inverse sigmoid-type function of the threshold T.

2.5. *Conditional MTT*

To get the conditional MTT, we first compute the mean time $\bar{\tau}(\sigma)$ for a trajectory parametrized by $\sigma = (i_0, i_1, i_2, \ldots, i_n)$ where $0 = i_0 \leq i_1 \leq \cdots i_{n-1} < i_n = T$ and $1 \leq n \leq M - T + 1$, which follows a path in the Markov diagram (see Fig. 2).

When there are k bound molecules and there remain $m - k$ free molecules, we shall estimate the mean transition time from this state, (k, m), to the state $(k + 1, m)$. This event is Poissonian with rate $\lambda_{k,m} = \lambda(S - k)(m - k)$, and the probability of binding before a molecule is killed is given by $\dfrac{\lambda_{k,m}}{\lambda_{k,m} + \mu(m - k)}$. Thus, the conditional mean binding time is

$$E(\tau^{MS}, |MS| = k, |M| = m - k) = \frac{1}{\lambda_{k,m}} \frac{\lambda_{k,m}}{\lambda_{k,m} + \mu(m - k)}$$

$$= \frac{1}{\lambda_{k,m} + \mu(m - k)}. \qquad (25)$$

Similarly, the mean time to killing is

$$E(\tau^K, |MS| = k, |M| = m - k) = \frac{1}{\lambda_{k,m} + \mu(m - k)}. \qquad (26)$$

The random times along the path σ are independent; thus, the total mean time $\bar{\tau}(\sigma)$ is the sum of all the mean times

$$\bar{\tau}(\sigma) = \begin{cases} \sum_{k=0}^{T-1} \frac{1}{(\lambda(S-k)+\mu)(M-k)} & \text{if } \sigma = (0,T), \\[3mm] \sum_{j=1}^{n-1} \frac{1}{(\lambda(S-i_j)+\mu)(M-j+1-i_j)} & \\[3mm] + \sum_{j=1}^{n} \sum_{\substack{k=i_{j-1} \\ i_{j-1} \neq i_j}}^{i_j-1} \frac{1}{(\lambda(S-k)+\mu)(M-j+1-k)} & \text{otherwise.} \end{cases}$$

The probability $P(\sigma)$ that the dynamics follows the path σ is

$$P(\sigma) = \begin{cases} \prod_{i=0}^{T-1} \frac{\lambda(S-i)}{\lambda(S-i)+\mu} & \text{if } \sigma = (0,T), \\[3mm] \left(\prod_{i=0}^{T-1} \frac{\lambda(S-i)}{\lambda(S-i)+\mu} \right) \left(\prod_{k=1}^{n-1} \frac{\mu}{\lambda(S-i_k)+\mu} \right) & \text{otherwise.} \end{cases}$$

$$(27)$$

Finally, for a forward binding constant λ and a killing rate μ, the conditional MTT $\bar{\tau}_T(\lambda, \mu, T)$ is

$$\bar{\tau}_T(\lambda, \mu, T) = \sum_{\sigma} \tau(\sigma) P(\sigma | T \text{ is reached}) = \frac{\sum_{\sigma} \tau(\sigma) P(\sigma)}{p(\frac{\mu}{\lambda}, T)},$$

$$(28)$$

where $p(\frac{\mu}{\lambda}, T)$ is the probability to reach the threshold computed in the previous subsection (formula (21)). We now approximate $\tau_T(\lambda, \mu, T)$ at first order in $\frac{\mu}{\lambda}$, which means that we neglect all the paths $\sigma = (i_0, i_1, i_2, \ldots, i_n)$ such that $n > 2$, (the probability for the other paths is at least of order $\left(\frac{\mu}{\lambda}\right)^2$). Consequently, consider $\sigma = (0, T)$ and $\sigma = (0, i, T)$ with $0 \leq i \leq T - 1$, at first order, using formula (22), $p\left(\frac{\mu}{\lambda}, T\right) = 1 + o(\frac{\mu}{\lambda})$. In addition, when $M \gg T$, we obtain from relations (27) and (27) the approximations for $0 \leq i \leq T - 1$

$$P(\sigma = (0, T))\tau(\sigma = (0, T)) \approx \left(1 - \sum_{k=0}^{T-1} \frac{\mu}{\lambda(S - k)} \right)$$

$$\sum_{k=0}^{T-1} \frac{1}{\lambda M(S - k)} \left(1 - \frac{\mu}{\lambda(S - k)} \right),$$

$$P(\sigma = (0, i, T))\tau(\sigma = (0, i, T)) \approx \left(\frac{\mu}{\lambda} \right) \frac{1}{\lambda M(S - i)}$$

$$\left(\sum_{k=0}^{T-1} \frac{1}{S - k} + \frac{1}{S - i} \right).$$

Finally, we obtain

$$\tau_T(\lambda, \mu, T) \approx \frac{1}{\lambda M} \log \left(\frac{S}{S - T} \right) \quad \text{when } M \gg T, \lambda \gg \mu. \quad (29)$$

Interestingly, in Eq. (29), the term in $\frac{\mu}{\lambda}$ vanishes, and thus we recover the zero-order approximation (Eq. (9)) for the MTT when particles cannot escape. In Fig. 3, the analytical formula (28) and the result of Brownian simulations show reasonable agreement. It might be tempting to believe that replacing degradation by a small absorbing window would give similar results. It does not. Indeed, the probability to reach a window in a sphere containing several others depends nonlinearly on their distribution, through their capacitance.[27,28] Thus, the rates $\lambda_{k,m}$ and $\mu_{k,m}$ will differ from the ones we obtained here. The geometrical configuration of the holes will now influence the escape and binding rates, and this effect should be studied carefully.

3. Competing Thresholds

When m molecules M can bind to sites S_1 (of number s_1) with a rate k_1 or to sites S_2 (of number s_2) with a rate k_2 (no backward rate),

$$M + S_1 \xrightarrow{k_1} MS_1, \quad (30)$$

$$M + S_2 \xrightarrow{k_2} MS_2, \quad (31)$$

we propose to estimate the probability that the threshold T_1 of MS_1 bindings is reached before that the number of bound MS_2 reaches the threshold T_2. We will also compute the corresponding conditional MTT.

3.1. Applications of MTT in cell biology

Several applications of the above Markovian approximation concern predictions of the rate of molecular dynamics that underlies the spindle assembly checkpoint during cell division and the probability that a messenger RNA (mRNA) escapes degradation through binding a certain number of microRNAs, and these are discussed in Ref. 16. During cell division, the spindle checkpoint prevents separation of the duplicated chromosomes until each chromosome is properly attached to an apparatus called the spindle apparatus. Another application concerns the ordered arrival a $T_1, \ldots, T_k$ molecules to a target. This situation appears following the induction of a dsDNA break, where diffusing repair proteins should arrive in the correct order.

3.2. Extreme statistics or the first arrival to a target and the paradigm shift

How is the theory of chemical reaction relevant for studying molecular signaling, and in particular the time scale of molecular process that sustain physiological functions?

The mass-action law (1) predicts the production of one species from another in a mixed compartment where the forward rate k_1 represents the flux of 3D Brownian particles arriving at a small ball of radius a. Smoluchowski's 1916 forward rate computation shows that

$$k_1 = 4\pi Dca, \tag{32}$$

where D is the diffusion coefficient, when the concentration c is maintained constant far away from the reaction site. When the binding site is modeled by an absorbing window, the forward rate k_1 is the reciprocal of the MFPT of a Brownian particle to the window. The precise geometry of the activating small windows has been captured by general asymptotics of the mean first arrival time at high activation energy. This mean time is, indeed, sufficient to characterize the rate, because the binding process is Poissonian and the rate is precisely the reciprocal of the MFPT. These computations are summarized in the narrow escape theory.[27,31,32] The forward rate is given by

$$k_1^{-1} = \frac{|\Omega|}{4aD \left[1 + \frac{L(\mathbf{0})+N(\mathbf{0})}{2\pi} a \log a + o(a \log a) \right]}, \tag{33}$$

with $|\Omega|$ being the volume of the domain of Brownian motion, a the radius of the absorbing window,[30] and $L(0)$ and $N(0)$ the principal mean curvatures of the surface at the small absorbing window. Formula 33 reveals that, on average, a particle explores the entire volume before exit (Fig. 4). In summary, in the diffusion theory, the rate k_1 is computed from the mean arrival rate of a single particle. However, this approach is insufficient to study the rate of activation, which needs a different formulation: in cell biology, the time scale of activation of a biochemical reaction is given by the first (fastest) particle that reaches a small binding target, so that the average arrival time of a single particle does not necessarily represent the time scale of activation. In particular, when there are initially no activating molecules, the Smoluchowski rate is not the correct description of the rate of arrival, because it relies precisely on the assumption of a constant concentration at infinity. Thus, even the Gillespie algorithm would give a rate, sampled from the mean k_1, the statistics of which will be different from the rate of the fastest arrival. This difference is the key to the determination of the time scale of cellular activation that can be computed from full Brownian simulations and/or asymptotics of the fastest particle. This is an important departure from the traditional paradigm, reviewed in detailed in Ref. 33.

Briefly, the statistics of the first particle to arrive at a target can be computed from the statistics of a single particle when all are independently and identically distributed.[34–37] With N non-interacting i.i.d. Brownian trajectories (ions) in a bounded domain Ω that bind at a site, the shortest arrival time τ^1 is by definition

$$\tau^1 = \min(t_1, \ldots, t_N), \tag{34}$$

where t_i are the independent arrival times of the N ions in the medium. The distribution of τ^1 is expressed in terms of a single particle,

$$\Pr\{\tau^1 > t\} = \Pr^N\{t_1 > t\}. \tag{35}$$

Here, $\Pr\{t_1 > t\}$ is the survival probability of a single particle prior to binding at the target. This probability is computed by solving the diffusion equation[40]

$$
\begin{aligned}
\frac{\partial p(\boldsymbol{x}, t)}{\partial t} &= D\Delta p(\boldsymbol{x}, t) && \text{for } \boldsymbol{x} \in \Omega, \ t > 0, \\
p(\boldsymbol{x}, 0) &= p_0(\boldsymbol{x}) && \text{for } \boldsymbol{x} \in \Omega, \\
\frac{\partial p(\boldsymbol{x}, t)}{\partial \boldsymbol{n}} &= 0 && \text{for } \boldsymbol{x} \in \partial\Omega_r, \\
p(\boldsymbol{x}, t) &= 0 && \text{for } \boldsymbol{x} \in \partial\Omega_a,
\end{aligned}
\tag{36}
$$

where the boundary $\partial\Omega$ contains N_R binding sites $\partial\Omega_i \subset \partial\Omega$ ($\partial\Omega_a = \cup_{i=1}^{N_R}\partial\Omega_i$, $\partial\Omega_r = \partial\Omega - \partial\Omega_a$). The single particle survival probability is

$$\Pr\{t_1 > t\} = \int_\Omega p(\boldsymbol{x}, t)\, d\boldsymbol{x}, \tag{37}$$

so that $\Pr\{\tau^1 = t\} = \frac{d}{dt}\Pr\{\tau^1 < t\} = N(\Pr\{t_1 > t\})^{N-1}\Pr\{t_1 = t\}$, where $\Pr\{t_1 = t\} = \oint_{\partial\Omega_a}\frac{\partial p(\boldsymbol{x}, t)}{\partial \boldsymbol{n}}\, dS_{\boldsymbol{x}}$. The pdf of the arrival time is

$$\Pr\{\tau^1 = t\} = NN_R \left[\int_\Omega p(\boldsymbol{x}, t)d\boldsymbol{x}\right]^{N-1} \oint_{\partial\Omega_1} \frac{\partial p(\boldsymbol{x}, t)}{\partial \boldsymbol{n}} dS_{\boldsymbol{x}}, \tag{38}$$

which gives the MFPT

$$\bar{\tau}^1 = \int_0^\infty \Pr\{\tau^1 > t\}dt = \int_0^\infty [\Pr\{t_1 > t\}]^N dt. \tag{39}$$

New physical laws have been recently derived for the shortest time.[38,39] These laws are expressed in terms of the shortest ray from the source to the absorbing window δ_{min}. This distance plays a key role, because the fastest trajectory is as close as possible to that ray. The diffusion coefficient is D, the size of the window is a, and there are N Brownian particles present. The asymptotic laws for the expected first arrival time of Brownian particles to a target for large N are

$$\bar{\tau}^{d1} \approx \frac{\delta_{min}^2}{4D \ln\left(\frac{N}{\sqrt{\pi}}\right)}, \quad \text{in dim 1} \tag{40}$$

$$\bar{\tau}^{d2} \approx \frac{\delta_{min}^2}{4D \log\left(\frac{\pi\sqrt{2N}}{8\log\left(\frac{1}{a}\right)}\right)}, \quad \text{in dim 2} \tag{41}$$

$$\bar{\tau}^{d3} \approx \frac{\delta_{min}^2}{2D\sqrt{\log\left(N\frac{4a^2}{\pi^{1/2}\delta_{min}^2}\right)}}, \quad \text{in dim 3.} \tag{42}$$

These formulas show that the expected arrival time of the fastest particle is in dimensions 1 and 2, $O(1/\log(N))$ (see Fig. 4). They should be used instead of the classical forward rate in models of activation in biochemical reactions. Interestingly, the path associated with the fastest particle is not necessarily the physical path, but is

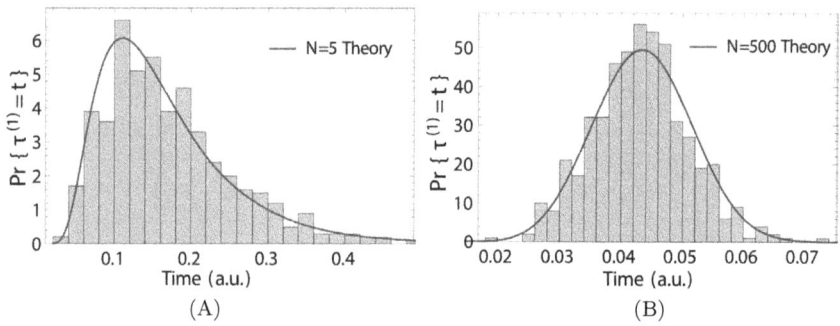

Fig. 4. Histograms of the arrival times to the boundary of the fastest particle, obtained from Brownian simulations with Euler's scheme. The number of Brownian particles is $N = 5$ in **A** and $N = 500$ in **B**.[40]

the shortest as chosen among many of them. Using the history of the trajectory, when n Brownian particles are generated, the path integral methodology[41] shows that the trajectories associated with $\tau^{(n)}$ are concentrated near the optimal trajectories of a minimization problem for the functional energy of the velocity, which gives the classical geodesics. When a disk obstacle is located between the initial position and the narrow exiting window, the optimal trajectories is again concentrated near the shortest geodesic as indicated by numerical simulations (Fig. 5). When there are no obstacles, the solution of the extreme narrow equation[41] is constructed by the ray method of the Green's function.[40] Shortest paths are strategically used in the context of Brownian molecular signaling in cell biology to define the time scale of activation. These questions of molecular activation have changed our concept of chemical reactions.

To conclude, large and disproportionate numbers of particles in natural processes should not be considered wasteful, but rather, they serve a purpose[33]: They are necessary for generating the fastest possible response or to find a key target under time constraints. This property is universal, ranging from the molecular scale to the population level. Interestingly, this strategy for optimizing the response time is not necessarily defined by the physics of the motion of an individual particle, but rather by the collective extreme statistics.

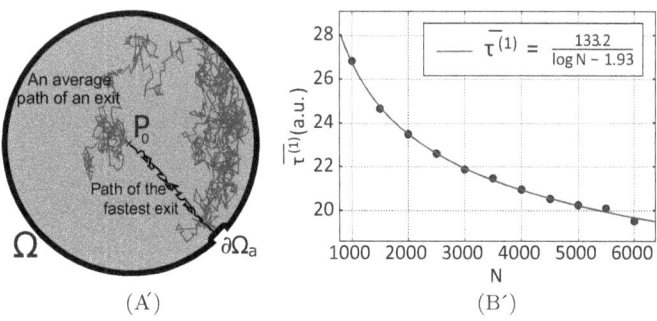

Fig. 5. Escape through a narrow opening in a planar disk. (A′) The geometry of the NEP for the fastest particle. (B′) Plot of the expected arrival time of the fastest path vs the number of particles N. The asymptotic solution (solid curve) is fit to $\frac{\alpha}{\log(N)+\beta}$.

This is precisely the strategy of finding a hidden target: the search process selects the one to arrive first: although these trajectories are rare, they are the ones that set the time scale. This result comes with two consequences: the search time decays with the reciprocal of log of the number and the optimal paths are concentrated along geodesic (for Brownian motion) and the optimal trajectories shoot directly toward the small target, ignoring obstacles.

In cell biology, transient biochemical processes are often activated by the first (fastest) particle that reaches a small binding target, so that the average arrival time of a single particle does not necessarily represent the time scale of activation. In particular, when there are no molecules at steady state, the Smoluchowski rate theory does not apply because it relies precisely on the assumption of constant concentration at infinity. Sampled with classical forward rate k_1, the Gillespie algorithm would give an incorrect description of the transient, because its statistics would be different from the rate of the fastest arrival. This difference is the key to the determination of the time scale of cellular activation that can be computed from full Brownian simulations and/or asymptotics of the fastest particle. This is an important departure from the traditional paradigm. With the increase of computer power, modeling stochastic chemical reactions can easily done by following individual trajectories even if they number thousand or more, but the geometrical organization of the cellular environment should be incorporated in the simulations.

4. Chemical Reaction for Brain Patterning

Brain patterning refers to the process by which neurons connect large regional territories. Neurons wire the brain by connecting regions that can be located far away (see Fig. 6(a)).[42,43] The general mechanism for such patterning involves neuronal migration, and the information about position is provided by morphogen gradients. The formation of these gradients remains unclear,[44] but consists of molecules that are secreted by neurons or glial cells. These secreted molecules or morphogens can lead to their own synthesis in cells because they are TFs that activate genes. When morphogens of two opposite gradients centered at different locations meet in an ensemble of cells, they form a boundary that separates two neuronal territories.[21,22] Many of the chemical reactions involved in such processes have been reviewed.[23]

We briefly discuss here how guidance cues,[43] that are secreted in gradient can be "read" by a cell so that the concentration is transformed into a positional information. Indeed, the tip of the neurons — called the growth cone — can travel far away (centimeters and more) from the place where they have been generated.[45] The path of the cell migration relies on a collection of specific cues and their associated receptors. The density of these receptors on each neuron is not uniform and depends on the initial location of neurons (retina–tectum, for example[46]).

A combination of cues such as two morphogens (e.g., Ehprin and Engrailed) together can lead to a more accurate positioning of the growth cone through a local molecular amplification chemical reaction,[47] reviewed in Ref. 23. In this section, we introduce recent modeling scenarios to investigate how a triangulation of the concentration at the tip of a neuron can be used to localize the gradient source. The classical paradigm was concerned with estimating the concentration in a ball of size a, for example by computing the difference of flux received between the south and north pole,[48,49] even when various binding properties[50] are accounted for. But this is clearly not enough to reconstruct the source location and thus to explain growth cone navigation.

In the limit of fast binding of morphogens to receptors, the problem of finding the location of a gradient source becomes one of finding the molecular fluxes through small receptor windows, modeled as absorbers, located on an otherwise reflecting cell membrane. We recently addressed this problem by solving the stationary diffusion equation (Poisson equation) in a domain with small absorbing windows situated on the boundary of a reflecting disk of radius R.[51]

The probability fluxes through the windows can be determined numerically by injecting Brownian particles at the location of the

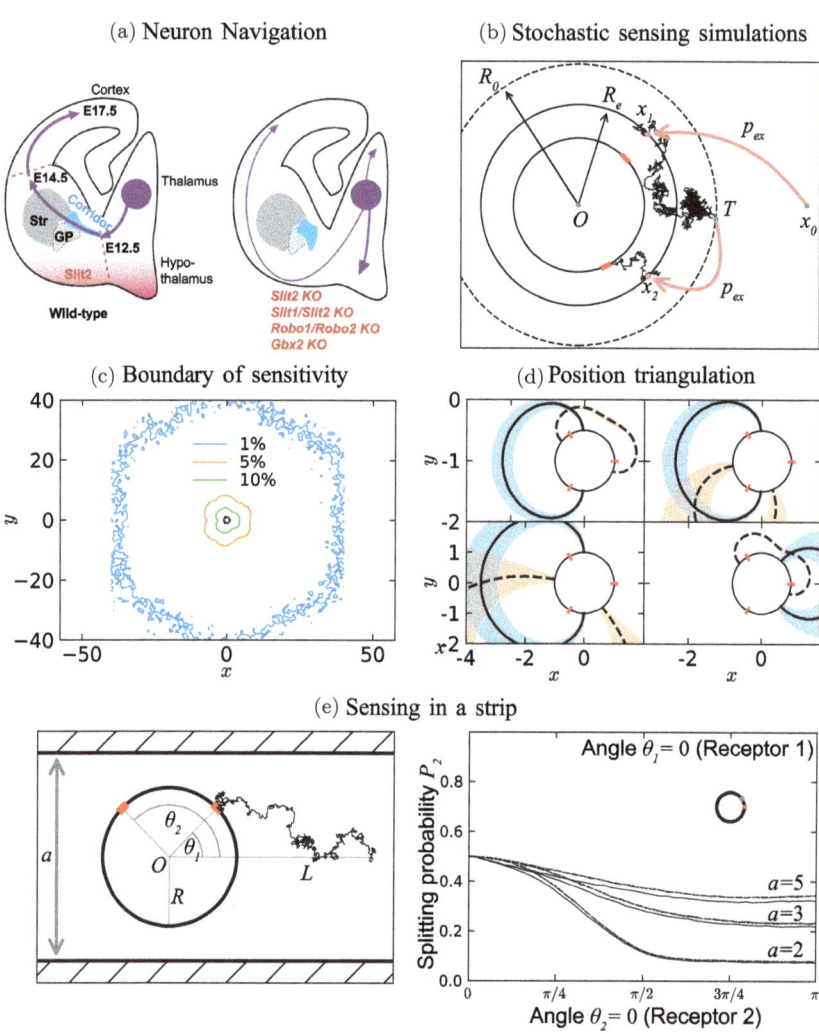

Fig. 6. Sensing gradient (a) Neuronal growth cones navigate over large distances from the mouse thalamus into the cortex. (Reproduced from Ref. 45 with the permission of Elsevier.) Slit acts as repulsive gradient to steer axons through permissive corridors (left). Knockout of various morphogen signals leads to misnavigation (right). (b) Hybrid simulation procedure for windows on a ball. Brownian particles started at x_0 are injected on the circle with radius R_e according to the exit map p_{ex} (red arrow). Trajectories are advanced via the Euler Scheme until the particle leaves the disk with radius $R_o > R_e$, whereupon the particle jumps from point T and to a new position

gradient source and using Euler's scheme to simulate their trajectories until absorbtion at one of the windows. However, this naive method is problematic because, in free space, particles can make large excursions, which renders this method computationally infeasible without an artificial cut-off boundary. To overcome this issue, a hybrid analytical–numerical simulation scheme has been developed in Ref. 52, see also Figs. 6(a) and (b). This method works by immediately mapping particles injected at x_0 to a new position x_1 on the boundary of a disk with radius $R_e > R$ according to the exit probability distribution p_{ex}. This exit distribution is simply given by the flux through the boundary of the absorbing disk with radius R_e,[52] leading to the expression

$$p_{ex}(\boldsymbol{x}_0, \boldsymbol{x}_1) = \left. \frac{\partial G(x_0, x_1)}{\partial n_1} \right|_{|x_1| = R_e} = \frac{1}{2\pi} \frac{\frac{r_0^2}{R_e^2} - 1}{\frac{r_0^2}{R_e^2} - 2\frac{r_0}{R_e} \cos(\theta) + 1},$$

(43)

where $r_0 = |x_0|$ and θ is the angle subtended by the vectors $\boldsymbol{x}_0$ and $\boldsymbol{x}_1$. The particle trajectories are generated using a Euler's scheme until it leaves the domain with radius $R_o \geq R_e$ at position T, upon which the particle is mapped to a new position x_2. This is repeated until the particle is absorbed at one of the windows. Using an analytical approach based on asymptotics and this numerical scheme,

---

Fig. 6. (*Continued*)
determined by the map p_{ex}. (c) Detectability contours for three windows on a disk with normalized flux difference larger than the threshold $T = 1\%$, 5%, and 10%. Units are given in cell radii.[51] (d) Four examples of the recovery of the unique source position as the intersection of two curves given the fluxes to three windows. Shaded areas indicate the uncertainty of the curve with a fixed uncertainty in the fluxes. The uncertainty of the recovered source position, shown as the overlap of shaded areas, is inhomogeneous. (e) Two receptor windows on a disk inside a reflecting strip. (left) Scheme of the domain. (right) Splitting probability (flux) of hitting the second receptor as a function of angle θ_2 for source distances $L = 1.3$ (solid line), 2 (long-dashed line), 10 (dash-dotted line), and 100 (short-dashed line) cell radii. The flux becomes largely independent of the source position in contrast to a disk in free space.

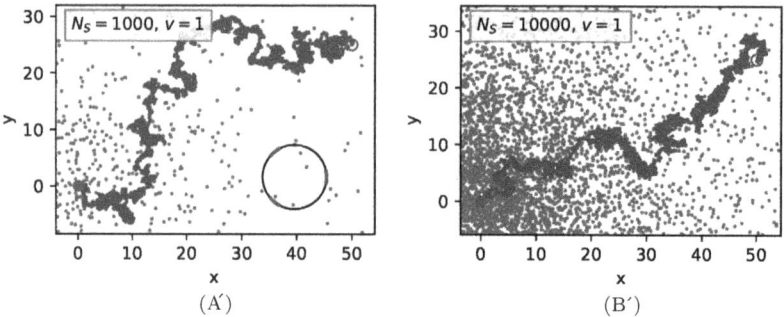

Fig. 7.   Navigating a gradient. A disk with three receptor windows is placed in a gradient. Each time step the disk samples (A′) $N_S = 1{,}000$ or (B′) $N_S = 10{,}000$ particles from the gradient to measure the receptor fluxes and measure the normalized direction of the gradient n. The disk is then advanced by vn and the process is repeated until the source is located. The grey dots illustrate example gradient molecule densities.

we compute the diffusion fluxes to at least three receptors located on the surface of a 2D disk and showed that the triangulation of the source can be resolved in the limit of very fast binding. Interestingly, the error of the position recovery depends on the number of receptors (Figs. 6(c) and (d)). However, the recovery of source located far away, although not realistic in a empty space, is possible in narrow strip Fig. 6(e). In general, a growth cone can find its way to a source (Fig. 7). More details about the resolution can be found in Refs. 51 and 52.

5.   Chemical Reactions in the Cell Nucleus Occurring through Chromatin Looping

There are several types of chemical reactions occurring in the nucleus: the ones related to gene activation–modulation and the other ones associated with gene maintenance, repair, and regulation of protein trafficking. Interestingly, in many cases, genes are activated by TFs, and in the past decade theoretical investigation was focused on the mean time for TFs to find a small activator target located on the multiscaled folded chromatin.[53–55] The total search time can be computed by considering separately the time a TF spends in 1D along the DNA and in 3D inside the nucleus,[56–58] but not by assuming that the two times are equal[59] based on assuming an optimal repartition principle. In reality, when several TFs are transiently generated, the activation time is not given by the mean time for a single one, but by the mean time for the first one to arrive to the target. In that case, physical estimation fall into the extreme narrow escape, discussed in Subsection 3.2.

In this subsection, we discuss the formula for polymer looping time: this time is related to the formation of a random loop between two loci. The formula allows to relate the parameters together and reveals the contribution of various physical parameters in the looping time, thus revealing the small and large ones responsible for the time scale. We first introduce various polymer models (Rouse, beta-polymer, Random cross-linked RCL) and summarize the formula for the first looping times.[60]

5.1. *Description of polymer models: Rouse, beta, and RCL*

5.1.1. *Rouse polymer model*

The Rouse polymer model is composed of N identical monomers at position $\boldsymbol{R} = [R_1, R_2, \ldots, R_N]$, connected sequentially by harmonic springs with a spring constant κ. Springs are allowed to cross each other, and no repulsion and hydrodynamic interactions are accounted for. Therefore, only nearest neighbor (NN) monomer interaction are considered. The Rouse model belongs to Gaussian chain family, in which the pdf for the vector $R_m - R_n$ is given by

$$P(R_m - R_n) = \frac{\exp\left(\frac{(R_m - R_n)^2}{2b^2|m-n|}\right)}{\sqrt{2\pi b^2|m-n|}}, \tag{44}$$

where b^2 is the mean square distance between adjacent monomers. The spring potential of the Rouse polymer is given by

$$\phi_R(\boldsymbol{R}) = \frac{\kappa}{2} Tr(\boldsymbol{R}^T \boldsymbol{M} \boldsymbol{R}) = \frac{\kappa}{2} \sum_{n=2}^{N} (R_n - R_{n-1})^2, \tag{45}$$

where Tr is the trace operator, and M is the Rouse matrix, defined as

$$M_{mn} = \begin{cases} -1, & |m-n] = 1; \\ -\sum_{j=1}^{N} M_{mj}, & m = n; \\ 0, & \text{else.} \end{cases} \tag{46}$$

Monomers of the Rouse polymer move under the influence of the force derived from the energy potential (45), and that of thermal fluctuations, and their dynamics is given by the Smoluchowski's limit of the Langevin equation[40]

$$\frac{d\boldsymbol{R}}{dt} = -\frac{dD}{b^2}\boldsymbol{M}\boldsymbol{R} + \sqrt{2D}\frac{d\boldsymbol{\omega}}{dt}, \tag{47}$$

where D is the diffusion constant, in units of $\mu m^2/s$, d is the dimension, and $\boldsymbol{\omega}$ are d-dimensional $N-$standard Brownian motions. In the Fourier space (normal or Rouse modes),[61] we recall

$$\boldsymbol{u_p} = \sum_{n=1}^{N} R_n \alpha_p^n, \tag{48}$$

where

$$\alpha_p^n = \begin{cases} \sqrt{\frac{1}{N}}, & p = 0 \\ \sqrt{\frac{2}{N}} \cos\left((n-1/2)\frac{p\pi}{N}\right), & \text{otherwise.} \end{cases} \tag{49}$$

$\boldsymbol{u_0}$ represents the motion of the center of mass and the potential ϕ defined in equation 45 is written in the new coordinates as

$$\phi(\boldsymbol{u_1}, \ldots, \boldsymbol{u_{N-1}}) = \frac{1}{2}\sum_{p=1}^{N-1} \kappa_p \boldsymbol{u}_p^2, \tag{50}$$

where

$$\kappa_p = 4\kappa \sin^2\left(\frac{p\pi}{2N}\right). \tag{51}$$

Equations (47) are now decoupled and we obtain a $(N-1)$ d-independent Ornstein–Uhlenbeck (OU) processes[62]

$$\frac{d\boldsymbol{u_p}}{dt} = -D_p \kappa_p \boldsymbol{u_p} + \sqrt{2D_p}\frac{d\tilde{\boldsymbol{w}}_p}{dt}, \tag{52}$$

where each $\tilde{\boldsymbol{w}}_p$ is an independent d-dimensional Brownian motion, with mean zero and variance 1 and $D_p = D$ for $p = 1, \ldots, N-1$, while $D_0 = D/N$ and the relaxation times are $\tau_p = 1/D\kappa_p$. The center of mass behaves as a freely diffusing particle. When initially the polymer is stretched, the characteristic relaxation time is the slowest time constant and is given by the reciprocal of the first eigenvalue

$$\tau_N = \frac{1}{D\kappa_1} = \frac{1}{4D\kappa \sin^2\left(\frac{\pi}{2N}\right)} \approx \frac{N^2}{D\kappa\pi^2}. \tag{53}$$

5.1.2. *The β-polymer model*

The Rouse polymer model can be modified so that the mean-square-displacement (MSD) for a monomer R_c behaves for small time as

$$\langle (\boldsymbol{R}_c(t_0 + t) - \boldsymbol{R}_c(t_0))^2 \rangle \propto t^\alpha, \tag{54}$$

where $\langle \cdot \rangle$ is the ensemble average over configurations and $\alpha > 0$. Thus, a β-polymer is a modified Rouse polymer model with a prescribed exponent α. Such a construction is possible,[60,63] and prescribing the anomalous exponent α imposes intrinsic long-range interactions between monomers, beyond the closest neighbors of the Rouse model. Starting with Rouse equation in the Fourier's space (Eq. (52)), the coefficients $\tilde{\kappa}_p$ are modified to

$$\tilde{\kappa}_p = 4\kappa \sin^\beta\left(\frac{p\pi}{2N}\right), \tag{55}$$

where $\beta > 1$, while $D_0 = D_{\text{cm}}$ and $D_p = D$ for $p > 0$ (see definition in the previous Subsection 5.1.1). When $\beta = 2$, we recover the Rouse model. The final polymer model is reconstructed by inverting the matrix Eq. (49) between the original and the Fourier space. In the procedure, the coefficients α_p^n are not changed, while the exponent in the eigenvalues of Eq. (55) is modified.

This procedure defines a unique ensemble of long-range interactions. Indeed, changing the eigenvalues results in long-range monomer–monomer interaction as revealed by computing the potential energy, which differs from (45). Using the reciprocal of the Fourier transform, we get

$$\tilde{\phi}(\boldsymbol{u}_1, \ldots, \boldsymbol{u}_N) = \frac{1}{2} \sum_p \tilde{\kappa}_p \boldsymbol{u}_p^2. \tag{56}$$

The Rouse transformation in Eq. (49) leads to an explicit expression for the interaction between each monomer

$$\tilde{\phi}(\boldsymbol{R}_1, \ldots, \boldsymbol{R}_N) = \frac{1}{2} \sum_p \tilde{\kappa}_p \left(\sum_{n=1}^{N} \boldsymbol{R}_n \alpha_p^n \right)^2$$

$$= \frac{1}{2} \sum_{l,m} \boldsymbol{R}_l \boldsymbol{R}_m \sum_p \tilde{\kappa}_p \alpha_p^l \alpha_p^m$$

$$= \frac{1}{2} \sum_{l,m} \boldsymbol{R}_l \boldsymbol{R}_m A_{l,m}. \tag{57}$$

The coefficients are

$$A_{l,m} = \sum_{p=1}^{N-1} \tilde{\kappa}_p \alpha_p^l \alpha_p^m = 4\kappa \frac{2}{N} \sum_{p=1}^{N-1} \sin^\beta \left(\frac{p\pi}{2N} \right)$$

$$\cos \left(\left(l - \frac{1}{2} \right) \frac{p\pi}{N} \right) \cos \left(\left(m - \frac{1}{2} \right) \frac{p\pi}{N} \right). \tag{58}$$

For $\beta \neq 2$, all monomers are now coupled and the strength of the interaction decays with the distance along the chain, as shown in Fig. 8(a) and 8(b). For example, the coefficient $A_{50,m}$ between monomer 50 and m depends on the position m (Fig. 8b for a polymer of length $N = 100$ and $\beta = 1.5$). The coefficients $A_{n,m}$ obtained for various β are summarized in Table 1. The long-range interactions between the monomers in the quadratic potential are linked to pair of monomers interacting by

$$\psi(\boldsymbol{R}_1, \ldots, \boldsymbol{R}_N) = \frac{1}{2} \sum_{ij;\, i>j} a_{i,j} (\boldsymbol{R}_i - \boldsymbol{R}_j)^2$$

$$= \sum_{i=1}^{N} \boldsymbol{R}_i^2 \sum_{j \neq i} \frac{a_{i,j}}{2} - \sum_{ij;\, i>j} a_{ij} \boldsymbol{R}_i \boldsymbol{R}_j. \tag{59}$$

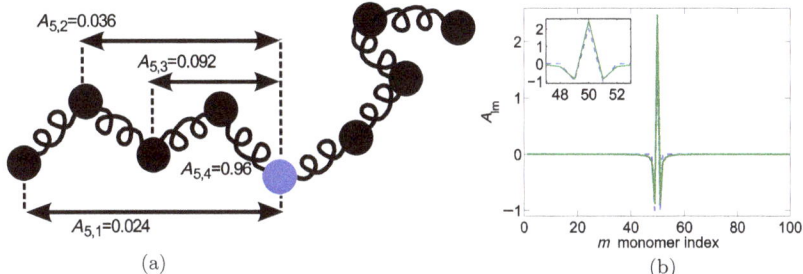

(a) (b)

Fig. 8. The β−polymer. (a) Representation of a β−polymer, where all monomers are connected together with a strength that decays with the distance along the chain. The central monomer (blue) interacts with all other monomers in a chain of length $N = 9$ for $\beta = 1.5$ (interaction unit $\kappa = 3/b^2$). (b) Monomer-monomer interactions in the modified Rouse polymer model (β−model). The coefficient A_{lm} (in units of κ) measures the strength of the interaction between two monomers. Shown are the coefficients A_{lm} for the polymer with $\beta = 1.1$, where $l = 50$ and $N = 100$. All monomers interact with each other and the strength of the interaction decays with the distance along the chain. (Reproduced from Ref. 63 with the permission of Natural Research.)

Table 1. Values of the coefficients $A_{l,m}$ in units of κ for the middle monomer in a polymer chain of length $N = 101$, for different values of β.

β	$A_{51,51}$	$A_{51,50}$	$A_{51,49}$	$A_{51,48}$	$A_{51,40}$	$A_{51,30}$
2	2	-1	0	0	0	0
1.5	2.22	-0.95	-0.087	-0.029	-1.07×10^{-3}	-2.2×10^{-3}
1.2	2.40	-0.90	-0.14	-0.054	-3.05×10^{-3}	-7.9×10^{-3}
1	2.55	-0.85	-0.17	-0.073	-5.48×10^{-3}	-16.68×10^{-3}

The coefficients $a_{i,j}$ are related to the β-polymer $A_{l,m}$, (see Eq. (57)) by the following formula

$$a_{l,m} = \begin{cases} \sum_{j \neq l} A_{jl}, & l = m, \\ -\frac{1}{2} A_{l,m}, & \text{otherwise.} \end{cases} \tag{60}$$

Expression (59) for the potential can be used to reconstruct the amplitude of the interactions from the coefficients of the β-model. The anomalous exponent α (Eq. (54)) is directly associated with the exponent β of the polymer by the relation $\alpha = 1 - \frac{1}{\beta}$. In summary, within a β-polymer model, all monomers are coupled and the strength of the interaction decays with the distance along the chain.

5.1.3. *The randomly cross-linked (RCL) polymer*

The RCL polymer can be used to represent chromatin: it is composed of a linear DNA backbone where additional cross-linkers are added, leading to chromatin folding. These cross-linkers can be mediated by binding proteins such as CTCF and cohesin. Binding positions for these molecules are distributed along the DNA, and cross-linkers are attached between any two binding sites. In a population of cells, cross-link positions are not preserved across cells and cell cycles. However, the concentration of binding molecules (number of cross-links) is expected to be constant. The RCL polymer accounts for such heterogeneity in the cross-linking over an ensemble of cells. RCL polymers are used to describe short-range ($<$Mbps) random cross-linking that give rise to chromatin folding patterns, such as TADs[63] and meta-TADs.[64,65] We now present the construction of RCL polymers.

We start with a Rouse polymer. The quadratic potential from the linear backbone is given by Eq. (45). Using harmonic springs, N_c randomly chosen non-NN monomer pairs are connected, such that the quadratic potential of the RCL polymer is now the sum

$$\phi_{RCL}(\boldsymbol{R}) = \phi_S(\boldsymbol{R}) + \phi_R(\boldsymbol{R}) = \frac{\kappa}{2}Tr(\boldsymbol{R}^T(\boldsymbol{M} + \boldsymbol{B}^{\mathcal{G}}(\xi))\boldsymbol{R}), \quad (61)$$

where $\boldsymbol{B}^{\mathcal{G}}(\xi)$ is the connectivity matrix of the added random connectors in realization $\mathcal{G}$, given by

$$\boldsymbol{B}^{\mathcal{G}}(\xi) = \begin{cases} -1, & |m - n| > 1, r_m, r_n \text{ are connected;} \\ -\sum_{j \neq m} B_{mj}^{\mathcal{G}}(\xi), & m = n; \\ 0, & \text{else.} \end{cases} \quad (62)$$

In each realization $\mathcal{G}$ of the RCL polymer, we randomize the choice of Nc monomer pairs to connect. The parameter ξ is the connectivity fraction, defined as the fraction of the non-NN connected monomers out of the total possible

$$\xi = \frac{2N_c}{(N-1)(N-2)}. \tag{63}$$

Monomers $\boldsymbol{R}$ of the RCL polymer move under the influence of the force derived from harmonic springs potential (61) and that of thermal fluctuations

$$\frac{d\boldsymbol{R}}{dt} = \frac{1}{\gamma}\nabla\phi_{RCL}(\boldsymbol{R}) + \sqrt{2D}\frac{d\boldsymbol{\omega}}{dt}$$

$$= -\frac{dD}{b^2}\left(\boldsymbol{M} + \boldsymbol{B}^{\mathcal{G}}(\xi)\right)\boldsymbol{R} + \sqrt{2D}\frac{d\boldsymbol{\omega}}{dt}, \tag{64}$$

where γ is the friction coefficient, D is the diffusion constant, b^2 is the mean square distance between adjacent monomers, and $\boldsymbol{\omega}$ are d dimensional standard Brownian motions.

To account for the heterogeneity in chromatin organization over cell population, we replace the added connectivity $\boldsymbol{B}^{\mathcal{G}}(\xi)$ in Eq. (64) by its averaging $\langle\boldsymbol{B}^{\mathcal{G}}(\xi)\rangle$ over all possible configurations when N_c monomer pairs are connected, and is given by

$$\langle\boldsymbol{B}^{\mathcal{G}}(\xi)\rangle_{mn} = \begin{cases} -\xi, & |m-n| > 1; \\ -\sum_{j=1}^{N}\langle\boldsymbol{B}^{\mathcal{G}}(\xi)\rangle_{mj}, & m = n; \\ 0, & \text{else.} \end{cases} \tag{65}$$

The mean-field equation

$$\frac{d\boldsymbol{R}}{dt} = -\frac{dD}{b^2}\left(\boldsymbol{M} + \langle\boldsymbol{B}^{\mathcal{G}}(\xi)\rangle\right)\boldsymbol{R} + \sqrt{2D}\frac{d\boldsymbol{\omega}}{dt}. \tag{66}$$

can be used to derive statistical properties, but simulations should be made from (64). In normal coordinates, the RCL-system (66) is decoupled, and we obtain

$$\frac{du_p}{dt} = -\frac{dD}{b^2}\chi_p u_p + \sqrt{2D}\frac{d\bar{\omega}}{dt}, \tag{67}$$

where $\chi_p(\xi)$ are the eigenvalues of the matrix $\boldsymbol{M}+\langle\boldsymbol{B}^{\mathcal{G}}(\xi)\rangle$, (Eq. (64)) given by

$$\chi_p(\xi) = N\xi + 4(1-\xi)\sin^2(p\pi/N), \tag{68}$$

and $\bar{\omega}$ are standard d-dimensional Brownian motions. The RCL polymer belongs to the class of generalized Gaussian chain models, studied in Refs. 66–69, for which the encounter probability (EP) between
any two monomers m and n at equilibrium is given by

$$P_{m,n}(\xi) = \left(\frac{d}{2\pi\sigma^2_{m,n}(\xi)}\right)^{\frac{d}{2}}. \tag{69}$$

Expression (69) is computed by first estimating the variance $\sigma^2_{m,n}(\xi)$
in normal coordinates (Eq. (48)):

$$\sigma^2_{m,n}(\xi) = \langle(R_m - R_n)^2\rangle = \sum_{p=0}^{N-1}(\alpha_p^m - \alpha_p^n)^2\langle u_p^2(\xi)\rangle. \tag{70}$$

Although there are several methods to compute the steady-sate
variance of Gaussian models,[69] we present here a direct computations using the normal coordinates (Eq. (48)) and the eigenvalues
(Eq. (68)).

Starting with the Ornstein–Uhlenbeck equations (67),[2] we obtain
the time-dependent variance

$$\langle u_p^2(\xi)\rangle = \frac{b^2}{\chi_p(\xi)}\left(1 - \exp\left(-\frac{2D\chi_p(\xi)t}{b^2}\right)\right). \tag{71}$$

The relaxation times $\tau_0 \geq \tau_1(\xi) \geq \cdots \tau_{N-1}(\xi)$ are given by

$$\tau_p(\xi) = \frac{b^2}{2D\chi_p(\xi)}, \tag{72}$$

and the slowest $\tau_0(\xi)$ corresponds to the diffusion of the center of
mass. At steady state, we obtain from Eq. (71)

$$\langle u_p^2(\xi)\rangle = \frac{b^2}{2(1-\xi)\left(y(N,\xi) - \cos\left(\frac{p\pi}{N}\right)\right)}, \tag{73}$$

where $y(N,\xi) = 1 + \frac{N\xi}{2(1-\xi)}$. Substituting (49) and (73) into (70), we
get

$$\sigma^2_{m,n}(\xi) = \frac{b^2}{N(1-\xi)}\sum_{p=0}^{N-1}\frac{\left(\cos\left(\frac{p(m-\frac{1}{2})\pi}{N}\right) - \cos\left(\frac{p(n-\frac{1}{2})\pi}{N}\right)\right)^2}{y(N,\xi) - \cos(\frac{p\pi}{N})}.$$

$$\tag{74}$$

For $N \gg 1$, the sum (74) can be approximated by the Euler–MacLauren integral, where $x = p\pi/N$. In the complex plan, using for the contour the unit disk parameterized by $z = e^{ix}$, we get

$$\sigma^2_{m,n}(\xi) = \frac{b^2}{2\pi(1-\xi)} \int_{-\pi}^{\pi} \frac{\left(\cos(x(m-\tfrac{1}{2})) - \cos(x(n-\tfrac{1}{2}))\right)^2}{y(N,\xi) - \cos(x)} dx$$

$$= \frac{-b^2}{4\pi i(1-\xi)} \oint_{|z|=1} \frac{(z - z^{m+n})^2(z^m - z^n)^2 dz}{(z - \zeta_0(N,\xi))(z - \zeta_1(N,\xi))z^{2(m+n)+1}},$$

(75)

where

$$\zeta_0(N,\xi) = y(N,\xi) + \sqrt{y^2(N,\xi) - 1},$$

$$\zeta_1(N,\xi) = y(N,\xi) - \sqrt{y^2(N,\xi) - 1}.$$

(76)

When $\zeta_0(N,0) = 1$, we recover from expression (74) the variance $\sigma^2_{m,n}(0) = b^2|m-n|$ of the Rouse chain $(N_c(\xi) = 0)$.[70] The integrand in (75) is symmetric in m and n and has a pole of order $2(m+n)+1$ at $z = 0$ and simple poles at $z = \zeta_0(N,\xi), z = \zeta_1(N,\xi)$. Because $y(N,\xi) \geq 1$, we have $\zeta_0(N,\xi) \geq 1$, which is outside the contour $|z| = 1$, and $\zeta_1(N,\xi) \leq 1$, for all N, $\xi \geq 0$. The pole $\zeta_0(N,\xi)$ is not in the disk and does not contribute to the residues of (75). For $\xi > 0$, we solve the integral (75) to obtain an exact expression for the variance

$$\sigma^2_{m,n}(\xi) = \begin{cases} \dfrac{b^2}{(1-\xi)\sqrt{y^2(N,\xi)-1}} \\ \quad \left(\dfrac{(\zeta_0^{m-n}(N,\xi)-1)^2 - 2\zeta_0^{m+n-1}(N,\xi)}{\zeta_0^{2m-1}(N,\xi)} + 2 \right), \quad m \geq n; \\[4mm] \dfrac{b^2}{(1-\xi)\sqrt{y^2(N,\xi)-1}} \\ \quad \left(\dfrac{(\zeta_0^{n-m}(N,\xi)-1)^2 - 2\zeta_0^{m+n-1}(N,\xi)}{\zeta_0^{2n-1}} + 2 \right), \quad m < n. \end{cases}$$

(77)

For $\xi \ll 1$, $k > 1$ we approximate the terms

$$\zeta_0^k(N, \xi) \approx \exp(k\sqrt{N\xi});$$
$$\zeta_1^k(N, \xi) \approx \exp(-k\sqrt{N\xi}); \tag{78}$$

and use (78) in expression (77) to extract the dominant terms. The variance is thus asymptotically given by

$$\sigma_{m,n}^2(\xi) \approx \frac{b^2}{\sqrt{N\xi}}(1 - \exp(-|m - n|\sqrt{N\xi})). \tag{79}$$

Expressions (79) or (77) can now be substituted into (69) to obtain the EP between monomer R_n and R_m.

5.2.   *Looping time formulas*

The asymptotic formulas for the looping time of polymer models are used to emphasize the role of various parameters. These formulas have been derived in Refs. 65, 72 and 73 and are reviewed in Ref. 74.

5.2.1.   *Mean first encounter time formulas in free space for a Rouse polymer*

Using the notations of the Rouse polymer introduced in Subsection 5.1.1, the time it takes for two monomers to meet for the first time in a ball of radius ε is given in dimensions two and three by

$$\langle\tau_\varepsilon\rangle_{2d} = \frac{N}{2D\kappa}\log\left(\frac{\sqrt{2}b}{\varepsilon}\right) + A_2\frac{b^2}{D}N^2 + \mathcal{O}(1), \tag{80}$$

$$\langle\tau_\varepsilon\rangle_{3d} = \left(\frac{N\pi}{\kappa}\right)^{3/2}\frac{\sqrt{2}}{D4\pi\varepsilon} + A_3\frac{b^2}{D}N^2 + \mathcal{O}(1), \tag{81}$$

where D is the diffusion coefficient, $\kappa = dk_BT/b^2$ is the spring constant with d the spatial dimension, k_B is the Boltzmann coefficient,

T is the temperature, and A_2 and A_3 are constants, fitted to simulations data. The distribution of looping time is

$$p(t) = \Pr\{\tau_\epsilon > t\} = \int_{\Omega_\epsilon} p(\boldsymbol{x}, t)dx = \sum_{i=0}^{\infty} C_i e^{-\lambda_i^\epsilon Dt}, \qquad (82)$$

where C_i are constants, and is well approximated by a sum of two exponentials. For N ($N = 16$ and 32) not too large, a single exponential is often sufficient for applications,

$$p_N(t) \sim e^{-\lambda_N t}, \qquad (83)$$

with $\varepsilon = 0.1b$. Here, $\lambda_{16} = 0.0125b^{-2}, \lambda_{32} = 0.0063b^{-2}$. For long polymers, a sum of two exponentials is more accurate

$$p_N(t) \sim C_0 e^{-\lambda_0^\epsilon t} + C_1 e^{-\lambda_1^\epsilon t}. \qquad (84)$$

For $N = 64$, the numerical values are $\lambda_0^\epsilon = 0.0012b^{-2}, \lambda_1^\epsilon = 0.0375b^{-2}, C_0 = 0.99$ and $C_1 = 0.28$. Although the two-exponential approximation works well for small $\varepsilon < 0.2b$, four exponents are needed for larger ε ($>0.4b$). For $N \in [4 - 64]$, $C_0 \approx 1$, while C_1 remains approximately constant for a given value of ε. For example, for $\varepsilon = 0.1b$, C_1 varied with N from 0.2 to 0.28.

5.2.2. *Mean first encounter time formula in a confined domain for a Rouse polymer*

The mean first encounter time (MFET) in a confined ball of radius A is estimated when the boundary is accounted for by replacing it by a parabolic potential, added to the Rouse potential ϕ_{Rouse}, such that the total energy is the sum

$$\phi_{\text{h}} = \phi_{\text{Rouse}} + \frac{\beta}{2} \sum_{n=1}^{N} R_n^2 = \frac{1}{2} \sum_{p=0}^{N-1} (\kappa_p + \beta) \boldsymbol{u}_p^2, \qquad (85)$$

where $\kappa_p = 4\kappa \sin^2(\frac{p\pi}{2N})$, $\boldsymbol{u}_p$ are the coordinates in which ϕ_{Rouse}, and the strength B is calibrated to the radius of the ball by Ref. 74

$$B = \frac{12}{A^4/b^2 + 2A^2}, \qquad (86)$$

so that the root mean square end-to-end distance of the polymer in the potential field is equal to the square radius of the confining ball domain A, that is

$$\langle (\mathbf{R}_N(B) - \mathbf{R}_1(B))^2 \rangle = A^2 \quad \text{for } N \gg 1. \tag{87}$$

In that case, the MFET in a confined ball of radius A $\langle \tau_{\mathrm{h}} \rangle$ for two end monomers of a Rouse polymer to meet is given for $\varepsilon \ll b$,

$$\langle \tau_{\mathrm{h}} \rangle \approx \frac{2^{1/2}}{4\pi\varepsilon D} \left[\frac{4\pi N}{N^2 B + \pi^2 \kappa} + \frac{4}{\sqrt{\kappa B}} \right.$$

$$\left. \times \left[\frac{\pi}{2} - \tan^{-1} \left(2\sqrt{\kappa/B} \tan (\pi/2N) \right) \right] \right]^{3/2} + \mathcal{O}(1). \tag{88}$$

Note that as N increases, the mean looping time $\langle \tau_{\mathrm{h}} \rangle$ does not diverge, but converges to an asymptotic value $\frac{1}{\pi\varepsilon D}(\frac{\pi}{\sqrt{\kappa B}})^{3/2}$. Finally, the distribution of looping times is always well approximated by a Poissonian law $p(t) \approx e^{-\lambda_0^\varepsilon D t}$ in contrast to looping in a free space.

5.2.3.   *Mean first encounter time formula for a β-polymer*

The mean encounter time for the two end monomers of a β-polymer is given by[74]

$$\langle \tau_\epsilon^B \rangle \approx \frac{1}{D\varepsilon(2\kappa)^{3/2}} \left[\frac{2}{3} \cos \left(\frac{\pi}{2N} \right)^3 {}_2F_1 \left(\frac{3}{2}, \frac{1+\beta}{2}, \frac{5}{2}, \cos \left(\frac{\pi}{2N} \right)^2 \right) \right.$$

$$\left. + \frac{\pi}{N} \frac{\cos^2 \left(\frac{\pi}{2N} \right)}{\sin^\beta \left(\frac{\pi}{2N} \right)} + \frac{\pi^2}{12N^2} \frac{\beta\pi \cos \left(\frac{\pi}{2N} \right)^3}{2 \sin(\frac{\pi}{2N})^{1+\beta}} \right]^{3/2} \tag{89}$$

and for large N, it is possible to use the estimation

$${}_2F_1 \left(\frac{3}{2}, \frac{1+\beta}{2}, \frac{5}{2}, \cos \left(\frac{\pi}{2N} \right)^2 \right)$$

$$= 3\pi^{3/2-\beta} 2^{-(1+\beta)} N^{\beta-1} + \mathcal{O}(1) + \mathcal{O}(N^{-2}), \tag{90}$$

where ${}_2F_1$ is the Gaussian hypergeometric function.[75] The two other terms in (89) scale as $N^{\beta-1}$. This asymptotic result is quite different for the Rouse polymer which scales as $N^{3/2}$, while for a β-polymer, for $N \gg 1$, the MFET scales as $N^{\frac{3}{2}(\beta-1)}$.

5.2.4. *Arrival time of a monomer to a small target located on the boundary of a bounded domain*

Although we are still missing an analytical derivation for the mean arrival time of a single monomer of a Rouse polymer to the boundary of a bounded domain, the approximation of this process as a single stochastic particle trapped in a single well leads to[19]

$$\tau(N) = \frac{(2\pi)^{3/2}\gamma\sqrt{kT}}{4a\omega_N^{3/2}} \exp\left[\frac{U_N}{kT}\right], \tag{91}$$

where a is the size of the small target, $U_N(r)$ is the energy barrier, generated by the polymer due to the presence of the boundary of the domain at the target site, ω_N is the frequency at the minimum $(U_N(r) \approx \frac{\omega_N}{2}r^2$ near 0),[71,76] and $\omega_N \sim N$. Additional research is expected to derive this statement from polymer model analysis.

5.2.5. *Mean first encounter time (MFET) $\langle\tau^\epsilon(\xi)\rangle$ for the RCL polymer*

The MFET for the Rouse and β-polymer are computed[72,77] from the first eigenvalue λ_0^ϵ of the Fokker–Planck operator associated with the stochastic equations of the polymer model. We use similar principle to derive the mean time for two monomers of the RCl polymer to encounter for the first time at a distance $\epsilon > 0$, which model possible interaction (Fig. 9(a)). We start with

$$\langle\tau^\epsilon(\xi)\rangle \approx \frac{1}{D\lambda_0^\epsilon(\xi)}. \tag{92}$$

The first-order approximation in ϵ is given by[77]

$$\lambda_0^\epsilon(\xi) = \frac{4\pi\epsilon \int_{C-P} e^{-\phi_\mathcal{G}(U)}dU}{|\tilde{\Omega}(\xi)|} + O(\epsilon^2), \tag{93}$$

where $\phi_\mathcal{G}(U)$ is the diagonalized potential (61) in normal coordinates $U = [u_0, \ldots, u_{N-1}]$, and $|\tilde{\Omega}(\xi)|$ is the integral over the entire RCL configuration space

$$|\tilde{\Omega}(\xi)| = \int e^{-\phi_g(U)} dU = \left(\frac{(2\pi)^{N-1}}{\prod_{p=1}^{N-1} \kappa \chi_p(\xi)} \right)^{\frac{d}{2}}. \tag{94}$$

The integral over $C - P$ in (93) is computed over the space of closed RCL polymer ensemble with fixed connector between monomers m and n and additional $N_c(\xi)$ random connectors. A direct computation gives[57]

$$\int_{C-P} e^{-\phi_g(U)} dU = (2\pi)^{\frac{(N-2)d}{2}} \left(\frac{\kappa b^2 \prod_{p=1}^{N-1} (\kappa \chi_p(\xi))^{-1}}{\sigma_{m,n}^2(\xi)} \right)^{\frac{d}{2}}, \tag{95}$$

where $\sigma^2(\xi)$ is the variance defined by (77). Using relations (94) and (95) in (92), we obtain the MFET between any two monomers m and n of the RCL polymer for a given connectivity fraction ξ in dimension $d = 3$:

$$\langle \tau_{m,n}^\epsilon(\xi) \rangle = \frac{1}{4\pi D\epsilon} \left(\frac{2\pi \sigma_{m,n}^2(\xi)}{\kappa b^2} \right)^{\frac{3}{2}}. \tag{96}$$

Using (79) into (96), we obtain the approximation

$$\langle \tau_{m,n}^\epsilon(\xi) \rangle \approx \frac{b^2 \left(1 - \exp(-|m-n|\sqrt{N\xi}) \right)^{d/2}}{4\sqrt{N\xi} \pi D\epsilon (\kappa b^2)^{d/2}} + \mathcal{O}(N\xi),$$

where $|m - n| \ll N$, and $\xi \ll 1$. The analytical formula (96) agrees with Brownian simulations of the MFET for the RCL polymer with $N = 20, 50$, and 100 monomers, and $N_c(\xi) = 25$ added random connectors (Fig. 9(b)). Other approaches and results are presented in Refs. 78 and 79.

5.3. Looping time for more elaborated polymer models

In a general polymer model framework, monomers can interact not only with their NN along the chain, but with everyone.[65,80,81] It is also possible to account for an impenetrable or exclusion volume

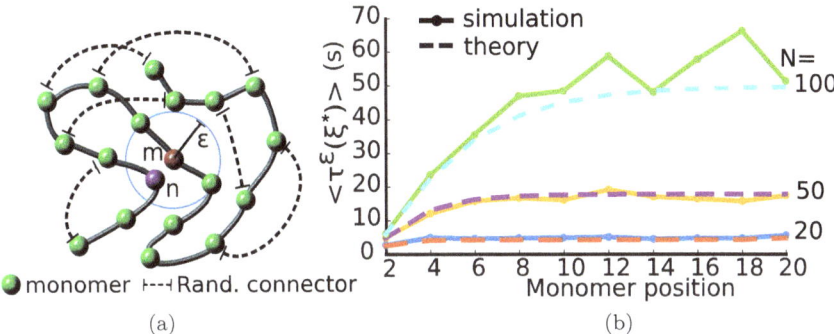

(a) (b)

Fig. 9. Transient RCL polymer properties: (a) Two monomers m (red) and n (purple) meet when they enter a ball of radius ϵ. Random connectors (dashed arrows) are added to a linear Rouse chain. (b) Stochastic simulations (dots) of the MFET between monomer 1 and monomers 2–20 of RCL polymers with $N = 20$ (blue), 50 (yellow) and 100 (green) monomers, with $N_c(\xi) = 25$ random connectors, agree with the formula (Eq. (96), dashed). Parameters: $\epsilon = b/10, D = 1, b = \sqrt{3}, \Delta t = 0.01s$, the RCL system is (64) (we used Eq. (96)).

around monomers, represented by a repulsion force. The associated excluded volume interaction energy ϕ_{EVI} is described by a short-range potential

$$\phi_{EVI}(\boldsymbol{R}_1, \ldots, \boldsymbol{R}_N) = \frac{1}{2}\sigma k_B T \sum_{n,m=1;n\neq m}^{N} \delta(\boldsymbol{R}_n - \boldsymbol{R}_m), \quad (97)$$

where σ is the excluded volume, which is computed from the radius of the monomer and δ is the classical Dirac function. There are almost no results for the asymptotic of the first looping time result in that case.

To conclude, due to thermal agitation, transient loops appear and disappear with a time scale summarized by the mean first passage time described above. Some loops can be stabilized by linker molecules such as CTCF or cohesin.[82] The principle of analyzing HiC data is to obtain statistical properties of these loop distributions over cell populations. These loops define 3D compartments of the scale of hundreds of nanometers, such as TAD, characterized by a block square matrix.[83,84] These compartments can also interact and intermingle partially, characterized by the radius of Gyration, the mean time for the two loci to meet, and the number of linker molecules[84] at a given scale.

6. Summary, Conclusion, Future Directions, and Perspectives

To conclude, the theory of chemical reactions at a molecular level has drastically changed in the past 20 years. The need to understand chemical reactions in a physiological context has created new demands, beyond studying the production rate of a continuum number. Novel concepts were developed such as the probability and first time a threshold for the number of bound molecules is reached. Time to threshold is a key feature because it reflects changing scale and the induction of physiological changes by molecular events, ultimately controlling the cell physiology.[8] An example is the induction of long-term changes at synapses that underlie learning and memory, which remains to this day an open question. To understand the fast time scale of organelle activation in a cell, we recently introduced the concepts of extreme statistics for the first molecules to find and activate a small target (receptor)[39,85,86]: It does not matter how many receptors are activated, but one must make sure that one is, because that one often triggers an avalanche that amplifies the signal. This mechanism explains fast calcium transients in neuronal and glial compartments.[85,87] This fast activation mechanism is more obvious to understand for the classical fertilization problem, where the first arriving sperms will have the chance to fertilize the ovule. The law of the mean arrival time depends on the log of the number in case of Brownian particles.[86] Having many sperms allows for the first ones to follow optimal paths, which define the search time scale. The question of redundancy, where there are many copies of the same object that could have been considered as waste, is not. Its purpose is to define time scale over a great expanse (due to the log effect).

Although we have not discussed much here on chemical networks, where compounds are activated in a partially ordered sequence, we emphasize that transient regimes should be carefully studied. Indeed, the classical approach to study the dynamics and interaction of tens of chemical compounds remains challenging, especially when their spatial geometrical organization is ignored. Ignoring the geometry distribution reduces chemical reactions to a forward and a backward rate. If there are many methods available to simulate the steady state with many compounds, the transient regimes should be studied much more carefully, especially if there are bottlenecks consisting of reactions that are activated successively in a given order. In that case, as discussed in the extreme statistics Subsection 3.2, traditional mass-action laws, reaction-diffusion equations, Gillespie's algorithm, Markov models, hybrid approaches, etc. should be replaced by simulations of stochastic trajectories. The number of initial molecules is a key factor in the simulations, and the first arrival time of the first particles should be tracked, because the extreme statistics predict that the time scale is not given by the mean first passage time nor by the forward rate.[85,88] The consequence is that the dynamics of protein and gene networks should probably be reconsidered, when estimating the rates of production, especially if the first arriving molecules are the activators.

Finally, we discussed an indirect mode of molecular or gene activation occurring through looping. This activation has connected polymer models with chemical reactions,[71,89] where rate constants have been computed for several polymer models.[59]

We are currently moving in a direction to include the cell or nucleus geometry at tens of nanometer resolution: a novel approach is to use stochastic chemical reaction simulations to predict the exact location of receptors.[85] Combining theoretical computations and Brownian simulation of trajectories is a powerful method to make predictions and to discover molecular organization with tens of nanometer precision, which can be compared to the latest super-resolution imaging data.

Appendix

MTT for stochastic chemical reactions

To compute the MTT for the number of bound molecules to reach the threshold T, we use a Markov chain for the substrate S and the diffusing molecules M interact as described by Eq. (1), which describes the pdfs

$$p_k(t) = \Pr\{|MS|(t) = k||MS|(0) = 0\}, \tag{A.1}$$

where the number of bound molecules $|MS|$ at time t is equal to k. Between time t and $t + \Delta t$, the transitions to the k state occur from states $k - 1$, k, and $k + 1$ with transition rates $\lambda_{k-1}, -(\lambda_k + k_{-1}k), k_{-1}(k+1)$.[11] Thus, p_k (for $0 \le k \le S_0$) satisfies

$$\dot{p}_0 = -\lambda_0 p_0 + k_{-1} p_1,$$
$$\dot{p}_k = -(\lambda_k + k_{-1}k)p_k + \lambda_{k-1}p_{k-1} + k_{-1}(k+1)p_{k+1}$$
$$\text{for } 0 < k < S_0, \tag{A.2}$$
$$\dot{p}_{S_0} = -(\lambda_{S_0} + k_{-1}S_0)p_{S_0} + \lambda_{S_0-1}p_{S_0-1},$$

where

$$\lambda_k = \lambda(S_0 - k)(M_0 - k), \tag{A.3}$$

which is the rate for one of the $M_0 - k$ free molecules to reach the $S_0 - k$ free binding sites. λ is the binding rate for one molecule M to a single target, it is the reciprocal of the NET $\bar{\tau}$.

To estimate the mean first time that the number of bound molecules reaches the threshold T, the state $|MS| = T$ in Eq. (A.2) is absorbing, which leads to the modified system

$$\dot{p}_0 = -\lambda_0 p_0 + k_{-1} p_1,$$
$$\dot{p}_k = -(\lambda_k + k_{-1}k)p_k + \lambda_{k-1}p_{k-1} + k_{-1}(k+1)p_{k+1}$$
$$\text{for } 0 < k < T - 1,$$
$$\dot{p}_{T-1} = -(\lambda_{T-1} + k_{-1}(T-1))p_{T-1} + \lambda_{T-2}p_{T-2},$$
$$\dot{p}_T = \lambda_{T-1}p_{T-1},$$

where $p_k(t)k = \Pr\{|MS(t)| = k\}$ with $0 \le k \le T$. By definition

$$p_T(t) = \Pr\{\tau_T \le t\}, \tag{A.4}$$

where τ_T is the first hitting time to the threshold T,

$$\tau_T = \inf\{t, |MS|(t) = T\}. \tag{A.5}$$

The MTT $\bar{\tau}_T$ is given by

$$\bar{\tau}_T = \int_0^{+\infty} \Pr\{\tau_T > t\}dt = \int_0^{+\infty} (1 - p_T(t))dt. \tag{A.6}$$

Equivalently using the normalization condition

$$\sum_0^T p_k(t) = 1, \tag{A.7}$$

we have the general expression

$$\bar{\tau}_T = \sum_0^{T-1} a_k, \tag{A.8}$$

where $a_k = \int_0^{+\infty} p_k(t)dt$. To obtain an analytical expression for $\bar{\tau}_T$, we integrate equations (A.4) between 0 and $+\infty$ with the initial conditions $p_0 = 1, p_k = 0$. Using $p_T(+\infty) = 1$ and $p_k(+\infty) = 0$ for $k < T$, we get

$$-1 = -\lambda_0 a_0 + k_{-1}a_1,$$
$$0 = -(\lambda_k + k_{-1}k)a_k + \lambda_{k-1}a_{k-1} + k_{-1}(k+1)a_{k+1}$$
$$\text{for } 0 < k < T-1,$$
$$0 = -(\lambda_{T-1} + k_{-1}(T-1))a_{T-1} + \lambda_{T-2}a_{T-2},$$
$$1 = \lambda_{T-1}a_{T-1},$$

equivalently

$$a_{T-1} = \frac{1}{\lambda_{T-1}}, \tag{A.9}$$

$$a_k = \frac{1}{\lambda_k} + (k+1)k_{-1}\frac{a_{k+1}}{\lambda_k} \text{ for } 0 \le k \le T-2. \tag{A.10}$$

When the binding is irreversible ($k_{-1} = 0$), the MTT $\bar{\tau}_T$ is the sum of the forward rates

$$\tau_T^{\text{irrev}} = \frac{1}{\lambda_0} + \frac{1}{\lambda_1} + \cdots + \frac{1}{\lambda_{T-1}} = \frac{1}{\lambda} \sum_{k=0}^{T-1} \frac{1}{(M_0 - k)(S_0 - k)}. \quad (A.11)$$

In particular, when $M_0 = S_0$ and $M_0 \gg 1$, the asymptotic formula for equation (A.11) becomes

$$\tau_T^{\text{irrev}} \approx \frac{T}{\lambda M_0 (M_0 - T)}. \quad (A.12)$$

In addition, when the diffusing molecules largely exceed the number of targets ($M_0 \gg S_0, T$), we further obtain from (A.11), the asymptotic formulas,

$$\tau_T^{\text{irrev}} \approx \begin{cases} \dfrac{1}{\lambda M_0} \log \dfrac{S_0}{S_0 - T} & \text{when } M_0 \gg S_0, T, \\[2ex] \dfrac{1}{\lambda S_0} \log \dfrac{M_0}{M_0 - T} & \text{when } S_0 \gg M_0, T, \\[2ex] \dfrac{T}{\lambda M_0 S_0} & \text{when } M_0, S_0 \gg T. \end{cases} \quad (A.13)$$

In Fig. 1, we plot the MTT τ_T^{irrev} for several values of the threshold T and we compare it with Brownian simulations performed in a circular disk $\Omega = D(R)$, the boundary of which is reflecting except at the targets. When $k_{-1} > 0$, the analytical expression for $\bar{\tau}_T$ is given by

$$\bar{\tau}_T = \sum_{k=0}^{T-1} \frac{1}{\lambda_k} + k_{-1} \sum_{k=1}^{T-1} \frac{k}{\lambda_k \lambda_{k-1}} + k_{-1}^2 \sum_{k=2}^{T-1} \frac{k(k-1)}{\lambda_k \lambda_{k-1} \lambda_{k-2}}$$

$$+ k_{-1}^3 \sum_{k=3}^{T-1} \frac{k(k-1)(k-2)}{\lambda_k \lambda_{k-1} \lambda_{k-2} \lambda_{k-3}} + \cdots$$

$$= \sum_{j=0}^{T-1} \left(k_{-1}^j \sum_{k=j}^{T-1} \frac{k!}{(k-j)! \prod_{i=k-j}^{k} \lambda_i} \right)$$

$$= \frac{1}{\lambda} \sum_{j=0}^{T-1} \left(\frac{k_{-1}}{\lambda} \right)^j \sum_{k=j}^{T-1} \frac{k!}{(k-j)!} \frac{(M_0 - k - 1)!}{(M_0 - k + j)!} \frac{(S_0 - k - 1)!}{(S_0 - k + j)!}.$$

Finally, we obtain

$$\bar{\tau}_T = \frac{1}{\lambda} \sum_{j=0}^{T-1} \left(\frac{k_{-1}}{\lambda}\right)^j \sum_{k=j}^{T-1} \frac{k!}{(k-j)!} \frac{(M_0 - k - 1)!}{(M_0 - k + j)!} \frac{(S_0 - k - 1)!}{(S_0 - k + j)!}.$$

(A.14)

This sum can be further approximated for the three following regimes $M_0 \gg S_0, T$, $S_0 \gg M_0, T$, and $S_0 = M_0 \gg T$, by using the first-order expansion in $\frac{k_{-1}}{\lambda}$ only, and the sum $\sum_1^N \frac{1}{k} = \log(N) + O(1)$. We obtain,

$$\bar{\tau}_T \approx \begin{cases} \tau_T^{irrev} + \frac{k_{-1}}{(\lambda M_0)^2} \left(\frac{T}{S_0 - T} - \log\left(1 + \frac{T}{S_0 - T}\right)\right) \\ \quad \text{when } M_0 \gg S_0, T, \\ \tau_T^{irrev} + \frac{k_{-1}}{(\lambda S_0)^2} \left(\frac{T}{M_0 - T} - \log\left(1 + \frac{T}{M_0 - T}\right)\right) \\ \quad \text{when } S_0 \gg M_0, T, \\ \tau_T^{irrev} + \frac{k_{-1}}{2\lambda^2} \left(\frac{T}{M_0^2}\right)^3, \quad \text{when } S_0 = M_0 \gg T. \end{cases}$$

(A.15)

Using the expression for τ^{irrev} (Eq. (A.13)), we finally obtain the asymptotic expressions:

$$\bar{\tau} \approx \begin{cases} \frac{1}{\lambda M_0} \left[\log\left(1 + \frac{T}{S_0 - T}\right) + \frac{k_{-1}}{\lambda M_0}\left(\frac{T}{S_0 - T} - \log\left(\frac{S_0}{S_0 - T}\right)\right)\right] \\ \quad \text{when } M_0 \gg S_0, T, \\ \frac{1}{\lambda S_0} \left[\log\left(\frac{M_0}{M_0 - T}\right) + \frac{k_{-1}}{\lambda S_0}\left(\frac{T}{M_0 - T} - \log\left(\frac{M_0}{M_0 - T}\right)\right)\right] \\ \quad \text{when } S_0 \gg M_0, T, \\ \frac{T}{\lambda M_0^2} \left[1 + \frac{k_{-1}}{2\lambda^2}\left(\frac{T}{M_0^2}\right)^2\right] \quad \text{when } S_0 = M_0 \gg T. \end{cases}$$

(A.16)

Furthermore, when $T \ll S_0$ and $T \ll M_0$, respectively, in the two first regimes (Eq. (A.16)), we obtain the refined estimates

$$
\bar{\tau} \approx
\begin{cases}
\frac{T}{\lambda M_0 S_0} \left[1 + \frac{1}{2} \left(1 + \frac{k_{-1}}{\lambda M_0} \right) \left(\frac{T}{S_0} \right) \right] \\
\quad \text{when } M_0 \gg S_0 \gg T, \\
\frac{T}{\lambda M_0 S_0} \left[1 + \frac{1}{2} \left(1 + \frac{k_{-1}}{\lambda S_0} \right) \left(\frac{T}{M_0} \right) \right] \\
\quad \text{when } S_0 \gg M_0 \gg T.
\end{cases}
\tag{A.17}
$$

References

1. D. Holcman and Z. Schuss, Commentary new mathematical physics needed for life sciences, *Phys. Today* **69**(1), 10–12 (2016).
2. Z. Schuss, *Theory and Applications of Stochastic Processes: An Analytical Approach*, Vol. 170. Springer Science & Business Media (2009).
3. J.-P. Gauthier and I. Kupka, *Deterministic Observation Theory and Applications*. Cambridge University Press (2001).
4. C. L. Worth, R. Preissner, and T. L. Blundell, Sdma server for predicting effects of mutations on protein stability and malfunction, *Nucleic Acids Res.* **39**(2), W215–W222 (2011).
5. W. R. Pitt, R. W. Montalvão, and T. L. Blundell, Polyphony: Superposition independent methods for ensemble-based drug discovery, *BMC Bioinformatics*, **15**, 324 (2014).
6. J. Reingruber and D. Holcman, Modeling and stochastic analysis of the single photon response, in: *Stochastic Processes, Multiscale Modeling, and Numerical Methods for Computational Cellular Biology*, Springer (2017), pp. 315–348.
7. J. Cartailler, T. Kwon, R. Yuste, and D. Holcman, Deconvolution of voltage sensor time series and electro-diffusion modeling reveal the role of spine geometry in controlling synaptic strength, *Neuron*, **97**(5), 1126–1136 (2018).
8. G. L. Fain, *Molecular and Cellular Physiology of Neurons*. Harvard University Press (1999).
9. D. Holcman and Z. Schuss, 100 years after smoluchowski: Stochastic processes in cell biology, *J. Phys. A. Math. Theor.* **50**(9), 093002 (2017).
10. Y. X. Bouchoucha, J. Reingruber, C. Labalette, M. A. Wassef, E. Thierion, C. D.-T. Dinh, D. Holcman, P. Gilardi-Hebenstreit, and P. Charnay, Dissection of a krox20 positive feedback loop driving cell fate choices in hindbrain patterning, *Mol. Syst. Biol.* **9**(1), 690 (2013).
11. D. Holcman and Z. Schuss, Stochastic chemical reactions in microdomains, *J. Chem. Phys.* **122**(11), 114710 (2005).

12. J. Reingruber, J. Pahlberg, M. L. Woodruff, A. P. Sampath, G. L. Fain, and D. Holcman, Detection of single photons by toad and mouse rods, *Proc. Nat. Acad. Sci.* **110**(48), 19378–19383 (2013).

13. D. Holcman and Z. Schuss, Control of flux by narrow passages and hidden targets in cellular biology, *Rep. Prog. Phys.* **76**(7), 074601 (2013).

14. K. D. Duc and D. Holcman, Threshold activation for stochastic chemical reactions in microdomains, *Phys. Rev. E.* **81**(4), 041107 (2010).

15. K. D. Duc and D. Holcman, Using default constraints of the spindle assembly checkpoint to estimate the associated chemical rates, *BMC Biophys.* **5**(1), 1 (2012).

16. D. Holcman and Z. Schuss, *Stochastic Narrow Escape in Molecular and Cellular Biology: Analysis and Applications.* Springer Verlag, NY (2015).

17. Z. Schuss, A. Singer, and D. Holcman, The narrow escape problem for diffusion in cellular microdomains, *Proc. Natl. Acad. Sci. USA.* **104**, 16098–16103 (2007).

18. D. Holcman and Z. Schuss, The narrow escape problem, *SIAM Rev.* **56**(2), 213–257 (2014).

19. Z. Schuss, *Diffusion and Stochastic Processes. An Analytical Approach.* Springer-Verlag, New York, NY (2009).

20. L. Wolpert, One hundred years of positional information, *Trends Genet.* **12**(9), 359–364 (1996).

21. V. Kasatkin, A. Prochiantz, and D. Holcman, Morphogenetic gradients and the stability of boundaries between neighboring morphogenetic regions, *Bull. Math. Biol.* **70**(1), 156–178 (2008).

22. D. Holcman, V. Kasatkin, and A. Prochiantz, Modeling homeoprotein intercellular transfer unveils a parsimonious mechanism for gradient and boundary formation in early brain development, *J. Theor. Biol.* **249**(3), 503–517 (2007).

23. J. Reingruber and D. Holcman, Computational and mathematical methods for morphogenetic gradient analysis, boundary formation and axonal targeting, in: *Seminars in Cell & Developmental Biology*, Vol. **35**, Elsevier (2014), pp. 189–202.

24. R. L. Huganir and R. A. Nicoll, Ampars and synaptic plasticity: The last 25 years, *Neuron* **80**(3), 704–717 (2013).

25. J. E. Lisman and K. M. Harris, Quantal analysis and synaptic anatomyintegrating two views of hippocampal plasticity, *Trends Neurosci.* **16**(4), 141–147 (1993).

26. K. DaoDuc and D. Holcman, Computing the length of the shortest telomere in the nucleus, *Phys. Rev. Lett.* **111**(22), 228104 (2013).

27. V. Kurella, J. C. Tzou, D. Coombs, and M. J. Ward, Asymptotic analysis of first passage time problems inspired by ecology, *Bull. Math. Biol.* **77**(1), 83–125 (2015).

28. M. I. Delgado, M. J. Ward, and D. Coombs, Conditional mean first passage times to small traps in a 3-d domain with a sticky boundary: Applications to t cell searching behavior in lymph nodes, *Multiscale Model Sim.* **13**(4), 1224–1258 (2015).

29. D. Holcman and Z. Schuss, Stochastic narrow escape in molecular and cellular biology. *Analysis and Applications*. Springer, New York (2015).

30. M. J. Ward, W. D. Heshaw, and J. B. Keller, Summing logarithmic expansions for singularly perturbed eigenvalue problems, *SIAM J. Appl. Math.* **53**(3), 799–828 (1993).

31. I. V. Grigoriev, Y. A. Makhnovskii, A. M. Berezhkovskii, and V. Y. Zitserman, Kinetics of escape through a small hole, *J. Chem. Phys.* **116**(22), 9574–9577 (2002).

32. Z. Schuss, K. Basnayake, and D. Holcman, "Redundancy principle for optimal random search in biology," *bioRxiv* (2017).

33. I. M. Sokolov, R. Metzler, K. Pant, and M. C. Williams, First passage time of n excluded-volume particles on a line, *Phys. Rev. E.* **72**(4), 041102 (2005).

34. S. B. Yuste and K. Lindenberg, Order statistics for first passage times in one-dimensional diffusion processes, *J. Stat. Phys.* **85**(3), 501–512 (1996).

35. T. Chou and M. Dorsogna, First passage problems in biology, *First-Passage Phenomena and Their Applications*, **35**, 306 (2014).

36. G. Schehr, S. N. Majumdar, G. Oshanin, and S. Redner, Exact record and order statistics of random walks via first-passage ideas, *First-Passage Phenomena and Their Applications*, **35**, 226 (2014).

37. Z. Schuss, K. Basnayake, C. Guerrier, and D. Holcman, Asymptotics of extreme statistics of escape time in 1,2 and 3-dimensional diffusions, *arXiv preprint arXiv:1711.01330*.

38. T. Antal, K. Blagoev, S. Trugman, and S. Redner, Aging and immortality in a cell proliferation model, *J. Theor. Biol.* **248**(3), 411–417 (2007).

39. K. Basnayake, Z. Schuss, and D. Holcman, Asymptotic formulas for extreme statistics of escape times in 1, 2 and 3-dimensions, *J. Nonlinear Sci.* (2018).

40. Z. Schuss, *Nonlinear Filtering and Optimal Phase Tracking*, Vol. 180. Springer Science & Business Media (2011).

41. A. Chédotal and L. J. Richards, Wiring the brain: The biology of neuronal guidance, in: *Cold Spring Harbor Perspectives in Biology.* (2010), p. a001917.

42. A. L. Kolodkin and M. Tessier-Lavigne, Mechanisms and molecules of neuronal wiring: A primer, *CSH. Perspect. Biol.* **3**(6), p. a001727 (2011).

43. J. G. Flanagan, Neural map specification by gradients, *Curr. Opinion Neurobiology*, **16**(1), 59–66 (2006).

44. H. Blockus and A. Chédotal, The multifaceted roles of slits and robos in cortical circuits: from proliferation to axon guidance and neurological diseases, *Curr. Opin. Neurobiol.* **27**, 82–88 (2014).

45. T. McLaughlin, R. Hindges, and D. D. OLeary, Regulation of axial patterning of the retina and its topographic mapping in the brain, *Curr. Opin. Neurobiol.* **13**(1), 57–69 (2003).

46. O. Stettler, R. L. Joshi, A. Wizenmann, J. Reingruber, D. Holcman, C. Bouillot, F. Castagner, A. Prochiantz, and K. L. Moya, Engrailed homeoprotein recruits the adenosine a1 receptor to potentiate ephrin a5 function in retinal growth cones, *Development*, **139**(1), 215–224 (2012).

47. G. Aquino, N. S. Wingreen, and R. G. Endres, Know the single-receptor sensing limit? think again, *J. Stat. Phys.* **162**(5), 1353–1364 (2016).

48. G. J. Goodhill, Can molecular gradients wire the brain?, *Trends Neurosci.* **39**(4), 202–211 (2016).

49. A. M. Berezhkovskii and A. Szabo, Effect of ligand diffusion on occupancy fluctuations of cell-surface receptors, *J. Chem. Phys.* **139**(12), 09B610_1 (2013).

50. U. Dobramysl and D. Holcman, Reconstructing the gradient source position from steady-state fluxes to small receptors, *Sci. Rep.* **8**(1), 941 (2018).

51. U. Dobramysl and D. Holcman, Mixed analytical-stochastic simulation method for the recovery of a brownian gradient source from probability fluxes to small windows, *J. Computat. Phys.* **355**, 22–36 (2018).

52. S. J. Greive and P. H. Von Hippel, Thinking quantitatively about transcriptional regulation, *Nat. Rev. Mol. Cell Biol.* **6**(3), 221 (2005).

53. J. Reingruber and D. Holcman, Gated narrow escape time for molecular signaling, *Phys. Rev. Lett.* **103**(14), 148102 (2009).

54. J. Reingruber and D. Holcman, Transcription factor search for a dna promoter in a three-state model, *Phys. Rev. E.* **84**, 020901 (2011).

55. G. Malherbe and D. Holcman, The search kinetics of a target inside the cell nucleus, *arXiv preprint arXiv:0712.3467*, 2007.

56. G. Malherbe and D. Holcman, Search for a dna target site in the nucleus, *Phys. Lett. A.* **374**, 466–471 (2010).

57. A. Amitai and D. Holcman, Polymer physics of nuclear organization and function, *bioRxiv*, 076661 (2016).

58. M. Slutsky and L. A. Mirny, Kinetics of protein-dna interaction: facilitated target location in sequence-dependent potential, *Biophys. J.* **87**(6), 4021–4035 (2004).

59. A. Amitai and D. Holcman, Polymer physics of nuclear organization and function, *Phys. Rep.* **678**, 1–83 (2017).

60. M. Doi and S. F. Edwards, *The Theory of Polymer Dynamics*. Clarendon Press, Oxford (1986).

61. G. E. Uhlenbeck and L. S. Ornstein, On the theory of the brownian motion, *Phys. Rev.* **36**, 823–841 (1930).

62. A. Amitai and D. Holcman, Polymer model with long-range interactions: Analysis and applications to the chromatin structure, *Phys. Rev. E.* **88**, 052604 (2013).

63. E. P. Nora, B. R. Lajoie, E. G. Schulz, L. Giorgetti, I. Okamoto, N. Servant, T. Piolot, N. L. van Berkum, J. Meisig, J. Sedat, *et al.*, Spatial partitioning of the regulatory landscape of the x-inactivation centre, *Nature* **485**(7398), 381–385 (2012).

64. J. Fraser, C. Ferrai, A. M. Chiariello, M. Schueler, T. Rito, G. Laudanno, M. Barbieri, B. L. Moore, D. C. Kraemer, S. Aitken, *et al.*, Hierarchical folding and reorganization of chromosomes are linked to transcriptional changes in cellular differentiation, *Mol. Syst. Biol.* **11**(12), 852 (2015).

65. O. Shukron and D. Holcman, Statistics of randomly cross-linked polymer models to interpret chromatin conformation capture data, *Phys. Rev. E.* **96**(1), 012503 (2017).

66. I. Sokolov, Cyclization of a polymer: first-passage problem for a non-markovian process, *Phys. Rev. Lett.* **90**(8), 080601 (2003).

67. A. A. Gurtovenko and A. Blumen, Generalized gaussian structures: Models for polymer systems with complextopologies, in: *Polymer Analysis Polymer Theory*, Springer (2005), pp. 171–282.

68. S. Jespersen, I. Sokolov, and A. Blumen, Small-world rouse networks as models of cross-linked polymers, *J. Chem. Phys.* **113**(17), 7652–7655 (2000).

69. B. Eichinger, Configuration statistics of gaussian molecules, *Macromolecules* **13**(1), 1–11 (1980).

70. M. Doi and S. Edwards, *The Theory of Polymer Dynamics Clarendon*. Oxford (1986).

71. A. Amitai, C. Amoruso, A. Ziskind, and D. Holcman, Encounter dynamics of a small target by a polymer diffusing in a confined domain, *J. Chem. Phys.* **137**(24), 244906 (2012).

72. A. Amitai and D. Holcman, Polymer model with long-range interactions: Analysis and applications to the chromatin structure, *Phys. Rev. E.* **88**(5), 052604 (2013).

73. A. Amitai, A. Seeber, S. M. Gasser, and D. Holcman, Visualization of chromatin decompaction and break site extrusion as predicted by statistical polymer modeling of single-locus trajectories, *Cell Rep.* **18**(5), 1200–1214 (2017).

74. A. Amitai, I. Kupka, and D. Holcman, Kinetics of diffusing polymer encounter in confined cellular microdomains, *J. Stat. Phys.* **153**, 1107–1131 (2013).

75. M. Abramowitz and I. A. Stegun, *Handbook of Mathematical Functions with Formulas, Graphs, and Mathematical Tables*. Dover, New York (1972).

76. A. Singer and Z. Schuss, Activation through a narrow opening, *SIAM J. Appl. Math.* **68**, 98 (2007).

77. A. Amitai, I. Kupka, and D. Holcman, Computation of the mean first-encounter time between the ends of a polymer chain, *Phys. Rev. Lett.* **109**(10), 108302 (2012).

78. O. Benichou, T. Guérin, and R. Voituriez, Mean first-passage times in confined media: from markovian to non-markovian processes, *J. Phys. A.: Math. Theor.* **48**(16), 163001 (2015).

79. T. Guérin, M. Dolgushev, O. Bénichou, R. Voituriez, and A. Blumen, Cyclization kinetics of gaussian semiflexible polymer chains, *Phys. Rev. E.* **90**(5), 052601 (2014).

80. W. Kuhn, Over the shape of threadlike molecules in solutions, *Kolloid* **68**, 2 (1934).

81. P. J. Flory, The configuration of real polymer chains, *J. Chem. Phys.* **17**, 303 (1949).

82. S. S. Rao, S.-C. Huang, B. G. St Hilaire, J. M. Engreitz, E. M. Perez, K.-R. Kieffer-Kwon, A. L. Sanborn, S. E. Johnstone, G. D. Bascom, I. D. Bochkov, *et al.*, Cohesin loss eliminates all loop domains, *Cell* **171**(2), 305–320 (2017).

83. O. Shukron and D. Holcman, Transient chromatin properties revealed by polymer models and stochastic simulations constructed from chromosomal capture data, *PLOS Comput. Biol.* **13**(4), e1005469 (2017).

84. O. Shukron and D. Holcman, Heterogeneous cross-linked polymers to reconstruct chromatin reorganization during cell differentiation, *bioRxiv* (2017).

85. E. Korkotian, K. Basnayake, and D. Holcman, The extreme statistics of fast calcium transient in dendritic spines with an endoplasmic reticulum, *Preprint*.

86. K. Basnayake, A. Hubl, Z. Schuss, and D. Holcman, Extreme narrow escape: shortest paths for the first particles to escape through a small window, *arXiv preprint arXiv:1804.10808* (2018).

87. P.-Y. Shih, L. P. Savtchenko, N. Kamasawa, Y. Dembitskaya, T. J. McHugh, D. A. Rusakov, R. Shigemoto, and A. Semyanov, Retrograde synaptic signaling mediated by K+ efflux through postsynaptic nmda receptors, *Cell Rep.* **5**(4), 941–951 (2013).

88. D. Holcman and Z. Schuss, *Asymptotics of Elliptic and Parabolic PDEs and their Applications in Statistical Physics, Computational Neuroscience, and Biophysics*, Vol. 199. Springer (2018).

89. M. Ptashne, Gene regulation by proteins acting nearby and at a distance, *Nature* **322**, 697 (1986).

Chapter 16

First-Passage Processes and Encounter-Controlled Reactions in Growing Domains

E. Abad[*,§], C. Escudero[†], F. Le Vot[‡], and S. B. Yuste[‡]

[*]*Departamento de Física Aplicada and Instituto de Computación Científica Avanzada (ICCAEx), Centro Universitario de Mérida, Universidad de Extremadura, E-06800 Mérida, Spain*
[†]*Departamento de Matemáticas, Universidad Autónoma de Madrid and Instituto de Ciencias Matemáticas, Consejo Superior de Investigaciones Científicas, E-28049 Madrid, Spain*
[‡]*Departamento de Física and Instituto de Computación Científica Avanzada (ICCAEx), Universidad de Extremadura, E-06071 Badajoz, Spain*
[§]*eabad@unex.es*

The theory of first-passage processes provides a very useful theoretical rationale for the kinetics of diffusion-controlled reactions, even in cases where the reactive events occur at random locations. While such reactions are usually considered in static domains, the case of a growing domain is highly interesting for applications, notably in biology and cosmology. A sufficiently fast domain growth may indeed have a strong impact on the kinetics, to the extent that the reaction is then no longer controlled by the diffusive transport. Here, we illustrate such effects by way of three examples that can be addressed with the theory of first-passage processes, namely, (a) the escape problem for a particle diffusing inside a growing hypersphere, (b) the calculation of the survival probability of a point trap diffusing in the presence of a growing spherical target, and (c) the $1d$ kinetics of the coalescence reaction $A + A \to A$ on a growing domain. For a sufficiently fast domain growth, all three cases lead to a non-vanishing survival probability of the diffusing particles in the long-time limit. In the case of the coalescence reaction, a fast domain growth leads to an enhancement of the memory of the initial condition. In this case, when the system starts from a random initial condition, the degree of self-ordering induced by the coalescence events is reduced drastically. A common feature of the systems considered is the fact that the domain grows uniformly, in which case analytical solutions can be obtained by means of a double transformation in time and space. The obtained results highlight the need to extend the existing first-passage theories of chemical kinetics in order to deal with the interplay between diffusional mixing and drift-dilution effects induced by the domain growth.

1. Introduction

In many situations, binary reactions occur with a high probability when a pair of co-reactants come sufficiently close to one another for the first time; the transport mechanism facilitating the first encounter between the two reactants then becomes the rate-limiting step. Diffusion is a very common mixing mechanism underlying such encounters, in which case one speaks of a diffusion-controlled reaction.[1] Here, the time needed to cover the typical interparticle separation distance by diffusive spreading is much larger than the time required for the formation of products via a binary collision. Examples are systems where the particles have a low diffusivity and a high probability to react upon encounter. However, even when the diffusivities are comparatively large and the efficiency of reactive collisions relatively low, the reaction may still be diffusion-controlled as a result of the geometric constraints imposed by the embedding substrate. A very common situation is the case of a low-dimensional substrate (see, e.g., Ref. 2), which strongly hinders the efficiency of diffusional mixing and amplifies the effect of local inhomogeneities, thereby entailing a certain persistence of the initial condition. This may result in the breakdown of the mean-field approximation and in the onset of fluctuation-dominated kinetics, thus leading to a slowed down time evolution of the reactant concentrations.

In this chapter, we shall deal with another mechanism which may also limit the efficiency of diffusional mixing, namely, the physical growth of the medium in which a collection of reactants diffuse. In a closed expanding system, volume growth increases the typical inter-particle distance and makes the spatial distribution of the reactants more sparse. One obvious conclusion is that, when the domain grows at a sufficiently fast rate, one may need to reexamine the recurrence and first-passage properties of Brownian motion. To avoid any mis-understanding, we emphasize that in the systems under consideration here, each volume element expands at a certain rate in the course of time; as a consequence of this expansion process, the boundary of a finite domain is also displaced. This is a fundamentally different situation from the case where the enclosed volume grows simply because the boundary is shifted. In the former case, a Brownian particle will move even when it does not jump, since it will be drifted as the volume element that contains the particle expands. Therefore, the first-passage properties characterizing the arrival at the boundary of the growing domain will in general be different in both cases.

Interestingly enough, works on reaction-diffusion systems in grow-ing domains as defined above remain relatively scarce in the litera-ture. The field of developmental biology appears to be one of the main sources of inspiration, and more specifically phenomena such as pattern formation and tissue growth.[3-5] In this context, cell divi-sion processes leading to a growth of the physical medium may occur on time scales that are relevant for the study of diffusion processes in the above systems.

From a more general perspective, despite the important advances made over the last decades by rate theories relying on the first-passage picture,[6-10] we are not aware of any work in which specific applications to real systems under growth are considered. In con-trast, some attempts have been recently made to consider idealized situations which will hopefully shed light on the behavior of real systems.[11–13] Taking this line of thought as a starting point, the aim of the present chapter is to familiarize the reader with the general formalism of diffusion in uniformly growing domains and to present him/her with three illustrative first-passage problems. In the first two problems, the reactions are localized at deterministically moving boundaries, whereas in the last problem, chemical reactions occur at random locations. Despite these differences, we study the three problems within a common theoretical framework. In all cases, a sufficiently fast domain growth entails a drastic modification of the long-time behavior, leading to an increased survival probability of the diffusing reactants and (in our last example) to an enhanced memory of the initial condition.

To illustrate the situation, let us consider the kinetics of the coalescence reaction $A + A \to A$ on a $1d$ growing domain, where A denotes a diffusing particle. In the well-studied case of a static domain, the system decays to the empty state at a slower rate than the mean-field prediction.[14] However, diffusional mixing is strong enough to sustain the decay to the empty state and even to erase the signature of the initial condition in the long-time limit (except for some very specific types of initial distributions[15]). In contrast, when the domain grows at a sufficiently large rate, it is clear that diffusional mixing will become negligible on relatively short time scales, implying that the kinetics may be strongly altered. The reason is that not all the particles may be able to meet and thereby react; therefore, the system will attain a non-empty final state associated with a finite survival probability that depends on the details of the initial particle distribution.[16,17] In the case of a random initial condition, the self-ordering properties of the system are also strongly affected by a fast domain growth.[17]

The remainder of this chapter is organized as follows. In Section 2, we give an overview of the formalism used to deal with diffusion in uniformly growing domains, which involves a double transformation in space and time. In Section 3, we take this formalism as a starting point to discuss escape problems and target problems in expanding domains. More specifically, we deal with the escape problem of a

Brownian particle initially located in the interior of a hypersphere and with the problem of the survival probability of a target in the presence of a diffusing point trap (the $1d$ extension to the case of multiple traps is also considered). As it turns out, the $1d$ version of the target problem with a single trap is closely related to the $1d$ kinetics of the $A + A \to A$ reaction on a growing domain, which is studied in Section 4. Finally, in Section 5 we recapitulate our main findings.

2. Free Diffusion in a Growing Domain: General Formalism

Consider a d-dimensional growing domain in which the evolution of the coordinate $\mathbf{y}(t)$ of a physical point with initial coordinate $\mathbf{x}$ is dictated by a uniform growth. The growth process is characterized by a time-dependent scale factor $a(t) > 0$, i.e., $\mathbf{y}(t) = a(t)\mathbf{x}$ with $a(t_0) = 1$ at the initial time t_0. The fixed coordinate $\mathbf{x}$ is termed Eulerian or comoving coordinate in the literature (see, e.g., Ref. 18). The case of a power-law scale factor

$$a(t) = (t/t_0)^\gamma, \quad t \geq t_0, \tag{1}$$

and the case of an exponential growth factor

$$a(t) = e^{H(t-t_0)}, \quad t \geq t_0, \tag{2}$$

are typical in cosmology, and also provide a suitable description of growth processes in many biological systems.[5,19,20]

In the uniformly expanding domain, the sojourn probability density function $P(\mathbf{y}, t)$ of a Brownian particle whose (unbiased) diffusive motion is intrinsic (in the sense that it is not coupled to the deterministic domain growth) evolves according to the following equation (see, e.g., Refs. 13 and 21):

$$\frac{\partial P(\mathbf{y}, t)}{\partial t} = -\frac{\dot{a}}{a} \nabla \cdot [\mathbf{y} P(\mathbf{y}, t)] + D\nabla^2 P(\mathbf{y}, t), \tag{3}$$

where (as usual) D denotes the diffusivity of the particle, hereafter assumed to be constant for simplicity. In the particular case of a $1d$

domain, one has

$$\frac{\partial P(y,t)}{\partial t} = -\frac{\dot{a}(t)}{a(t)}P(y,t) - \frac{\dot{a}(t)}{a(t)}y\,\frac{\partial P(y,t)}{\partial y}P(y,t) + D\frac{\partial^2 P(y,t)}{\partial y^2},$$

$$(4)$$

where the first two terms on the right-hand side arise from the domain growth. The first one describes the particle dilution effect, whereas the second term accounts for the drift experienced by the particle even in the absence of diffusion.

Next, let us define $Q(x,t) = a(t)P(y = a(t)x,t)$ and use this definition in Eq. (4) to obtain

$$\frac{\partial Q(x,t)}{\partial t} = \frac{D}{a^2}\frac{\partial^2 Q(x,t)}{\partial x^2},$$

$$(5)$$

i.e., the original problem is hereby cast into a pure diffusion problem with a time-dependent diffusivity (scaled Brownian motion). This time dependence can be removed by introducing the transformed time variable

$$\tau(t) = \int_{t_0}^{t}\frac{ds}{a^2(s)}.$$

$$(6)$$

By analogy with the (standard or ballistic) conformal time defined by the equation $\dot{\tau}_c = 1/a$ in the context of cosmology, the term "Brownian conformal time" was coined to denote τ in Ref. 13. In what follows, we shall use this term (or the equivalent term "scaled time") to refer to this quantity.

As discussed in Refs. 13, 16, 17 and 22, for a sufficiently fast expansion $\tau_\infty \equiv \tau(t \to \infty)$ becomes *finite*. Consider, e.g., the case of an exponentially growing scale factor. One has

$$\tau(t) = \frac{1}{2H}[1 - e^{-2H(t-t_0)}].$$

$$(7)$$

For a power-law scale factor one obtains

$$\tau(t) = t_0^{2\gamma}\frac{t^{-2\gamma+1} - t_0^{-2\gamma+1}}{1 - 2\gamma}, \qquad \gamma \neq 1/2,$$

$$(8a)$$

$$\tau(t) = t_0 \ln\left(\frac{t}{t_0}\right), \quad \gamma = 1/2. \tag{8b}$$

As we can see, one has $\tau_\infty < \infty$ in the exponential case with $H > 0$ and in the power-law case when $\gamma > 1/2$.

In terms of the Brownian conformal time τ and of the comoving coordinate x, Eq. (5) becomes identical with the standard diffusion equation

$$\frac{\partial Q(x,\tau)}{\partial \tau} = D\frac{\partial^2 Q(x,\tau)}{\partial x^2}. \tag{9}$$

The solution of Eq. (9) is well-known for a large family of boundary value problems.[23] The corresponding solution in physical space can be obtained by means of the inverse transformation $P(y,t) = Q[y/a(t), \tau(t)]/a(t)$. The so-called propagator solution corresponds to free boundary conditions and to the initial condition $P(y, t_0) = \delta(y)$. Taking into account that $a(t_0) = 1$ and $\tau(t_0) = 0$, one finds $Q(x, 0) = \delta(x)$, corresponding to the well-known Gaussian solution

$$Q_G(x, \tau) = \frac{1}{\sqrt{4\pi D\tau}} e^{-x^2/4D\tau} \tag{10}$$

in the comoving space and to the solution $G(y, t)$

$$G(y, t) = \frac{1}{a(t)} Q_G\left[\frac{y}{a(t)}, \tau(t)\right] = \frac{1}{\sqrt{4\pi D a^2(t)\tau(t)}} e^{-y^2/[4Da(t)^2\tau(t)]} \tag{11}$$

in physical space.

Note that a first consequence of the solution given by Eq. (10) is that, in the case of a sufficiently fast expansion (that is, an expansion with finite τ_∞), the overlap of two Brownian pulses initially separated by a distance $x_0 \gg \sqrt{2D\tau_\infty}$ becomes negligible as both pulses tend to their final frozen states.[13] If one switches to a comoving reference frame, one of the pulses becomes a stationary point, whereas the broadening of the other pulse in this reference frame is increased due to the fact that its diffusivity is doubled $(D \to 2D)$. Computing the probability of first encounter between the two Brownian particles represented by the pulses amounts to making the aforementioned stationary point absorbing, and then solving the corresponding first-passage problem for the remaining pulse. However, it is clear from the present discussion that such an encounter may never happen, as opposed to the standard case of a $1d$ static domain. We shall revisit this problem in more detail at the end of Subsection 3.2.1. But first, we shall study the problem of escape of a Brownian particle from a growing hypersphere.

3. Escape and Target Problems in Growing Domains

3.1. *Escape from a growing hypersphere*

3.1.1. *Survival probability*

Let us assume that a Brownian particle is placed at the center of a hypersphere of expanding radius $R_y = R_y(t) = a(t)R_0$, where R_0 denotes the initial radius. We seek the probability $S(t)$ that the particle has *not* escaped from the expanding region defined by the hypersphere up to time t. We can solve this problem by making the surface of the hypersphere fully absorbing and by identifying the escape process (surface crossing) with absorption. This justifies the use of the term "survival probability" for $S(t)$.

Because of the symmetry of the problem, Eq. (3) [formulated in polar coordinates for $P(r_y, \Omega_y, t)$] can be immediately integrated over the angular variables Ω_y to obtain a diffusion equation describing the time evolution of the density $P(r_y, t)$ associated with the probability of finding the particle within a shell of inner radius r_y and outer radius $r_y + dr_y$. The resulting diffusion equation is then subject to the delta-peaked initial condition $P(r_y, t = t_0) = s_d(r_y)^{-1}\delta_+(r_y)$ [where $s_d(r_y) = 2\pi^{d/2}r_y^{d-1}/\Gamma(d/2)$ is the surface of a hypersphere of radius r_y] and to the absorbing boundary condition $P(r_y = R_y, t) = 0$. Additionally, one has the implicit condition that $P(r_y, t)$ must remain finite everywhere at all times. The notation $\delta_+(\cdot)$ used above stands for the slightly modified delta-function with the property $\int_0^R \delta_+(r)dr = 1$ for any $R > 0$. From the solution $P(r_y, t)$ of the above problem, the survival probability follows immediately as $S(t) = \int_0^{R_y} P(r_y, t)\, s_d(r_y)dr_y$.

Proceeding as in the previous section, it is possible to cast the integrated version of Eq. (3) into a simpler form by introducing comoving coordinates $|\mathbf{x}| \equiv r_x = r_y/a(t)$, as well as a new function Q defined by the substitution $P(r_y, t) = Q[r_x, \tau(t)]/a(t)$. The resulting equation, $\partial Q/\partial \tau = D\nabla^2 Q$, is the d-dimensional generalization of Eq. (9) for the case of a hyperspherical geometry, i.e.,

$$\frac{\partial Q(r, \tau)}{\partial \tau} = D \left\{ \frac{\partial^2}{\partial r^2} + \frac{d-1}{r} \frac{\partial}{\partial r} \right\} Q(r, \tau). \tag{12}$$

In Eq. (12), we have set $r_x \equiv r$; this notation will be used throughout the remainder of the present section. Taking into account the equations $\tau(t_0) = 0$ and $a(t_0) = 1$ as well as the initial and boundary conditions for $P(r_y, t)$, one finds $Q(r, \tau = 0) = s_d(r)^{-1}\delta_+(r)$ and $Q(r = R_0, \tau) = 0$. In addition, the normalization condition $\int_0^{R_0} P(r, t_0) s_d(r) dr = \int_0^{R_0} Q(r, \tau = 0) s_d(r) dr \equiv 1$ must be satisfied. The well-known solution of the above problem (easily found by separation of variables) reads as follows[24,25]:

$$Q(r, \tau) = \sum_{n=1}^{\infty} \left(\frac{j_n}{2R_0} \right)^{d/2-1} \frac{r^{1-d/2}}{\pi^{d/2} R_0^2 J_{d/2}^2(j_n)} J_{d/2-1} \left(\frac{j_n r}{R_0} \right) e^{-j_n^2 D\tau/R_0^2}, \tag{13}$$

where $j_n \equiv j_{d/2-1,n}$ is the nth positive zero of the Bessel function of order $d/2 - 1$, i.e., $J_{d/2-1}(j_{d/2-1,n}) = 0$. For simplicity, we use the short-hand notation j_n; however, the reader should keep in mind that j_n depends on the spatial dimension d.

In terms of Q, the survival probability $S(t) = S[\tau(t)]$ is expressed as

$$S(\tau) = \int_0^{R_0} Q(r, \tau) s_d(r) dr = \frac{2^{2-d/2}}{\Gamma(d/2)} \sum_{n=1}^{\infty} \frac{j_n^{d/2-2}}{J_{d/2}(j_n)} e^{-j_n^2 D\tau/R_0^2}. \tag{14}$$

We see that the behavior of $S(t)$ depends on how $\tau(t)$ behaves. For example, if $\tau(t \to \infty) \equiv \tau_\infty \neq 0$, then there is a non-zero probability $S(\tau_\infty)$ that a particle is never trapped. This is the case for the previously defined power-law expansion with $\gamma > 1/2$, which gives $\tau_\infty = t_0/(2\gamma - 1)$ (cf. Eq. (8a)). On the other hand, if $\tau_\infty = \infty$, then $S(\tau_\infty) = 0$, i.e., the particle is trapped with certainty. This is the case, for example, for the power-law expansion with $\gamma \le 1/2$. A previous derivation of the above results for $d = 1, 2, 3$ has been given in Refs. 11 and 12. It is interesting to note the similarity of these results with the case of an expanding $1d$ cage,[26] where the boundaries of a symmetric interval $[-L(t), L(t)]$ are shifted outwards according to the long-time power law $L(t) \propto t^\gamma$. In this case, the physical space does not expand, only the boundaries move. However, the crossover exponent between certain death and a finite probability of eternal survival is also $\gamma = 1/2$. For a rather general treatment of problems involving the first-passage time of a Markov process to moving barriers the interested reader is referred to Ref. 27.

3.1.2. Power-law expansion: Moments of the first-passage time

For a power-law domain growth, the moments of the first-passage time $\langle t^n \rangle$ display an interesting behavior. These moments can be easily computed from the first-passage time distribution $F(t) = -dS(t)/dt$. The mth-order moment is

$$\langle t^m \rangle = \int_{t_0}^\infty dt\, F(t) t^m, \quad m = 0, 1, 2, \ldots, \tag{15}$$

or, integrating by parts,

$$\langle t^m \rangle = -t^m S(t)|_{t_0}^\infty + m \int_{t_0}^\infty dt\, t^{m-1} S(t). \tag{16}$$

Depending on the value of γ, we shall consider different subcases for the power-law scale factor $a(t) = (t/t_0)^\gamma$.

- Case $\gamma < 1/2$

    In this case, Eq. (14) gives

$$S(t) = \frac{2^{2-d/2}}{\Gamma(d/2)} \sum_{n=1}^\infty \frac{j_n^{d/2-2}}{J_{d/2}(j_n)} \exp\left[-j_n^2 D t_0^{2\gamma} \frac{t^{1-2\gamma} - t_0^{1-2\gamma}}{(1-2\gamma) R_0^2} \right]. \tag{17}$$

This can be expressed in a more compact way as follows:

$$S(t) = \sum_{n=1}^\infty \rho_n e^{-\alpha_n (t^{1-2\gamma} - t_0^{1-2\gamma})}, \tag{18}$$

where the quantities $\rho_n = 2^{2-d/2} j_n^{d/2-2}/[\Gamma(d/2) J_{d/2}(j_n)]$ and $\alpha_n = j_n^2 D R_0^{-2} t_0^{2\gamma}/(1 - 2\gamma)$ have been introduced. For $t = t_0$, the series $S(t_0) = \sum_{n=1}^{\infty} \rho_n$ is divergent for $d \geq 3$. This singularity in the initial condition is well-known from the analogous diffusion problem in static domains; however, we know that the normalization condition $\sum_{n=1}^{\infty} \rho_n \equiv 1$ must be satisfied in all dimensions. Fortunately, one can recover the physical value of the above divergent series by a suitable regularization procedure. To this end, a technique akin to Abel summation[28] can be applied. This technique relies on suitable regulator functions involving Bessel functions.[25] In particular, $\sum_{n=1}^{\infty} \rho_n$ is just $2^{2-d/2}/\Gamma(d/2)$ times the series denoted by $S(d/2 - 1, 0)$ in Ref. 25, which is equal to $2^{d/2-2}\Gamma(d/2)$ [see the result below Eq. (26) in Ref. 25].

When $\gamma < 1/2$, $S(t) \to 0$ for $t \to \infty$ and Eq. (16) becomes

$$\langle t^m \rangle = t_0^m + m \int_{t_0}^{\infty} t^{m-1} S(t) dt, \tag{19}$$

since, by construction, $S(t_0) \equiv 1$. Taking Eq. (18) into account, one obtains

$$\langle t^m \rangle = t_0^m + m \sum_{n=1}^{\infty} \rho_n e^{\alpha_n t_0^{1-2\gamma}} \int_{t_0}^{\infty} t^{m-1} e^{-\alpha_n t^{1-2\gamma}} dt. \tag{20}$$

The above expression for $\langle t^m \rangle$ can be rewritten in terms of incomplete Gamma functions. One has

$$\langle t^m \rangle = t_0^m + \frac{m}{1-2\gamma} \sum_{n=1}^{\infty} \rho_n \alpha_n^{-\frac{m}{1-2\gamma}} e^{\alpha_n t_0^{1-2\gamma}} \Gamma\left(\frac{m}{1-2\gamma}, \alpha_n t_0^{1-2\gamma}\right). \tag{21}$$

Note that, since we have assumed $\gamma < 1/2$, one has $\alpha_n > 0$. On the other hand, for a fixed value of the spatial dimension one has $j_n \to [n + (d-3)/4]\pi$ for large n according to McMahon's asymptotic expansion.[29] Furthermore, for fixed order and large values of the argument, the following asymptotic expansion of the Bessel function holds[29]:

$$J_\nu(z) \sim \sqrt{\frac{2}{\pi z}} \cos\left(z - \frac{\nu\pi}{2} - \frac{\pi}{4}\right), \quad |z| \to \infty, \tag{22}$$

implying that $J_{d/2}(j_n) \to (-1)^{n-1}[2/(\pi^2 n)]^{1/2}$ for $n \to \infty$. Using the large-x approximation $\Gamma(a,x) \sim x^{a-1}e^{-x}$, the series expansion (21) is found to converge for arbitrary $m > 0$ in one, two, and three dimensions.

- *Case $\gamma = 1/2$*

Taking $\tau = t_0 \ln(t/t_0)$ in Eq. (14) and using the definition of ρ_n, we obtain

$$S(t) = \sum_{n=1}^{\infty} \rho_n \left(\frac{t_0}{t}\right)^{\eta_n}, \tag{23}$$

with $\eta_n \equiv j_n^2 D t_0 / R_0^2$. For $t > t_0$, this series tends to zero as $t \to \infty$ in any spatial dimension. Let us now examine the behavior of the moments of the first-passage time. In this case, Eq. (16) becomes

$$\langle t^m \rangle = -\sum_{n=1}^{\infty} \left(1 + \frac{m}{\eta_n - m}\right) \rho_n t_0^{\eta_n} t^{m-\eta_n} \Bigg|_{t=t_0}^{t=\infty}, \tag{24}$$

when $m \neq \eta_n$. Since the η_n's increase monotonically with n, it is necessary and sufficient that the condition $\eta_1 = j_1 D t_0 / R_0^2 > m$ holds for $\langle t^m \rangle$ to be finite. When this is the case, the upper boundary term vanishes and one finally obtains

$$\langle t^m \rangle = t_0^m + m \sum_{n=1}^{\infty} \rho_n \frac{t_0^m}{\eta_n - m}. \tag{25}$$

For large n, one has $\eta_n \propto j_n \propto n$, and $\rho_n \propto (-1)^{n-1} n^{(d-3)/2}$. Hence, the above series converges in one, two, and three dimensions for any $m > 0$. The condition $\eta_1 > 1$ implies that the diffusion coefficient must exceed a threshold value,

$$D > \frac{1}{j_{d/2-1,1}} \frac{R_0^2}{t_0} \tag{26}$$

for the mean first-passage time $\langle t \rangle$ to exist. We have restored the full notation for the j_n's to emphasize the dependence on dimensionality. In more general terms, if the condition $m + 1 \geq \eta_1 > m$ holds, i.e., if

$$\frac{m+1}{j_{d/2-1,1}} \frac{R_0^2}{t_0} \geq D > \frac{m}{j_{d/2-1,1}} \frac{R_0^2}{t_0} \tag{27}$$

holds, the mth moment of the diffusion coefficient is still finite, but neither the $m+1$th moment nor higher order moments exist in one, two, or three dimensions. Note that, when $\eta_1 = m$, the m-th moment diverges logarithmically.

• *Case $\gamma > 1/2$*

As we have seen in 3.1.1, one has $S(\tau_\infty) > 0$ with $\tau_\infty = t_0/(2\gamma - 1)$. Thus, since $S(t \to \infty) \neq 0$, one concludes that neither the mean first-passage time nor any moments of positive integer order exist.

3.2. *The target problem*

In this subsection, we shall study the survival probability of an immobile target surrounded by one or more traps. Traditionally, the term "target problem" makes reference to the case of multiple traps, but here we shall also apply it to the single-trap case, which is the first setting we shall be dealing with.

3.2.1. *The target problem with a single trap*

Consider a hyperspherical, impenetrable target centered at the origin whose radius $R_y = a(t)R_0$ grows at the same rate as the embedding domain. A point trap is initially placed at a distance $r_0 > R_0$ from the target. At time $t = t_0$, the trap starts moving diffusively, and if it ever hits the target surface both the target and the trap are assumed to disappear. We seek the probability that the trap has not collided with the target up to time t, i.e., the survival probability $S_{\text{target}}(t) = S_{\text{trap}}(t) \equiv S(t)$.

If one assumes that the initial distribution of the trap position does not depend on the angular variables, Eq. (12) for the radial motion of the trap in comoving space holds, i.e.,

$$\frac{\partial Q(r, r_0, \tau)}{\partial \tau} = D \left\{ \frac{\partial^2}{\partial r^2} + \frac{d-1}{r} \frac{\partial}{\partial r} \right\} Q(r, r_0, \tau), \qquad (28)$$

where the additional argument r_0 in Q emphasizes the dependence on the initial position of the particle. The initial condition is formally written as $Q(r, r_0, \tau = 0) = s_d(r)^{-1} \delta(r - r_0)$. Additionally, one has the boundary conditions

$$Q(r = R_0, r_0, \tau) = 0, \qquad (29a)$$

$$\lim_{r \to \infty} Q(r, r_0, \tau) = 0. \qquad (29b)$$

The survival probability of both target and trap is given by the formula $S(R_0, r_0, \tau) = \int_{R_0}^{\infty} Q(r, r_0, \tau) s_d(r) \, dr$. In terms of the adjoint diffusion equation, the relevant boundary value problem is formulated as follows:

$$\frac{\partial S(R_0, r_0, \tau)}{\partial \tau} = D \nabla_{r_0}^2 S(R_0, r_0, \tau), \qquad (30a)$$

$$S(R_0, r_0, 0) = 1, \qquad (30b)$$

$$S(R_0, R_0, \tau) = 0, \qquad (30c)$$

$$\lim_{r_0 \to \infty} S(R_0, r_0, \tau) = 1. \qquad (30d)$$

The solution of the above problem in the case of a static domain has a closed form in Laplace space, which also leads to exact expressions valid for arbitrary times in $d = 1$ and $d = 3$.[30–32] For the case of a growing domain, the solution is obtained by performing the replacement $t \to \tau$ in these expressions. One then has

$$S(R_0, r_0, \tau) = 1 - \left(\frac{R_0}{r_0} \right)^{\alpha} \operatorname{erfc} \left(\frac{r_0 - R_0}{2\sqrt{D\tau}} \right), \qquad (31)$$

where $\alpha = 0$ for $d = 1$ and $\alpha = 1$ for $d = 3$. In the $d = 1$ case, one has

$$S(0, r_0, \tau) \equiv S(r_0, \tau) = \text{erf}\left(\frac{r_0}{2\sqrt{D\tau}}\right) \tag{32}$$

in the limit $R_0 \to 0$ of a point target. Note that, for a sufficiently fast expansion ($\tau_\infty < \infty$), the trap is spared with a finite probability $\text{erf}\left(\frac{r_0}{2\sqrt{D\tau_\infty}}\right)$. In contrast, the death of the trap is of course a certain event in the case of a static domain.

Alternatively, Eq. (32) can be obtained by noting that the trap density is given by the Green's function

$$Q_{1d}^{R_0=0}(r, r_0, \tau) = G_D(r, r_0, \tau)$$
$$\equiv \frac{1}{(4\pi D\tau)^{1/2}} \left\{\exp\left[-\frac{(r-r_0)^2}{4D\tau}\right] - \exp\left[-\frac{(r+r_0)^2}{4D\tau}\right]\right\}, \quad r > 0, \tag{33}$$

which is easily obtained by the method of images.[7] The integral $S(r_0, \tau) = \int_0^\infty G_D(r, r_0, \tau)\, dr$ then yields the desired probability. Alternatively, one may write

$$S(r_0, \tau) = 1 - \int_0^\tau d\tau'\, F(0, r_0, \tau') = 1 - \int_0^\tau d\tau'\, D\, \partial_r G_D(r, r_0, \tau')\big|_{r=0}, \tag{34}$$

where $F(0, r_0, \tau')$ stands for the first-passage density.

3.2.2. The target problem with multiple traps

One can now take advantage of the $1d$ result (32) to study the survival probability of a point target surrounded by a sea of diffusing, non-interacting point traps that are initially randomly scattered on the real axis. Assuming that the trap concentration is equal to c, the survival probability $S_{mt}(\tau)$ of the target in the multiple-trap problem is[32–36]:

$$S_{mt}(\tau) = \exp\left[c\int_0^\infty dr_0\, (1 - S(r_0, \tau))\right] = \exp[-4(Dc^2\tau/\pi)^{1/2}]. \tag{35}$$

The above result implies that, for a fast expansion, a non-zero final survival probability $\exp[-4(Dc^2\tau_\infty/\pi)^{1/2}]$ is obtained even in the case of multiple traps.

4. A First-Passage Problem Approach to the Kinetics of the $A + A \to A$ Reaction on a $1d$ Growing Domain

In what follows we shall take advantage of the results in 3.2.1 to deal with the problem of the coalescence reaction $A + A \to A$ on a $1d$ growing domain. The following discussion is based on the initial work by Doering and ben-Avraham,[14] and on a subsequent summary of the main results given by Redner in his monograph.[7] These two works refer to the case of a static domain, but we shall see that the applied methodology also works in the case of a growing domain.

Let us first recall the method of solution for the case of a static domain. The main idea is to reduce to the multiple-body problem to an effective one-body problem. This is easily done without loss of generality by considering an asymmetric version of the model in which a diffusing particle is killed when it meets its left nearest neighbor, whereas the particle continues to survive if it meets its right neighbor. To obtain the survival probability of an arbitrary particle, one only needs to consider a particle and its left neighbor, since the former will remain unaffected by collisions with particles to its right. Because both the particle and its left neighbor have diffusivity D, computing the survival probability of the particle amounts to solving the problem of a single particle with diffusivity $2D$ that moves in the presence of a stationary absorbing point.

The above arguments remain valid for a growing domain; the only difference is that the above first-passage problem must be formulated in terms of the comoving coordinate x and of the scaled time τ (see discussion at the end of Section 2 and see also Section 3.2.1). We now proceed to give some details.

4.1. *General expressions*

The density function $\rho(x, \tau)$ associated with the probability that the separation between the two particles (measured in comoving coordinates) equals x at a scaled time τ obeys an ordinary diffusion equation with diffusivity $2D$. Besides, it satisfies the boundary conditions $\rho(0, \tau) = \rho(\infty, \tau) = 0$. The solution of this problem is easily extrapolated from the static case and involves the Green's function already introduced in Section 3.2.1:

$$\rho(x, \tau) = \int_0^\infty dx' G_{2D}(x, x', \tau)\rho(x', 0). \tag{36}$$

The survival probability $S(\tau(t))$ is just the probability that the interparticle distance remains positive, i.e.,

$$S(\tau(t)) = \int_0^\infty dx\, \rho(x, \tau) = \int_0^\infty dx' \left[\int_0^\infty dx\, G_{2D}(x, x', \tau) \right] \rho(x', 0)$$

$$= \int_0^\infty dx'\, \mathrm{erf}(x'/(2\sqrt{2D\tau}))\, \rho(x', 0), \tag{37}$$

where Eq. (32) has been used in the last step. The average concentration (number of particles per unit length in comoving space) is then simply

$$c(\tau(t)) = c_0\, S(\tau(t)), \tag{38}$$

whereas the IPDF or interparticle distribution function $p(x, \tau)$ (distance distribution of the surviving particles in the many-body problem) is

$$p(x, \tau) = \rho(x, \tau)/S(\tau). \tag{39}$$

4.2. *Discussion of the main results*

4.2.1. *Behavior of the survival probability*

For the case of a periodic initial distribution $p(x, 0) = \delta(x - 1/c_0)$, Eq. (37) trivially yields

$$S_{\mathrm{per}}(\tau(t)) = \mathrm{erf}\left(\frac{1}{\sqrt{8Dc_0^2\tau(t)}} \right), \tag{40}$$

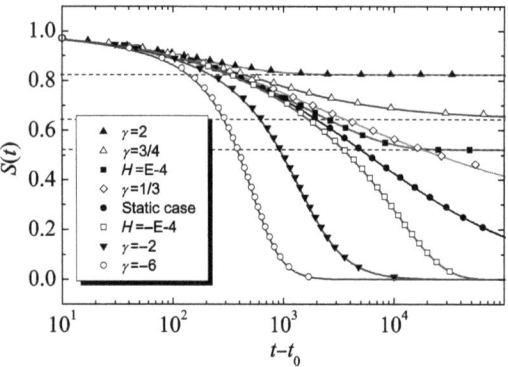

Fig. 1. Survival probability versus $t - t_0$ for the coalescence reaction. Solid lines represent theoretical results (given by Eq. (41)), whereas symbols are simulation results. Here we have used 10^5 particles with $c_0 = 1/100$ and $D = 1/2$. Different uniform expansions of the domain are considered, i.e., power-law expansions with $t_0 = 10^3$ and $\gamma = \{2, 3/4, 1/3, -2, -6\}$, exponential expansions with $H = \{10^{-4}, -10^{-4}\}$. Results for the static case ($\gamma = 0$) are also given. Dashed lines correspond to the asymptotic value S_∞ for $\gamma = 2$ ($S_\infty \approx 0.823$), $\gamma = 3/4$ ($S_\infty \approx 0.644$) and $H = 10^{-4}$ ($S_\infty \approx 0.523$).

whereas for a fully random initial condition $\rho(x,0) = p(x,0) = c_0 \exp(-c_0 x)$, one finds

$$S_{\text{rand}}(\tau(t)) = \exp\left[2c_0^2 D\,\tau(t)\right]\text{erfc}(\sqrt{2c_0^2 D\,\tau(t)}). \qquad (41)$$

Figure 1 shows the behavior of the survival probability for a given initial concentration c_0 of randomly distributed particles and different types of exponential and power-law expansions. In those cases where the expansion is sufficiently fast (finite τ_∞), the survival probability tends asymptotically to a constant value S_∞ which strongly depends on the domain growth rate. This is completely at odds with the fluctuation-dominated long-time decay to the empty state observed in the case of a static domain ($S(t) \approx (2\pi Dt)^{-1/2}$). The expressions derived previously remain valid in the case of a contracting domain ($\gamma < 0$ or $H < 0$); as one can easily anticipate, a contracting domain leads to an accelerated decay to the empty state.

Displayed in Fig. 2 is the time evolution of the survival probability for the case of an exponential domain growth. Both a periodic and a

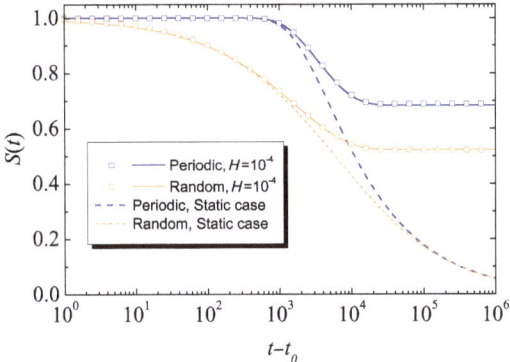

Fig. 2. Survival probability for the case of an exponential domain growth. Here, we have taken $c_0 = 0.01$ and $D = 1/2$.

fully random initial condition with the same value of c_0 are considered. As one can see, the final value of the survival probability turns out to depend not only on c_0, but also on the details of the initial particle arrangement. Most reactive events take place in the short-time regime, and here the high particle clustering induced by a fully random initial condition increases the encounter rate and decreases the survival probability with respect to the case of a periodic initial distribution. Therefore, $S_{\mathrm{per}} > S_{\mathrm{rand}}$. These strong memory effects arising from the lack of mixing are in sharp contrast with the results for the static case, where the asymptotic long-time decay completely loses the memory of the initial condition (see both dashed lines in Fig. 2). The above results do not come entirely as a surprise, since for a static domain, similar memory effects are observed for immobile reactants with short-range interactions, again as a consequence of poor mixing.[37]

4.2.2. *Behavior of the IPDF*

Let us now briefly discuss the behavior of the IPDF $p(x, \tau(t))$ when $\tau_\infty < \infty$. We first recall the main findings for the static case. Starting from the exponential form characteristic of a random initial condition, a gap of the scaled IPDF $c(\tau(t))^{-1} p(x, \tau(t))$ as a function of the rescaled distance $z = c(\tau(t)) x$ progressively develops

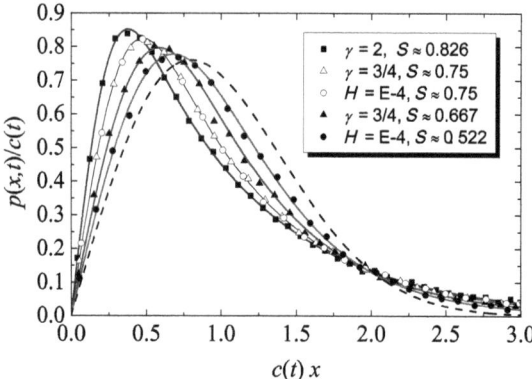

Fig. 3.   Scaled IPDF versus $c(t)x$ for coalescence and three different expansions
with $\tau_\infty < \infty$. The three thick lines from left to right near the origin depict
the limiting IPDFs, respectively, obtained for $t \to \infty$ in two different cases of
power-law expansion, i.e., $\gamma = 2$ ($S_\infty \approx 0.823$), $\gamma = 3/4$ ($S_\infty \approx 0.645$), as
well as for an exponential expansion with $H = 10^{-4}$ ($S_\infty \approx 0.523$). The thin
line is the IPDF for a time at which $S(t) = 0.75$, whereas the dashed line
is the long-time asymptotic IPDF for the static case. The symbols are simulation results for (i) power-law expansions with $\gamma = 2$ and a time t such
that $S(t) \approx 0.826$, $\gamma = 3/4$ and $S(t) \approx 0.75$, $\gamma = 3/4$ and $S(t) \approx 0.667$
(with $t_0 = 10^3$ in all cases). (ii) Exponential expansions where $H = 10^{-4}$ and
$S(t) \approx 0.75$, and $H = 10^{-4}$ and $S(t) \approx 0.522$. In all the cases depicted by
thick lines, the curves are very close to the theoretical prediction for the corresponding stationary long-time curves; we have also checked within our numerical
accuracy that the corresponding values of $S(t)$ are very close to the theoretical
values of S_∞.

as the reactions take place, indicating a dynamic self-ordering process characterized by the long-time asymptotic form $p(z, t \to \infty) \approx$
$\pi(z/2) \exp\{-\pi(z/2)^2\}$. This scaling form has a maximum at a characteristic value of the scaling variable z and is represented by the
dashed line in Fig. 3. However, if one now considers sufficiently fast
expansions with $\tau_\infty < \infty$, the scaling form is never reached, and the
IPDF freezes into a final state characterized by a displacement of
the maximum of the IPDF to shorter distances (smaller z-values)
and thus by a smaller gap (*cf.* Fig. 3). In other words, the deviation from the initial exponential distribution (which corresponds to
maximum disorder) becomes less pronounced.

4.2.3.   Power-law growth: Exact versus approximate solution

In the case of a power-law growth rate [*cf.* Eqs. (8a) and (8b)], it is interesting to consider the behavior of the concentration $\hat{c}(t) = c(t)/a(t)$ in physical space. For a random initial condition, Eq. (41) yields

$$\hat{c}(t) = \begin{cases} b_<(\gamma, D)\, t^{-1/2} & \text{if} \quad 0 < \gamma < 1/2, \\ b_=(t_0, D)\, [t \ln(t)]^{-1/2} & \text{if} \quad \gamma = 1/2, \\ b_>(c_0, \gamma, t_0, D)\, t^{-\gamma} & \text{if} \quad \gamma > 1/2, \end{cases} \tag{42}$$

with suitably defined constants $b_<$, $b_>$ and $b_=$. At the level of the decay exponent, agreement with the case of a static domain holds only as long as $\gamma \le 1/2$, with a logarithmic correction for $\gamma = 1/2$. In contrast, for $\gamma > 1/2$, the behavior is dominated by the dilution effect of the expansion.

This behavior can now be compared with an approximation obtained by adding to the right-hand side of the standard mean-field equation a term describing the global dilution effect induced by the domain growth. The resulting equation is

$$\frac{d\hat{c}_{\text{app}}(t)}{dt} = -\frac{\dot{a}(t)}{a(t)}\,\hat{c}_{\text{app}}(t) - \alpha[\hat{c}_{\text{app}}(t)]^2. \tag{43}$$

In the long-time limit, this yields

$$\hat{c}_{\text{app}}(t) = \begin{cases} b'_>(\gamma, \alpha)\, t^{-1} & \text{if} \quad 0 < \gamma < 1, \\ [\alpha t \ln(t)]^{-1} & \text{if} \quad \gamma = 1, \\ b'_<(c_0, \gamma, t_0, \alpha)\, t^{-\gamma} & \text{if} \quad \gamma > 1, \end{cases} \tag{44}$$

with suitably defined coefficients $b'_>$ and $b'_<$, implying that the agreement of the decay exponent with that of the exact solution is only reached for $\gamma \ge 1$.

5.   Summary and Outlook

The first-passage properties of diffusing particles in the presence of absorbing boundaries can be used to study the kinetics of chemical reactions localized at interfaces, or even to study the kinetics of cooperative processes characterized by the occurrence of reactive events at random locations. As we have seen, such properties are strongly altered when the embedding domain grows at a sufficiently fast rate.

In the first example considered (a diffusing particle initially localized at the center of a growing hypersphere with a fully absorbing surface), the behavior of the long-time survival probability and of the moments of the first-passage time distribution depends critically on the growth rate of the sphere. A rich phenomenology is observed in the case of a power-law growth, implying that the particle has a finite survival probability when $\gamma > 1/2$ (or, equivalently, when $\tau_\infty < \infty$). In this case, the moments of positive integer order do not exist, whereas in the marginal case $\gamma = 1/2$ the existence of the mth order moment depends on the diffusivity D of the particle, on the initial radius R_0 of the growing sphere, and on the initial time t_0.

The second example considered is the computation of the survival probability of a hyperspherical target which expands with the growing domain. The target is killed when hit by a diffusing point trap initially located at a certain distance (for practical purposes, the trap is also assumed to die). Already in one dimension, the target has a finite probability of eternal survival for a sufficiently fast domain growth (once again implying $\tau_\infty < \infty$). The behavior is thus markedly different from the case of a static $1d$ domain, in which the target is known to die with certainty due to the recurrence properties of the Brownian motion performed by the trap. Moreover, when $\tau_\infty < \infty$, the target still has a non-zero survival probability when it is initially surrounded by a sea of non-interacting diffusing traps scattered at random along the real axis.

The result for the target problem with a single trap turns out to have important implications for the $1d$ kinetics of the $A + A \to A$ reaction, which in the case of an expanding domain may still be effectively reduced to a first-passage problem for a single particle in the presence of a trapping point. Once again, the behavior depends on whether the expansion is fast enough to make τ tend to a finite value at long times. If that is the case, the behavior is radically different from the case of a static domain; the system tends to a non-empty state characterized by a survival probability which depends on the details of the initial condition, reflecting an enhancement of memory effects induced by the domain growth. In the case of a random initial condition, the self-ordering processes induced by the reactions stop prematurely, and the IPDF does not necessarily reach the asymptotic long-time scaling form characteristic of the static case.

In the specific case of a power-law growth rate, when the exponent γ is gradually increased, the $A + A$ system undergoes a dynamical phase transition from a slow long-time decay of the concentration in physical space to a fast decay essentially driven by the dilution effect associated with the domain growth. The decay for $\gamma < 1/2$ is proportional to $t^{-1/2}$, and the system is essentially diffusion-controlled as in the static case. For $\gamma > 1/2$, reactions become less frequent, and the dilution effect prevails. In the marginal case $\gamma = 1/2$, a logarithmic correction appears as a signature of the ultraslow decay of the survival probability.

In view of the above findings, one can easily anticipate that a subtle trade-off between diffusional mixing and particle separation effects induced by the domain growth may give rise to a very rich behavior in systems where both effects occur on comparable time scales. To account for such effects, an extension of the current "first-passage" picture of chemical kinetics will be needed.

Acknowledgments

This work was partially funded by MINECO (Spain) through Grant No. MTM2015-72907-EXP (C. E.) and No. FIS2016-76359-P (partially financed by FEDER funds) (S. B. Y. and E. A.), and by the Junta de Extremadura through Grant No. GR15104 and GR18079 (S. B. Y. and E. A.). F. L. V. acknowledges financial support from the Junta de Extremadura through Grant. No. PD16010 (FSE funds).

References

1. S. A. Rice, *Diffusion-limited Reactions*. Elsevier, Amsterdam (1985).
2. V. Privman, (ed.), *Nonequilibrium Statistical Mechanics in One Dimension*. Cambridge University Press, Cambridge (1997).

3. E. J. Crampin and P. K. Maini, Modelling biological pattern formation: The role of domain growth, *Comments Theor. Biol.* **6**(3), 229–249 (2001).

4. B. J. Binder, K. A. Landman, M. J. Simpson, M. Mariani, and D. F. Newgreen, Modeling proliferative tissue growth: A general approach and an avian case study, *Phys. Rev. E.* **78**(3), 031912 (2008).

5. J. D. Murray, *Mathematical Biology. Vol II: Spatial Models and Biomedical Applications.* Springer, Berlin (2003).

6. P. Hänggi, P. Talkner, and M. Borkovec, Reaction-rate theory: Fifty years after Kramers, *Rev. Mod. Phys.* **62**(2), 251–342 (1990).

7. S. Redner, *A Guide to First-passage Processes.* Cambridge University Press, New York (2001).

8. O. Bénichou, C. Chevalier, J. Klafter, B. Meyer, and R. Voituriez, Geometry-controlled kinetics, *Nat. Chem.* **2**, 472–477 (2010).

9. R. Metzler, S. Redner, and G. Oshanin, (eds.), *First-passage Phenomena and their Applications.* World Scientific, Singapore (2014).

10. D. S. Grebenkov, First exit times of harmonically trapped particles: A didactic review, *J. Phys. A.: Math. Theor.* **48**(1), 013001 (2015).

11. M. J. Simpson, J. A. Sharp, and R. E. Baker, Survival probability for a diffusive process on a growing domain, *Phys. Rev. E.* **91**(4), 042701 (2015).

12. M. J. Simpson and R. E. Baker, Exact calculations of survival probability for diffusion on growing lines, disks, and spheres: The role of dimension, *J. Chem. Phys.* **143**(9), 094109 (2015).

13. S. B. Yuste, E. Abad, and C. Escudero. Diffusion in an expanding medium: Fokker-Planck equation, Green's function, and first-passage properties, *Phys. Rev. E.* **94**(3), 032118 (2016).

14. C. Doering and D. ben-Avraham, Interparticle distribution functions and rate equations for diffusion-limited reactions, *Phys. Rev. A.* **38**(6), 3035–3042 (1988).

15. P. A. Alemany, Novel decay laws for the one-dimensional reaction-diffusion model as consequence of initial distributions, *J. Phys. A.: Math. Gen.* **30**(10), 3299–3311 (1997).

16. C. Escudero, S. B. Yuste, E. Abad, and F. Le Vot, *Reaction-diffusion kinetics in growing domains*, in Integrated Population Biology and Modeling, edited by A. S. R. Srinivasa Rao and C. R. Rao, Handbook of Statistics, Part A, Vol. 39 (Elsevier, Amsterdam, 2018). ISBN: 9780444640727.

17. F. Le Vot, C. Escudero, E. Abad, and S. B. Yuste, Encounter-controlled coalescence and annihilation on a one-dimensional growing domain, *Phys. Rev. E.* **98**, 032137, 1–15 (2018).

18. V. Berezinsky and A. Z. Gazizov, Diffusion of cosmic rays in the expanding universe, *I. Astrophys. J.* **643**(1), 8–13 (2006).

19. B. Ryden, *Introduction to Cosmology.* Addison-Wesley, Reading, PA (2003).

20. P. Gerlee, The model muddle: In search of tumor growth laws, *Cancer Res.* **73**(8), 2407–2411 (2013).

21. C. N. Angstmann, B. I. Henry, and A. V. McGann, Generalized fractional diffusion equations for subdiffusion in arbitrarily growing domains, *Phys. Rev. E.* **96**(4), 042153 (2017).

22. F. Le Vot, E. Abad, and S. B. Yuste, Continuous-time random-walk model for anomalous diffusion in expanding media, *Phys. Rev. E.* **96**(3), 032117 (2017).

23. H. S. Carslaw and J. C. Jaeger, *Conduction of Heat in Solids.* Oxford University Press, Oxford (1959).

24. R. Borrego, E. Abad, and S. B. Yuste, Survival probability of a subdiffusive particle in a d-dimensional sea of mobile traps, *Phys. Rev. E.* **80**(6), 061121 (2009).

25. S. B. Yuste, R. Borrego, and E. Abad, Divergent series and memory of the initial condition in the long-time solution of some anomalous diffusion problems, *Phys. Rev. E.* **81**(2), 021105 (2010).

26. P. L. Krapivsky and S. Redner, Life and death in an expanding cage and at the edge of a receding cliff, *Am. J. Phys.* **64**(5), 546–552 (1996).

27. H. C. Tuckwell and F. Y. M. Wan, First-passage time of Markov processes to moving barriers, *J. Appl. Prob.* **21**(4), 695–709 (1984).

28. G. H. Hardy, *Divergent Series.* Oxford University Press, Oxford (1949).

29. M. Abramowitz and I. A. Stegun, *Handbook of Mathematical Functions.* Dover, New York (1965).

30. A. V. Barzykin and M. Tachiya, Diffusion-influenced reaction kinetics on fractal structures, *J. Chem. Phys.* **99**(12), 9591–9597 (1993).

31. A. M. Berezhkovskii, D.-Y. Yang, S. H. Lin, Yu. A. Makhnovskii, and S.-Y. Sheu, Smoluchowski-type theory of stochastically gated diffusion-influenced reactions, *J. Chem. Phys.* **106**(17), 6985–6998 (1997).

32. S. B. Yuste and K. Lindenberg, Subdiffusive target problem: Survival probability, *Phys. Rev. E.* **76**(5), 051114 (2007).

33. R. A. Blythe and A. J. Bray, Survival probability of a diffusing particle in the presence of Poisson-distributed mobile traps, *Phys. Rev. E.* **67**(4), 041101 (2003).

34. A. J. Bray, S. N. Majumdar, and R. A. Blythe, Formal solution of a class of reaction-diffusion models: Reduction to a single-particle problem, *Phys. Rev. E.* **67**(6), 060102(R) (2003).

35. G. Oshanin, O. Bénichou, M. Coppey, and M. Moreau, Trapping reactions with randomly moving traps: Exact asymptotic results for compact exploration, *Phys. Rev. E.* **66**(6) 060101(R) (2002).

36. S. B. Yuste, G. Oshanin, K. Lindenberg, O. Bénichou, and J. Klafter, Survival probability of a particle in a sea of mobile traps: A tale of tails, *Phys. Rev. E.* **78**(2), 021105 (2008).

37. E. Abad, P. Grosfils, and G. Nicolis, Nonlinear reactive systems on a lattice viewed as Boolean dynamical systems, *Phys. Rev. E.* **63**(4), 041102 (2001).

Chapter 17

A Case Study of Thermodynamic Bounds for Chemical Kinetics

K. Proesmans*, L. Peliti†, and D. Lacoste‡,§

*Hasselt University, B-3590 Diepenbeek, Belgium
Collège de France, 75005 Paris, France
†Santa Marinella Research Institute,
00058 Santa Marinella (RM), Italy
‡Laboratoire de Physico-Chimie Théorique —
UMR CNRS Gulliver 7083,
PSL Research University, ESPCI,
10 rue Vauquelin, F-75231 Paris, France
§david.lacoste@espci.fr

In this chapter, we illustrate recently obtained thermodynamic bounds for a number of enzymatic networks by focusing on simple examples of unicyclic or multicyclic networks. We also derive complementary relations which constrain the fluctuations of first-passage times to reach a threshold current.

1. Introduction

There are generally several parameters which determine the performance of a thermodynamic system. One parameter is the average output flux delivered by the system, which is related to its output power. Another parameter is the dissipation, which may be viewed as a cost for operating the system. For the behavior of the system to be reliable and robust, one wants small fluctuations in the output flux while at the same time a small cost of operation. These goals are generally incompatible as emphasized by a trade-off known as thermodynamic uncertainty relations.[1,2] This trade-off constrains the fluctuations in product formation in enzyme kinetics[3] and can thus be used to infer information on the topology of the underlying chemical network[4,5] or to estimate the dissipation from the fluctuations of observed fluxes.[6] Suppressing fluctuations of an output flux is required to achieve some accuracy with brownian clocks,[7] while suppressing dissipation leads to an improvement in the thermodynamic efficiency of machines.[8-12] This trade-off, originally obtained for non-equilibrium steady states, holds in fact for systems at finite time[13,14] evolving in either continuous or discrete time.[15-17] It has also been adapted to Brownian motion,[18] non-equilibrium self-assembly,[19] active matter,[20] equilibrium order parameter fluctuations,[21] phase transitions,[22] and first-passage-time fluctuations,[23,24] and it continues today to generate many new applications or extensions.[25-30]

In this chapter, we focus on the implications of thermodynamic uncertainty relations for chemical kinetics. We illustrate the theoretical predictions by studying particle conversion fluxes and their fluctuations for both unicyclic and multicyclic chemical reactions. We also show the connection with the statistics of first-passage time, defined as the first time that a given number of particles has been converted.

This review is organized as follows. In Section 2, we illustrate bounds on the Fano factor for three examples of unicyclic networks, namely, the isomerization reaction, the Michaelis–Menten reaction, and the active catalysis. In Section 3, we study one example of a multicyclic network containing two cycles, which we call the misfolding reaction. In Section 4, we study complementary relations for the fluctuations of first-passage times. We conclude in Section 5.

2.   Bounds for Unicyclic Networks

2.1.   *The isomerization reaction*

Let us consider a single enzyme which can catalyze the transition between two isomers E_1 and E_2,

$$E_1 \underset{k^-}{\overset{k^+}{\rightleftharpoons}} E_2$$

with constant transition rates k^+ and k^-. This model can be mapped on a biased random walk, with a rate of forward jumps k^+ and of backward jumps k^-. The total displacement of this walker corresponds to the difference between the number of isomers E_1 converted into E_2 minus the number of reverse conversions. The master equation of this system is given by

$$\frac{d}{dt}P_n(t) = k^- P_{n+1}(t) - (k^+ + k^-)P_n(t) + k^+ P_{n-1}(t), \qquad (1)$$

where $P_n(t)$ is the probability to be at time t at the position n. The detailed balance relation takes the form

$$\frac{k^+}{k^-} = e^{\mathcal{A}}, \qquad (2)$$

where $\mathcal{A}$ is the dimensionless affinity given our choice of units, $k_B T = 1$.

In order to characterize the fluctuations of the variable n, we introduce the *generating function*

$$\Psi(\lambda, t) = \sum_n e^{\lambda n} P_n(t). \qquad (3)$$

Using the master equation (1), this generating function evolves according to the equation

$$\frac{d\Psi(\lambda, t)}{dt} = \theta(\lambda)\Psi(\lambda, t), \qquad (4)$$

where we have defined

$$\theta(\lambda) \equiv \lim_{t \to \infty} \frac{1}{t} \ln\langle e^{n\lambda} \rangle = k^- e^{\lambda} - (k^+ + k^-) + k^+ e^{-\lambda}. \qquad (5)$$

The mean and the variance of n can be expressed in terms of the derivatives of $\theta(\lambda)$ at $\lambda = 0$ as follows:

$$J = \lim_{t \to \infty} \frac{\langle n \rangle}{t} = \theta'(0) = k^+ - k^-, \tag{6}$$

$$D = \lim_{t \to \infty} \frac{\langle n^2 \rangle - \langle n \rangle^2}{2t} = \frac{\theta''(0)}{2} = \frac{k^+ + k^-}{2}. \tag{7}$$

Using these expressions, Barato *et al.*[1] have derived an "uncertainty relation" involving the following measure of the precision of the fluctuating variable n:

$$\epsilon^2 = \frac{\langle n^2 \rangle - \langle n \rangle^2}{\langle n \rangle^2} = \frac{k^+ + k^-}{(k^+ - k^-)^2 t}. \tag{8}$$

For a duration t, the total energy cost C is the product of the entropy production rate by the time t, so $C = \mathcal{A}Jt$, where J is the average conversion rate introduced above. Then the product of this cost C by the relative uncertainty ϵ^2 can be expressed by means of the detailed balance relation, Eq. (2), as

$$C\epsilon^2 = \frac{2D\mathcal{A}}{J} = \mathcal{A} \coth\left(\frac{\mathcal{A}}{2}\right) \geq 2, \tag{9}$$

where the last inequality follows from a well-known property of hyperbolic tangent. Importantly, this relation expresses a trade-off between the precision quantified by ϵ and the cost quantified by C. Note that the inequality in Eq. (9) holds arbitrarily far from equilibrium and becomes saturated only in the linear regime close to equilibrium when $\mathcal{A} \to 0$.

2.2.   *The reversible Michaelis–Menten reaction*

We now consider another important unicyclic network, namely, the well-known Michaelis–Menten kinetics.[3] In this chemical network, a substrate S is transformed into a product P due to the presence of an enzyme E via the formation of an unstable complex ES

$$S + E \underset{k_1^-}{\overset{k_1^+}{\rightleftarrows}} ES \underset{k_2^-}{\overset{k_2^+}{\rightleftarrows}} P + E,$$

where k_1^+ is proportional to the substrate concentration and k_2^- is proportional to the product concentration. The local detailed balance relation is now given by

$$\frac{k_1^+ k_2^+}{k_1^- k_2^-} = e^{\mathcal{A}}. \tag{10}$$

We introduce $p_{\alpha,n}(t)$ as the probability to have the enzyme in the state $\alpha = 0, 1$ with $\alpha = 0$ representing the free state and $\alpha = 1$ the bound state, and with n molecules of P produced. This probability satisfies the master equations

$$\frac{dp_{0,n}}{dt} = k_1^- p_{1,n} + k_2^+ p_{1,n-1} - (k_1^+ + k_2^-)p_{0,n}, \tag{11}$$

$$\frac{dp_{1,n}}{dt} = k_1^+ p_{0,n} + k_2^- p_{1,n+1} - (k_1^- + k_2^+)p_{1,n}. \tag{12}$$

We again introduce generating functions associated with these probability distributions by

$$\Psi_\alpha(\lambda, t) = \sum_{n=-\infty}^{+\infty} e^{\lambda(n+\alpha/2)} p_{\alpha,n}(t). \tag{13}$$

By convention, a half integer value of n is assigned to states where the enzyme is bound, and an integer number when the enzyme is free. By transforming the master equation into an evolution equation for the generating function, we find

$$\frac{d}{dt}\begin{pmatrix}\Psi_0\\\Psi_1\end{pmatrix} = L(z)\begin{pmatrix}\Psi_0\\\Psi_1\end{pmatrix}, \tag{14}$$

with the evolution matrix

$$L(z) = \begin{pmatrix} -(k_1^+ + k_2^-) & z^{-1}k_1^- + zk_2^+ \\ zk_1^+ + z^{-1}k_2^- & -(k_1^- + k_2^+) \end{pmatrix}, \tag{15}$$

where $z = e^{\lambda/2}$. By the Perron–Frobenius theorem, there is a non-degenerate positive leading eigenvalue of L, which we denote $\theta(z(\lambda))$.

Explicit evaluation of this function yields

$$J = \left. \frac{d\theta}{d\lambda} \right|_{\lambda=0} = \frac{k_1^+ k_2^+ - k_1^- k_2^-}{k_1^+ + k_1^- + k_2^+ + k_2^-}, \tag{16}$$

$$D = \left. \frac{1}{2} \frac{d^2\theta}{d^2\lambda} \right|_{\lambda=0} = \frac{k_1^+ k_2^+ + k_1^- k_2^- - 2J^2}{2(k_1^+ + k_1^- + k_2^+ + k_2^-)}. \tag{17}$$

In the context of enzymatic kinetics, the *Fano factor* characterizes the fluctuations in the formation of product by the enzyme. It is defined as

$$F = \frac{2D}{J}. \tag{18}$$

Using the above expressions for J and D together with the detailed balance condition (10), we find that

$$F = \coth\left(\frac{\mathcal{A}}{2}\right) - \frac{2k_2^- k_1^- (e^{\mathcal{A}} - 1)}{(k_2^- + k_2^+ + k_1^- + k_1^+)^2}. \tag{19}$$

A lower bound for the Fano factor can be obtained by minimizing the right-hand side of Eq. (19) with respect to all the transition rates. The minimum is obtained when $k_1^+ = k_2^+ \equiv k^+$, and $k_1^- = k_2^- \equiv k^-$. Then, using the detailed balance condition, one proves the inequality[3]

$$F \geq \frac{1}{2} \coth\left(\frac{\mathcal{A}}{4}\right) \geq \frac{2}{\mathcal{A}}. \tag{20}$$

Since the relation $C\epsilon^2 = F\mathcal{A}$ holds generally, both Eqs. (9) and (20) are particular cases of a general inequality for $C\epsilon^2$ valid for a unicyclic enzyme containing N states. In their original work,[1] Barato *et al.* provided an inequality involving $\mathcal{A}$ and N/n_c where n_c, was defined as the number of consumed substrate molecules in each cycle. If one defines the Fano factor per molecule instead of per cycle, as we do here, there is no need for n_c, and the result is simpler to state: the Fano factor of a unicyclic enzyme containing N states is bound by an expression that only depends on $\mathcal{A}$ and N[7]

$$F \geq \frac{1}{N} \coth\left(\frac{\mathcal{A}}{2N}\right). \tag{21}$$

The example we gave previously of an isomerization reaction satisfies this relation with $N = 1$, while the Michaelis–Menten scheme does so with $N = 2$. In Fig. 1, we verify this bound by computing the Fano factor for 1000 random transition rates of the form $k_1^+ = 10^{2r_1 - 1}$ where r_1 is a uniform random number in $[0, 1]$, while keeping a fixed value of the affinity $\mathcal{A}$.

An affinity-independent bound follows from Eq. (21) by letting $\mathcal{A} \to \infty$[31,32]

$$F \geq \frac{1}{N}. \tag{22}$$

The limit $\mathcal{A} \to \infty$ is realized in practice as soon as at least one transition contributing to the affinity $\mathcal{A}$ becomes irreversible. Note that Eq. (22) can also be derived from an analysis of current fluctuations in a periodic 1D lattice.[33] It represents a central result of statistical kinetics, since it allows to estimate the number of states in an enzymatic cycle from measurements of fluctuations.[34]

As shown in Fig. 1, all the points are indeed above the bound $F_{\min} = 0.5$ in the case of Michaelis–Menten kinetics.

In order to explore further Fano factor bounds, we now move to more complex examples.

2.3.   *The active catalysis*

In this type of chemical reaction, the folding of the substrate molecule A into the product molecule B is accompanied by the hydrolysis of an ATP molecule. This reaction can be represented as a unicyclic network with four intermediate states E, E_1, E_2, and E_3 for the enzyme and two substrates A and B as shown in Fig. 2. The various

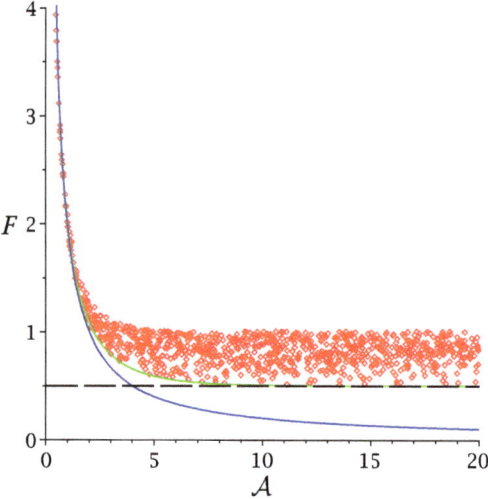

Fig. 1.   The Fano factor F as a function of the affinity $\mathcal{A}$ in the Michaelis–Menten kinetics. The black horizontal dashed line represents $F_{\min} = 0.5$, the blue line the bound $2/\mathcal{A}$, the green line is the hyperbolic bound of Eq. (20).

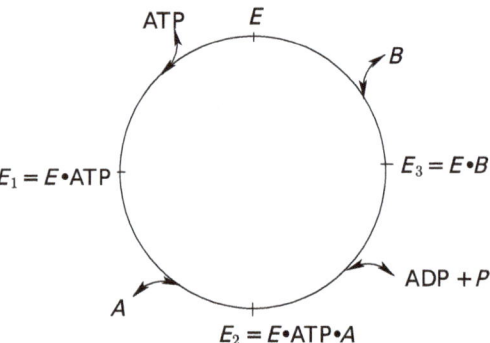

Fig. 2.   Cycle representation of the reactions involved in the active catalysis.

reactions are

$$\text{ATP} + E \mathrel{\substack{k_1^+ \\ \rightleftharpoons \\ k_1^-}} E_1, \; E_1 + A \mathrel{\substack{k_2^+ \\ \rightleftharpoons \\ k_2^-}} E_2,$$

$$E_2 \mathrel{\substack{k_3^+ \\ \rightleftharpoons \\ k_3^-}} E_3 + \text{ADP} + P, \; E_3 \mathrel{\substack{k_4^+ \\ \rightleftharpoons \\ k_4^-}} E + B \cdot$$

The local detailed balance condition takes the form

$$\frac{k_1^+ k_2^+ k_3^+ k_4^+}{k_1^- k_2^- k_3^- k_4^-} = e^{\mathcal{A}}, \qquad (23)$$

where the affinity of the cycle now reads $\mathcal{A} = \mu_A - \mu_B + \Delta\mu$, in terms of μ_A (resp. μ_B) the chemical potential of the substrate A (resp. B) and $\Delta\mu$ the chemical potential difference associated with the ATP hydrolysis reaction.

The framework of the previous section applies again here: now the evolution of the generating function is governed by a 4×4 matrix, and for this reason there is no simple analytic expression for its eigenvalues. Nevertheless, it is still possible to compute the corresponding currents and diffusion coefficients without having to obtain these explicitly by exploiting a method due to Koza as explained in Refs. 3, 35.

By following this method, we obtain the plot shown in Fig. 3. As shown in this figure, the bound satisfies the general property expected for a unicyclic enzyme given in Eq. (21) with $N = 4$ intermediate states.

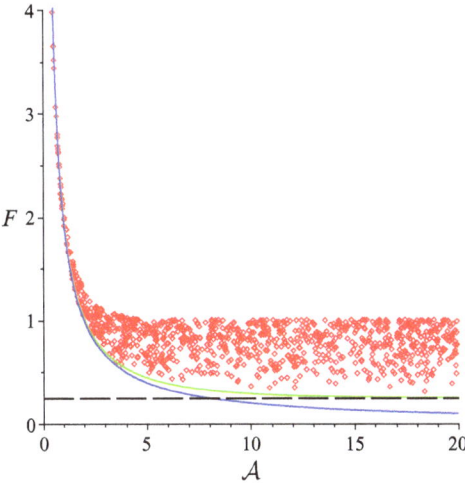

Fig. 3. The Fano factor F as a function of the affinity $\mathcal{A}$ for the active catalyst kinetics. The black horizontal dashed line represents $F_{\min} = 0.25$, the blue line the bound $2/\mathcal{A}$, the green line is the hyperbolic bound of Eq. (21) with $N = 4$.

3. Bounds for Multicyclic Networks: The Misfolding Reaction

We now switch to multicyclic reaction networks. As a simple example, we consider the misfolding reaction, which describes an enzyme that can make errors. More precisely, the enzyme can bind a molecule A and lead to the production of the "correct" molecule, say, B, or a "wrong" one, say C. This scheme represents a network with two cycles, characterized by the same free and bound state of the enzyme. The two possible reactions that can occur are

$$
A + E \underset{k_1^-}{\overset{k_1^+}{\rightleftharpoons}} E^* \underset{k_2^-}{\overset{k_2^+}{\rightleftharpoons}} B + E,
$$

$$
A + E \underset{k_1^-}{\overset{k_1^+}{\rightleftharpoons}} E^* \underset{k_3^-}{\overset{k_3^+}{\rightleftharpoons}} C + E.
$$

We now have two different affinities driving each one of these reactions, defined by

$$
\frac{k_1^+ k_2^+}{k_1^- k_2^-} = e^{A_1}, \quad \frac{k_1^+ k_3^+}{k_1^- k_3^-} = e^{A_2}. \tag{24}
$$

The master equations describe the evolution of the probability of being in the two different enzyme states (bound and free) as a function with two integer chemical variables n (resp. m), which represent the number of B (resp. C) produced since an arbitrary time. We then have

$$
\frac{dp_0(n, m, t)}{dt} = k_1^- p_1(n, m, t) + k_2^+ p_1(n - 1, m, t) + k_3^+ p_1(n, m - 1, t)
$$
$$
- (k_1^+ + k_2^- + k_3^-) p_0(n, m, t),
$$
$$
\frac{dp_1(n, m, t)}{dt} = k_1^+ p_0(n, m, t) + k_2^- p_0(n + 1, m, t) + k_3^- p_0(n, m + 1, t)
$$
$$
- (k_1^- + k_2^+ + k_3^+) p_1(n, m, t).
$$
$$
\tag{25}
$$

The generating function for this system can be defined by

$$
\Psi_\alpha(\overline{\lambda}, t) = \sum_{n,m} e^{\overline{\lambda} \cdot (n + \alpha/2, m + \alpha/2)} p_\alpha(n, m, t), \tag{26}
$$

where $\bar{\lambda}$ is a vector containing the two variables λ_1 and λ_2 associated to the degrees of freedom n and m, respectively. The evolution matrix that governs the dynamics of the generating function is given by

$$L(z_1, z_2) = \begin{pmatrix} -(k_1^+ + k_2^- + k_3^-) & (z_1 z_2)^{-1} k_1^- + \frac{z_2}{z_2} k_2^+ + \frac{z_2}{z_1} k_3^+ \\ z_1 z_2 k_1^+ + \frac{z_2}{z_1} k_2^- + \frac{z_1}{z_2} k_3^- & -(k_1^- + k_2^+ + k_3^+) \end{pmatrix}.$$

$$(27)$$

Again, the leading eigenvalue of $L(z_1, z_2)$, called $\Theta(z_1, z_2)$, allows us to obtain the currents J_1 and J_2 and their diffusion coefficients D_1 and D_2.

Using these parameters, we can compute two Fano factors F_1 and F_2, defined by $F_i = 2D_i/J_i$, where $i = 1, 2$. Similarly, we define the corresponding cost-fluctuations parameters $C\epsilon_i^2 = F_i \Sigma / J_i$, which instead contain the total entropy production rate $\Sigma = A_1 J_1 + A_2 J_2$.

The error-cost parameters $C\epsilon_i^2$ are constrained by the uncertainty relation[3]

$$C\epsilon_i^2 \geq \frac{\bar{A}}{2} \coth\left(\frac{\bar{A}}{4}\right) \geq \max\left(2, \frac{\bar{A}}{2}\right), \qquad (28)$$

where $\bar{A}$ is the minimum of the two cycle affinities A_1 and A_2.

We have verified numerically this inequality in the left panel of Fig. 4, which was constructed by generating random transition rates for 5000 iterations and then evaluating the error-cost parameters and the two different cycle affinities A_1 and A_2 using the detailed balance conditions (24).

In this case where the affinities A_1 and A_2 differ, we observe that a bound of the form of Eq. (21) does not hold in general for the two Fano factors F_i in terms of the affinities A_i or $\bar{A}$. In contrast to that, the two Fano factors F_i are always bound by $1/2$, which is the limit expected for unicycles with $N = 2$ states. This confirms that the bound in $1/N$ for the Fano factor — a central result in statistical kinetics — holds in general and is not just a consequence of the thermodynamic uncertainty relation.

In the particular case where the two affinities A_1 and A_2 are equal, then the bound of the form of Eq. (21) holds again for the two

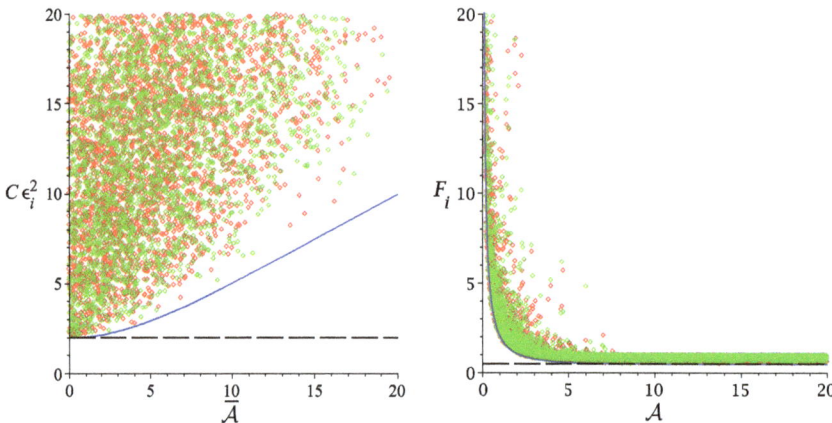

Fig. 4.   Left: Error-cost parameters $C\epsilon_i^2$ as a function of the minimum $\bar{\mathcal{A}}$ of the two different affinities $\mathcal{A}_1$ and $\mathcal{A}_2$ of the two cycles. The blue solid and black dashed line illustrate the bound of Eq. (28). Right: Fano factor F_i as function of the single affinity $\mathcal{A} = \mathcal{A}_1 = \mathcal{A}_2$. The blue solid and black dashed line illustrate the bound of Eq. (21). In both figures, red points represent $i = 1$ and green points $i = 2$.

Fano factors F_i as shown in the right panel of Fig. 4. Thus, we have illustrated the bound for the error-cost parameter which is linked to the uncertainty relation, but importantly we find that no affinity-dependent bound of this type exists for the two Fano factors except in the particular case where all the cycles have the same affinity.

4.   Bound on the Fluctuations of First-Passage Times

The Fano factor bounds derived above represent a general property of current fluctuations probed for a fixed observation time. It is also interesting to look at these results from a different point of view, which focuses instead on fluctuations of first-passage times to reach a threshold of time-integrated current.[24]

In this section, we derive the corresponding relations for the examples we have studied above. We rely on the so-called renewal equation to analyze first-passage times for Markovian jump processes. This equation connects the propagator $p(n, t)$, i.e., the probability of being in n at time t given that one starts from 0 at time 0, to the probability $F(n, t)$ to reach the state n for the first time at time t [36, p. 307]

$$p(n, t) = \delta_{n,0}\delta_{t,0} + \int_0^t d\tau\, F(n, \tau)p(0, t - \tau), \qquad (29)$$

where one assumes that the initial state is at $n = 0$ and that the process is translation invariant. Taking the Laplace transform of this equation gives

$$\tilde{F}(n, s) = \frac{\tilde{p}(n, s)}{\tilde{p}(0, s)}, \qquad (30)$$

with $\tilde{F}(n, s)$ (resp. $\tilde{p}(x, s)$) the Laplace transforms of $F(n, t)$ (resp. $p(n, t)$). Since $\tilde{F}(x, s)$ is the moment generating function of the first-passage time to reach n, which we now denote T, the first and second cumulants of T are

$$\langle T \rangle = -\frac{d}{ds} \ln\left(\tilde{F}(n, s)\right)\Big|_{s=0},$$

$$\mathrm{Var}(T) = \frac{d^2}{ds^2} \ln\left(\tilde{F}(n, s)\right)\Big|_{s=0}. \qquad (31)$$

The connection between current fluctuations probed for a fixed observation time and first-passage times fluctuations goes in fact beyond the first and second moments. Indeed, let us evaluate the cumulant generating function of the first-passage time T

$$g(s) = \lim_{n\to\infty} \frac{1}{n} \ln\left\langle e^{-sT} \right\rangle = \lim_{n\to\infty} \frac{1}{n} \ln \tilde{F}(n, s). \qquad (32)$$

One can then show that $g(s)$ is related to the cumulant generating function of the flux of n, which is the function we have denoted earlier by $\theta(s)$, by the following relation:[24]

$$g(s) = \theta^{-1}(s). \qquad (33)$$

4.1. *Isomerization relation*

Let us now verify these new relations, starting with the isomerization reaction. In this case, the equation obtained by Laplace transforming the master equation Eq. (1) admits a solution of the form $\tilde{p}_n(s) = \mathcal{N}\lambda(s)^n$ when $W^+ \geq W^-$ and $n \geq 0$, with

$$\lambda(s) = \frac{s + W^+ + W^- - \sqrt{(s + W^+ + W^-)^2 - 4W^+W^-}}{2W^+}. \tag{34}$$

Using Eq. (30), one obtains $\tilde{F}(n, s) = \lambda(s)^n$. Then with Eq. (31), one finds

$$\langle T \rangle = \frac{n}{W^+ - W^-}, \qquad \mathrm{Var}(T) = \frac{n(W^+ + W^-)}{(W^+ - W^-)^3}. \tag{35}$$

We recall that the energy cost C is related to the entropy production rate Σ by $C = t\Sigma = \mathcal{A}Jt$, with the affinity $\mathcal{A}$ defined in Eq. (2) and the average current J defined in Eq. (6). Then, one obtains the uncertainty relation complementary to Eq. (9)

$$\Sigma \frac{\mathrm{Var}(T)}{\langle T \rangle} = \mathcal{A}\coth\left(\frac{\mathcal{A}}{2}\right) \geq 2. \tag{36}$$

This relation means that fluctuations in first-passage times can be reduced only at the price of an increase of dissipation. Note that fluctuations of first-passage times are not constrained by dissipation when $W^+ \leq W^-$, because in this case the mean first-passage time is infinite.

The relation between the generating functions of first-passage time and current is also easily verified. Indeed, since $\tilde{F}(n, s) = \lambda(s)^n$,

$$g(s) = \ln\lambda(s). \tag{37}$$

Then, from the definition of the cumulant generating function of n introduced in Eq. (4), one finds

$$s = \theta[\theta^{-1}(s)] = W^+ e^{\theta^{-1}(s)} + W^- e^{-\theta^{-1}(s)} - (W^+ + W^-), \tag{38}$$

therefore,

$$\theta^{-1}(s) = \ln\left(\frac{s + W^+ + W^- - \sqrt{(s + W^- + W^+)^2 - 4W^-W^+}}{2W^+}\right),$$

$$\tag{39}$$

which is clearly equivalent to plugging Eq. (34) into Eq. (37) in agreement with Eq. (33).

4.2. *Michaelis–Menten reaction*

The first-passage time uncertainty relation can be validated in an analogous way for the Michaelis–Menten reaction. We first need to determine $\tilde{p}_{0/1,n}(n, s)$. This can be done by taking the Laplace transform of Eq. (12)

$$(s + k_1^+ + k_2^-)\tilde{p}_{0,n} = k_1^- \tilde{p}_{1,n} + k_2^+ \tilde{p}_{1,n-1}, \tag{40}$$

$$(s + k_1^- + k_2^+)\tilde{p}_{1,n} = k_1^+ \tilde{p}_{0,n} + k_2^- \tilde{p}_{0,n+1}. \tag{41}$$

Along the lines of the isomerization reaction, we assume that $\tilde{p}_{0/1,n}(s) = \mathcal{N}_{0/1}\lambda(s)^n$, leading to

$$\lambda(s) = \frac{s^2 + Ks + k^+ + k^- - \sqrt{(s^2 + Ks + k^+ + k^-)^2 - 4k^+k^-}}{2k^-},$$
$$\tag{42}$$

where we have introduced

$$K = k_1^+ + k_2^+ + k_1^- + k_2^-, \qquad k^- = k_1^- k_2^-, \qquad k^+ = k_1^+ k_2^+ \tag{43}$$

to simplify notations. This again leads to the first two cumulants

$$\langle T \rangle = \frac{Kn}{k^+ - k^-}, \quad \mathrm{Var}(T) = \frac{\left(K^2(k^+ + k^-) - 2(k^+ - k^-)^2\right)n}{(k^+ - k^-)^3}. \tag{44}$$

The thermodynamic uncertainty relation for first-passage times, as in Eq. (36), can now be verified easily.

As mentioned before, the cumulant generating function associated with the first-passage time is given by $g(s) = \ln \lambda(s)$, where $\lambda(s)$ is given by Eq. (42). One can invert this expression to determine the cumulant generating function $\theta(\mu)$ associated with the number of produced particles

$$\theta(s) = g^{-1}(s) = \frac{\sqrt{K^2 + 4(e^s - 1)(k^- - k^+e^{-s})} - K}{2}. \tag{45}$$

4.3.   *Misfolding reaction*

As a final example, we shall now turn to the misfolding reaction. This reaction network can be decomposed into two independent fluxes: the production of B molecules and the production of C molecules. Let us focus on the first-passage time to produce n molecules of B type. This problem can be mapped on the Michaelis–Menten reaction: indeed B is produced from E^* at a rate k_2^+ and produces E^* at a rate k_2^-. On the other hand, E^* is constructed from some other source (either A or C) at the rate $k'^+_1 = k_1^+ + k_3^-$ and deconstructed at rate $k'^-_1 = k_1^- + k_3^+$. Therefore, the system can be mapped onto a Michaelis–Menten system with k_1 replaced by k'_1. One concludes that Eqs. (42–44) also hold for the misfolding reaction, with the appropriate change of rates. Using the expression for the entropy production rate Σ determined in Section 3 leads to the thermodynamic uncertainty relation in the form

$$\Sigma \frac{\mathrm{Var}(T)}{\langle T \rangle} \geq 2. \qquad (46)$$

5.   Conclusion

In this chapter, we have illustrated a number of thermodynamic bounds for chemical kinetics and particularly for chemical cycles. In both unicyclic and multicyclic networks, we have confirmed the thermodynamic uncertainty relation which limits the precision that a chemical system can achieve for a given cost in terms of chemical dissipation. We have pointed out that only in unicyclic networks or in multicyclic networks subjected to a single affinity, there is a simple affinity-dependent bound. In contrast to that, there is always an affinity-independent bound for the Fano in terms of the inverse number of states, but this bound does not contain any trade-off.

Very recently, Gingrich and Horowitz reported a relation between the large deviation functions for currents and first-passage times in general Markov chains.[24] They also made an interesting connection between the thermodynamic uncertainty relation and first-passage time statistics. In this chapter, we have verified their result on our examples. In future work, we would like to explore this connection further, because it could be used in both ways: on the one hand, one could gain insights into currents fluctuations using results on first-passage time statistics, and on the other hand, one can understand better first-passage time statistics using large-deviation techniques, originally introduced for the analysis of current fluctuations in non-equilibrium systems.

Acknowledgments

LP acknowledges support from a Chair of the Labex CelTisPhysBio (Grant No. ANR-10-LBX-0038). KP was supported by the Flemish Science Foundation (FWO-Vlaanderen) travel grant (Grant No. V436217N).

References

1. A. C. Barato and U. Seifert, Thermodynamic uncertainty relation for biomolecular processes, *Phys. Rev. Lett.* **114**, 158101 (2015).
2. T. R. Gingrich, J. M. Horowitz, N. Perunov, and J. L. England, Dissipation bounds all steady-state current fluctuations, *Phys. Rev. Lett.* **116**, 120601 (2016).
3. A. C. Barato and U. Seifert, Universal bound on the Fano Factor in enzyme kinetics, *J. Phys. Chem. B.* **119**(22), 6555–6561 (2015).
4. P. Pietzonka, A. C. Barato, and U. Seifert, Universal bounds on current fluctuations, *Phys. Rev. E.* **93**(5), 052145 (2016).
5. P. Pietzonka, A. C. Barato, and U. Seifert, Affinity- and topology-dependent bound on current fluctuations, *J. Phys. A.: Math. Gen.* **49**(34), 34LT01 (2016).
6. T. R. Gingrich, G. M. Rotskoff, and J. M. Horowitz, Inferring dissipation from current fluctuations, *J. Phys. A.: Math. Gen.* **50**(18), 184004 (2017).
7. A. C. Barato and U. Seifert, Cost and precision of brownian clocks, *Phys. Rev. X.* **6**, 041053 (2016).
8. P. Pietzonka, A. C. Barato, and U. Seifert, Universal bound on the efficiency of molecular motors, *J. Stat. Mech.* **2016**(12), 124004 (2016).
9. M. Polettini, A. Lazarescu, and M. Esposito, Tightening the uncertainty principle for stochastic currents, *Phys. Rev. E.* **94**, 052104 (2016).
10. C. Maes, Frenetic bounds on the entropy production, *Phys. Rev. Lett.* **119**(16), 160601 (2017).

11. W. Hwang and C. Hyeon, Energetic costs, precision, and transport efficiency of molecular motors, *J. Phys. Chem. Lett.* **3**, 513–520 (2018).

12. H. Vroylandt, D. Lacoste, and G. Verley, Degree of coupling and efficiency of energy converters far-from-equilibrium, *J. Stat. Mech.* 023205 (2018).

13. P. Pietzonka, F. Ritort, and U. Seifert, Finite-time generalization of the thermodynamic uncertainty relation, *Phys. Rev. E.* **96**(1), 012101 (2017).

14. J. M. Horowitz and T. R. Gingrich, Proof of the finite-time thermodynamic uncertainty relation for steady-state currents, *Phys. Rev. E.* **96**, 020103 (2017).

15. S. Pigolotti, I. Neri, É. Roldán, and F. Jülicher, Generic properties of stochastic entropy production, *Phys. Rev. Lett.* **119**(14), 140604 (2017).

16. K. Proesmans and C. Van den Broeck, Discrete-time thermodynamic uncertainty relation, *Europhys. Lett.* **119**(2), 20001 (2017).

17. D. Chiuchiù and S. Pigolotti, Mapping of uncertainty relations between continuous and discrete time, *Phys. Rev. E.* **97**, 032109 (2018).

18. C. Hyeon and W. Hwang, Physical insight into the thermodynamic uncertainty relation using brownian motion in tilted periodic potentials, *Phys. Rev. E.* **96**(1), 012156 (2017).

19. M. Nguyen and S. Vaikuntanathan, Design principles for nonequilibrium self-assembly, *Proc. Natl. Acad. Sci. USA.* **113**(50), 14231–14236 (2016).

20. G. Falasco, R. Pfaller, A. P. Bregulla, F. Cichos, and K. Kroy, Exact symmetries in the velocity fluctuations of a hot brownian swimmer, *Phys. Rev. E.* **94**, 030602 (2016).

21. J. Guioth and D. Lacoste, Thermodynamic bounds on equilibrium fluctuations of a global or local order parameter, *Europhys. Lett.* **115**(6), 60007 (2016).

22. A. P. Solon and J. M. Horowitz, Phase transition in protocols minimizing work fluctuations, *Phys. Rev. Lett.* **120**, 180605 (2018).

23. J. P. Garrahan, Simple bounds on fluctuations and uncertainty relations for first-passage times of counting observables, *Phys. Rev. E.* **95**(3), 032134 (2017).

24. T. R. Gingrich and J. M. Horowitz, Fundamental bounds on first passage time fluctuations for currents, *Phys. Rev. Lett.* **119**, 170601 (2017).

25. G. M. Rotskoff, Mapping current fluctuations of stochastic pumps to nonequilibrium steady states, *Phys. Rev. E.* **95**(3), 030101 (2017).

26. A. Dechant and S.-i. Sasa, Current fluctuations and transport efficiency for general Langevin systems, *J. Stat. Mech.* 063209 (2018).

27. K. Brandner, T. Hanazato, and K. Saito, Thermodynamic bounds on precision in ballistic multiterminal transport, *Phys. Rev. Lett.* **120**(9), 090601 (2018).

28. S. K. Manikandan and S. Krishnamurthy, Exact results for the finite time thermodynamic uncertainty relation, *J. Phys. A.: Math. Gen.* **51**, 116701 (2018).

29. H. Wierenga, P. R. T. Wolde, and N. B. Becker, Quantifying fluctuations in reversible enzymatic cycles and clocks, *Phys. Rev. E.* **97**, 042404 (2018).

30. A. Dechant and S. Sasa, Entropic bounds on currents in langevin systems, *Phys. Rev. E.* **97**, 062101 (2018).

31. J. R. Moffitt, Y. R. Chemla, and C. Bustamante, Mechanistic constraints from the substrate concentration dependence of enzymatic fluctuations, *Proc. Natl. Acad. Sci. USA.* **107**(36), 15739–15744 (2010).

32. G. Knoops and C. Vanderzande, On the motion of kinesin in a viscoelastic medium, *Phys. Rev. E.* **97**, 052408 (2018).

33. B. Derrida, Velocity and diffusion constant of a periodic one-dimensional hopping model, *J. Stat. Phys.* **31**, 433 (1983).

34. J. R. Moffitt and C. Bustamante, Extracting signal from noise: Kinetic mechanisms from a Michaelis–Menten like expression for enzymatic fluctuations, *FEBS J.* **281**, 498 (2014).

35. Z. Koza, Maximal force exerted by a molecular motor, *Phys. Rev. E.* **65**(3), 031905 (2002).

36. N. Van Kampen, *Stochastic Processes in Physics and Chemistry*. North-Holland Personal Library, Amsterdam (2007).

Chapter 18

The Essential Role of Thermodynamics in Metabolic Network Modeling: Physical Insights and Computational Challenges

A. De Martino[*], D. De Martino[†], and E. Marinari[‡,§]

[*]*Soft & Living Matter Lab, CNR-NANOTEC, Rome, Italy*
Italian Institute for Genomic Medicine, Turin, Italy
[†]*Jozef Stefan Institute, Jamova 39, SI-1000, Ljubljana, Slovenia*
Institute of Science and Technology Austria, Am Campus 1,
A-3400 Klosterneuburg, Austria
[‡]*Sapienza Università di Roma, INFN Sezione di Roma 1 and*
CNR-NANOTECH, UOS di Roma, P.le A. Moro 2, 00185 Roma, Italy
[§]*enzo.marinari@roma1.infn.it*

Quantitative studies of cell metabolism are often based on large chemical reaction network models. A steady-state approach is suited to analyze phenomena on the timescale of cell growth and circumvents the problem of incomplete experimental knowledge on kinetic laws and parameters, but it should be supported by a correct implementation of thermodynamic constraints. In this chapter, we review the latter aspect, highlighting its computational challenges and physical insights. The simple introduction of Gibbs inequalities avoids the presence of unfeasible loops allowing for correct timescale analysis, but leads to possibly non-convex feasible flux spaces whose exploration needs efficient algorithms. We briefly review the implementation of thermodynamics through variational principles in constraint-based models of metabolic networks.

1. Introduction

Because of its uniquely universal nature, thermodynamics has been linked to physiology from its very inception, both to rationalize observations and to elucidate fundamental limits to physiological functions. With the advent of genetics and molecular biology, the discovery of the molecular mechanisms underlying physiology became the primary challenge. However, as the molecular actors and their interactions were mapped out at increasingly fine resolution, the focus gradually shifted to understanding their system-level organization.[1] And, perhaps unsurprisingly, it has become more and more clear that thermodynamic aspects are crucial for the emergent large-scale behavior of these systems. Currently, renewed interest has flourished around the thermodynamics of cellular processes, only this time with the possibility of relying on a host of data at various scales for quantitative analyses.[2]

In no area of physiology is thermodynamic analysis more central than in metabolic network modeling.[3] In brief, metabolic networks encode the set of chemical reactions that, in any cell, break down nutrients and harvest free energy to synthesize the macromolecular building blocks essential to life (amino acids, nucleotides, fatty acids, etc.) and, ultimately, biomass. Their structures can be inferred by combining gene–enzyme reaction associations with regulatory information and transcriptional data. The availability of detailed metabolic network reconstructions for a large number of organisms and cell types is, in our view, among the most significant successes that computational methods have reaped in biology to date.[4]

Building reliable and predictive dynamical models of metabolism based on this information is however challenging, mainly due to our vastly incomplete knowledge about intracellular enzyme kinetics, transport mechanisms, and rate constants. On the other hand, non-equilibrium steady-state approaches appear to be more feasible. Such methods are perhaps best represented by the broad class of computational schemes known as "constraint-based models".[5,6] From a physical viewpoint, such models should essentially rely on two "Kirchhoff-type" assumptions regarding (a) mass balance for chemical species (i.e., metabolic homeostasis) and (b) energy balance for reactions (i.e., thermodynamic feasibility of material fluxes).[7] In the most basic setup, energy balance simply requires reaction fluxes at steady state to proceed downhill in free energy, in accordance with the second law of thermodynamics. Unfortunately, implementing this constraint in genome-scale models is drastically harder than enforcing the stationarity of metabolite concentrations. Inclusion of thermodynamic constraints is however essential not only to obtain physically viable flux patterns, but also to highlight timescales and turnover rates and to allow for the estimation of metabolite concentrations. Needless to say, the range of applications of such results, from biotechnology to pharmacology, would be enormous.

Our main goal here is to present this problem and its multiple ramifications, which span from basic biochemistry to some fundamental algorithmic challenges, under a statistical physics lens. We shall discuss what, in practice, makes it so hard to solve efficiently and review some of the alternative approaches that have been attempted. Finally, we will point to some recent developments that may hold some promising keys to finally unlock the puzzle of metabolic network thermodynamics at the genome scale.

2. Background

In the most simple setting, metabolism can be modeled in terms of the dynamics of the chemical compounds concentration levels.[8] Upon assuming well-mixing and neglecting noise, we still have a large possibly nonlinear dynamical system whose parameters could be not known in their entirety. For a chemical reaction network in which M metabolites participate in N reactions with the stoichiometry encoded in a matrix $\mathbf{S} = \{S_{\mu i}\}$, the concentrations c_μ change in time according to mass-balance equations

$$\dot{\mathbf{c}} = \mathbf{S} \cdot \mathbf{v}, \tag{1}$$

where a component of the vector $\mathbf{v}$, v_i, is the flux of the reaction i that is in turn a (possibly unknown) function of the concentration levels

$v_i(\mathbf{c})$ (and several other parameters, like enzyme copy number, etc).
On the other hand, in order to analyze phenomena with timescales
longer than diffusion and typical turnover times (like cell growth),
it is possible to assume a steady state, i.e., a flux configuration
satisfying

$$\mathbf{S} \cdot \mathbf{v} = 0. \tag{2}$$

In so-called constraint-based modeling, apart from mass balance,
fluxes are bounded in certain ranges $v_r \in [v_r^{\min}, v_r^{\max}]$ that take into
account thermodynamic irreversibility, kinetic limits, and physiological constraints. The set of constraints

$$\mathbf{S} \cdot \mathbf{v} = 0, \quad v_r \in [v_r^{\min}, v_r^{\max}] \tag{3}$$

defines a convex closed set in the space of reaction fluxes: the
polytope of feasible steady states. The productive capabilities of the
network can be investigated computationally by maximizing suited
linear objective functions in the aforementioned space,[9] in particular
the biomass growth itself (flux balance analysis,[10] based on linear
programming). On the other hand, more generic inference problems
can be afforded quite efficiently with Monte Carlo methods given
the convexity of the space.[11] It should be noted that thermodynamics is implemented in a very simple way, i.e., by setting reaction
reversibility, i.e., $f_i \geq 0$ for some reactions. When flux bounds are
not provided, it is customary to set them to an arbitrary large number, $f_i \in [-C, C]$ (typically $C = 10^3, 10^4$), that, for a meaningful
model, will not influence the results. On the other hand, a more rigorous yet simple approach consists in postulating that fluxes shall
follow a free energy gradient. If $f_i \neq 0$, then $f_i \Delta G_i < 0$ (Gibbs
inequality), where ΔG_i is the free energy change of reaction i. The
ΔG's can be written as the difference between the chemical potentials g_μ of products and substrates through the stoichiometric matrix

$\Delta G_i = \sum_{\mu} S_{i\mu} g_{\mu}$. In terms of chemical potentials, we thus have a system of linear inequalities ($\xi_{i\mu} = -sign(f_i)S_{i\mu}$)

$$\sum_{\mu} \xi_{i\mu} g_{\mu} > 0 \quad \forall i, \tag{4}$$

whose feasibility (existence of a solution g_{μ}) is necessary for the thermodynamic feasibility of the flux configuration. Even in this basic approach, the addition of free energy variables (g_{μ} and ΔG_i) makes the problem nonlinear, in particular quadratic and possibly nonconvex. On the other hand, upon conditioning on flux variables, we can get useful hints from duality theorems of the alternatives that characterize thermodynamic feasibility in terms of the unfeasibility of particular flux configurations.[12] Specifically, according to the Gordan theorem, we have that system (4) has a feasible solution if and only if

$$\sum_{i} S_{i\mu} k_i = 0 \quad k_i \geq 0 \quad \forall \mu \tag{5}$$

has no non-trivial solutions (unfeasible loops). Such a duality can be exploited in order to define efficient algorithms, as we discuss in the following section.

3. Relaxational Algorithms

Reconstruction of complete metabolic networks is becoming, thanks to new, modern, and accurate experiments, a possible option for many simple organisms. What is important in the present context is that this reconstruction needs to be compatible with thermodynamic principles, and that implementing thermodynamic requirements can help in an accurate reconstruction of the network. This is sometimes a difficult computational problem, and we describe here a useful algorithm to implement thermodynamic consistency and use this to help in (correct) network reconstruction.[13]

In this way, we will gather information about Gibbs free energy and about their landscape, which will have to be compatible with the selected vector of reaction directions. We use stoichiometric information via a constructive algorithm inspired by perceptron[14] learning approach. In the method we use a preliminary reconstruction of the network structure to iteratively build up correlations between the chemical potentials of the chemical species, until we reach a thermodynamically consistent profile. The algorithm is nicely scalable, and it can allow the crucial result of guaranteeing the feasibility of flux configurations, or of identifying and removing unfeasible cycles. The algorithm can also be useful to get an estimate of reaction affinities, and it can be used to derive bounds for concentrations.

We consider the Gibbs energy at temperature T and volume V

$$G \equiv E - PV - TS, \tag{6}$$

where E is the internal energy of the systems, P its pressure and S the entropy. Let us call δ_i the direction of chemical reaction i: $\delta_i = \pm 1$, i.e., the reaction can proceed in the "forward" direction or in the reverse direction. If ΔG_i is the Gibbs energy difference induced by reaction i, one needs that $\delta_i \Delta G_i \leq 0 \ \forall i$ (the Gibbs energy cannot increase in the direction where the chemical reaction operates). Let us consider now the stoichiometric coefficients, that are $S_{\mu,j} < 0$ for substrates and $S_{\mu,j} > 0$ for products. Let us also define the vector $\boldsymbol{g} \equiv \{g_\mu\}$ of the Gibbs energies per mole of species μ. In terms of $\boldsymbol{g}$, we have that $\Delta G_i = S_{j,\mu}^T \, g_\mu$. Now we can discuss the following problem. Given a set of reaction directions $\{\delta_i\}$ (that have been inferred in the first step of the procedure) determine, *if it exists*, $\boldsymbol{g}$ such that

$$\Delta_i \equiv -\delta_i \sum_\alpha S_{j,\mu} g_\mu \geq 0 \, \forall i. \tag{7}$$

For fixed δ_i, the solution space is convex. Relaxation methods are a typical, and potentially effective choice, for solving a problem of this kind. If a solution does not exist, the reconstructed network is not consistent and it has, at best, to be cured.

To solve this problem, we have introduced[13] an algorithm based on the so-called MinOver approach,[14] which was originally developed for neural network learning. The algorithm starts from a configuration of the g_μ, that is extracted under the probability distribution $P_0(g)$. P_0 is selected a priori after phenomenological considerations. All the experimental input to the algorithm is indeed in the choice of P_0. We assume the simple ansatz

$$P_0(g) = \prod_{\alpha=1}^{M} P_0^\mu(g), \tag{8}$$

where P_0^μ is uniform around the values estimated from experiments with a range also suggested from experimental data. We now generate a random vector under $P_0(g)$ and we compute the vector $\Delta = \{\Delta_i\}$ from 7. Let us call i_0 the index of the most broken constraint, i.e., let us set

$$i_0 = \arg \min_i \Delta_i. \tag{9}$$

Now if $\Delta_{i_0} \geq 0$ g is a thermodynamically consistent chemical potential, and solves our problem, we can accept it and exit, or look for more solutions (including the one already found) by restarting the algorithm from a different seed. If instead $\Delta_{i_0} < 0$ g, we do not have a solution and we update g by setting

$$g \longrightarrow g - \lambda \delta_{i_0} S_{i_0}, \tag{10}$$

with λ an appropriate constant. One iterates till convergence, that is guaranteed (maybe after a very long, unpractical time) if a solution exists.

If the problem has no solutions, the assignment of the directions is not consistent. This happens if and only if there is at least one unfeasible loop. The main problem at this point is that it is not easy to find a loop that can be, in principle, also very long. We can phrase better the problem by saying that there is an unfeasible loop if there is a set $\mathcal{I}$ of reactions such that a set of positive constants $k_i > 0$ exists such that

$$\sum_{i \in \mathcal{I}} k_i \delta_i S_{\mu,i} = 0 \ \forall \mu.$$

If there is a loop, the algorithm does not converge, since the least satisfied constraint rotates on the loop and does not get fixed. If this is happening, we can start from our algorithm to localize and kill not too long loops in the following way. After discarding the first part of the iteration steps (where we typically are in a transient region) we start storing the values of $i_0(t)$ (where t labels the iterations of the procedure), i.e., the value of the most broken constraint at iteration t of the MinOver procedure. Now we look among the reactions appearing in the set of the $\{i_0(t)\}$ for $t > \tilde{T}$, where we can vary the minimum iteration $\tilde{T}$, and we search for loops of length $\mathcal{L}$, starting from $\mathcal{L} = 3$, and increasing it if needed. When we find a loop, we change one of the directions and try the MinOver procedure again. The loops in the dual space can be searched exhaustively or with aid of Monte Carlo methods.[15]

The possibility of improving the reconstruction of a network, making it compatible with thermodynamical basic principles, is important, and our algorithm helps in this direction.

4.   Flux Scales

It is interesting to note that scales analysis in metabolic networks need implementation of thermodynamics constraints beyond reversibility assignment. This leads to a geometrically more complex picture of the flux space, but its lack leads possibly to wrong conclusions. Previous work on sampling the flux space seemed to show *scale free* distributions,[16] in contradiction with the existence of physical limiting factors, e.g., resources availability[17] or maximum ribosome elongation rate.[18] As we stressed in the background, models of metabolic networks come with arbitrary bounds on the fluxes, upon which the solution space could depend, in turn hampering scaling analysis, in particular in presence of the aforementioned unfeasible loops. Suppose in fact that $f_{i,0}$ is a feasible state and k_i is a solution of (5), the line $f_i(L) = f_{i,0} + k_i s_i L$ verifies the steady-state mass balance constraints by construction and it will be inside the polytope till L possibly reaches the arbitrary bounds. A simple illustration is

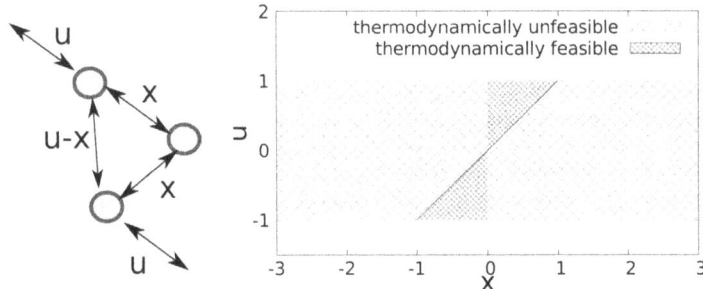

Fig. 1.   A simple three reaction network and its stationary flux space (x, u), $u \in [-1, 1]$ and $x \in [-3, 3]$: thermodynamics constraints impose $ux \geq 0$ $|x| \leq |u|$.

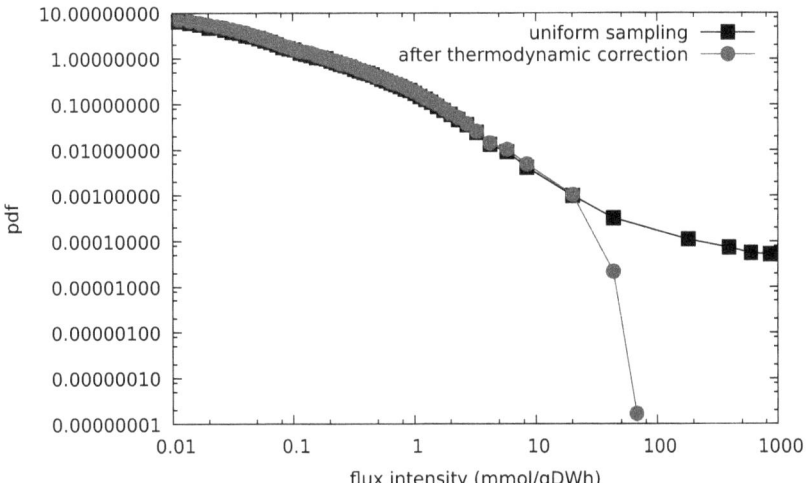

Fig. 2.   Flux intensity distribution (log-log) before and after removing thermo-dynamically unfeasible loops (*E.coli* model iJR904,[19] $R = 10^5$ configurations).

depicted in Fig. 1: A is injected with rate u, it can be transformed either in B and subsequently in C with rate x or directly in C with rate $u - x$, that is consumed with rate u. (if $u < 0$, C is injected and A consumed). The variable x is unbounded unless closed loops are forbidden, leading to the nonlinear constraints $ux \geq 0$, $|x| \leq |u|$. In general, thermodynamics forbids the orthants in the flux space that include closed loops: the remaining feasible space is not convex any-more (see Fig. 1), but its scales now reflect true physical constraints.

These issues have been studied in a genome scale metabolic network model, specifically the typical steady states of the *E. coli* metabolic network iJR904[19] in a glucose limited minimal medium in aerobic conditions.[20] Flux configurations have been uniformly sampled and corrected from unfeasible loops with the methods described in the previous section. Results for the distribution of flux intensities $|f_i|$ are shown in Fig. 2, before and after correcting for unfeasible loops. The long tail corresponding to the uniform sampling depends on the arbitrary constant fixed for flux bounds. After thermodynamic correction, the flux intensity distribution has a cut-off that scales simply with the glucose input.[20]

5.   Enhanced Turnover

What is the overall turnover time of metabolism? This question receives useful hints from thermodynamic information even if the knowledge of reactions kinetics is lacking. In particular, fluctuation analysis upon application of the fluctuation theorem returns a faster picture of metabolism with respect to standard turnover estimates.[21] In general and formally, relaxation times are calculated from a linear stability analysis of the steady states, but this requires detailed knowledge of reactions kinetics. On the other hand, if at least fluxes and concentrations are experimentally known, it is possible to calculate the metabolites turnover times τ, i.e., the ratio between the concentration c and the net flux of production P (or equivalently consumption D, given the steady state), schematically

$$\dot{c} = P - D = 0, \quad \tau = \frac{c}{P}, \tag{11}$$

where, e.g., the flux P can be calculated from the network and the flux configuration for a given metabolite μ (θ is the Heaviside step function)

$$P_\mu = \sum_i \theta(S_{i\mu} v_i) S_{i\mu} v_i. \tag{12}$$

Table 1. Turnover times of selected compounds in *E. coli* metabolism simulated for a genome scale model in aerobic glucose limited minimal environment, from net fluxes and corrected for fluctuations.

Metabolic compound	Turnover time estimated from net flux (s)	Turnover time corrected for fluctuations (s)
ATP	2.0 ± 0.1	0.4 ± 0.2
ADP	0.120 ± 0.005	0.02 ± 0.01
AMP	0.5 ± 0.1	0.11 ± 0.06
NAD	1.1 ± 0.1	0.3 ± 0.1
NADH	$3.5 \pm 0.2 \cdot 10^{-2}$	$1.0 \pm 0.4 \cdot 10^{-2}$
NADP	$1.6 \pm 0.2 \cdot 10^{-3}$	$2 \pm 1 \cdot 10^{-4}$
NADPH	$9 \pm 1 \cdot 10^{-2}$	$2 \pm 1 \cdot 10^{-2}$
Glutammate	90 ± 20	16 ± 8
3-Phosphoglycerate	2.0 ± 0.2	0.12 ± 0.6

Such turnover time is the typical time it takes to fully replenish a given metabolic pool. On the other hand, net fluxes result from the difference between forward and backward contributions $\nu = \nu^+ - \nu^-$, and the latter can be estimated by the fluctuation theorem if the free energy ΔG change is known

$$\frac{\nu^+}{\nu^-} = e^{-\Delta G/RT}. \tag{13}$$

Upon taking into account the backward contribution, the turnover time can be shorter, i.e., schematically

$$\dot{c} = (P^+ + D^-) - (D^+ + P^-) = 0,$$

$$\tau = \frac{c}{P^+ + D^-}. \tag{14}$$

For instance, consider Glucose-6-phosphate in the human red blood cell. This is produced by the Hexokinase enzyme ($\Delta G_1 \simeq -29$ KJ/mol), and consumed by the phosphoglucoisomerase enzyme ($\Delta G_2 \simeq -2.9$ KJ/mol). At $RT = 2.5$ KJ/mol, the turnover time τ_0 estimated only from net fluxes overestimates the one τ that takes into account backward contribution by a factor

$$\frac{\tau_0 - \tau}{\tau} \simeq \frac{1}{e^{-\frac{\Delta G_2}{RT}} - 1} \simeq 45\%. \tag{15}$$

Such analysis has been performed for the genome scale *E.coli* metabolic network iJR904[19] in a glucose limited minimal medium in aerobic conditions,[21] returning a faster picture of intermediate metabolism, that we summarize in Table 1, reporting the turnover times estimate from net and total fluxes of the metabolites ruling the energetics of the network.

6.   Variational Principles

An alternative route to implementing thermodynamic feasibility at network scale consists in devising global variational principles ensuring that optimal mass-balanced flux patterns are void of cycles. The simplest such principle is perhaps given by the minimization of the total flux

$$Q(\mathbf{v}) = \frac{1}{N} \sum_{i=1}^{N} v_i^2, \tag{16}$$

where it is understood that flux vectors $\mathbf{v} = \{v_i\}$ satisfy the mass balance conditions $\mathbf{Sv} = \mathbf{0}$ with predefined ranges of variability for fluxes (as well as the additional constraints that may be required on a case-by-case basis). An argument proving that Q is minimum for flux configurations that are thermodynamically viable (assuming such configurations exist for the network under study) is as follows. Consider a mass-balanced flux configuration $\mathbf{v}$ and assume it contains an infeasible cycle. Such a cycle must be described by a non-zero solution $\mathbf{k}$ to the system

$$\sum_i \Omega_i^\mu k_i = 0, \tag{17}$$

where $\Omega_i^\mu = -v_i S_i^\mu$. Now consider the flux configuration $\mathbf{w}$ defined by

$$w_i = v_i + \alpha k_i v_i, \tag{18}$$

with α a constant. Clearly, if $\mathbf{v}$ is mass-balanced, so is $\mathbf{w}$. However, $Q(\mathbf{v}) > Q(\mathbf{w})$ provided α is such that $\frac{\partial Q(\mathbf{w})}{\partial \alpha} = 0$. In particular, one finds

$$Q(\mathbf{v}) = Q(\mathbf{w}) + \frac{\left(\sum_i k_i v_i^2\right)^2}{\sum_i k_i^2 v_i^2}. \tag{19}$$

In other terms, given a thermodynamically infeasible mass-balanced flux configuration, it is always possible to construct another mass-balanced flux configuration whose total flux is lower. In turn, the resulting flux pattern has to be thermodynamically feasible when Q is minimized.

This idea, originally put forward in Ref. 22, has been applied in various computational schemes for genome-scale metabolism, such as pFBA (parsimonious Flux Balance Analysis),[23] CycleFreeFlux,[24] the global method to remove infeasible cycles from NESS flux configurations introduced in Ref. 15. Clearly, it is useful in practice whenever one is interested in finding a single feasible flux pattern as long as the minimum of Q lies within the solution space defined by mass balance constraints. A more integrative principle has been proposed in Ref. 25. It is most easily expressed by distinguishing forward (F) and reverse (R) directions for each flux, so that the net flux v_i can be written as $v_i = v_{i,F} - v_{i,R}$ (with $v_{i,F} \geq 0$ and $v_{i,R} \geq 0$), as well as exchange fluxes corresponding to sources or sinks of the reaction network $(v_{i,E})$. In brief, it states that the triplet $(\mathbf{v}_F^\star, \mathbf{v}_R^\star, \mathbf{v}_E^\star)$ satisfies a Flux Balance Analysis-like problem (i.e., is mass balanced and maximizes a given linear objective function) thermodynamically, provided it minimizes the functional

$$\mathcal{F}(\mathbf{v}_F, \mathbf{v}_R) = \mathbf{v}_F \cdot [\log(\mathbf{v}_F + \mathbf{c} - \mathbf{1})] + \mathbf{v}_R \cdot [\log(\mathbf{v}_R + \mathbf{c} - \mathbf{1})], \tag{20}$$

subject to

$$\mathbf{S}_I(\mathbf{v}_F - \mathbf{v}_R) + \mathbf{S}_E \mathbf{v}_E^\star = \mathbf{0}, \tag{21}$$

where $\mathbf{S}_I$ and $\mathbf{S}_E$ stand for the intracellular and exchange parts of the stoichiometric matrix, $\mathbf{c}$ is a generic vector in $\mathbb{R}^N$ and $\mathbf{1}$ is the vector with all entries equal to 1. In other terms, if $\mathbf{v}_E^\star$ is the vector of optimal exchange fluxes for the solution of an FBA problem, one can obtain a thermodynamically viable solution to the same FBA problem by minimizing $\mathcal{F}$. Importantly, the vector $\mathbf{g}$ of chemical potentials are related to the vector $\boldsymbol{\lambda}$ of Lagrange multipliers enforcing (21) by

$$\mathbf{g} = -2RT\boldsymbol{\lambda}. \tag{22}$$

This result follows from standard convex analysis (see Ref. 25 for details) and benefits from the standard conceptual and computational advantages of convex optimization problems with linear constraints (uniqueness of solution, efficient computational implementation). In addition, it is fully generic, in the sense that each thermodynamically viable mass-balanced flux vector must minimize $\mathbf{F}$ for a certain choice of $\mathbf{c}$. It therefore provides a rather transparent description of thermodynamic feasibility (as far as optimality is concerned). On the other hand, the existence of a free parameter constitutes a limitation, at least in part. The most serious drawback, in our view, however, derives from the fact that the above formulation does not allow to account for explicit constraints on *net* fluxes (as discussed in Ref. 25).

Thermodynamic arguments have also inspired different types of variational principles that effectively extend the reach of flux-based models by allowing to account for concentrations. For instance, in Ref. 26, feasible NESS are assumed to minimize the function

$$H = \sum_{\mu \in \text{Ext}} \frac{(u^\mu)^2}{c_{\text{ext}}^\mu} \tag{23}$$

over exchange fluxes $\mathbf{u}$ and intracellular fluxes $\mathbf{v}$, subject to

$$\mathbf{Sv} = \mathbf{u} \tag{24}$$

and with prescribed bounds of the form $u^\mu \in [u_{\text{min}}^\mu, u_{\text{max}}^\mu]$ and $v_i \in [v_{i,\text{min}}, v_{i,\text{max}}]$ for each extracellular compound μ and each intracellular reaction i. The explicit dependence of H on external concentrations (considered as fixed parameters) makes it possible to use the above principle to infer intracellular reaction rates given the levels of a set of extracellular metabolites. On the other hand, the minimization of (23) does not ensure that the resulting flux pattern is thermodynamically viable, unless in specific cases. The reader is referred to Ref. 27 for details.

7. Conclusions

Integrating thermodynamics with genome-scale biochemical reconstructions is perhaps the central theoretical open challenge of metabolic network modeling. Basically, two classes of approaches are currently being attempted that can mutually benefit from each other. On the one hand, novel empirical data and biochemical methods are employed to estimate more accurate standard-free energies for reactions and chemical potentials for metabolites.[28,29] Having better estimates of such quantities is crucial, especially if they cover a larger part of the reactome and of the metabolome than the ones currently available. On the other hand, optimization principles based on different physico-chemical arguments can be used to obtain approximate (but genome-scale) estimates, whose accuracy depends on the prior biochemical information as well as on the underlying assumptions. The key issue to be faced here is computational, relating both to the scalability of the algorithms required for the study of genome-scale networks and to the fact that thermodynamics may require the study of non-convex optimization problems. At the same time, ongoing work is uncovering how thermodynamics constrains flux patterns at steady state starting from the kinetics of individual enzymes, with results which suggest that the problem of computing thermodynamic potentials described here might have to be modified as more is known about individual reaction mechanisms, at least to some degree.[30] In this respect, it is the convergence of novel statistical physics,[31–34] biochemical and algorithmic ideas that will likely provide the tools to effectively tackle this challenge.

References

1. H. Kacser, The control of flux. *Symp. Soc. Exp. Biol.* **27**, 65–104, (1973).
2. R. A. Alberty, *Thermodynamics of Biochemical Reactions*. Wiley (2003).
3. M. Ataman and V. Hatzimanikatis, Heading in the right direction: thermodynamics-based network analysis and pathway engineering, *Curr. Opin. Biotech.* **36**, 176–182 (2015).
4. K. R. Patil, M. Åkesson, and J. Nielsen. Use of genome-scale microbial models for metabolic engineering, *Curr. Opin. Biotech.* **15**(1), 64–69 (2004).
5. B. O. Palsson, *Systems Biology: Properties of Reconstructed Networks*. Cambridge University Press, Cambridge, UK, 1st edn, (2006).
6. D.-A. Beard and H. Qian, *Chemical Biophysics*, Cambridge University Press, Cambridge, UK (2008).
7. D. Beard, S. Liang, and H. Qian, Energy balance for analysis of complex metabolic networks. *Biophys. J.* **83**(1), 79 (2002).
8. R. Heinrich and S. Schuster, *The Regulation of Cellular Systems*. Chapman & Hall, New York (1996).
9. R. A. Majewski and M. M. Domach, Simple constrained-optimization view of acetate overflow in *E. coli*. *Biotechnol. Bioeng.* **35**(7), 732–738 (1990).
10. J. Orth, I. Thiele, and B. O. Palsson, What is flux balance analysis? *Nat. Biotechnol.* **28**(3), 245–248 (2010).
11. D. De Martino, M. Mori, and V. Parisi, Uniform sampling of steady states in metabolic networks: Heterogeneous scales and rounding, *PLoS ONE* **10**(4), e0122670 (2015).
12. D. De Martino, Thermodynamics of biochemical networks and duality theorems. *Phys. Rev. E.* **87**(5), 052108 (2013).
13. D. De Martino, M. Figliuzzi, A. De Martino, and E. Marinari, A scalable algorithm to explore the gibbs energy landscape of genome-scale metabolic networks, *PLoS Comput Biol.* **8**(6), e1002562 (2012).
14. W. Krauth and M. Mezard, Learning algorithms with optimal stability in neural networks, *J. Phys. A.: Math. Gen.* **20**, L745 (1987).
15. D. De Martino, F. Capuani, M. Mori, A. De Martino, and E. Marinari, Counting and correcting thermodynamically infeasible flux cycles in genome-scale metabolic networks, *Metabolites* **3**(4), 946–966 (2013).
16. E. Almaas, B. Kovács, T. Vicsek, Z. N. Oltvai, and A.-L. Barabási, Global organization of metabolic fluxes in the bacterium *Escherichia coli*. *Nature* **427**(6977), 839 (2004).
17. J. Monod, The growth of bacterial cultures, *Ann. Rev. Microbiol.* **3**(1), 371–394 (1949).
18. H. Bremer and P. P. Dennis, Modulation of chemical composition and other parameters of the cell by growth rate. *Ecosal Plus.* **3**(1), (2008), doi: 10.1128/ecosal.5.2.3 (see http://www.asmscience.org/content/journal/ecosalplus/10.1128/ecosal.5.2.3).
19. J. L. Reed, T. D. Vo, C. H. Schilling, and B. O. Palsson, An expanded genome-scale model of *Escherichia coli* k–12 (i jr904 gsm/gpr), *Genome Biol.* **4**(9), R54, (2003).

20. D. De Martino, Scales and multimodal flux distributions in stationary metabolic network models via thermodynamics, *Phys. Rev. E.* **95**(6), 062419 (2017).

21. D. De Martino, Genome-scale estimate of the metabolic turnover of e. coli from the energy balance analysis, *Phys. Biol.* **13**(1), 016003 (2016).

22. H. G. Holzhütter, The principle of flux minimization and its application to estimate stationary fluxes in metabolic networks, *Eur. J. Biochem.* **271**(14), 2905–2922 (2004).

23. N. E. Lewis, K. K. Hixson, T. M. Conrad, J. A. Lerman, P. Charusanti, A. D. Polpitiya, J. N. Adkins, G. Schramm, S. O. Purvine, D. Lopez-Ferrer, *et al.* Omic data from evolved *E. coli* are consistent with computed optimal growth from genome-scale models, *Mol. Syst. Biol.* **6**(1), 390 (2010).

24. A. A. Desouki, F. Jarre, G. Gelius-Dietrich, and M. J. Lercher, Cyclefreeflux: Efficient removal of thermodynamically infeasible loops from flux distributions, *Bioinformatics* **31**(13), 2159–2165 (2015).

25. R. M. T. Fleming, C. M. Maes, M. A. Saunders, Yinyu Ye, and B. Ø. Palsson, A variational principle for computing nonequilibrium fluxes and potentials in genome-scale biochemical networks, *J. Theor. Bio.* **292**, 71–77 (2012).

26. D. De Martino, F. Capuani, and A. De Martino, Inferring metabolic phenotypes from the exometabolome through a thermodynamic variational principle, *New J. Phys.* **16**(11), 115018 (2014).

27. A. De Martino, D. De Martino, R. Mulet, and G. Uguzzoni, Reaction networks as systems for resource allocation: A variational principle for their non-equilibrium steady states, *PLoS ONE* **7**(7), e39849 (2012).

28. M. Jankowski, C. Henry, L. Broadbelt, and V. Hatzimanikatis, Group contribution method for thermodynamic analysis of complex metabolic networks, *Biophys. J.* **95**(3), 1487 (2008).

29. E. Noor, H. S. Haraldsdóttir, R. Milo, and R. M. T. Fleming. Consistent estimation of gibbs energy using component contributions, *PLoS Comput. Biol.* **9**(7), e1003098 (2013).

30. E. Noor, A. Bar-Even, A. Flamholz, E. Reznik, W. Liebermeister, and R. Milo, Pathway thermodynamics highlights kinetic obstacles in central metabolism, *PLoS Comput. Biol.* **10**(2), e1003483 (2014).

31. M. Polettini and M. Esposito, Irreversible thermodynamics of open chemical networks. i. emergent cycles and broken conservation laws, *J. Chem. Phys.* **141**(2), 07B610_1 (2014).

32. T. Schmiedl and U. Seifert, Stochastic thermodynamics of chemical reaction networks, *J. Chem. Phys.* **126**, 044101 (2007).

33. A. Wachtel, R. Rao, and M. Esposito, Thermodynamically consistent coarse graining of biocatalysts beyond Michaelis–Menten, *New J. Phys.* **20**(4), 042002 (2018).

34. R. Rao and M. Esposito, Conservation laws shape dissipation. *New J. Phys.* **20**(2), 023007 (2018).

Chapter 19

Kinetics of Protein–DNA Interactions: First-Passage Analysis

Maria P. Kochugaeva[*,§], Alexey A. Shvets[†,¶],
and Anatoly B. Kolomeisky[‡,∥]

[*]*Department of Biomedical Engineering and
System Biology Institute, Yale University,
West Haven, CT 06516, USA*
[†]*Institute for Medical Engineering and Science,
Massachusetts Institute of Technology,
Cambridge, MA 02142, USA*
[‡]*Department of Chemistry and Center for
Theoretical Biological Physics, Rice University,
Houston, TX 77005, USA*
[§]*maria.kochugaeva@yale.edu*
[¶]*shvets@mit.edu*
[∥]*tolya@rice.edu*

All living systems can function only far away from equilibrium, and for this reason chemical kinetic methods are critically important for uncovering the mechanisms of biological processes. Here we present a new theoretical method of investigating dynamics of protein–DNA interactions, which govern all major biological processes. It is based on a first-passage analysis of biochemical and biophysical transitions, and it provides a fully analytic description of the processes. Our approach is explained for the case of a single protein searching for a specific binding site on DNA. In addition, the application of the method to investigations of the effect of DNA sequence heterogeneity and the role of multiple targets and traps in the protein search dynamics are discussed.

1.   Introduction

One of the most striking features of living systems is their dynamic nature.[1,2] Biological processes involve time-dependent fluxes of energy and materials, which makes them strongly deviate from the equilibrium. This implies that concepts of equilibrium thermodynamics have limited applications for biological systems, while the role of chemical kinetic methods that study the dynamical transformations is significant.[3] In this chapter, we present a new theoretical method of investigating the complex mechanisms of chemical and biological phenomena, which is based on explicit calculations of dynamic properties via a first-passage analysis. The first-passage ideas have been already widely utilized in studies of various processes in chemistry, physics, and biology.[4,5] Our method employs these ideas in developing a discrete-state stochastic framework for analyzing the dynamics of protein–DNA interactions.

It is known that many biological processes start when protein molecules bind to specific target sequences on DNA to initiate cascades of biochemical reactions.[1,2] This fundamental aspect of protein–DNA interactions has been studied extensively by various experimental and theoretical methods.[6–8] Although a significant progress in understanding the protein search phenomena has been achieved, the detailed molecular mechanisms remain not well clarified. Furthermore, there are extensive theoretical discussions on how to explain the fast dynamics of the protein search for the targets on DNA, which is also known as a facilitated diffusion.[8]

A large amount of experimental observations, coming mostly from the single-molecule measurements, suggests that the protein search is a complex dynamic phenomenon which combines 3D (in the bulk solution) and 1D (on DNA) motions.[6–8] But the most paradoxical observation is that the protein molecules spend most of the search time ($\geq$90–99%) on the DNA chain where they diffuse very slowly, and they still find the targets very fast, in some cases faster than the bulk diffusion would allow.[6–8] Several theoretical ideas discussing the role of lowering of dimensionality, electrostatic effects, correlations between 3D and 1D motions, conformational transitions, bending fluctuations, and hydrodynamics effects have been proposed. However, the analysis shows that none of these mechanisms can fully explain the facilitated diffusion.[9] To investigate the mechanisms of protein–DNA interactions, we developed a discrete-state stochastic framework to take into account the most relevant chemical states and transitions in the system. The application of first-passage approach allows us to explicitly evaluate the dynamic properties and to clarify several important molecular aspects of protein–DNA interactions.

2. Discrete-State Stochastic Model of Protein Search for the Specific Target on DNA

To explain our method in detail, let us consider a simple model where one protein molecule diffuses through the bulk solution around the DNA molecule with occasional non-specific bindings to DNA (when the scanning along the chain is taking place) until the specific target site is located, as shown in Fig. 1.[9] In the discrete-state model the

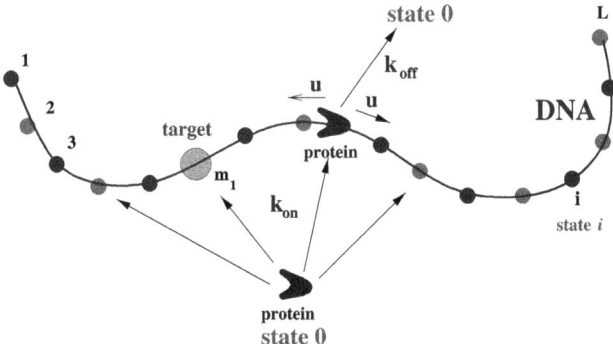

Fig. 1. A schematic view of the discrete-state stochastic model of the protein search. DNA chain has $L-1$ non-specific binding sites and one specific site. A protein molecule can diffuse along the DNA segment with a rate u in both directions. It can also associate to DNA with a rate k_{on} or dissociate with the a k_{off}. The search is finished when the protein binds to the target site at the position m.

DNA chain has L binding sites, and one of them at the position m is the target for the protein molecule. Because the diffusion of the proteins in the bulk is very fast, all solutions states for the protein are combined into one state that we label as a state 0: see Fig. 1. It is assumed then that from the bulk solution the protein can bind with equal probability to any site on DNA, and the total association rate to DNA is equal to k_{on}, while the dissociation rate from DNA is k_{off}. The non-specifically bound proteins can diffuse along the DNA contour with a rate u (see Fig. 1).

Since the search process ends as soon as the protein molecule arrives to the specific site for the first time, we introduce a function $F_n(t)$, which is defined a probability density function of reaching the site m (target site) at time t if at $t = 0$ the protein started in the state n ($n = 0$ is the bulk solution, and $n = 1, \ldots, L$ are protein–DNA bound states). This function is also known as a first-passage probability density function.[4,5] To determine these first-passage probabilities, we utilize backward master equations that describe the temporal evolution of these quantities,[4,5,9]

$$\frac{F_n(t)}{dt} = u\left[F_{n+1}(t) + F_{n-1}(t)\right] + k_{off}F_0(t) - (2u + k_{off})F_n(t), \quad (1)$$

for $2 \leq n \leq L - 1$, while at the boundaries ($n = 1$ or $n = L$) we have

$$\frac{F_1(t)}{dt} = uF_2(t) + k_{off}F_0(t) - (u + k_{off})F_1(t), \quad (2)$$

and

$$\frac{F_L(t)}{dt} = uF_{L-1}(t) + k_{off}F_0(t) - (u + k_{off})F_L(t). \quad (3)$$

For the state $n = 0$, the backward master equation is

$$\frac{F_0(t)}{dt} = \frac{k_{on}}{L} \sum_{n=1}^{L} F_n(t) - k_{on}F_n(t). \quad (4)$$

In the last equation, we used the fact that the rate to bind to any site on DNA is k_{on}/L, while the total association rate to DNA is k_{on}. In addition, the initial conditions require that $F_m(t) = \delta(t)$ and $F_{n \neq m}(t = 0) = 0$.

It is important to explain the physical meaning of the backward master equations because they differ from classical forward master equations widely employed in chemical kinetics. All trajectories that start at the state n and end the target at the state m can be divided into several groups. For example, for $2 \leq n \leq L$ all trajectories starting at n will go to the state $n - 1$, $n + 1$, or to the state 0 in the next time step, and the fraction of those trajectories is given by $u/(2u + k_{\text{off}})$, $u/(2u + k_{\text{off}})$ and $k_{\text{off}}/(2u + k_{\text{off}})$, respectively. Equation (1) describes this partition of the trajectories in a time-dependent manner. Thus, the backward master equations reflect the temporal evolution of the first-passage probabilities.

The most convenient way to analyze the dynamics in the system is to use Laplace representations of the first-passage probability functions, $\widetilde{F_n(s)} \equiv \int_0^\infty e^{-st} F_n(t) dt$. Then, Equations (1),(2), (3), and (4) can be written as simpler algebraic expressions

$$(s + 2u + k_{\text{off}})\widetilde{F_n(s)} = u \left[\widetilde{F_{n+1}(s)} + \widetilde{F_{n-1}(s)} \right] + k_{\text{off}} \widetilde{F_0(s)}, \quad (5)$$

$$(s + u + k_{\text{off}})\widetilde{F_1(s)} = u\widetilde{F_2(s)} + k_{\text{off}} \widetilde{F_0(s)}, \quad (6)$$

$$(s + u + k_{\text{off}})\widetilde{F_L(s)} = u\widetilde{F_{L-1}(s)} + k_{\text{off}} \widetilde{F_0(s)}, \quad (7)$$

$$(s + k_{\text{on}})\widetilde{F_0(s)} = \frac{k_{\text{on}}}{L} \sum_{n=1}^{L} \widetilde{F_n(s)}. \quad (8)$$

In addition, we have $\widetilde{F_m(s)} = 1$. These equations are solved assuming that the general form of the solution is $\widetilde{F_n(s)} = Ay^n + B$, where the unknown coefficients A, y, and B are determined from the initial and boundary conditions.[9] One could argue that the target site m divides the DNA molecule into two homogeneous segments ($1 \leq n \leq m$ and $m \leq n \leq L$), at which these parameters should have different values. It was shown[9] that this approach leads to explicit expressions for the first-passage probability functions. Specifically, one obtains

$$\widetilde{F_0(s)} = \frac{k_{\text{on}}(k_{\text{off}} + s)S_1(s)}{Ls(k_{\text{off}} + k_{\text{on}} + s) + k_{\text{off}} k_{\text{on}} S_1(s)}, \quad (9)$$

with an auxiliary function $S_1(s)$ defined as

$$S_1(s) = \frac{y(1+y)(y^{-L} - y^L)}{(1-y)(y^{1-m} + y^m)(y^{m-L} + y^{1+L-m})}, \tag{10}$$

and with the parameters y and B given by

$$y = \frac{s + 2u + k_{off} - \sqrt{(s + 2u + k_{off})^2 - 4u^2}}{2u}, \tag{11}$$

$$B = \frac{k_{off}\widetilde{F_0(s)}}{(k_{off} + s)}. \tag{12}$$

Explicit expressions for the first-passage probabilities provide a full dynamic description of the protein search processes. For example, the mean search time from the bulk solution, which is inversely proportional to the chemical association rate for the specific target site, can be found from[9]

$$T_0 \equiv -\frac{\partial \widetilde{F_0}(s)}{\partial s}\bigg|_{s=0} = \frac{1}{k_{on}}\frac{L}{S_1(0)} + \frac{1}{k_{off}}\frac{L - S_1(0)}{S_1(0)}. \tag{13}$$

This result has a clear physical meaning. Here the parameter $S_1(0)$ describes the average number of sites that the protein molecule scans during each visit to DNA while searching for a single specific site. Then, on average, to find the target the protein must make $L/S_1(0)$ visits to DNA. Each visit, on average, lasts $1/k_{on}$ while the protein scans for the target diffusing along the DNA chain. The protein also makes $L/S_1(0) - 1$ dissociations back into the solution. The number of dissociation events is smaller by one than the number of association events because the last binding to DNA leads to finding the specific site.

The results of our calculations for the mean search times are presented in Fig. 2. Three dynamic search regimes are predicted depending on the values of kinetic parameters. If the protein molecule has a strong affinity to bind non-specifically the the DNA molecule (small k_{off}), then there will be only one searching cycle. After binding to DNA the protein will not dissociate until it finds the target. In this case, the mean search time scales $\sim L^2$ because the DNA-bound protein does a simple unbiased random walk. We call this dynamic phase

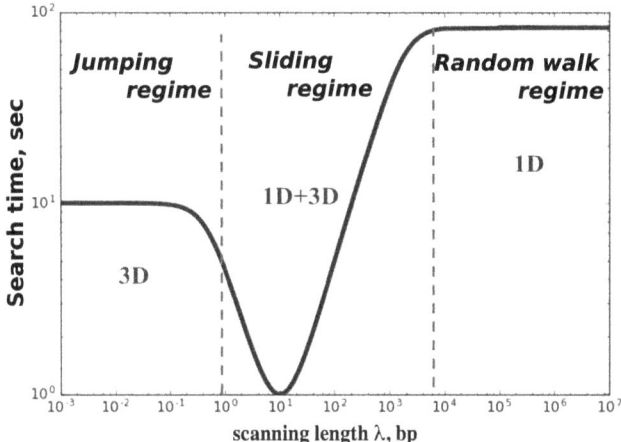

Fig. 2. Average times to find the target on DNA for the protein molecule starting from the solution as a function of the scanning length $\lambda = \sqrt{u/k_{\text{off}}}$. The parameters are $L = 10^3$ bp, $u = k_{\text{on}} = 10^5$ s^{-1}, and $m = L/2$. The transition rate k_{off} is varied to change λ.

a random walk regime. In the opposite limit of weak attractions between DNA and protein molecules (large k_{off}), the protein can bind to DNA but it cannot slide. So that the scanning length is of order of 1, and the protein makes L searching cycles ($T_0 \sim L$). This dynamic regime is called a jumping regime. The most interesting behavior is observed for intermediate interactions, which we label as a sliding regime. Here the scanning lengths are larger than one but smaller than the length of DNA L, and the number of searching cycles is also proportional to L. But in this regime the system can reach the most optimal dynamic behavior with the smallest search time. This search facilitation is achieved due to the fact that the fluxes to the target are coming from both the bulk solution and from the DNA chain.

3. Multiple Targets and Traps

The advantage of our method is the ability to extend and generalize it to more complex situations that can better describe the real biological systems. Let us present several examples to illustrate this.

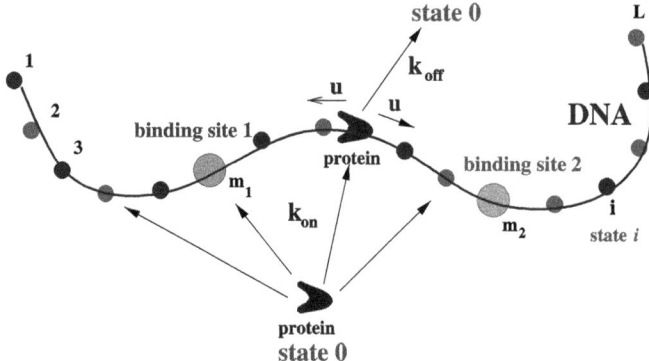

Fig. 3.   A discrete-state stochastic model of the protein search for multiple targets on DNA. The search ends when the protein finds one of the targets located at the sites m_1 and m_2. (Reproduced from Ref. 10 with the permission of AIP Publishing.)

In eukaryotic cells there are multiple target sites on accessible DNA fragments,[1,2] which raises a question how long does it take for a protein to find *any* specific binding site on DNA. It has been argued that the mean search time in this system might not decrease proportionally to the number of targets, as one would naively expect, due to the complex mechanism of the search that involves both 3D and 1D motions.[10] So we extended our discrete-state stochastic framework to consider a model with multiple targets as presented in Fig. 3. To describe the search dynamics in this system, we again introduce a first-passage probability function $F_n(t)$ of finding *any* of the target at time t if the process started at $t = 0$ at the site n. Solving the corresponding backward master equations leads to the explicit expression for the mean search time for any number of targets,[10]

$$T_0 = \frac{1}{k_{\text{on}}} \frac{L}{S_i(0)} + \frac{1}{k_{\text{off}}} \frac{L - S_i(0)}{S_i(0)}, \qquad (14)$$

with the function $S_i(0)$ describing the average number of scanned sites on DNA with i targets. This formula clearly is a generalization of Eq. (13) when there is only one target ($i = 1$). Specific expressions for $S_i(0)$ for various numbers of randomly distributed targets have been obtained.[10] For example, for $i = 2$ it was shown that

$$S_2 = \frac{(1+y)[2(1-y^{2L+m_1-m_2}) + (1-y^{m_2-m_1})(y^{2m_1-1} + y^{1+2(L-m_2)})]}{(1-y)(1+y^{2m_1-1})(1+y^{1+2(L-m_2)})(1+y^{m_2-m_1})},$$

$$(15)$$

where the parameter y is given in Eq. 11.

The results of explicit calculations for the mean search times are presented in Fig. 4. The presence of multiple targets does not affect the overall dynamic phase diagram: three search regimes are observed depending on the size of the scanning length and the size of the DNA segment, and in most cases the search is faster. However, increasing the number of specific sites might not always accelerate the search. To quantify this effect, we introduced an acceleration parameter, $a_n = T_0(1)/T_0(n)$, where $T_0(n)$ is the mean search for the system with n targets. This ratio gives a numerical value of how faster the search in the presence of n targets in comparison with a single-target system. It is interesting to analyze the results given in Fig. 5. One can see that there is a range of parameters when, surprisingly, the search dynamics in the system with two targets can be slower than the dynamics in the system with one target. This happens in the effectively 1D search regime when the single target is located in the middle of the DNA chain, while two targets are close to each other and located near one of the ends of the DNA segment. Our theoretical analysis predicts that the degree of acceleration due to the presence of multiple targets depends on the nature of the dynamic search phase and on the location of the specific sites.[10]

Another important factor in the protein search is the existence of DNA sequences that have chemical compositions similar to the specific targets. Because the protein molecule can be trapped in these sites, the dynamics might be strongly affected. To analyze this effect, we extended the original model to include the possibility of traps, assuming that associations to these semi-specific sites are irreversible.[11] This assumption is reasonable because the search times

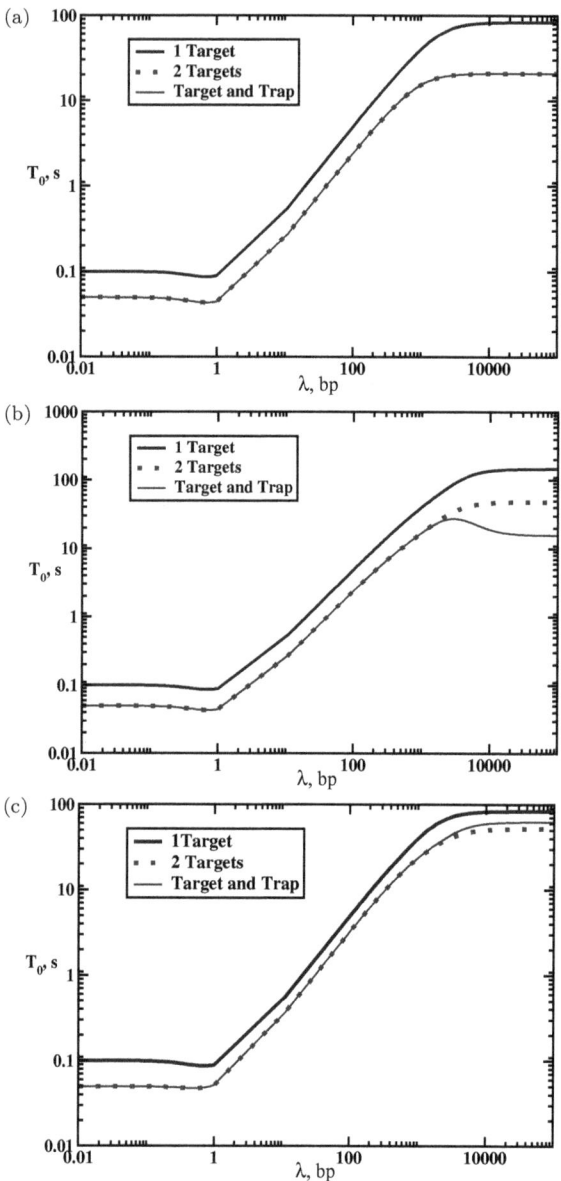

Fig. 4. Dynamic phase diagrams for the protein search on DNA with one target at the position m, with two targets at the positions m_1 and m_2 and with the target and the trap at the positions m_1 and m_2. Parameters used for calculations are: $k_{on} = u = 10^5$ s^{-1} and $L = 10,000$. (a) $m = L/2$, $m_1 = L/4$, and $m_2 = 3L/4$, (b) $m = L/4$, $m_1 = L/4$, and $m_2 = L/2$, and (c) $m = L/2$, $m_1 = L/2$, and $m_2 = L$. (Reproduced from Ref. 11 with the permission of AIP Publishing.)

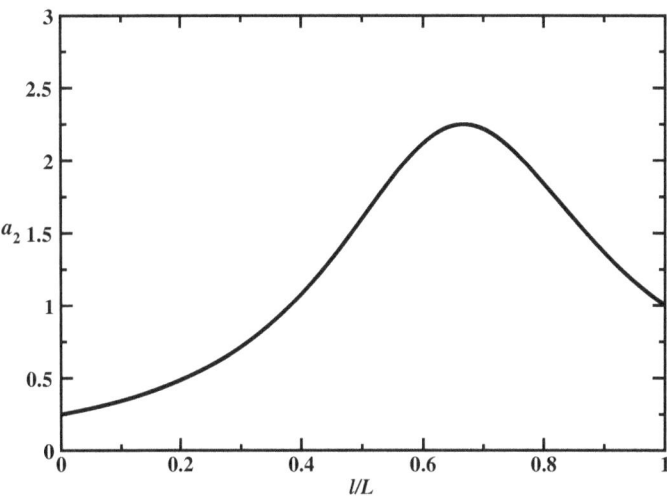

Fig. 5. Ratio of the mean search times as a function of the normalized distance between the targets for single-target and two-target systems (l is the distance between between targets, L is the DNA length). The single target is in the middle of the chain. In the two-target system, one of the specific sites is fixed at the end and the position of the second one is varied. The parameters used in calculations are: $u = k_{on} = 10^6$ s^{-1}, $k_{off} = 10^{-4}$ s^{-1}, and $L = 10,000$. (Reproduced from Ref. 10 with the permission of AIP Publishing.)

in many systems are relatively short and the experimental observations also have limited duration times. Our first-passage method can be easily applied here, but we have to notice that only a fraction of trajectories will reach the target site. Then the main quantity of our calculations, the first-passage probability function $F_n(t)$, is now a *conditional* probability for the protein molecules not captured by the trap to find the target site.

Let us consider a system consisting of a single target at the site m_1 and a single trap at the site m_2 on the DNA molecule with L sites.[11] Solving the corresponding backward master equations yields the Laplace transform of the first-passage probability function to find the target if the protein starts from the bulk solution,[11]

$$\widetilde{F_0(s)} = \frac{k_{on}(k_{off} + s)S_0(s)}{Ls(k_{off} + k_{on} + s) + k_{off}k_{on}S_2(s)}, \qquad (16)$$

with

$$S_0(s) = \frac{(1+y)(1 - y^{m_1+m_2-1})}{(1-y)(1 + y^{2m_1-1})(1 + y^{m_1-m_2})} \tag{17}$$

and the parameters y and S_2 given in Eqs. (11) and (15), respectively. This allows us to evaluate all dynamic properties in the system. The probability to reach the target (the fraction of successful trajectories) is given by the so-called splitting probability function,

$$\Pi = \widetilde{F_0(s = 0)} = \frac{S_0(0)}{S_2(0)}. \tag{18}$$

The mean search time, which is the conditional mean first-passage time to reach the target, can be estimated by averaging over the successful trajectories, producing

$$T_0 \equiv -\frac{\frac{\partial \widetilde{F_0}(s)}{\partial s}\big|_{s=0}}{\Pi} = \frac{1}{k_{on}}\frac{L}{S_2(0)} + \frac{1}{k_{off}}\frac{L - S_2(0)}{S_2(0)}$$
$$+ \Pi \frac{\partial}{\partial s}\left[\frac{S_2(s)}{S_0(s)}\right]\bigg|_{s=0}. \tag{19}$$

It is interesting to note that the first two terms in this expression is exactly the mean search time for the system with two targets (at the sites m_1 and m_2) as we discussed above,[10] while the third term is a correction which accounts for the fact that the site at m_2 is a trap. The main reason for this is the observation that the sites m_1 and m_2 are special locations where all trajectories are end up in both systems, with two targets and with the target and the trap. For the two-target case the mean search times are averaged over all trajectories, while for the target and the trap system the mean search times are obtained only by considering the trajectories finishing at the target.[11]

Theoretical calculations for dynamic properties of the protein search in the presence of traps are illustrated in Figs. 4 and 6. Again three dynamic search phases are predicted, but adding the trap generally facilitates the search dynamics: see Fig. 4. However, this acceleration (in comparison with the single-target system) is associated with lowering of the probability of reaching the specific target, as

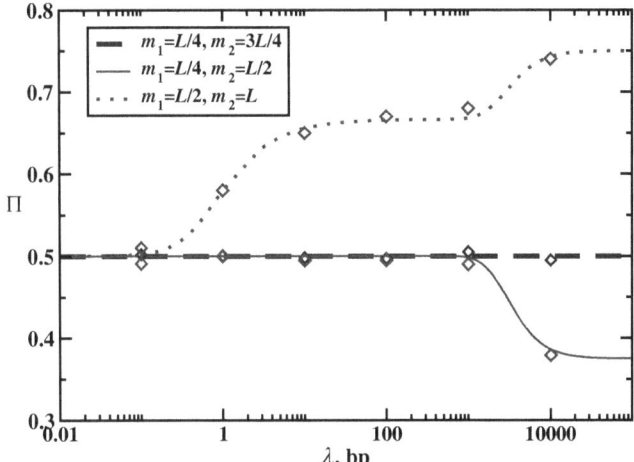

Fig. 6. Probability to reach the target as a function of the scanning length for different distributions of the target and trap sites. Parameters used for calculations are: $k_{on} = u = 10^5 \text{ s}^{-1}$, $L = 10,000$, and k_{off} is changing. Symbols are from Monte Carlo computer simulations. (Reproduced from Ref. 11 with the permission of AIP Publishing.)

shown in Fig. 6. In addition, the search dynamics depends on the nature of the dynamic phase. The strongest effect is observed in the random-walk regime (because it has only one searching cycle) where the locations of the target and the trap strongly influence the search.

4. Sequence Heterogeneity

It is known that real DNA are heterogeneous polymer molecules consisting of several types of subunits, and the interactions between protein and DNA molecules depend on the DNA sequence at the interaction site. Obviously, this should affect the protein search dynamics because the diffusion rate for the non-specifically bound proteins must be position-dependent. Our theoretical framework is a convenient tool to investigate this problem.[12]

To model the role of DNA sequence heterogeneity on protein search dynamics, we assume a simplified picture in which each monomer in DNA can be in one of two chemical species, A or B.[12]

When the protein is bound to the subunit $A(B)$, it interacts with energy $\varepsilon_A(\varepsilon_B)$, and the difference between interaction energies is given by $\varepsilon = \varepsilon_A - \varepsilon_B \geq 0$. This means that the protein attracts stronger to the B sites than to the A sites. The protein molecule can diffuse along DNA with a rate $u_A \equiv u$ or $u_B = ue^{-\varepsilon}$, where ε is measured in $k_B T$ units. In addition, we assume that, independently of the chemical nature of neighboring sites, sliding out of the sites A is characterized by the rate u_A, while the diffusion out of the sites B is given by u_B. From the solution the protein associates with any site A or B on DNA with the corresponding rates $k_{on}^A = k_{on}$ or $k_{on}^B = k_{on}e^{-\theta\varepsilon}$. Note that for convenience the on-rates defined here as the rates per unit site, in contrast to our definitions in the previous sections. Similarly, the dissociations from the DNA chain are described by the rates $k_{off}^A = k_{off}$ and $k_{off}^B = k_{off}e^{(\theta-1)\varepsilon}$. Here, the parameter $0 \leq \theta \leq 1$ specifies how the protein–DNA interaction energy is distributed between the association and dissociation transitions.[12] The physical meaning of this parameter is that the protein molecule tends to bind faster and to dissociate slower from the stronger attracting sites B, as compared with weaker attracting A sites. The parameter θ accounts for these effects. To quantify the role of sequence heterogeneity, we consider the DNA molecule with a fixed chemical composition (the fractions of A and B monomers are the same), but with different arrangements of subunits. Two limiting cases are specifically analyzed. One of them views the DNA molecule as two homogeneous segments of only A and only B subunits separated by the target in the middle of the chain. Another one is the DNA chain with the alternating A and B sites. The block copolymer has a more homogeneous sequence, while the alternating polymers are more heterogeneous. It is important to note that in both cases, the overall interaction between the protein and DNA is the same (because the overall chemical composition in both cases is identical), and thus our analysis probes only the effect of the heterogeneity.

Applying the first-passage approach to calculate the dynamic properties in this system leads to the explicit expressions for mean search times for all situations.[12] For example, for the block copolymer sequences, we obtain

$$T_0 = \frac{k_{\text{off}} + k_{\text{on}}\left[(L/2 - P_A) + e^{\varepsilon}(L/2 - P_B)\right]}{k_{\text{on}}k_{\text{off}}(1 + P_A + e^{\theta\varepsilon}P_B)}, \tag{20}$$

where

$$P_i = \frac{x_i^{1-L/2} - x_i^{1+L/2}}{(1 - x_i)(x_i^{1+L/2} + x_i^{L/2})}, \tag{21}$$

$$x_i = \frac{2u_i + k_{\text{off}}^{(i)} - \sqrt{(2u_i + k_{\text{off}}^{(i)})^2 - 4u_i^2}}{2u_i}, \tag{22}$$

for $i = A$ or B. The expressions for the mean search time for alternating sequences are more bulkier and they can be found in Ref. 12. The results of our calculations are presented in Fig. 7 where the ratio of the mean search times for the block copolymer and alternating sequences are plotted.

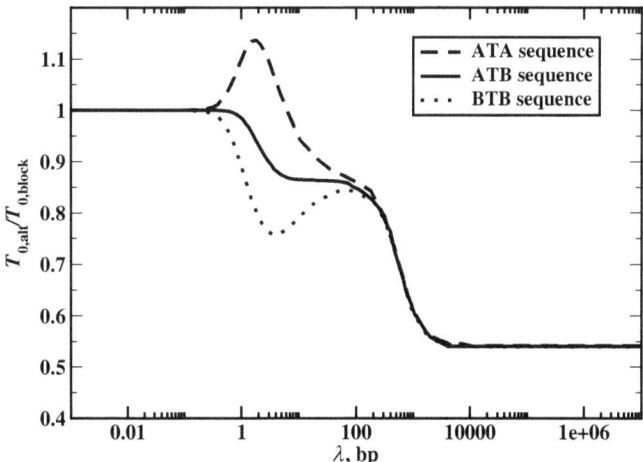

Fig. 7. The ratio of the mean search times for the alternating DNA sequences and for the block copolymer DNA sequences as a function of the scanning length $\lambda = \sqrt{u/k_{\text{off}}}$. Three different chemical compositions near the target (T) are distinguished, namely, ATA, ATB, BTB. The transition rates are $u = 10^5\ s^{-1}$ and $k_{\text{on}} = 0.1\ s^{-1}$. The DNA length is $L = 1,000$, the loading parameter is $\theta = 0.5$, and the energy difference of interactions for the protein with A and B sites is $\varepsilon = 5\ k_B T$. (Reproduced from Ref. 12 with the permission of AIP Publishing.)

One can see that the effect of the sequence heterogeneity on protein search dynamics depends on the nature of the dynamic phase. In the jumping regime when the protein does not slide along the DNA contour ($\lambda < 1$, where the parameter λ is proportional to the number of visited sites during each binding to DNA), the symmetry of the sequence does not play any role. This is because in this case, the process is taking place only via associations and dissociations (3D search), and the structure of the DNA chain is not important. The situation is different for the intermediate sliding regime (3D+1D search, $1 < \lambda < L$) where in most cases, the search on alternating sequences is faster. This can be explained by noticing that the search time in this dynamic phase is proportional to L/λ, which gives the average number of cycles before the protein can find the target. In the block copolymer sequence, the protein mostly comes to the target from the B segment because of stronger interactions with these sites. In the alternating sequences, the protein can reach the target from both sides of DNA, and this lowers the overall search time. It can be shown analytically that the scanning length on the alternating segment is larger than the scanning length for the B segment, i.e., $\lambda_{AB} > \lambda_B$.[12] Then the search is faster for the alternating sequence because $L/\lambda_{AB} < L/\lambda_B$, i.e., the number of searching cycles is lower for the alternating sequences. The only deviation from this picture is found in ATA sequences where for the small range of parameters the search is slower than in the block copolymer sequence. This effect can be explained by the fact that the protein does not sit at A sites for the long time and it moves quickly away, effectively increasing the barrier to enter the target via DNA.[12]

In the random-walk regime (1D search, $\lambda > L$), the effect of the sequence heterogeneity is even stronger: the protein molecule finds the specific binding site up to 2 times faster for more heterogeneous alternating DNA sequences. To understand this behavior, we note that in this case the mean first-passage time to reach the target is a sum of residence times on the DNA sites since the protein will not dissociate until the target is located. Because the target is in the middle of the chain, the mean time to reach the target from the block copolymer sequence will be $T_0 \simeq (L/4)\tau_B$, where τ_B is the average residence time at any site B. The protein prefers to start the search at any position on the B segment with equal probability, i.e., the distance to the target varies from 0 to $L/2$. Then, the average starting position of the protein is $L/4$ sites away from the target. For the alternating sequences, the average distance to the target is approximately the same $(L/4)$, but the chemical composition of intermediate sites on the path to the target is different, yielding, $T_0 \simeq (L/8)\tau_A + (L/8)\tau_B$ (τ_A is the residence time on A sites). The protein spends much less time on A subunits, and this leads to faster search for the alternating DNA sequences. For $\tau_A \ll \tau_B$, this also explains the factor of 2 in the search speed. In this case, the B subunits can be viewed as effective traps that slow down the search dynamics. Thus, our theoretical calculations make surprising predictions that the sequence heterogeneity almost always accelerates the protein search for targets on DNA. And the stronger the contribution of 1D search modes, the stronger will be the effect of the sequence heterogeneity.

5. Other Problems

In addition to problems discussed above, we extended and generalized the first-passage method to a variety of problems associated with protein–DNA interactions. More specifically, the role of crowding on DNA during the protein search was explicitly investigated.[13] It was found that the mobility of crowding agents (other DNA bound proteins) is a key factor affecting the facilitated diffusion: highly mobile crowders do not affect the search, while slow crowders inhibit the search dynamics. Similar analysis have been done for the protein search in the presence of static and dynamic obstacles (particles that occupy specific sites on DNA).[14] In this case, it was found that the key properties determining the search dynamics are size of the obstacle, the distance between the obstacles and the target, the dynamics of the obstacle and the nature of the dynamic search phase. We also analyzed the role of DNA looping for the target search by multisite proteins.[15] In addition, the effect of protein conformational transitions on the target search has been also explored.[16] Furthermore, we studied the surface-assisted dynamic processes by generalizing our method to 2D surfaces.[17] It is also important to note that our theoretical method has been successfully applied for analyzing of several experimental observations, including the kinetic studies of inducible transcription factor Egr-1,[18] the mechanism of the homology search by RecA protein filaments,[19] and the dynamics of genome interrogation by CRISPS-Cas9 protein–RNA systems.[20]

6.   Conclusions

A new theoretical method to investigate the dynamics of protein–DNA interactions is presented. It utilizes a discrete-state stochastic framework where dynamic properties are explicitly evaluated using the first-passage approach. Our approach takes into account the most relevant biochemical states and transitions. Because of exact calculations, it allows us to clarify many features of the complex mechanisms in these biological systems. The method is also successfully applied for understanding various experimental results. Thus, our theoretical approach is a powerful tool in studying protein–DNA interactions. There are several directions that we are planning to follow in the future studies. They include the role of coupling between 1D, 2D, and 3D motions, the effect of crowing in the bulk solutions, heterogeneity of the transition rates, DNA topological effects, and many others.

Acknowledgments

A.B.K. acknowledges the support from the Welch Foundation (C-1559), the NSF (CHE-1664218) and the Center for Theoretical Biological Physics sponsored by the NSF (PHY-1427654).

References

1. B. Alberts *et al.*, *Molecular Biology of Cell.* Garland Science, New York, 6th edn. (2014).
2. H. Lodish *et al.*, *Molecular Cell Biology.* W. H. Freeman, New York, 6th edn. (2007).
3. R. Phillips, J. Kondev, and J. Theriot, *Physical Biology of the Cell.* Garland Science, New York, 2nd ed. (2012).
4. N. G. Van Kampen, *Stochastic Processes in Physics and Chemistry.* North Holland, Amsterdam, 3rd edn. (2007).
5. S. Redner, *A Guide to First-Passage Processes.* Cambridge University Press, Cambridge (2001).
6. S. E. Halford and J. F. Marko, How do site-specific DNA-binding proteins find their targets? *Nucl. Acids Res.* **32**, 3040–3052 (2004).
7. L. Mirny, M. Slutsky, Z. Wunderlich, A. Tafvizi, J. Leith, and A. Kosmrlj, How a protein searches for its site on DNA: The mechanism of facilitated diffusion. *J. Phys. A.: Math. Theor.* **42**, 434019 (2009).
8. A. B. Kolomeisky, Physics of protein–DNA interactions: Mechanisms of facilitated target search. *Phys. Chem. Chem. Phys.* **13**, 2088–2095 (2011).
9. A. Veksler and A. B. Kolomeisky, Speed-selectivity paradox in the protein search for targets on DNA: Is it real or not? *J. Phys. Chem. B.* **117**, 12695–12701 (2013).
10. M. Lange, M. Kochugaeva, and A. B. Kolomeisky, Protein search for multiple targets on DNA. *J. Chem. Phys.* **143**, 105102 (2015).
11. M. Lange, M. Kochugaeva, and A. B. Kolomeisky, Dynamics of the protein search for targets on DNA in the presence of traps. *J. Phys. Chem. B.* **119**, 12410–12416 (2015).
12. A. A. Shvets and A. B. Kolomeisky, Sequence heterogeneity accelerates protein search for targets on DNA. *J. Chem. Phys.* **143**, 245101 (2015).
13. A. A. Shvets and A. B. Kolomeisky, Crowding on DNA in protein search for targets. *J. Phys. Chem. Lett.* **7**, 2502–2506 (2016).
14. A. A. Shvets, M. Kochugaeva and A. B. Kolomeisky, The role of static and dynamic obstacles in the protein search for targets on DNA. *J. Phys. Chem. B.* **120**, 5802–5809 (2015).
15. A. A. Shvets and A. B. Kolomeisky, The role of DNA looping in the search for specific targtes on DNA by multisite proteins. *J. Phys. Chem. Lett.* **7**, 5022–5027 (2016).
16. M. P. Kochugaeva, A. A. Shvets, and A. B. Kolomeisky, How conformational dynamics influences the protein search for targets on DNA. *J. Phys. A.: Math. Theor.* **49**, 444004 (2016).
17. J. Shin and A. B. Kolomeisky, Surface-assisted dynamic search processes. *J. Phys. Chem. B.* **122**, 2243–2250.
18. A. Esadze, C. A. Kemme, A. B. Kolomeisky, and J. Iwahara, Positive and negative impacts of nonspecific sites during targte location by a sequence-specific DNA-binding protein: Origin of the optimal search at physiological ionic strength. *Nucl. Acids Res.* **42**, 7039–7046 (2014).

19. M. P. Kochugaeva, A. A. Shvets, and A. B. Kolomeisky, On the mechanism of homology search by ReacA protein filaments. *Biophys. J.* **112**, 859–867 (2017).
20. A. A. Shvets and A. B. Kolomeisky, Mechanism of genome interrogation: How CRISPR RNA-guided Cas9 proteins locate specific targets on DNA. *Biophys. J.* **112**, 1416–1424 (2017).

Chapter 20

Modeling Chemotaxis of Microswimmers: From Individual to Collective Behavior

B. Liebchen[*] and H. Löwen[†]

*Institut für Theoretische Physik II: Weiche Materie,
Heinrich-Heine-Universität Düsseldorf,
Universitätsstr. 1, D-40225 Düsseldorf, Germany*
*liebchen@hhu.de
†Hartmut.Loewen@uni-duesseldorf.de

We discuss recent progress in the theoretical description of chemotaxis by coupling the diffusion equation of a chemical species to equations describing the motion of sensing microorganisms. In particular, we discuss models for autochemotaxis of a single microorganism which senses its own secretion, leading to phenomena such as self-localization and self-avoidance. For two heterogeneous particles, chemotactic coupling can lead to predator–prey behavior, including chase and escape phenomena, and to the formation of active molecules, where motility spontaneously emerges when the particles approach each other. We close this chapter with some remarks on the collective behavior of many particles where chemotactic coupling induces patterns involving clusters, spirals, or traveling waves.

1. Introduction

Chemotaxis plays a crucial role in the life of many microorganisms. It allows them to navigate toward food sources and away from toxins, but it is also used for signaling underlying self-organization in multicellular communities. Here, microorganisms sense the concentration of a chemical and adjust their motion to the chemical gradient:[1–3] if the corresponding chemical signal is externally imposed and, say, a food source, microorganisms will try to move up the gradient towards the food source ("chemoattraction", or "positive chemotaxis"). In the opposite case of a toxin, microorganisms migrate down the gradient ("chemorepulsion", or "negative chemotaxis").[4]

Many microorganisms can produce the chemicals to which they respond themselves and use chemotaxis for signaling. Here, chemotactic behavior strongly couples to *chemical kinetics*, which is the main topic of the present book.

In this chapter, we review recent progress in the theoretical description of chemotaxis beyond present textbook knowledge. Modeling of chemotaxis concerns the coupling between the dynamics of the chemical described by the diffusion equation together with appropriate source and sink terms, and the motion of microorganisms. Therefore, we discuss the basic equations for the chemical diffusion and the motion of bacteria (or other microorganisms) which is coupled to the chemical field. We then proceed step by step from few to many bacteria, where the chemotactic response of a bacterium to chemicals produced by another one lead to chemical interactions, or signaling, among microorganisms.

The simplest case of a single bacterium (or "particle") which senses its own secretion is discussed first. This case, also called *autochemotaxis*, is both of biological relevance and of fundamental importance as it may lead to effects such as self-localization. This in turn leads to dynamical scaling laws for the mean-square-displacement of the particle which are different from ordinary diffusion and are therefore of general interest.[5] We then proceed to a two-particle predator and prey system governed by chemotactic sensing and finally discuss the general case of many (more than two) particles which probably play a key role for dynamical cluster formation and other patterns. Since chemotaxis allows microorganisms to navigate, and also allows to steer synthetic microswimmers, it is linked to the rapidly expanding research field of active particles; for recent reviews, see Refs. 6–9.

2. Basics: Diffusion of Chemicals in Different Spatial Dimensions and Chemotactic Coupling

2.1. *Diffusing chemicals around static (non-moving) point sources*

To set a theoretical framework for chemotaxis, we first explore the kinetics of the chemical that constitutes the chemotactic signal. We start with the diffusion equation for a chemical concentration field $c(\vec{r}, t)$ in solution with a point source emitting the chemical with a rate $\lambda_e(t)$, which may generally depend on time, at fixed position $\vec{r}_0$

$$\frac{\partial c(\vec{r}, t)}{\partial t} = D_c \Delta c(\vec{r}, t) - \mu c(\vec{r}, t) + \lambda_e(t)\delta(\vec{r} - \vec{r}_0). \tag{1}$$

Here, D_c is the diffusion coefficient of the chemical in solution. The chemical may also evaporate (or disappear) with a rate μ, e.g., due to another chemical reaction. In the following, we focus on constant emission rates. The diffusion equation (1) can be considered in $d = 1, 2, 3$ spatial dimensions, where $d = 1$ corresponds to an effective slab and $d = 2$ to an effective cylindrical geometry. Accordingly, Δ denotes the Laplacian operator in d spatial dimensions. For an instantaneous onset of chemical emission at $t = 0$, i.e., $\lambda_e(t) = \lambda_e \Theta(t)$, where $\Theta(t)$ denotes the unit step function and $\mu = 0$, the solution of Eq. (1) is given in d dimensions by[10]

$$c(\vec{r}, t) = \lambda_e \int_0^t dt' \frac{1}{(4\pi D_c |t - t'|)^{\frac{d}{2}}} \exp\left(-\frac{(\vec{r} - \vec{r}_0)^2}{4D_c |t - t'|}\right). \tag{2}$$

By substituting $t' \to s := (\vec{r} - \vec{r}_0)^2/(4D_c |t - t'|)$, this expression can also be written in terms of the upper incomplete Gamma function $\Gamma(a, b) = \int_b^\infty e^{-x} x^{a-1} dx$,

$$c(\vec{r}, t) = \frac{\lambda |\vec{r} - \vec{r}_0|^{2-d}}{4\pi^{d/2} D_c} \Gamma\left(\frac{d}{2} - 2, \frac{(\vec{r} - \vec{r}_0)^2}{4D_c t}\right). \tag{3}$$

Expression (2) can be generalized for $\mu \neq 0$ to

$$c(\vec{r}, t) = \lambda_e \int_0^t dt' \frac{1}{(4\pi D_c |t - t'|)^{\frac{d}{2}}} \exp\left(-\frac{(\vec{r} - \vec{r_0})^2}{4D_c |t - t'|} - \mu |t - t'|\right).$$

(4)

In many cases, the dynamics of the chemical is fast compared to all other relevant time scales in a given system (e.g., the response time of a microorganism). In these cases, we are mainly interested in the chemical steady-state profile, corresponding to $\dot{c} = 0$ in Eq. (1). This steady-state problem is formally equivalent to screened electrostatics, or in other words, to linear Debye–Hückel theory of screening[11] with an inverse screening length κ now given by

$$\kappa = \sqrt{\mu/D_c},$$

(5)

while without evaporation ($\mu = 0$) we have an analogy to the Poisson's equation of ordinary (unscreened) electrostatics. Therefore, in various spatial dimensions the solutions d, for a localized initial state, are as follows:

(i) In $d = 1$, there is an "exponential orbital" around the secreting source fixed at the origin of the coordinate system such that for a spatial coordinate x, the concentration field is for $\mu > 0$

$$c(x) = \frac{\lambda_e}{\sqrt{4D_c \mu}} \exp(-\kappa |x|).$$

(6)

For $\mu = 0$, the steady-state solution becomes unphysical; in that case, the time-dependent solution does not converge to a steady state but increases forever.

(ii) For $d = 2$, there is a "Macdonald orbital"

$$c(r) = \frac{\lambda_e}{D_c} K_0(\kappa r),$$

(7)

with r denoting the radial distance in two dimensions from the source. Here, $K_0(x)$ is a Macdonald function (or modified Bessel function) (see e.g., Eq. (39) in[12]). Like in $d = 1$, for $\mu = 0$ the chemical density does not converge and the steady-state solution becomes unphysical.

(iii) Finally, for $d = 3$, there is a radial-symmetric Debye–Hückel (or Yukawa) orbital around the point source

$$c(r) = \frac{\lambda_e}{4\pi D_c r} \exp(-\kappa r), \tag{8}$$

which reduces for $\mu = 0$ to the classical Coulomb solution

$$c(r) = \frac{\lambda_e}{4\pi D_c r}, \tag{9}$$

again with r denoting the radial distance from the point source.

2.2. *Moving point sources*

When the point source is moving with a constant velocity $\vec{v}$, the general diffusion equation for constant emission rate is

$$\frac{\partial c(\vec{r}, t)}{\partial t} = D_c \Delta c(\vec{r}, t) - \mu c(\vec{r}, t) + \lambda_e \delta(\vec{r} - \vec{r}_0 - \vec{v}t). \tag{10}$$

By a Galilean transformation from the laboratory frame into the moving particle frame, this equation can be transformed such that it reads under steady-state conditions as follows:

$$-D_c \Delta c(\vec{r}) + (\vec{v} \cdot \vec{\nabla}) c(\vec{r}) + \mu c(\vec{r}) = \lambda_e \delta(\vec{r}). \tag{11}$$

Solutions of Eq. (11) go beyond textbook knowledge and have not been discussed yet in this context. In general, the Green's function associated with Eq. (11) can be expressed as a Fourier integral as

$$c(\vec{r}) = \frac{\lambda}{(2\pi)^d} \int_{-\infty}^{\infty} d^d k \frac{e^{-i\vec{k} \cdot \vec{r}}}{D_c \vec{k}^2 + i\vec{v} \cdot \vec{k} + \mu}. \tag{12}$$

Evaluating this integral in one spatial dimension ($d = 1$) yields a solution consisting of two exponentials with different decay lengths in the front and in the rear of the moving source

$$c(x) = \frac{\lambda_e}{\sqrt{4D_c\mu}} \begin{cases} e^{-\chi_+|x|} & \text{for} \quad x \geq 0 \\ e^{-\chi_-|x|} & \text{for} \quad x < 0 \end{cases} \tag{13}$$

$$\text{with} \quad \chi_\pm = \sqrt{\frac{\mu}{D_c} + \frac{v^2}{4D_c^2}} \pm \frac{v}{2D_c}. \tag{14}$$

This solution is plotted in (1). Clearly, increasing the speed of the source enhances the front-rear asymmetry while at $v = 0$ we recover the solution (6) of a static point source. Increasing the evaporation rate basically decreases the range of the chemical concentration around the source.

For the corresponding solution in two dimensions, we find

$$c(\vec{r}) = \frac{\lambda_e}{D_c} K_0(\tilde{\kappa}r) \exp\left(-\frac{\vec{v} \cdot \vec{r}}{2D_c}\right); \quad \text{with} \quad \tilde{\kappa} = \sqrt{\frac{\mu}{D_c} + \frac{v^2}{4D_c^2}} \tag{15}$$

and in three dimensions

$$c(\vec{r}) = \frac{\lambda_e}{4\pi D_c r} \exp\left(-\tilde{\kappa}r - \frac{\vec{v} \cdot \vec{r}}{2D_c}\right). \tag{16}$$

For $\mu = 0$, in a coordinate system whose x-axis points along $\vec{v}$, the three-dimensional solution reduces to

$$c(\vec{r}) = \frac{\lambda_e}{4\pi D_c r} \exp\left[-\frac{|\vec{v}|(x + |\vec{r}|)}{2D_c}\right]. \tag{17}$$

Remarkably, in the rear of the moving source ($x < 0$ at $y = z = 0$ such that $x + r = 0$) the chemical concentration decays algebraically as $1/|x|$ while in all other directions it decays with a Yukawa-behavior as $\exp(-vx/D_c)/x$, i.e., algebraically for $r \ll \sqrt{D_c/\mu}$ and exponentially at longer distances. This asymmetry indicates the significance of the source trail and a memory effect about the past of the secreting particle (Fig. 1).

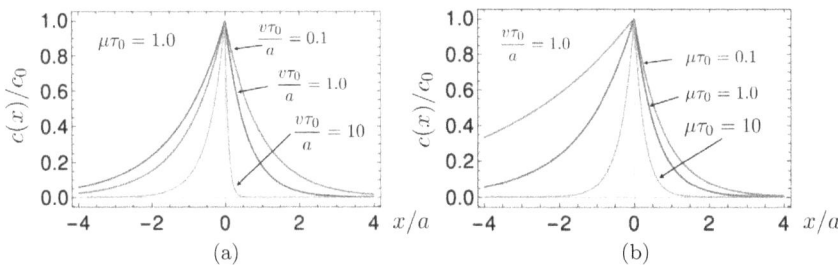

Fig. 1. Reduced concentration field $c(x)/c_0$ as a function of reduced distance x/a to the moving source. This is the steady-state solution for the 1D diffusion equation of a point source moving with a velocity v with constant emission rate λ_e for various relative speeds (a) and relative evaporation rates (b). All length and time scales are given in terms of $a = \sqrt{D_c/\lambda_e}$ and $\tau_0 = 1/\lambda_e$, and $c_0 = 1/(2a\sqrt{\mu\tau_0})$. The parameters are: (a) $v\tau_0/a = 0.1, 1.0, 10$ at $\mu\tau_0 = 1$ and (b) $\mu\tau_0 = 0.1, 1.0, 10$ at $v\tau_0/a = 1$.

2.3. *Chemotactic coupling and secreting particle dynamics*

For chemotaxis in its simplest form, the particle directs its motion according to the gradient of the chemical field. We describe this coupling to the chemical concentration as an effective force

$$\vec{F} = \alpha \nabla c(\vec{r}, t) \tag{18}$$

acting on the particle. Here positive α values represent "positive" chemotaxis or "chemoattraction", whereas negative α values represent "negative" chemotaxis or "chemorepulsion". The linear coupling to the gradient is the simplest possible form, but other couplings like logarithmic ones are also conceivable and probably relevant for microbiological systems[13] where

$$\vec{F} = \alpha \vec{\nabla} \ln c(\vec{r}, t) = \alpha \frac{\vec{\nabla} c(\vec{r}, t)}{c(\vec{r}, t)}. \tag{19}$$

A more complicated coupling involves a concentration-dependent prefactor α as proposed in Ref. 14. We will basically use and discuss formula (18) in the sequel.

The effective chemotactic force typically acts on a completely overdamped particle with position $\vec{r}_p(t)$, leading to the following equation of motion

$$\gamma \frac{d}{dt}\vec{r}_p(t) = \alpha \nabla c(\vec{r}_p(t), t).\tag{20}$$

Here, γ is the Stokes drag coefficient. If necessary, additional noise terms can be added to Eq. (20) in order to model the stochastic collisions of the particle with the solvent molecules. Here, we neglect any hydrodynamic flow effects stemming from a finite radius of the point source. Their inclusion would require a more sophisticated analysis.

3. Autochemotaxis for a Single Particle

If a single particle emits a chemical to which it responds itself, we call this "autochemotaxis". Here, Tsori and de Gennes[15] have coupled the chemical diffusion equation to an equation of motion for a particle. For the chemoattractant case at vanishing evaporation rate $\mu = 0$, they have found "*self-trapping*" of the particle in a spatial dimension $d = 1, 2$ but not for $d = 3$. This implies that a particle traps itself if it moves towards the chemical which it has secreted in the past. The concept of "perfect" self-trapping was subsequently questioned by Grima[16,17] in a model with a positive evaporation rate $\mu > 0$. Here, it turned out that self-trapping is a transient phenomenon at $\mu > 0$ and crosses over to normal diffusion at very long times even for $d = 1, 2$. For sufficiently strong negative chemotaxis, Grima found long-time diffusive or *ballistic* motion depending on the secretion rate λ_e. This result which was obtained for any dimensionality d suggests that a particle might *self-propel* if it avoids the region where it has been in the past, and in some sense constitutes a link between repulsive autochemotaxis and the rapidly growing research field of active particle or microswimmers.[7,9,18] (Note, however, that while self-propulsion due to autochemorepulsion might apply to particles on a surface, for microorganisms in bulk, which oblige momentum conservation, self-propulsion based on autochemorepulsion is conceptually not immediate and would probably need to involve some combination of parity-symmetry breaking and phoresis.)

Now, we shall mainly review subsequent studies which include Brownian noise due to solvent molecules acting on the self-driven particles. Noise statistics is a relevant part of the actual trajectories of self-propelled particles[8] both for microorganisms and synthetic microswimmers.[19-21] It is expected that noise will destroy the perfect localization for positive autochemotaxis as well as the ballistic long-time motion for negative autochemotaxis. Indeed, this was confirmed by numerical work and theoretical analysis in a subsequent paper of Sengupta *et al.*[22] which we shall discuss in the following in more detail.

The governing equations of the Brownian noise model introduced by Sengupta *et al.*[22] describe a coupling between the diffusion equation of the chemical concentration field $c(\vec{r}, t)$ and the trajectory of the secreting particle $\vec{r}_p(t)$. At vanishing evaporation rate μ, the chemical is emitted with a constant rate λ_e and is diffusing in d spatial dimensions according to

$$\frac{\partial c(\vec{r}, t)}{\partial t} = D_c \nabla^2 c(\vec{r}, t) + \lambda_e \delta(\vec{r} - \vec{r}_p(t)). \tag{21}$$

The equation of motion for the emitting particle is given by

$$\gamma \dot{\vec{r}}_p(t) = \vec{F}(\vec{r}_p, t) + \vec{\eta}(t). \tag{22}$$

Here, $\vec{\eta}(t)$ is an effective Gaussian white noise with zero mean and variance

$$\langle \eta_i(t) \eta_j(t') \rangle = 2\gamma \beta^{-1} \delta_{ij}(t - t'), \tag{23}$$

with i and j denoting the Cartesian spatial components and β an effective inverse thermal energy. In Ref. 22, the chemotactic force $\vec{F}(\vec{r}_p, t)$ depends on the history of the particle trajectory $\vec{r}_p(t)$ apart from a delay (or memory) time t_0 which takes into account that a finite time is needed for sensing the chemical. Integrating or superimposing over the Green's function of chemical diffusion, Eq. (4), the chemotactic force $\vec{F}(\vec{r}_p, t)$, Eq. (18), is modeled as

$$\vec{F}(\vec{r}, t) = -2\alpha\lambda_e \int_0^{t-t_0} dt' \frac{(\vec{r} - \vec{r}_p(t'))}{4D_c|t - t'|} \frac{\exp\left[\frac{-(\vec{r} - \vec{r}_p(t'))^2}{(4D_c|t-t'|)}\right]}{(4\pi D_c|t - t'|)^{d/2}}. \tag{24}$$

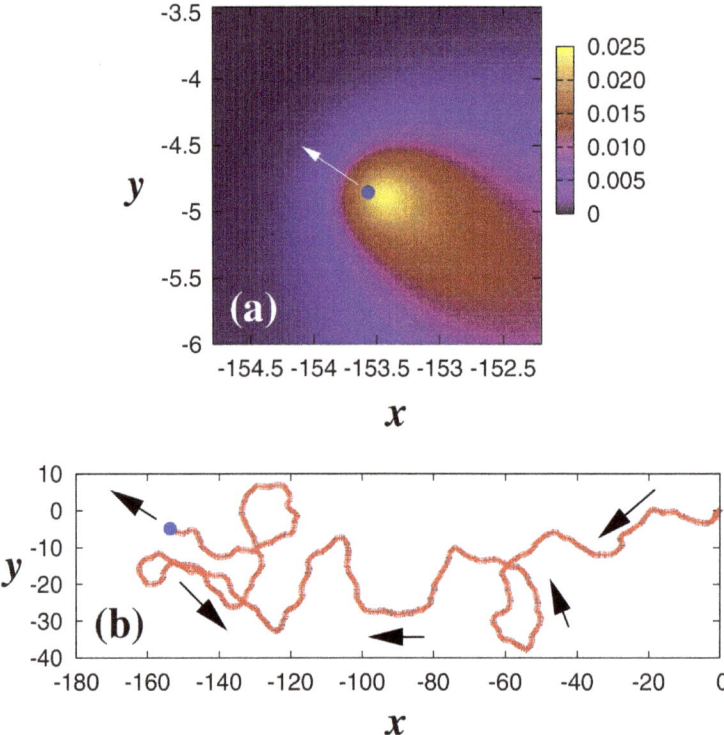

Fig. 2. (a) Snapshot of the instantaneous density profile $c(x, y)$ of the chemore-pellent released by the microorganism moving in two dimensions, obtained from simulation, at time instant $t = 10$ (in units of λ_e^{-1}). (b) The entire trajec-tory, shown as the red (thick) curve, of the microorganism. The current posi-tion of the microorganism $\vec{r}_p$ is indicated by the blue (black) dot in both the figures, and the direction of motion is indicated by arrows along the trajec-tory. The corresponding coupling strength is $|\lambda| = 10{,}000$. The parameters are $D \equiv 1/\beta\gamma = 0.1\ell_o^2/\tau_0$, $D_c/D = 100$, $t_0 = 0.001\tau_0$ with length, time, and energy scales of $\ell_0 = \sqrt{\sqrt{D_c D}/\lambda_e}$, $\tau_0 = 1/\lambda_e$, $1/\beta$. (Reproduced from Ref. 22 with the permission of American Physical Society.)

For $d = 2$, a typical particle trajectory in the chemorepellent case and the associated chemical density field are shown in Fig. 2. Clearly, the particle avoids its own trail and where it had been in the past, giv-ing rise to a persistent random walk along the arrow shown in Fig. 2.

Results for the "exact" numerical solution of these governing equations are presented in Fig. 3 for the chemoattractive case.

There is long-time diffusive behavior with a long-time diffusion coefficient D_l but for stronger couplings α, an intermediate transient time region shows up where the particle is quasi-localized. This localization is most pronounced in low spatial dimensions d. The long-time self-diffusion coefficient D_l drops strongly with the coupling α for any d (see Fig. 3(d)) and scales with $1/\alpha^2$ in agreement with scaling arguments proposed in Ref. 22.

Figure 4 shows the chemorepulsive case. Here, we again have a transient ballistic transient regime which is most pronounced in low

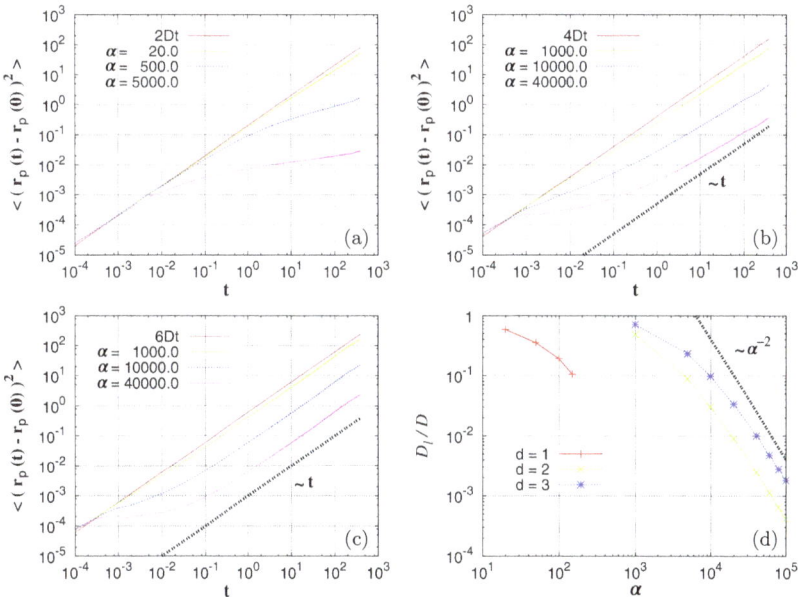

Fig. 3. Mean-square displacement $\langle [\vec{r}_b(t) - \vec{r}_b(0)]^2 \rangle$ of the microorganism as a function of time t with chemoattractant in (a) $d = 1$ with $\alpha = 20{,}500{,}5{,}000$, (b) $d = 2$ with $\alpha = 1{,}000{,}10{,}000{,}40{,}000$, (c) $d = 3$ with $\alpha = 1{,}000$, $10{,}000$, $40{,}000$. The non-chemotactic diffusion reference lines are also indicated as $2Dt, 4Dt$, and $6Dt$ correspondingly for $d = 1, 2, 3$. Reference lines (thick dotted) are used to indicate the long-time diffusive behavior (t) wherever possible. The relative long-time diffusivity D_l/D is shown as a function of α in (d) for $d = 1, 2, 3$. The reference line (thick dotted) shows a power-law scaling behavior $1/\alpha^2$ (see text). The parameters are as in 2, $D = 1/\beta\gamma$ is the short-time particle diffusivity and the coupling parameter α is measured in terms of ℓ_0^d/β. (Reproduced from Ref. 22 with the permission of American Physical Society.)

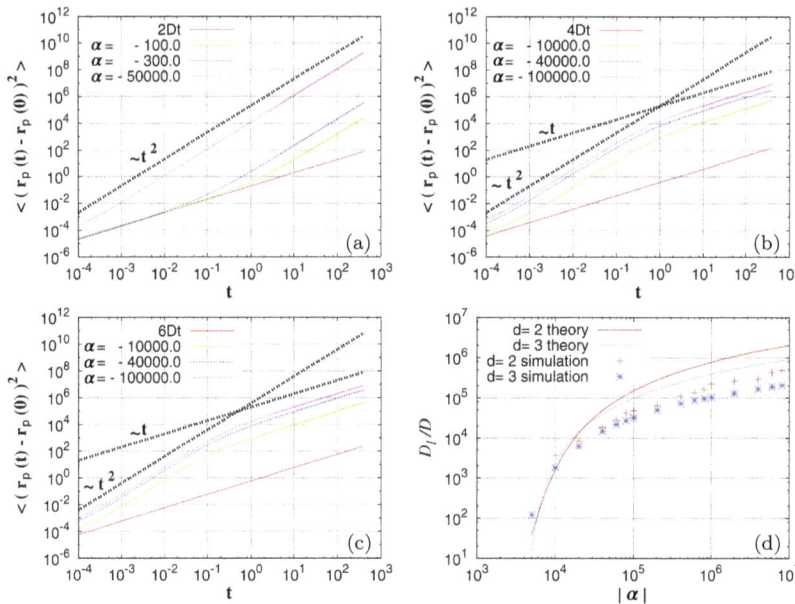

Fig. 4. Mean-square displacement $\langle[\vec{r}_b(t) - \vec{r}_b(0)]^2\rangle$ of the microorganism as a function of time t with chemorepellent in (a) $d = 1$ with $\alpha = -100, -300, -50,000$; (b) $d = 2$ with $\alpha = -1,000, -40,000, -100,000$; (c) $d = 3$ with $\alpha = -1,000, -40,000, -100,000$. The non-chemotactic diffusion reference lines are also indicated as $2Dt, 4Dt$, and $6Dt$ correspondingly for $d = 1, 2, 3$. Reference lines (thick dotted) indicating the ballistic (t^2) and the long-time diffusive (t) dynamics is shown as guide to the eye. The relative long-time diffusivity D_l/D is shown as a function of $|\alpha|$ in (d) for $d = 2, 3$. The points represent the actual data obtained from simulations, the lines correspond to a semiquantitative theory (see text). The parameters are as in Figure 2, $D = 1/\beta\gamma$ is the short-time particle diffusivity and the coupling parameter α is measured in terms of ℓ_0^d/β. (Reproduced from Ref. 22 with the permission of American Physical Society.)

spatial dimensions. A simple theory put forward in Ref. 22 describes the strong increase of the long-time particle diffusivity with the coupling strength $|\lambda|$.

At this stage, we mention that it is now possible to create synthetic particles which react in principle such that they avoid their own secretion trail. One important example discussed recently is an oil droplet in an aqueous surfactant solution which "remembers" the surfactant concentration which constitutes its self-propagation.[23]

Another idea is to dynamically control the motion of colloidal particles by optical fields which are dynamically adapted (programmed) such that the colloids avoid positions where they have been at earlier times; this leads to self-propulsion.[24] A third realization in the macroscopic world are robots which can be programmed at wish.[25] Moreover, there are further but related theoretical models for particles avoiding their own past trails[26,27] or that involve memory effects leading to similar phenomena.[28,29]

4. Chemotactic Predator–Prey Dynamics

Next, we shall explore two particles which are sensing each other via chemotaxis mimicking signaling among microorganisms. One particle ("predator") is attracted by the chemical secreted by the second particle and the latter ("prey") is repelled by the chemical secreted by the first one. Here, we follow the model of Ref. 10. Now we have two chemicals characterized by concentration fields $c_i(\vec{r}, t)$ ($i = 1, 2$) and two trajectories $\vec{r}_i(t)$. In the absence of chemical evaporation ($\mu = 0$), the concentration fields read

$$c_i(\vec{r}, t) = \lambda_i \int_0^t dt' \frac{1}{(4\pi D_{ci}|t - t'|)^{\frac{d}{2}}} \exp\left(-\frac{[\vec{r} - \vec{r}_i(t')]^2}{4D_{ci}|t - t'|}\right), \qquad (25)$$

which is the solution of the chemical diffusion equation for given trajectories $\vec{r}_i(t)$ with D_{ci} denoting the diffusion coefficient of the two chemicals and λ_i being the production rate of chemical species i. The equation of motion determining the predator trajectory reads

$$\gamma_1 \ddot{\vec{r}}_1 = +\alpha_1 \nabla c_2(\vec{r}_1, t) + \vec{\eta}_1(t), \qquad (26)$$

and the equation of motion for the prey is

$$\gamma_2 \ddot{\vec{r}}_2 = -\alpha_2 \nabla c_1(\vec{r}_2, t) + \vec{\eta}_2(t), \qquad (27)$$

where $\gamma_{1,2}$ are friction coefficients and $\alpha_{1,2}$ are chemotactic coupling coefficients.

As in Eq. (22), we generally allow for Gaussian white noise, represented by $\vec{\eta}_{1,2}(t)$.

It is important to remark here that the chemically mediated inter-action between the predator and the prey is non-reciprocal, i.e., the force exerted by the predator acting on the prey is unequal to the force acting on the predator due to the prey particle. This violation of Newton's third law stems from the non-equilibrium conditions and applies to the *effective interaction* among predator and prey; the microscopic interactions among all solvent molecules, and the preda-tor and the prey particle are of course reciprocal so that momentum conservation applies and any net motion of the predator–prey-pair (in bulk) will be generally balanced by a counter-propagating flow of solvent (and or chemicals). Such non-reciprocal interactions are frequently encountered in situations away from equilibrium, e.g., in dusty plasmas.[30]

A typical snapshot in the noise-free case is shown in Fig. 5 which highlights the two concentration fields and the predator (left particle) following the prey (right particle).

As a result of the analysis performed in Ref. 10, there are basically two dimensionless parameters which govern the escape and chase scenario, still in the case of vanishing noise. The first parameter is

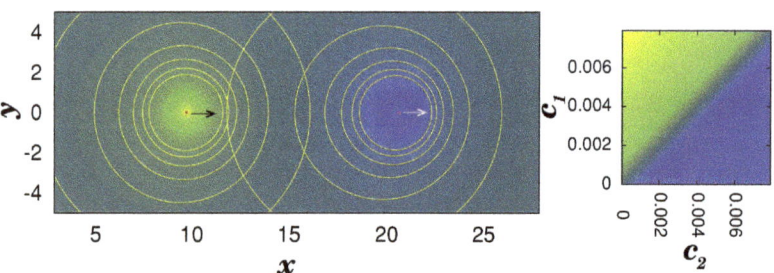

Fig. 5. (Left) A predator (red dot on left) chases a prey (red not visible dot on right), while the latter tries to escape through chemotactic gradient sensing of the diffusing chemicals. The arrows indicate their respective directions of motion in the absence of fluctuations. The contours around each microbe represent the equiconcentration lines of the secreted chemicals in a 2D projected plane in this case, indicating the asymmetry of the distribution. The color code used here for the spatial distribution of the secreted chemorepellant (c_1) and the chemoattrac-tant (c_2), as they mingle in space, is shown in the right panel. (Reproduced from Ref. 10 with the permission of American Physical Society.)

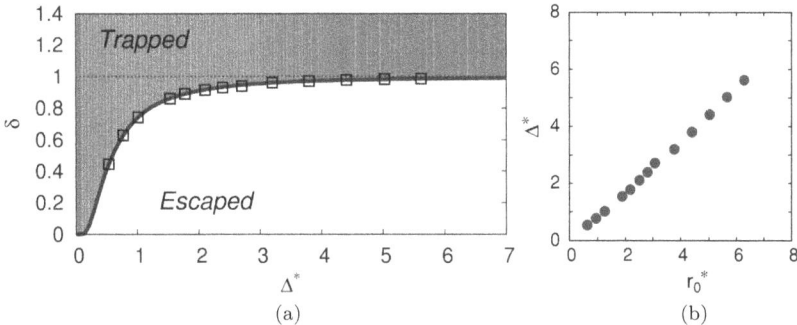

Fig. 6. (a) Dynamical phase diagram of the chemotactic predator–prey system, constructed in the $\Delta^* - \delta$ parameter space, showing the trapped (shaded) and escaped phases. The phase boundary (thick solid line) is obtained analytically and matches the simulation data (boxes). The horizontal thin dotted line ($\delta = 1$) represents the upper bound for the trapped-to-escaped dynamical phase transition (see text). (b) The dependence of the catching range (Δ^*) on the initial separation (r_0^*), as obtained from simulations. (Reproduced from Ref. 10 with the permission of American Physical Society.)

the reduced length scale Δ^* depending on r_{12} which is the steady-state distance between the particles; the second parameter is a sensibility ratio δ; see Ref. 10 for details. The state diagram is shown in Fig. 6 in the parameter plane, spanned by Δ^* and δ, and shows regions of escape and capture. For $\delta > 1$, there is always trapping, independent of the initial conditions. For $\delta < 1$, it depends on the initial particle separation r_0^* (see Fig. 6(b)) in which effective Δ^* is realized. On the separation line, there is steady-state motion with a constant particle separation, i.e., both particles move with the same speed. This line can be calculated analytically and is given by

$$\delta(\Delta^*) = (1 + \Delta^{*-1})\exp(-\Delta^{*-1}). \tag{28}$$

The particular form of the diffusing chemical field in the absence of evaporation, Eq. (17), allows the prediction of two scaling laws. The first one applies to escape situations and shows that the distance x between the two particles increases subdiffusively with time t as $t^{1/3}$, while the second one applies to trapping situations and predicts that the distance between predator and prey decreases as $(t_{\text{trap}} - t)^{1/3}$ with a finite trapping time t_{trap} where $x = 0$.

As a final remark, there are also other predator–prey models which model the predator and prey dynamics either on a lattice[31] or with more complex models designed for real bacteria.[32] A marvellous realization of a synthetic predator–prey system employs a pair of an ion-exchange resin and a passive charged colloidal particle which moves autonomously as a "modular swimmer"[33]; another interesting realization of a predator–prey-like pair is based on two different particles, which are actuated as a pair.[34]

5. Collective Behavior of Few and Many Active Particles

Tsori and de Gennes[15] have pointed out the analogy between chemoattractive matter and *gravity* in three spatial dimensions. In fact, the chemotactic coupling (18) together with the Coulomb orbitals (9) and the linearity of the diffusion equations implies that a one-component system is identical to gravitating particles. The dynamics will lead to clustering and finally to a collapse ("black hole"). Hence, one can study aspects of the dynamics of a black hole collapse scaled down in a Petri dish, see e.g., Ref. 35 for a similar idea.

Binary mixtures of particles with chemotactic coupling coefficients of opposite sign lead to a similar physical behavior as oppositely electrically charged mixtures. These systems form interesting cluster structures and lead to effectively non-reciprocal forces, i.e., they break Newton's third law, actio=reactio. These non-reciprocal forces can lead to self-propulsion and self-rotation,[36] which only emerge if different colloids closely approach each other and form "active molecules" appearing in a broad variety of shapes as movers, rotators and circle swimmers.[34,37,38] The simplest example of such an active molecule is a moving dimer similar to the predator–prey system discussed in the previous section. A direct experimental confirmation of active molecules consisting of two species was found in Ref. 39 where two different types of ion exchange resins provide the active constituents.

The full many-body behavior of a binary mixture interacting with chemotactic-based non-reciprocal interactions was studied in Ref. 40, however, in the different context of complex plasmas. In Ref. 36, the connection to chemotaxis was worked out explicitly. One example for the collective behavior in a binary system of chemotactically coupled species is shown in Fig. 7 at vanishing noise. On the x-axis, a relative wake charge is plotted which corresponds to

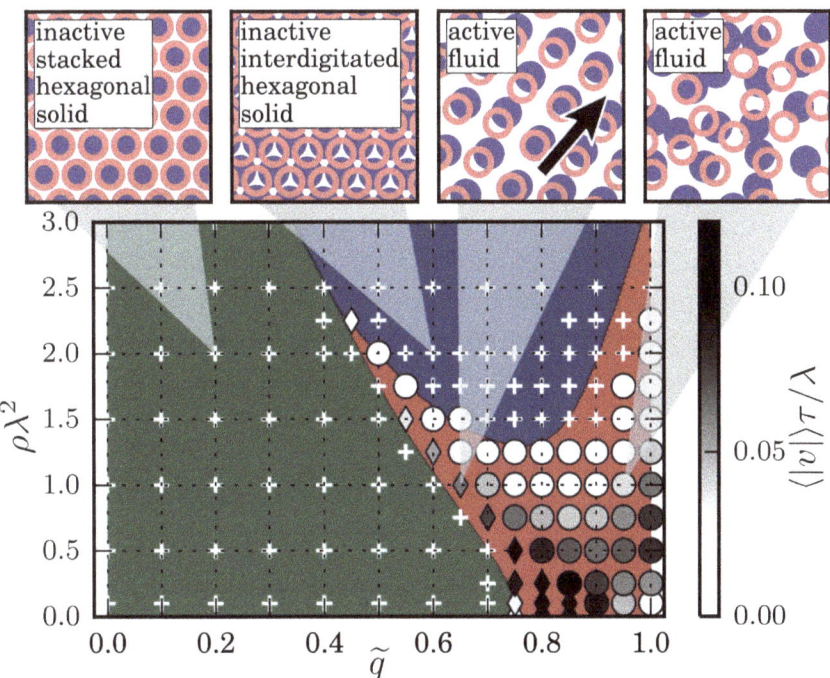

Fig. 7. State diagram in the zero-temperature limit, plotted in the plane of a reduced number density ρ and relative wake charge $\tilde{q}$. Color coding depicts results obtained from the stability analysis, symbols show numerical results. Inactive systems (+) can be either *stacked hexagonal solid* (green background) or *interdigitated hexagonal solid* (blue background). For *active fluid* regimes (○, red background), the average particle velocities are indicated by a gray scale. Diamonds (◇) are used instead of circles if active doublets emerge whose decay time τ_D exceeds a threshold of $10^3\tau$. The states are illustrated by typical snapshots. (Reproduced from Ref. 40 with the permission of AIP Publishing.)

the non-reciprocity governed by the asymmetry in the sensing mechanisms of the two particles. The y-axis shows the two-dimensional particle density. There is a rich steady-state diagram with four different dynamical states involving inactive (i.e., non-moving) states and active ones. The latter are either swarms or orientationally disordered active fluids.

A further interesting setup of particles interacting via chemotaxis is provided by autophoretic Janus colloids. These particles catalyze a chemical reaction on part of their surface only, resulting in a chemical gradient across their own surface. This self-produced gradient sets them into motion via diffusiophoresis or a similar mechanism. Remarkably, a Janus colloid does not only respond to self-produced gradients, but also to gradients produced by other Janus colloids, essentially by chemotaxis (or taxis with respect to another phoretic field). Thus, phoretic Janus colloids interact "chemically" and provide a synthetic analogon to microbiological signaling.[14,41,42] These chemical interactions play a crucial role for the collective behavior of large suspensions of Janus colloids[43] and can generate patterns including clusters,[41–45] traveling waves[43] and continuously moving patterns[42] and, in case of chiral active particles, also spiral patterns and phase separation with traveling waves emerging within the dense phase.[46]

6.   Conclusions

In conclusion, we have discussed models for chemotactic behavior of microorganisms and synthetic particles in a diffusing chemical concentration field focusing on three different scenarios: (i) autochemotaxis of a single particle, (ii) predator–prey models arising from chemotactic coupling to two different chemicals secreted by the predator and the prey, (iii) clusters and collective behavior of many particles coupled via their chemotactic response to chemical fields produced by other particles. Here, attractive autochemotaxis may lead to self-localization, while repulsive autochemotaxis leads to trajectories avoiding their own past. We have also discussed that a moving chemotactic source leads to a front-rear asymmetry in the chemical field resulting in marked scaling laws for predator–prey systems. Finally, chemotaxis in multispecies systems provides an avenue towards a new world of active molecules where we have just started to tap the full potential of the novel cluster formation processes. Collective behavior includes a swarming of chemotactically coupled particles at finite concentrations.

We close with an outlook to future problems. First, at high chemical concentration or strongly coupled chemical fields, the simple diffusion equation picture will break down, calling for new models. This is in particular important for multivalent microions at high concentrations. Recent developments have considered these effects of strong coupling in using nonlinear diffusion equations. These equations can be based on the dynamical version of classical density functional theory, so-called dynamical density functional theory (DDFT).[47–50] In this framework, the equations of motion of a chemical around a point source are given by

$$\frac{\partial c(\vec{r}, t)}{\partial t} = D\vec{\nabla}c(\vec{r}, t)\vec{\nabla}\frac{\delta\mathcal{F}\left[c(\vec{r}, t)\right]}{\delta c(\vec{r}, t)} - \mu c(\vec{r}, t) + \lambda_e\delta(\vec{r}), \qquad (29)$$

where $\mathcal{F}[n([\vec{r}]$ is the equilibrium free energy density functional.[51–53] For a non-interacting system (ideal gas), the functional is known explicitly and in this limit the traditional diffusion equation (1) is recovered. Non-trivial particle correlations as arising from interactions among the chemical species are contained in the functional in the general case and make the diffusion equation nonlinear. In certain cases, linearization is possible and corresponding analytical solutions for the Green's function of diffusing interacting particles can indeed be found within DDFT.[54]

A second important generalization concerns time-dependent secreting rates as embodied in a non-constant function $\lambda_e(t)$ such as, e.g., an emission rate that is periodic in time. This situation has recently been considered[54] and leads to propagating density waves of the chemical around the emitting source. Again, in some special cases, the Green's function can be found analytically within DDFT.[54]

Third, in terms of predator–prey models, the situation of a single predator and a single prey can be generalized towards a herd of prey and to a group of chasers. This has been discussed in the literature within different models, see e.g.,[55–58] but needs to be extended within the chemotactic context. Efficient chase and escape strategies[59] may depend on the details of predator/prey perception.

Fourth, most of our considerations were done in the bulk. Confinement near system walls and crowding situations will change both the diffusion of the chemical as well as the chemotactic dynamics. Similarly, chemotaxis in complex environments, such as traveling waves, may lead to interesting transport effects.[60] We are just at the beginning of a systematic understanding of chemotaxis in complex environments.

Finally, we have considered the evaporation of different chemicals by a constant rate in our modeling. If two different chemicals which, e.g., govern a predator–prey system or the formation of the active molecule, react between themselves this would constitute a more complicated and highly interesting problem with new scenarios induced by coupling nonlinear reaction-diffusion equations to the chemical kinetics and the particle motions.

Acknowledgment

H.L. gratefully acknowledges support by the Deutsche Forschungsgemeinschaft (DFG) through grant LO 418/19-1.

References

1. M. Kollmann, L. Lovdok, and K. Bartholome, Design principles of a bacterial signalling network, *Nature* **438**, 504 (2005).
2. U. B. Kaupp, N. D. Kashikar, and I. Weyand, Mechanisms of sperm chemotaxis, *Annu. Rev. Physiol.* **70**, 93 (2008).
3. W. C. K. Poon, Soft matter: From synthetic to biological materials, *Lecture Notes of the 39th IFF Spring School, Forschungszentrum Julich GmbH.* **11**, 1 (2008).

4. C. Hoell and H. Löwen, Theory of microbe motion in a poisoned environment, *Phys. Rev. E.* **84**, 42903 (2011).

5. R. Metzler and J. Klafter, The random walk's guide to anomalous diffusion: a fractional dynamics approach, *Phys. Rep.* **339**, 1 (2000).

6. E. J., R. G. Winkler, and G. Gompper, Physics of microswimmers–single particle motion and collective behavior: a review, *Rep. Prog. Phys.* **78**, 56601 (2015).

7. A. Zoettl and H. Stark, Emergent behavior in active colloids, *J. Phys. Condens. Matter.* **28**, 253001 (2016).

8. C. Bechinger, R. Di Leonardo, H. Löwen, C. Reichhardt, and G. Volpe, Active particles in complex and crowded environments, *Rev. Mod. Phys.* **88**, 045006 (2016).

9. A. M. Menzel, Tuned, driven, and active soft matter, *Phys. Rep.* **554**, 1 (2015).

10. A. Sengupta, T. Kruppa, and H. Löwen, Chemotactic predator-prey dynamics, *Phys. Rev. E.* **83**, 31914 (2011).

11. J.-P. Hansen and H. Löwen, Effective interactions between electric double-layers, *Annu. Rev. Phys. Chem.* **51**, 209 (2000).

12. H. Löwen, Charged rodlike colloidal suspensions: An ab initio approach, *J. Chem. Phys.* **100**, 6738 (1994).

13. J. D. Murray, *Mathematical Biology. II Spatial Models and Biomedical Applications Interdisciplinary Applied Mathematics.* Springer-Verlag Berlin Heidelberg (2003).

14. M. Meyer, L. Schimansky-Geier, and P. Romanczuk, Active Brownian agents with concentration-dependent chemotactic sensitivity, *Phys. Rev. E.* **89**, 022711 (2014).

15. Y. Tsori and P.-G. de Gennes, Self-trapping of a single bacterium in its own chemoattractant, *EPL* **66**, 599 (2004).

16. R. Grima, Strong-coupling dynamics of a multicellular chemotactic system, *Phys. Rev. Lett.* **95**, 128103 (2005).

17. R. Grima, Phase transitions and superuniversality in the dynamics of a self-driven particle, *Phys. Rev. E.* **74**, 011125 (2006).

18. G. Gompper, C. Bechinger, S. Herminghaus, R. E. Isele-Holder, U. B. Kaupp, H. Löwen, H. Stark, and R. G. Winkler, Microswimmers — from single particle motion to collective behaviour, *Eur. Phys. J. ST.* **226**, 2061 (2016).

19. B. ten Hagen, S. van Teeffelen, and H. Löwen, Brownian motion of a self-propelled particle, *J. Phys. Cond. Matt.* **23**, 194119 (2011).

20. X. Zheng, B. ten Hagen, A. Kaiser, M. Wu, H. Cui, Z. Silber-Li, and H. Löwen, Non-Gaussian statistics for the motion of self-propelled Janus particles: Experiment versus theory, *Phys. Rev. E.* **88**, 032304 (2013).

21. F. Kümmel, B. ten Hagen, R. Wittkowski, I. Buttinoni, R. Eichhorn, G. Volpe, H. Löwen, and C. Bechinger, Circular motion of asymmetric self-propelling particles, *Phys. Rev. Lett.* **110**, 198302 (2013).

22. A. Sengupta, S. van Teeffelen, and H. Löwen, Dynamics of a microorganism moving by chemotaxis in its own secretion, *Phys. Rev. E.* **80**, 31122 (2009).

23. C. Jin, C. Krüger, and C. C. Maass, Chemotaxis and autochemotaxis of self-propelling droplet swimmers, *Proc. Natl. Acad. Sci. USA.* **114**, 5089 (2017).

24. J. Bewerunge, A. Sengupta, R. F. Capellmann, F. Platten, S. Sengupta, and S. U. Egelhaaf, Colloids exposed to random potential energy landscapes: From particle number density to particle–potential and particle–particle interactions, *J. Chem. Phys.* **145**, 044905 (2016).

25. M. Mijalkov, A. McDaniel, J. Wehr, and G. Volpe, Engineering sensorial delay to control phototaxis and emergent collective behaviors, *Phys. Rev. X.* **6**, 011008 (2016).

26. W. T. Kranz, A. Gelimson, K. Zhao, G. C. L. Wong, and R. Golestanian, Effective dynamics of microorganisms that interact with their own trail, *Phys. Rev. Lett.* **117**, 038101 (2016).

27. A. Gelimson and R. Golestanian, Collective dynamics of dividing chemotactic cells, *Phys. Rev. Lett.* **114**, 028101 (2015).

28. J. Taktikos, V. Zaburdaev, and H. Stark, Modeling a self-propelled autochemotactic walker, *Phys. Rev. E.* **84**, 041924 (2011).

29. C. Valeriani, R. J. Allen, and D. Marenduzzo, Non-equilibrium dynamics of an active colloidal "chucker", *J. Chem. Phys.* **132**, 204904 (2010).

30. A. V. Ivlev, J. Bartnick, M. Heinen, C.-R. Du, V. Nosenko, and H. Löwen, Statistical mechanics where Newton's third law is broken, *Phys. Rev. X.* **5**, 011035 (2015).

31. G. Oshanin, O. Vasilyev, P. L. Krapivsky, and J. Klafter, Survival of an evasive prey, *Proc. Natl. Acad. Sci. USA.* **106**, 13696 (2009).

32. H. Jashnsaz, G. G. Anderson, and S. Presse, Statistical signatures of a targeted search by bacteria, *Phys. Biol.* **14**, 065002 (2017).

33. R. Niu, D. Botin, J. Weber, A. Reinmüller, and T. Palberg, Assembly and speed in ion-exchange-based modular phoretic microswimmers, *Langmuir* **33**, 3450–3457 (2017).

34. F. Schmidt, B. Liebchen, H. Löwen, and G. Volpe, Light-controlled assembly of active colloidal molecules, *arXiv: 1801.06868* (J. Chem. Phys., in press) (2018).

35. J. Bleibel, S. Dietrich, A. Domínguez, and M. Oettel, Shock waves in capillary collapse of colloids: A model system for two-dimensional screened Newtonian gravity, *Phys. Rev. Lett.* **107**, 128302 (2011).

36. J. Bartnick, M. Heinen, A. V. Ivlev, and H. Löwen, Structural correlations in diffusiophoretic colloidal mixtures with nonreciprocal interactions, *J. Phys. Cond. Matt.* **28**, 025102 (2016).

37. R. Soto and R. Golestanian, Self-assembly of catalytically active colloidal molecules: Tailoring activity through surface chemistry, *Phys. Rev. Lett.* **112**, 068301 (2014).

38. R. Soto and R. Golestanian, Self-assembly of active colloidal molecules with dynamic function, *Phys. Rev. E.* **91**, 052304 (2015).

39. R. Niu, T. Palberg, and T. Speck, Self-assembly of colloidal molecules due to self-generated flow, *Phys. Rev. Lett.* **119**, 028001 (2017).

40. J. Bartnick, A. Kaiser, H. Löwen, and A. V. Ivlev, Emerging activity in bilayered dispersions with wake-mediated interactions, *J. Chem. Phys.* **144**, 224901 (2016).

41. S. Saha, R. Golestanian, and S. Ramaswamy, Clusters, asters, and collective oscillations in chemotactic colloids, *Phys. Rev. E.* **89**, 062316 (2014).

42. B. Liebchen, D. Marenduzzo, I. Pagonabarraga, and M. E. Cates, Clustering and pattern formation in chemorepulsive active colloids, *Phys. Rev. Lett.* **115**, 258301 (2015).

43. B. Liebchen, D. Marenduzzo, and M. E. Cates, Phoretic interactions generically induce dynamic clusters and wave patterns in active colloids, *Phys. Rev. Lett.* **118**, 268001 (2017).

44. O. Pohl and H. Stark, Dynamic clustering and chemotactic collapse of self-phoretic active particles, *Phys. Rev. Lett.* **112**, 238303 (2014).

45. O. Pohl and H. Stark, Self-phoretic active particles interacting by diffusiophoresis: A numerical study of the collapsed state and dynamic clustering, *Euro. Phys. J. E.* **38**, 93 (2015).

46. B. Liebchen, M. E. Cates, and D. Marenduzzo, Pattern formation in chemically interacting active rotors with self-propulsion, *Soft Matter.* **12**, 7259 (2016).

47. U. M. B. Marconi and P. Tarazona, Dynamic density functional theory of fluids, *J. Chem. Phys.* **110**, 8032 (1999).

48. A. J. Archer and R. Evans, Dynamical density functional theory and its application to spinodal decomposition, *J. Chem. Phys.* **121**, 4246 (2004).

49. P. Espanol and H. Löwen, Derivation of dynamical density functional theory using the projection operator technique, *J. Chem. Phys.* **131**, 244101 (2009).

50. H. Löwen. Dynamical density functional theory for Brownian dynamics of colloidal particles, In: *Variational Methods in Molecular Modeling*, Chapter 9, (J. Wu, (ed.)), Springer (2017), p. 255.

51. R. Evans, The nature of the liquid-vapour interface and other topics in the statistical mechanics of non-uniform, classical fluids, *Adv. Phys.* **28**, 143 (1979).

52. Y. Singh, Density-functional theory of freezing and properties of the ordered phase, *Phys. Rep.* **207**, 351 (1991).

53. H. Löwen, Melting, freezing and colloidal suspensions, *Phys. Rep.* **237**, 249 (1994).

54. H. Löwen and M. Heinen, Dynamical density functional theory for the diffusion of injected brownian particles, *Eur. Phys. J. ST.* **223**, 3113 (2014).

55. M. Schwarzl, A. Godec, G. Oshanin, and R. Metzler, A single predator charging a herd of prey: Effects of self volume and predator–prey decision-making, *J. Phys. A.* **49**, 225601 (2016).

56. C. Mejía-Monasterio, G. Oshanin, and G. Schehr, First passages for a search by a swarm of independent random searchers, *J. Stat. Mech.* **2011**, P06022 (2011).

57. R. Nishi, A. Kamimura, K. Nishinari, and T. Ohira, Group chase and escape with conversion from targets to chasers, *Physica A.* **391**, 337 (2012).

58. T. Saito, T. Nakamura, and T. Ohira, Group chase and escape model with chasers interaction, *Physica A.* **447**, 172 (2016).

59. O. Bénichou, C. Loverdo, M. Moreau, and R. Voituriez, Intermittent search strategies, *Rev. Mod. Phys.* **83**, 81 (2011).

60. A. Geiseler, P. Hänggi, F. Marchesoni, C. Mulhern, and S. Savel'ev, Chemotaxis of artificial microswimmers in active density waves, *Phys. Rev. E.* **94**, 012613 (2016).

Chapter 21

Adopting the Boundary Homogenization Approximation from Chemical Kinetics to Motile Chemically Active Particles

M. N. Popescu[*,‡] and W. E. Uspal[†]

*Max-Planck-Institut für Intelligente Systeme,
Heisenbergstr. 3, D-70569, Stuttgart, Germany
IV. Institut für Theoretische Physik,
Universität Stuttgart Pfaffenwaldring 57,
D-70569 Stuttgart, Germany
†Department of Mechanical Engineering
University of Hawai'i at Mānoa, USA
‡popescu@is.mpg.de

Chemically active particles, which are capable of moving within a fluid environment by promoting at (parts of) their surfaces chemical reactions in the surrounding solution, are of significant interest both for basic science, due to their intrinsic non-equilibrium nature, as well as for applied science, due to their envisioned role as carriers operating in novel lab-on-a-chip devices. Irrespective of the exact mechanism through which the non-equilibrium distribution of molecular species ("chemical field") in the suspending medium is transduced to motility of the particle and hydrodynamic flow of the solution, that distribution must be explicitly determined. Here we discuss the use of the so-called boundary homogenization approximation, developed in the calculations of the reaction rate in chemical kinetics, in simple models of chemically active particles. The analytical results obtained based on this approximation capture qualitatively, and even semi-quantitatively, the far-field distribution of chemical species and hydrodynamic flow induced by the active particle.

1. Introduction

Recently, there has been increasing interest in the development of chemically active particles which are capable of moving within a fluid environment by promoting at (parts of) their surfaces chemical reactions involving their surrounding solution.[1–5] Such particles exhibit motility in the absence of external forces or torques acting on them or on the fluid. Their motion is intrinsically connected with a "chemical field", i.e., the distribution of number densities of the various chemical species present in the solution, and with hydrodynamic flow of the solution around the particle. Typical experimental realizations of such systems involve aqueous suspending solutions, molecularly sized, fast diffusing reactants and products, and micrometer-sized particles moving at speeds of the order of a couple of particle diameters per second. Therefore, they operate at very small Reynolds number in Newtonian fluids,[1,4,5] a regime in which viscous friction dominates over inertial effects as far as hydrodynamics is concerned. Consequently, achieving motility is not a straightforward issue: It requires the development of ways of breaking the time-reversibility of the Stokes equations (i.e., to avoid the "scallop theorem" of Purcell, Edward M.).[6,7] These systems are thus valuable both from a theoretical perspective, serving as paradigms for non-equilibrium processes, as well as from an application viewpoint, according to which active particles are envisioned to play the role of carriers ("engines") in novel lab-on-a-chip devices.

A wide variety of such particles has been proposed and studied experimentally.[8–28] The mechanisms of motility have been the topic of numerous theoretical studies.[29–47] Irrespective of the exact mechanism through which the non-equilibrium chemical field is transduced to the motility of the particle, the chemical field must be determined in order to find the (translational and rotational) velocity of the particle at steady state.[30,34,37,48] Certain significant simplifications are possible in many cases by noting that the aforementioned available experimental realizations typically deal — as discussed above — with molecularly sized, fast diffusing reactants and products and with hydrodynamic flow velocities of the order of micrometer per second. Therefore, typically the Péclet numbers of the reactant and product species are very small; accordingly, transport of the reactant and product species by diffusion dominates that by advection.[1,2,5,30,36] In this case, the chemical field decouples from the hydrodynamics and, at the steady state, it is determined as the solution of Laplace equation subject to suitable boundary conditions.[2,5,30,31,34,37,39] The latter account for the reaction(s) taking place at the surface of the particle and, typically, for prescribed bulk (i.e., far from the particle) values of the number densities of the various chemical species present in the solution.

In such cases, it was noticed by Ref. 44 that the problem of determining the chemical field of an active particle is similar to the one in the context of chemical kinetics concerning the calculation of the reaction rate,[49–52] formulated as a diffusion problem by Smoluchowski, Marian in the classic theory of kinetics of coagulation[53] (and later generalized by Collins and Kimball[54]). Regarding the reaction rate, the case of spherical objects with anisotropic reactivity, which is most relevant for the discussion here, has been the subject of numerous studies, both concerning the steady state (including the cases in which the reactants experience an interaction field or in which the diffusion constant varies spatially)[55–68] as well as the time dependence.[69–72] The theory or reaction rate has been further extended to account for the effect of hydrodynamic interactions between the reacting species (see, e.g., Ref. 73) as well as for competitive effects between spatially distributed reactive particles.[56,74–77]

In determining the steady-state chemical field of an active particle that promotes on its surface, e.g., a first, order chemical reaction,[31,36,37,44,78] one of the significant difficulties in theoretical analyses is represented by the mixed boundary conditions at the surface of the particle: chemical reaction on some parts, chemically inert on the rest. There are cases in which the solution of such equations can be obtained analytically (see Ref. 79 for a rather exhaustive list), but these cases are the exception rather than the rule; therefore, in general, the theoretical analysis is restricted to approximate solutions (see, e.g., Refs. 69 and 76). In view of the aforementioned similarity with the study of reaction rate in chemical kinetics, one can adapt and make use of the wealth of results and techniques available in that field for studies of the motility of chemically active particles.

Here we discuss the use of the so-called boundary homogenization method,[65–68] which is known to provide very good approximations for the reaction rate, to the case of chemically active spherical particles.[44] We focus on the analytical estimates for the first four coefficients in the multipole expansion of the chemical field. (These coefficients are important because, e.g., for simple models of self-motility they determine the far-field hydrodynamic flow induced by the particle[78,80] and the leading order contributions to the interaction of such particles with boundaries.[30,80–82]) We show that for simple geometries, such as those shown in Fig. 1, the results obtained within this approach compare very well with the exact ones obtained by numerically solving the corresponding mixed boundary problem.

2. Chemical Field of Model Active Particle

The system of interest in this work is a spherical particle of radius R, with inhomogeneous surface properties, immersed in a well-stirred bath of solvent, S, and reactant, A, molecular species. On parts Σ_a of the surface of the particle (the dark gray areas in Fig. 1), an "annihilation" chemical reaction $A \to S$ is promoted; the rest, Σ_i, of the surface (light gray areas in Fig. 1) is chemically *inert*. We shall consider various geometries with axial symmetries for the reactive regions[30] (see Fig. 1): (a) a spherical cap of opening angle θ_0; (b) two spherical caps of same opening angle θ_0; and (c) a ring, centered on the equator (the complement of the two caps of opening angle θ_0 from (b)). The particle is impenetrable to both solvent and reactant molecules. Furthermore, we shall consider the (dilute) limit of small number density $c(\mathbf{r})$ of reactant molecules, i.e., we effectively

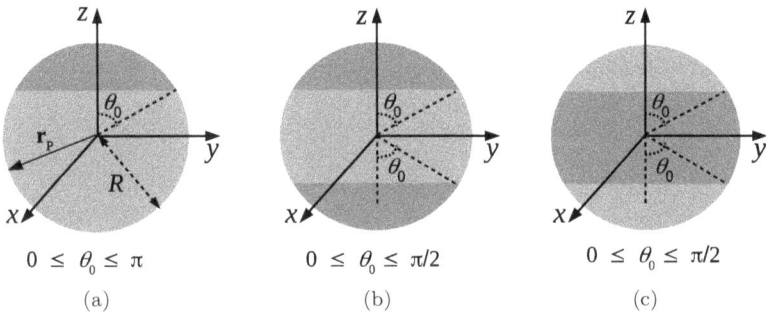

Fig. 1. (a)–(c) Spherical particles with inhomogeneous surface properties immersed in a well stirred bath of reactants (not shown); dark gray areas denote the regions Σ_a on the particle where a reaction $A \to S$ can take place, while gray depicts the inert (non-reactive) areas Σ_i. In all three geometries shown here the particles have axial symmetry and the angles θ_0 indicate the opening angle of the corresponding spherical caps.

approximate the reactant as an ideal gas of non-interacting, point-like molecules freely diffusing in the solution with diffusion constant D. The system is unbounded and is in contact with a reservoir of particles which fixes the chemical potentials of the solvent and reactant species; thus, implicitly the value $c_\infty > 0$ of the number density of reactant molecules in the bulk (far away from the particle) is also fixed. For the calculation of the chemical field, it is convenient to consider the particle to be immobile (e.g., by being held within an optical trap); the system of coordinates is chosen such that the origin coincides with the center of the particle, and the z-axis is along the axis of symmetry of the particle (see Fig. 1). The excess interaction of the reactant molecules, over that of the solvent molecules, with the surface of the particle gives rise to body forces on the surrounding solution; this induces hydrodynamic flow of the solution. We further assume, motivated by the typical experimental realizations, that the diffusion of the reactant dominates the advective transport by such flows (see the corresponding discussion in Section 1). We shall focus on the case in which the annihilation reaction obeys first-order kinetics, i.e., the rate of conversion of A to S on the small area element centered at the point $\mathbf{r}_P \in \Sigma_a$ is proportional to the local number density $c(\mathbf{r}_P)$.

2.1. *Mathematical formulation*

Under these assumptions, at steady state the number density $c(\mathbf{r})$ of reactant molecules is the solution of the Laplace equation

$$\nabla^2 c(\mathbf{r}) = 0 \quad \text{for } |\mathbf{r}| > R, \tag{1}$$

subject to boundary conditions (BCs) at infinity (far from the particle)

$$c(|\mathbf{r}| \to \infty) \to c_\infty \tag{2a}$$

and at the surface of the particle

- on the reactive part

$$\mathbf{n} \cdot [-D\nabla c(\mathbf{r})] = -kc(\mathbf{r}) \quad \text{for } \mathbf{r} \in \Sigma_a, \tag{2b}$$

- on the inert part

$$\mathbf{n} \cdot [-D\nabla c(\mathbf{r})] = 0 \quad \text{for } \mathbf{r} \in \Sigma_i, \tag{2c}$$

where $\mathbf{n}$ denotes the normal to the surface of the particle pointing into the solution and the constant k (units of m/s) is positive (consistent with an "annihilation" reaction).[a]

The solution $c(\mathbf{r})$ of Eq. (1), which has axial symmetry and satisfies the boundary condition at infinity (Eq. (2a)), can be conveniently expressed as a multipole series expansion in terms of the spherical coordinates (r, θ, ϕ) (defined in the usual way)

$$c(r, \theta) = c_\infty \left[1 + \sum_{n=0}^{\infty} a_n \left(\frac{R}{r} \right)^{n+1} P_n(\cos \theta) \right], \tag{3}$$

where P_n denotes the Legendre polynomial of degree n and the dimensionless coefficients $\{a_n\}_{n \geq 0}$ are to be determined from the BC at the surface of the particle, Eqs. (2b) and (2c).

---

[a]In principle, this description holds only in the region outside the range of the interaction between the reactant molecules and the surface of the particle. Since these interactions are short-ranged, with a range λ much smaller then the typical values (micrometers) of the radius R of the particle, the error made by imposing the boundary conditions at R rather than at $R + \lambda$ is negligibly small.

Before proceeding with the calculation of the coefficients $\{a_n\}_{n\geq0}$, we note that the particle geometries in the cases (b) and (c) are symmetric with respect to the transformation $z \mapsto -z$, i.e., $\theta \mapsto \pi-\theta$. Therefore, in these cases the solution $c(\mathbf{r})$ must have the additional property $c(r,\theta) = c(r, \pi - \theta)$. By noting that $\cos(\pi - \theta) = -\cos\theta$, and that the Legendre polynomials posses parity, i.e., $P_n(-x) = (-1)^n P_n(x)$, one concludes that for the geometries in (b) and (c) all the coefficients a_n of odd index n vanish.

2.2. Dual series equation for the coefficients $\{a_n\}_{n\geq0}$

By plugging Eq. (3) into Eqs. (2b) and (2c), and defining the Damköhler number (dimensionless) as

$$\mathrm{Da} := \frac{kR}{D}, \tag{4}$$

one arrives at the following so-called "dual series" equation[69,76,79] for the yet to be determined coefficients $\{a_n\}_{n\geq0}$

$$-\sum_{n=0}^{\infty}(n+1)a_n P_n(\cos\theta)$$

$$= \mathrm{Da} \times \begin{cases} 1 + \sum_{n=0}^{\infty}a_n P_n(\cos\theta) & \text{for } \theta \in \Sigma_a, \\ 0 & \text{for } \theta \in \Sigma_i. \end{cases} \tag{5}$$

After multiplying both sides of Eq. (5) with $\sin\theta\, P_k(\cos\theta)$, integrating over $\theta \in [0, \pi]$, exploiting the orthogonality of the Legendre polynomials,

$$\int_0^\pi d\theta\, \sin\theta\, P_n(\cos\theta) P_k(\cos\theta) = \frac{2}{2n+1}\delta_{n,k} \tag{6}$$

and defining

$$H_{n,k} := \int_{\Sigma_a} d\theta\, \sin\theta\, P_n(\cos\theta) P_k(\cos\theta), \tag{7a}$$

$$b_n := -\int_{\Sigma_a} d\theta\, \sin\theta\, P_n(\cos\theta), \tag{7b}$$

one arrives at the following infinite system of linear equations for the coefficients $\{a_n\}_{n \geq 0}$

$$\sum_{k \geq 0} \left(H_{n,k} + \frac{1}{\mathrm{Da}} \frac{2(k+1)}{2k+1} \delta_{n,k} \right) a_k = b_n, n = 0, 1, \ldots . \quad (8)$$

In general, the above system of equations cannot be analytically solved (i.e., the operator acting on the infinitely dimensional vector $(a_0, a1, \ldots)^T$ cannot be inverted analytically); for the geometries of interest here (Fig. 1), various analytical systematic approximations, allowing one to obtain accurate estimates for (a number of) the coefficients a_n, have been proposed and discussed in detail (see, e.g., Refs. 69 and 76). The alternative path, which will be followed here (see, c.f., Section 3), is to obtain numerical solutions by truncating the system at a sufficiently large order $n_{\max} = N$.

2.3. *Approximate analytical solution via boundary homogenization*

The so-called "boundary homogenization" approximation was introduced by Ref. 65 (see also Ref. 67) as a mean to obtain explicit analytical estimates for the reaction rate of spherical objects with anisotropic reactivity. The method was later refined and further extended to more complex shapes (e.g., patchy planes or stripped rods).[68,83–88] In the context of chemically active colloids, this approximation has been first used recently by Ref. 44.

The approximation consists in replacing the position dependent right-hand side (rhs) of the boundary condition at the reactive part of the surface of the particle, Eq. (2b), with a constant one, i.e.,

$$\mathbf{n} \cdot [-D\nabla c(\mathbf{r})] = -kc_\infty Q, \quad \text{for } \mathbf{r} \in \Sigma_a. \quad (9)$$

The dimensionless constant $Q > 0$ (consistent with "annihilation" reaction) is determined from the self-consistency condition that the former Eq. (2b) holds *only* as an integral over the active region Σ_a, i.e.,

$$kc_\infty Q \mathcal{A}(\Sigma_a) = - \int_{\Sigma_a} ds\, k\, c(R, \theta), \tag{10}$$

where $\mathcal{A}(\Sigma_a)$ denotes the area of the surface Σ_a and ds denotes the area element on the surface of the particle.

By plugging Eq. (3) into the left-hand side (lhs) of Eqs. (9) and (2c), multiplying both sides of the equations with $\sin\theta\, P_k(\cos\theta)$, integrating over $\theta \in [0, \pi]$, and using the definition of b_n (Eq. (7b)) one arrives at

$$a_n = \frac{2n+1}{2(n+1)} b_n \, \mathrm{Da}\, Q. \tag{11}$$

Combining Eqs. (3), (11), and (10), and introducing the geometric factor[65,73]

$$f_g := \frac{\mathcal{A}(\Sigma_a)}{4\pi R^2}, \tag{12}$$

i.e., the proportion of the area of the particle taken by the active region, one arrives at

$$Q := \left(1 + \frac{1}{4}\mathrm{Da} \sum_{n \geq 0} \frac{2n+1}{n+1} \frac{b_n^2}{f_g} \right)^{-1}. \tag{13}$$

Equations (13) and (11) provide the approximate solution for the coefficients $\{a_n\}_{n \geq 0}$.

3. Results and Discussion

The choice of the geometries shown in Fig. 1 allows one to explore the dependence of the chemical field both on the extent of the reactive part (i.e., varying θ_0) as well as on the spatial distribution of the reactive parts at fixed total extent (i.e., compare the three cases at the same value of the area of the dark gray patches).[30] In what concerns the coefficients $\{a_n\}$, the extent of the reactive part is characterized by the geometrical factor f_g (Eq. ((12)), while the geometry of reactive area is encoded in the parameters $H_{n,k}$ and b_n (Eq. (7)).

For the geometries shown in Fig. 1, most of these parameters can be computed analytically[80]; in terms of $w_0 := \cos\theta_0$, they are given by

- **geometry (a)** ($\theta_0 \in [0, \pi]$, $w_0 \in [-1, 1]$):

$$f_g^{(a)}(w_0) = \frac{1 - w_0}{2},$$

$$b_n^{(a)}(w_0) = \begin{cases} -1 + w_0 & \text{for } n = 0, \\ \dfrac{(1 - w_0^2)^{1/2}}{n(n+1)} P_n^1(w_0) & \text{for } n \geq 1, \end{cases} \tag{14}$$

$$H_{n,k}^{(a)}(w_0) = \begin{cases} \int_{w_0}^1 dx\, [P_n(x)]^2 & \text{for } k = n, \\ -(1 - w_0^2)^{1/2} \dfrac{P_k^1(w_0)P_n(w_0) - P_n^1(w_0)P_k(w_0)}{k(k+1) - n(n+1)} & \text{for } k \neq n, \end{cases}$$

where P_n^1 is the associated Legendre polynomial of degree n and order 1.

- **geometry (b)** ($\theta_0 \in [0, \pi/2]$, $w_0 \in [0, 1]$):

$$\begin{aligned} f_g^{(b)}(w_0) &= 2f_g^{(a)}(w_0), \\ b_n^{(b)}(w_0) &= [1 + (-1)^n]\, b_n^{(a)}(w_0), \\ H_{n,k}^{(b)}(w_0) &= [1 + (-1)^{n+k}]\, H_{n,k}^{(a)}(w_0). \end{aligned} \tag{15}$$

- **geometry (c)** ($\theta_0 \in [0, \pi/2]$, $w_0 \in [0, 1]$):

$$\begin{aligned} f_g^{(c)}(w_0) &= 1 - 2f_g^{(a)}(w_0), \\ b_n^{(c)}(w_0) &= -2\delta_{n,0} - [1 + (-1)^n]\, b_n^{(a)}(w_0), \\ H_{n,k}^{(c)}(w_0) &= \frac{2}{2n+1}\delta_{n,k} - [1 + (-1)^{n+k}]\, H_{n,k}^{(a)}(w_0). \end{aligned} \tag{16}$$

In what concerns the calculation of the coefficients a_0–a_3 by solving the corresponding linear system, Eq. (8), it turns out that the truncation converges very fast for all values w_0 in the corresponding ranges, as well as for values of Da in the broad range $10^{-2} \leq \text{Da} \leq 10^2$ (see also the discussion of the dependence on Da below). This is in agreement with the findings of previous reports.[69,76] All the results reported here correspond to the choice $N = 30$ for the truncation-order; although this is a relatively large value, we note that solving such system on modern computers is a very quick and simple task[b]. Regarding the calculations within the boundary homogenized approximation, the series in the expression for Q, Eq. (13), was calculated by truncating it to 100 terms at small values of θ_0 ($\theta_0 \leq 0.1$) and to 30 terms otherwise. These cut-offs ensured the convergence of the series while maintaining the CPU-time demands within reasonable limits.

In addition to the dependence on the geometry of the reactive areas, the number density of reactant molecules (i.e., the coefficients $\{a_n\}$) depends also on the relative importance of the rate of annihilation by the chemical reaction at the surface of the particle, as compared with the transport of reactant molecules to the particle by diffusion; this comparison is encoded by the Damköhler number Da (see Eqs. (8), (11), and (13)).

The case in which the Damköhler number is very small, $\mathrm{Da} \ll 1$, corresponds to the so-called reaction-limited regime. In this case, the rhs of the BC in Eq. (2b) can be accurately approximated by kc_∞, i.e., at the reactive part of the particle one has effectively a "constant flux" boundary condition (see, e.g., Ref. 89). This implies that in this regime the boundary homogenization approximation, Eq. (9), is practically *exact*. Furthermore, from Eq. (13) one infers that $Q(\mathrm{Da} \ll 1) \simeq 1$; therefore, Eq. (11) (which for $\mathrm{Da} \ll 1$ is also practically an exact result) renders

$$a_n(\mathrm{Da} \ll 1) \propto \mathrm{Da}, \tag{17}$$

where the n-dependent proportionality factor follows immediately from Eq. (11) and the parameter b_n of the corresponding geometry, Eqs. (14)–(16). Therefore, in the followings we focus the discussion to the cases $\mathrm{Da} \sim \mathcal{O}(1)$ and $\mathrm{Da} \gg 1$ (the latter corresponds to

---

[b]For example, by taking advantage of the coefficients as given by Eqs. (14)–(16) and employing Mathematica 11 under Linux on a laptop, we obtained the corresponding solutions for all 3 geometries, for 50 values of the parameter θ_0 and another 50 for the parameter Da, in less than an hour.

the so-called diffusion-limited regime). In the diffusion-limited regime $Da \gg 1$, in order for the number density $c(\mathbf{r})$ to remain bounded, the BC on the reactive part of the particle, Eq. (2b), turns into that of a perfectly absorbing sink, i.e., $c(\mathbf{r} \in \Sigma_a) \approx 0$. This implies that in this limit the coefficients $\{a_n\}$ become practically independent of Da. This qualitative feature is well captured by the boundary homogenization approximation: since $Q(Da \gg 1) \simeq Da^{-1}$ (Eq. (13)), from Eq. (11) it follows that $a_n(Da \gg 1) \propto Da^0$.

After these general considerations, we turn now specifically to the discussion of the coefficients a_0–a_3. Their importance for the motility of the chemically active particles can be intuitively grasped in the context of self-phoresis of particles with phoretic-slip hydrodynamic boundary condition at their surface, i.e., tangential relative velocity proportional to the gradient along the surface of the number density $c(\mathbf{r})$ in the particular case that the proportionality factor (the so-called "phoretic mobility") does not vary over the surface of the particle.[30,39,80] In this case, the coefficient a_1 determines the velocity of the (force- and torque-free) motile particle,[30,48] the coefficient a_2 determines the magnitude of the leading order (in the far-field) hydrodynamic flow of the particle, decaying as r^{-2}, while a_3 and a_1 determine the amplitude of the first subleading term in the hydrodynamic flow (decaying as r^{-3}).[40,80] The monopole coefficient a_0 (which from the perspective of chemical kinetics is directly connected with the effective reaction rate, see below), does not contribute to the motility in an unbounded suspension. However, its significance arises from that it contributes to the leading order (in the far-field) interaction of such an active particle with a confining boundary.[30,80–82,90–93]

3.1. *The monopole amplitude a_0 and the effective reaction rate*

In chemical kinetics, the effective reaction rate $\mathcal{K}$ (units of m^3/s) is defined as the flux of reactants that would pass through the surface of the particle, as if the particle were porous to the reactant, divided by the bulk density. For the system under consideration here, by using

Eq. (3) one arrives at

$$K := \frac{1}{c_\infty} \int_{\Sigma_a \cup \Sigma_i} -\mathbf{n} \cdot [-D\nabla c(\mathbf{r})]ds = -\underbrace{4\pi RD}_{K_D} a_0, \qquad (18)$$

where K_D is the classic Smoluchowski expression of the rate constant; i.e., the dependences of the effective rate constant on the geometry and on Da are encoded in the monopole amplitude a_0 of the multipole expansion of the number distribution of the reactant.

As discussed above, in the limit of small Da numbers the boundary homogenization approximation becomes quasi-exact. Combining Eqs. (17) (with $Q \simeq 1$) and (14)–(16) for the corresponding values of b_0 and f_g, it follows from Eq. (11) that for Da $\ll 1$ one has $a_0(\omega_0) \simeq -\mathrm{Da}f_g(\omega_0)$ irrespective of the geometry of the reactive part of the particle. Thus, for Da $\ll 1$ the effective reaction rate is proportional to Da; factoring this out, the remainder is decreased, with respect to the Smoluchowski value, solely by the geometric factor f_g, consistent with the discussion above of an effective "constant flux" BC at the surface of the particle.

In Fig. 2 we show the parametric dependence of $-a_0(\omega_0) = K/K_D$ on the geometric factor $f_g(\omega_0)$ (which can vary from $f_g = 0$, for completely inert particles, to $f_g = 1$, when the whole surface is reactive) for all three geometries from Fig. 1 and for several values Da ≥ 1. (The results at small values of Da are not shown because, as discussed above, in that regime the analytical theory is practically exact.) This representation allows one to easily read-off the influence of the exact geometry of the reactive part of the particle for a given total area of the reactive part (i.e., a fixed value of f_g). It can be seen that in all cases the boundary homogenization approximation provides very accurate estimates for a_0, which is consistent with previous reports.[68,83,85,87] The results show that for each geometry, and at each value of Da, the monopole amplitude depends significantly on the extent of the part that is active, i.e., on f_g, and it increases with f_g, as expected. For large Da, the dependence on f_g is clearly non-linear in f_g, which reflects that the reaction rate is reduced from the Smoluchowski value beyond the trivial decrease due to the active area

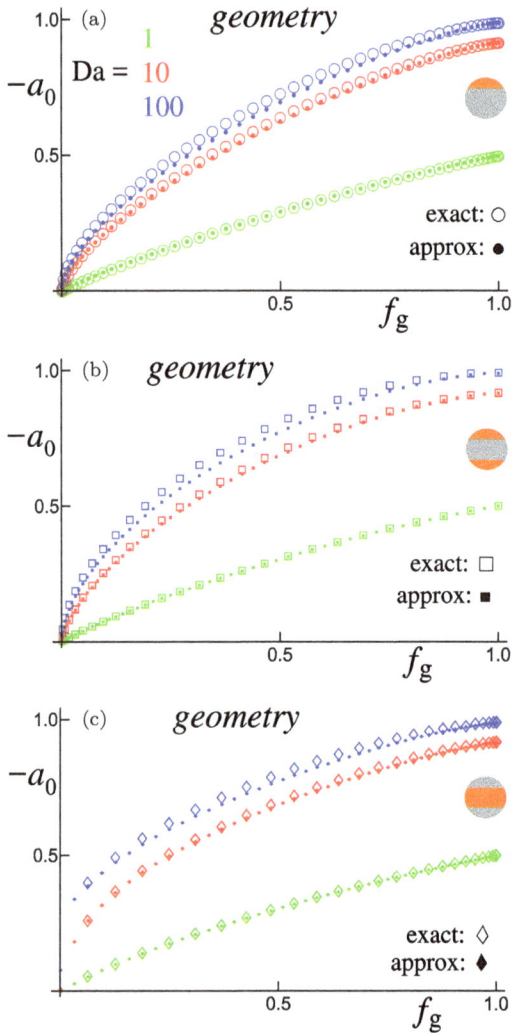

Fig. 2. The dependence of the negative of the monopole amplitude, $-a_0$ (which equals $\mathcal{K}/\mathcal{K}_D$), on the geometric factor f_g for various values $\mathrm{Da} \geq 1$. From top to bottom, the panels correspond to the geometries shown in Figs. 1(a)–(c).

being just a fraction of the whole surface (a behavior well known in the chemical kinetics literature).[56,71,73] Furthermore, at large Da, the monopole amplitude $-a_0$ approaches the value 1 at $f_g = 1$, which is the expected behavior since this is the limit of a perfectly absorbing sphere (for which the reaction rate is $\mathcal{K} = \mathcal{K}_D$). At fixed geometry, i.e., one of the panels, the curves $-a_0(f_g)$ indeed seem to approach a limiting form with increasing value of Da, in agreement with the independent of Da behavior predicted for Da $\gg 1$ within the boundary homogenization approximation. Finally, we note that the dependence on the exact geometry of the reactive part is weak (compare the three panels at a given value of f_g), but particularly noticeable when one compares the ring geometry, (c), with the cap geometry, (a), and the two-cap geometry, (b).

3.2. The dipole (a_1) and octopole (a_3) amplitudes

As noted in Subsection 2.1, the additional "top-bottom" symmetry of the two caps and of the ring geometry (Figs. 1(b) and 1(c)), respectively, implies that for these systems the dipole and octopole amplitudes vanish: $a_1 = a_3 = 0$. This is correctly captured both by the numerical solution of the dual series equation (results not shown), which provides a cross-check of the numerics, as well as by the boundary homogenization approximation, see Eqs. (11), (15), and (16) (for the latter two, the second line of each). Therefore, these amplitudes will be discussed only for the single cap geometry of the active particle (Fig. 1(a)).

For both amplitudes, the boundary homogenization approximation perfectly captures the qualitative behavior, and it remains quantitatively accurate over the whole range of f_g at Da $= 1$, i.e., well above the reaction-controlled kinetics regime (see Fig. 3). At large values of Da, i.e., upon approaching the limit of diffusion-controlled kinetics, quantitative discrepancies — more pronounced in the case of the amplitude a_3 — are noticeable once the active cap occupies more than 20–30% of the surface. While for a_1 the discrepancies are always in the direction of an underestimate of the exact value by the approximation (i.e., a larger magnitude $|a_1|$, which for the motile particle translates into a larger velocity), for a_3 both under- and over-estimates occur in different ranges of f_g. It should be also noted that the boundary homogenization approximation predicts that the

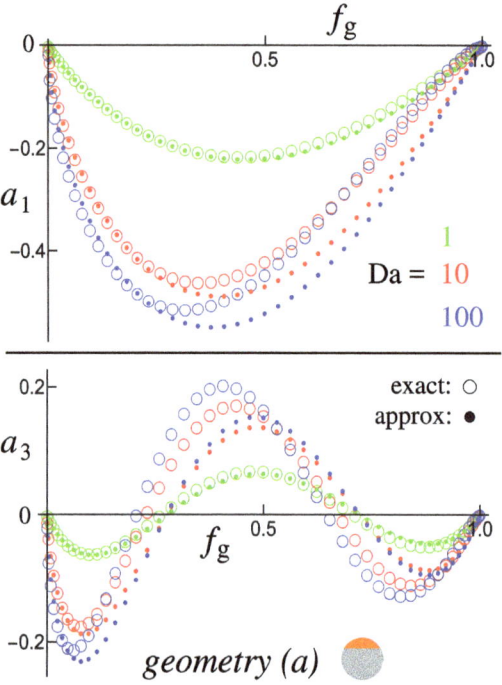

Fig. 3. The dependence of the dipole, a_1 (top panel), and octopole a_3 (bottom panel) amplitudes on the geometric factor f_g for various values Da ≥ 1 and for the case of the geometry shown in Fig. 1(a).

zero-crossings by a_3 are independent of Da (these are determined by $b_1 = 0$ and $b_3 = 0$, respectively; see Eq. (11)), which contrasts with the exact result of weak, but noticeable, shifts of these crossings toward smaller values of f_g with increasing Da.

3.3. The quadrupole amplitude (a_2)

For simple models of motile particles, the quadrupole amplitude a_2 determines the magnitude of the leading-order (in the far-field) hydrodynamic flow induced by the particle. Its sign distinguishes between "puller" and "pusher" type of motile particles (or "neutral", if the amplitude a_2 vanishes). As shown in Fig. 4, similarly to the conclusions drawn for the other amplitudes considered in the

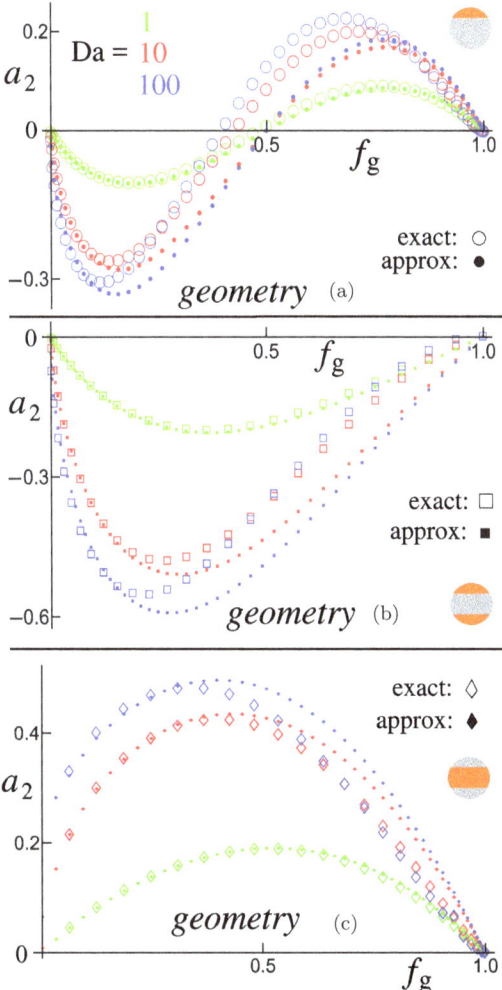

Fig. 4. The dependence of the quadrupole amplitude, a_2, on the geometric factor f_g for various values Da $\geq$ 1. From top to bottom, the panels correspond to the geometries shown in Figs. 1(a)–(c).

previous subsections, for all three geometries considered the boundary homogenization approximation leads to excellent qualitative predictions, and it is quantitatively accurate over the whole range of f_g even for Da values as large as Da = 1. (Note that the quite distinct qualitative behaviors exhibited by the amplitude a_2 in the three model geometries — compare, e.g., the top and the middle panels — provide a solid test for the approximation.) Finally, we remark that (as also noted for a_3 in the previous subsection) the zero-crossing occurring in the geometry (a) near $f_g = 0.5$ cannot be captured accurately by the approximate theory (which predicts the crossing to be set by $b_2^{(a)} = 0$, independent of Da). Although the quantitative differences are not very large, yet they should still be approached cautiously if the sign of a_2 is used, as noticed above, to infer a "puller" or "pusher" character.

4. Conclusions

The use of the so-called boundary homogenization approximation in studies of chemically active spherical particles was recently proposed by Ref. 44. Here we have studied in detail, for a simple model of a spherical particle and an annihilation reaction of first-order kinetics, the accuracy of the approximate analysis in terms of the dependences of the first four coefficients in the multipole expansion of the chemical field on the geometry of the reactive part and on the Damköhler number Da. Our results show that for simple geometries of the reactive part, such as those shown in Fig. 1, the results obtained within the confines of this approximation provide estimates that are in excellent qualitative agreement with the exact solutions. Furthermore, even for Damköhler numbers as large as Da $= 1$ (a range that compares favorably with estimates, based on[43,92] $k \sim 10^{-6}$ to 10^{-4} m/s, of the values in experimental studies of chemically active particles) the results obtained via the boundary homogenization approximation are quantitatively accurate over the whole range of geometries (configurations (a)–(c) in Fig. 1) and values of the geometric factor f_g (see Section 3.1).

In view of these promising outcomes, it would be interesting to further asses the robustness of the boundary homogenization approximation by testing more complex cases, such as particles with ellipsoidal shapes or particles with reactive zones that do not posses axial symmetry.

References

1. Y. Hong, D. Velegol, N. Chaturvedi, and A. Sen, Biomimetic behavior of synthetic particles: From microscopic randomness to macroscopic control, *Phys. Chem. Chem. Phys.* **12**, 1423–1435 (2010).

2. S. J. Ebbens and J. R. Howse, In pursuit of propulsion at the nanoscale, *Soft Matter* **6**, 726–738 (2010).

3. J. Elgeti, R. G. Winkler, and G. Gompper, Physics of microswimmers — single particle motion and collective behavior: A review, *Rep. Prog. Phys.* **78**, 056601 (2015).

4. C. Bechinger, R. Di Leonardo, H. Löwen, C. Reichhardt, G. Volpe, and G. Volpe, Active particles in complex and crowded environments, *Rev. Mod. Phys.* **88**, 045006 (2016).

5. J. L. Moran and J. D. Posner, Phoretic self-propulsion, *Ann. Rev. Fluid Mech.* **49**, 511–540 (2016).

6. E. Lauga and T. Powers, The hydrodynamics of swimming microorganisms, *Rep. Prog. Phys.* **72**, 096601 (2009).

7. E. M. Purcell, Life at low Reynolds number, *Am. J. Phys.* **45**, 3–11 (1977).

8. R. F. Ismagilov, A. Schwartz, N. Bowden, and G. M. Whitesides, Autonomous movement and self-assembly, *Angew. Chem. Int. Ed.* **41**, 652–654 (2002).

9. W. F. Paxton, K. C. Kistler, C. C. Olmeda, A. Sen, S. K. St. Angelo, Y. Y. Cao, T. E. Mallouk, P. E. Lammert, and V. H. Crespi, Catalytic nanomotors: Autonomous movement of striped nanorods, *J. Am. Chem. Soc.* **126**, 13424–13431 (2004).

10. G. A. Ozin, I. Manners, S. Fournier-Bidoz, and A. Arsenault, Dream nanomachines, *Adv. Mater.* **17**, 3011–3018 (2005).

11. W. F. Paxton, S. Sundararajan, T. E. Mallouk, and A. Sen, Chemical locomotion, *Angew. Chem. Int. Ed.* **45**, 5420–5429 (2006).

12. W. F. Paxton, P. T. Baker, T. R. Kline, Y. Wang, T. E. Mallouk, and A. Sen, Catalytically induced electrokinetics for motors and micropumps, *J. Am. Chem. Soc.* **128**, 14881–14888 (2006).

13. A. A. Solovev, Y. F. Mei, E. B. Urena, G. S. Huang, and O. G. Schmidt, Catalytic microtubular jet engines self-propelled by accumulated gas bubbles, *Small.* **5**, 1688–1692 (2009).

14. T. Mirkovic, N. S. Zacharia, G. D. Scholes, and G. A. Ozin, Nanolocomotion — catalytic nanomotors and nanorotors, *Small.* **6**, 159–167 (2010).

15. S. Fournier-Bidoz, A. C. Arsenault, I. Manners, and G. A. Ozin, Synthetic self-propelled nanorotors, *Chem. Commun.* **0**, 441–443 (2005).

16. J. R. Howse, R. A. L. Jones, A. J. Ryan, T. Gough, R. Vafabakhsh, and R. Golestanian, Self-motile colloidal particles: From directed propulsion to random walk, *Phys. Rev. Lett.* **99**, 048102 (2007).

17. G. Volpe, I. Buttinoni, D. Vogt, H. J. Kümmerer, and C. Bechinger, Microswimmers in patterned environments, *Soft Matter.* **7**, 8810–8815 (2011).

18. S. J. Ebbens, M. H. Tu, J. R. Howse, and R. Golestanian, Size dependence of the propulsion velocity for catalytic Janus-sphere swimmers, *Phys. Rev. E.* **85**, 020401(R) (2012).

19. F. Kümmel, B. ten Hagen, R. Wittkowski, I. Buttinoni, R. Eichhorn, G. Volpe, H. Löwen, and C. Bechinger, Circular motion of asymmetric self-propelling particles, *Phys. Rev. Lett.* **110**, 198302 (2013).

20. I. Buttinoni, J. Bialké, F. Kümmel, H. Löwen, C. Bechinger, and T. Speck, Dynamical clustering and phase separation in suspensions of self-propelled colloidal particles, *Phys. Rev. Lett.* **110**, 238301 (2013).

21. T. C. Lee, M. Alarcón-Correa, C. Miksch, K. Hahn, J. G. Gibbs, and P. Fischer, Self-propelling nanomotors in the presence of strong Brownian forces, *Nano Lett.* **14**, 2407–2412 (2014).

22. S. J. Ebbens, D. A. Gregory, G. Dunderdale, J. R. Howse, Y. Ibrahim, T. B. Liverpool, and R. Golestanian, Electrokinetic effects in catalytic Pt-insulator Janus swimmers, *EPL.* **106**, 58003 (2014).

23. B. ten Hagen, F. Kümmel, R. Wittkowski, D. Takagi, H. Löwen, and C. Bechinger, Gravitaxis of asymmetric self-propelled colloidal particles, *Nature Comm.* **5**, 4829 (2014).

24. X. Ma, S. Jang, M. N. Popescu, W. E. Uspal, A. Miguel-López, K. Hahn, D. P. Kim, and S. Sánchez, Reversed Janus micro/nanomotors with internal chemical engine, *ACS Nano.* **10**, 8751–8759 (2016).

25. S. Herminghaus, C. C. Maas, C. Krüger, S. Thutupalli, L. Goehring, and C. Bahr, Interfacial mechanisms in active emulsions, *Soft Matter.* **10**, 7008–7022 (2014).

26. R. Seemann, J. B. Fleury, and C. C. Maas, Self-propelled droplets, *Eur. Phys. J. Special Topics.* **225**, 2227–2240 (2016).

27. K. Kroy, D. Chakraborty, and F. Cichos, Hot microswimmers, *Eur. Phys. J. Special Topics.* **225**, 2207–2226 (2016).

28. C. Lozano, B. ten Hagen, H. Löwen, and C. Bechinger, Phototaxis of synthetic microswimmers in optical landscapes, *Nature Comm.* **7**(12828), 1–10 (2016).

29. R. Golestanian, T. B. Liverpool, and A. Ajdari, Propulsion of a molecular machine by asymmetric distribution of reaction products, *Phys. Rev. Lett.* **94**, 220801 (2005).

30. R. Golestanian, T. B. Liverpool, and A. Ajdari, Designing phoretic micro- and nano-swimmers, *New J. Phys.* **9**(126), 1–8 (2007).

31. G. R. Rückner and R. Kapral, Chemically powered nanodimers, *Phys. Rev. Lett.* **98**, 150603 (2007).

32. F. Jülicher and J. Prost, Generic theory of colloidal transport, *Eur. Phys. J. E.* **29**, 27–36 (2009).

33. M. N. Popescu, M. Tasinkevych, and S. Dietrich, Pulling and pushing a cargo with a catalytically active carrier, *EPL.* **95**, 28004 (2011).

34. B. Sabass and U. Seifert, Dynamics and efficiency of a self-propelled, diffusiophoretic swimmer, *J. Chem. Phys.* **136**, 064508 (2012).

35. B. Sabass and U. Seifert, Nonlinear, electrocatalytic swimming in the presence of salt, *J. Chem. Phys.* **136**, 214507 (2012).

36. R. Kapral, Nanomotors without moving parts that propel themselves in solution, *J. Chem. Phys.* **138**, 202901 (2013).

37. N. Sharifi-Mood, J. Koplik, and C. Maldarelli, Diffusiophoretic self-propulsion of colloids driven by a surface reaction: The sub-micron particle regime for exponential and van der Waals interactions, *Phys. Fluids.* **25**, 012001 (2013).

38. B. ten Hagen, S. van Teeffelen, and H. Löwen, Brownian motion of a self-propelled particle, *J. Phys. Cond. Matt.* **23**, 194119 (2011).

39. M. N. Popescu, S. Dietrich, M. Tasinkevych, and J. Ralston, Phoretic motion of spheroidal particles due to self-generated solute gradients, *Eur. Phys. J. E.* **31**, 351–367 (2010).

40. S. Michelin and E. Lauga, Autophoretic locomotion from geometric asymmetry, *Eur. Phys. J. E.* **38**(7), 1–16 (2015).

41. J. Hu, A. Wysocki, R. G. Winkler, and G. Gompper, Physical sensing of surface properties by microswimmers — directing bacterial motion via wall slip, *Sci. Rep.* **5**, 9586 (2015).

42. A. Zöttl and H. Stark, Emergent behavior in active colloids, *J. Phys.: Condens. Matter.* **28**, 253001 (2016).

43. J. de Graaf, G. Rempfer, and C. Holm, Diffusiophoretic self-propulsion for partially catalytic spherical colloids, *IEEE Trans. NanoBiosci.* **14**, 272–288 (2015).

44. G. Oshanin, M. N. Popescu, and S. Dietrich, Active colloids in the context of chemical kinetics, *J. Phys. A.* **50**, 134001 (2017).

45. P. E. Lammert, V. H. Crespi, and A. Nourhani, Bypassing slip velocity: rotational and translational velocities of autophoretic colloids in terms of surface flux, *J. Fluid Mech.* **802**, 294–304 (2016).

46. A. T. Brown, W. C. K. Poon, C. Holm, and J. de Graaf, Ionic screening and dissociation are crucial for understanding chemical self-propulsion in polar solvents, *Soft Matter* **13**, 1200–1222 (2017).

47. D. G. Crowdy, Wall effects on self-diffusiophoretic Janus particles: A theoretical study, *J. Fluid Mech.* **735**, 473–498 (2013).

48. J. L. Anderson, Colloid transport by interfacial forces, *Ann. Rev. Fluid Mech.* **21**, 61–99 (1989).

49. G. H. Weiss, Overview of theoretical models for reaction rates, *J. Stat. Phys.* **42**, 3–36 (1986).

50. G. H. Weiss, (ed.), *Contemporary Problems in Statistical Physics*. SIAM, Philadelphia (1994).

51. S. Rice, *Diffusion-limited Reactions*. Elsevier, Amsterdam (1985).

52. A. Ovchinnikov, S. Timashev, and A. Belyy, *Kinetics of Diffusion-controlled Chemical Processes*. Nova Science, New York (1989).

53. M. Smoluchowski, Versuch einer mathematischen Theorie der Koagulationskinetik kolloider Lösungen, *Z. Phys. Chem.* **92**, 129–168 (1917).

54. F. Collins and G. Kimball, Diffusion-controlled reaction rates, *J. Colloid Sci.* **4**, 425–437 (1949).

55. R. Samson and J. M. Deutch, Diffusion-controlled reaction rate to a buried active site, *J. Chem. Phys.* **68**, 285–290 (1978).

56. S. D. Traytak, The diffusive interaction in diffusion-limited reactions: the steady-state case, *Chem. Phys. Lett.* **197**, 247–254 (1992).

57. K. Šolc and W. Stockmayer, Kinetics of diffusion-controlled reaction between chemically asymmetric molecules. II. Approximate steady-state solution, *Int. J. Chem. Kinet.* **5**, 733–752 (1973).

58. S. D. Traytak, Diffusion-controlled reaction rate to an active site, *Chem. Phys.* **192**, 1–7 (1995).

59. S. D. Traytak and M. Tachiya, Diffusion-controlled reaction rate to an active site in an external electric field, *J. Chem. Phys.* **102**, 2760–2771 (1995).

60. S. D. Traytak and M. Tachiya, Diffusion-controlled reaction rate to asymmetric reactants under Coulomb interaction, *J. Chem. Phys.* **102**, 9240–9247 (1995).

61. S. D. Traytak and W. Price, Exact solution for anisotropic diffusion-controlled reactions with partially reflecting conditions, *J. Chem. Phys.* **127**, 184508 (2007).

62. C. Eun, Effect of surface curvature on diffusion-limited reactions on a curved surface, *J. Chem. Phys.* **147**, 184112 (2017).

63. M. Baldo, A. Grassi, and A. Raudino, Exact analytical solution of the rotational-translational diffusion equation with mixed boundary conditions. An application to diffusion-controlled enzyme reactions, *J. Chem. Phys.* **91**, 4658–4663 (1989).

64. J. M. Schurr and K. S. Schmitz, Orientation constraints and rotational diffusion in bimolecular solution kinetics. A simplification, *J. Phys. Chem.* **80**, 1934–1936 (1976).

65. D. Shoup, G. Lipari, and A. Szabo, Diffusion-controlled bimolecular reaction rates. The effect of rotational diffusion and orientation constraints, *Biophys. J.* **36**, 697–714 (1981).

66. D. Shoup and A. Szabo, Role of diffusion in ligand binding to macromolecules and cell-bound receptors, *Biophys. J.* **40**, 33–39 (1982).

67. V. Berdnikov and A. Doktorov, Steric factor in diffusion-controlled chemical reactions, *Chem. Phys.* **69**, 205–212 (1982).

68. L. Dagdug, M. Vázquez, A. M. Berezhkovskii, and V. Y. Zitserman, Boundary homogenization for a sphere with an absorbing cap of arbitrary size, *J. Chem. Phys.* **145**, 214101 (2016).

69. S. D. Traytak, The steric factor in the time-dependent diffusion-controlled reactions, *J. Phys. Chem.* **98**, 7419–7421 (1994).

70. K. Ivanov, N. Lukzen, and A. Doktorov, On the time dependence of rate coefficients of irreversible reactions between reactants with anisotropic reactivity in liquid solutions, *J. Chem. Phys.* **145**, 064104 (2016).

71. K. Šolc and W. Stockmayer, Kinetics of diffusion-controlled reaction between chemically asymmetric molecules. I. General theory, *J. Chem. Phys.* **54**, 2981–2988 (1971).

72. Z. Schulten and K. Schulten, The generation, diffusion, spin motion, and recombination of radical pairs in solution in the nanosecond time domain, *J. Chem. Phys.* **66**, 4616–4634 (1977).

73. J. M. Deutch and B. U. Felderhof, Hydrodynamic effect in diffusion-controlled reaction, *J. Chem. Phys.* **59**, 1669–1671 (1973).

74. D. Calef and J. M. Deutch, Diffusion-controlled reactions, *Ann. Rev. Phys. Chem.* **34**, 493–524 (1983).

75. J. Keizer, Diffusion effects on rapid bimolecular chemical reactions, *Chem. Rev.* **87**, 167–180 (1987).

76. S. D. Traytak, The diffusive interaction in diffusion-limited reactions: The time-dependent case, *Chem. Phys.* **193**, 351–366 (1995).

77. S. D. Traytak and D. S. Grebenkov, Diffusion-influenced reaction rates for active "sphere-prolate spheroid" pairs and Janus dimers, *J. Chem. Phys.* **148**, 024107 (2018).

78. S. Michelin and E. Lauga, Phoretic self-propulsion at finite Peclét numbers, *J. Fluid Mech.* **747**, 572–604 (2014).

79. I. Sneddon, *Mixed boundary Value in Potential Theory.* North-Holland, Amsterdam, The Netherlands (1966).

80. M. N. Popescu, W. E. Uspal, Z. Eskandari, M. Tasinkevych, and S. Dietrich. Effective squirmer models for self-phoretic chemically active spherical colloids, *Eur. Phys. J. E.* **41**, 145 (2018).

81. W. E. Uspal, M. N. Popescu, S. Dietrich, and M. Tasinkevych, Self-propulsion of a catalytically active particle near a planar wall: From reflection to sliding and hovering, *Soft Matter* **11**, 434–438 (2015).

82. Y. Ibrahim and T. B. Liverpool, How walls affect the dynamics of self-phoretic microswimmers, *Eur. Phys. J. Special Topics.* **225**, 1843–1874 (2016).

83. A. Berezhkovskii, Y. Makhnovskii, M. Monine, V. Zitserman, and S. Y. Shvartsman, Boundary homogenization for trapping by patchy surfaces, *J. Chem. Phys.* **121**, 11390 (2004).

84. Y. Makhnovskii, A. Berezhkovskii, and V. Zitserman, Homogenization of boundary conditions on surfaces randomly covered by patches of different sizes and shapes, *J. Chem. Phys.* **122**, 236102 (2005).

85. A. Berezhkovskii, M. Monine, C. Muratov, and S. Y. Shvartsman, Homogenization of boundary conditions for surfaces with regular arrays of traps, *J. Chem. Phys.* **124**, 036103 (2006).

86. C. Muratov and S. Y. Shvartsman, Boundary homogenization for periodic arrays of absorbers, *Multiscale Model. Simul.* **7**, 44–61 (2008).

87. L. Dagdug, A. Berezhkovskii, and A. Skvortsov, Trapping of diffusing particles by striped cylindrical surfaces. Boundary homogenization approach, *J. Chem. Phys.* **142**, 234902 (2015).

88. A. Skvortsov, A. Berezhkovskii, and L. Dagdug, Note: Boundary homogenization for a circle with periodic absorbing arcs. Exact expression for the effective trapping rate, *J. Chem. Phys.* **143**, 226101 (2015).

89. J. F. Brady, Particle motion driven by solute gradients with application to autonomous motion: continuum and colloidal perspectives, *J. Fluid Mech.* **667**, 216–259 (2011).

90. W. E. Uspal, M. N. Popescu, S. Dietrich, and M. Tasinkevych, Guiding catalytically active particles with chemically patterned surfaces, *Phys. Rev. Lett.* **117**, 048002 (2016).

91. W. E. Uspal, M. N. Popescu, M. Tasinkevych, and S. Dietrich, Shape-dependent guidance of active Janus particles by chemically patterned surfaces, *New J. Phys.* **20**, 015013 (2018).

92. A. Domínguez, P. Malgaretti, M. N. Popescu, and S. Dietrich, Effective interaction between active colloids and fluid interfaces induced by Marangoni flows, *Phys. Rev. Lett.* **116**, 078301 (2016).

93. A. Mozaffari, N. Sharifi-Mood, J. Koplik, and C. Maldarelli, Self-diffusiophoretic colloidal propulsion near a solid boundary, *Phys. Fluids.* **28**, 053107 (2016).

Chapter 22

Modeling Active Emulsions

Holger Stark* and Maximilian Schmitt

*Institute of Theoretical Physics, Technische Universität Berlin,
Hardenbergstr. 36, D-10623 Berlin, Germany*

*Holger.Stark@tu-berlin.de

We present a diffusion–advection-reaction equation to describe the dynamics of a mixture of two types of surfactants at the interface of an emulsion droplet. The advective current follows from a non-uniform surface-tension field at the interface, which induces Marangoni flow. We apply this theory to two examples for producing active emulsions.

First, bromine dissolved in water droplets reacts with the surfactants and locally enhances surface tension. For sufficiently strong gradients quantified by the Marangoni number M, a symmetry breaking transition toward a swimming droplet occurs followed by stopping and oscillating states with increasing M. Taking into account thermal fluctuations of the surfactant mixture, the coarsening dynamics toward the phase-separated profile is governed by an initially slow growth of domain size followed by a nearly ballistic regime. On larger time scales, thermal fluctuations cause random changes in the swimming direction, and the droplet performs a persistent random walk.

Second, we consider photosensitive surfactants, which can be switched between *trans* and *cis* isomers by light. Depending on the wavelength of the exciting light beam and the surfactant type in the outer bulk fluid, one can either push droplets along unstable trajectories or pull them along straight or oscillatory trajectories. We demonstrate these cases for strongly absorbing and for transparent droplets.

1. Introduction

This chapter talks about creating motion in the specific system of emulsion droplets, where fluid droplets are suspended in a fluid phase. Within the field of active motion, active emulsions have established an interesting research direction since they provide mechanisms to produce autonomous motion.[1–5] Understanding swimming mechanisms of different types of microswimmers ranging from Janus particles driven by diffusiophoresis to microorganisms, which use beating and rotating flagella, has always initiated theoretical research in order to obtain a qualitative understanding and perform a quantitative description of the underlying processes.[6–13]

Following this route, our work[2,5,14] summarized in this chapter was motivated by a realization of active emulsions. Bromine water droplets with a typical radius of $80\,\mu\text{m}$ placed into a surfactant-rich oil phase spontaneously start to swim with a typical speed of about $15\,\mu\text{m/s}$.[1] Surfactants move to the droplet interface and form a dense monolayer, thereby lowering the surface tension σ of the interface. Bromine from within the droplet chemically reacts with surfactants and generates weaker surfactants, which give a higher surface tension. As a consequence, local gradients in σ induce flow at and close to the interface toward regions of higher surface tension. This phenomenon is called Marangoni effect. It advects surfactants and thereby further enhances the gradients of σ. If the advective current exceeds the smoothing diffusion current, the surfactant mixture phase-separates. Thus, the original isotropic symmetry of the droplet interface is broken. The droplet develops a polar symmetry and starts to move in a random direction, which fluctuates around such that the droplet performs a persistent random walk. In parallel, brominated surfactants are constantly replaced by non-brominated surfactants from the oil phase by means of desorption and adsorption. The swimming motion comes to an end when the fueling bromine is exhausted.

Another means to generate a surfactant mixture is by using photoresponsive surfactants. They contain stereoisomers like azobenzene,[15] spiropyran,[16] or stilbene.[17] They reversibly change shape between *trans* and *cis* isomers under illumination with specific wavelengths of light and thereby change surface tension of a liquid interface.[18] Consequently, Marangoni flow can be induced between illuminated and dark regions of the interface.[14,19–21] Through this mechanism, photoresponsive surfactants offer a unique way to control fluid flow by shining static or changing patterns of light on an interface. This allows to move and position individual surfactant-covered droplets,[22] to turn droplets into self-propelled particles,[14] to move oil droplets[20] or marbles[23] on a photoresponsive fluid surface, to pump fluid near a planar water surface,[19] to photocontrol the coffee-ring effect,[24] and to explore photoresponsive fluid interfaces for feedback control.[21]

The chapter is based on three papers published by the authors, where we formulated the theory of a surfactant mixture, which fully covers the interface of an emulsion droplet.[2,5] The theory also needs the Marangoni flow field due to an arbitrary field of surface tension, which we derived in Ref. 14. After reviewing the complete theory in Section 2, we address in Section 3 the active emulsion droplets that swim based on the bromination reaction of surfactants. We characterize the different motional states induced by the spontaneous symmetry breaking and present them in a state diagram.[2] Taking into account thermal fluctuations in the surfactant mixture,[5] we study its coarsening dynamics toward full, motion-induced phase separation at the interface of the swimming droplet and also address rotational diffusion of the swimming direction. In Section 4, we explore light-induced motion of emulsion droplets using a beam of laser light and illustrate possible trajectories for adsorbing and transparent emulsion droplets initially covered either by *trans* or *cis* surfactants.[14]

2. Theory

Active droplets are driven by Marangoni flow initiated by a non-uniform surfactant mixture at the droplet interface. To model their dynamics, we formulate a diffusion–advection-reaction equation for the mixture order parameter ϕ. We assume that the surfactants completely cover the droplet interface without any intervening solvent. The head area ℓ^2 of both types of surfactant molecules (brominated and non-brominated or surfactants in the *trans* or *cis* state) are assumed to be the same so that they occupy the same area at the droplet interface. Their densities c_1 and c_2 in units of ℓ^{-2} then satisfy $c_1 + c_2 = 1$ and the mixture order parameter is the concentration difference $\phi = c_1 - c_2$. Local coverage with one type of surfactant is indicated by $\phi = 1$ and -1, respectively, and the surfactant densities are $c_1 = (1 + \phi)/2$ and $c_2 = (1 - \phi)/2$. Note that a theory without the constraint of full coverage is formulated in Ref. 21.

2.1. *Diffusion–advection-reaction equation*

We express the dynamics of the order-parameter field ϕ at the droplet interface in the form of a continuity equation with an additional source and thermal noise (ζ) term,

$$\partial_t \phi = -\nabla_s \cdot (\mathbf{j}_D + \mathbf{j}_A) - \tau_R^{-1}(\phi - \phi_{\text{eq}}) + \zeta(\mathbf{r}, t). \tag{1}$$

Here, $\nabla_s = \mathbf{P}_s \nabla = (1 - \mathbf{n} \otimes \mathbf{n})\nabla$ stands for the directional gradient on a sphere with radius R, where ∇ is the nabla operator, $\mathbf{P}_s$ the surface projector, and $\mathbf{n}$ a unit vector along the surface normal. The current is split into a diffusive part $\mathbf{j}_D$ and an advective part $\mathbf{j}_A$, which arises due to the Marangoni effect. We explain these in Sections 2.2 and 2.3, respectively. The source term is a simplified phenomenological description for the adsorption and desorption of surfactants from and to the surrounding fluid. Without diffusive and advective currents, it gives a relaxation of the surfactant mixture to an equilibrium composition ϕ_{eq} during the relaxation time τ_R. A more detailed model would include fluxes from and to the bulk fluid.[25] In our first example, the source term describes the bromination reaction as well as desorption of brominated and adsorption of non-brominated surfactants to and from the outer fluid. Adsorption and desorption dominate for $\phi_{\text{eq}} < 0$, while bromination dominates for $\phi_{\text{eq}} > 0$. In the second example of photoisomeric surfactants, we set $\phi_{\text{eq}} = -1$ or $+1$ to indicate that the outer fluid is laden with either *trans* or *cis* isomers, respectively. Finally, the thermal noise term is due to compositional fluctuations in the surfactant mixture. It is related to the diffusive current by the fluctuation–dissipation theorem as explained at the end of Subsection 2.2.

2.2. *Diffusive current*

To derive the diffusive current $\mathbf{j}_D$, we start from a Flory–Huggins type free energy density f for the surfactant mixture at the droplet interface in terms of the densities c_1 and c_2. It is composed of the mixing entropy plus terms mimicking attractive interactions between surfactants,

$$f = \frac{k_B T}{\ell^2}[c_1 \ln c_1 + c_2 \ln c_2 - b_1 c_1^2 - b_2 c_2^2 - b_{12} c_1 c_2]. \qquad (2)$$

Here, ℓ^2 denotes the area occupied by a surfactant in the interface, and b_1 and b_2 are dimensionless parameters characterizing the interaction between surfactant molecules of one type, while b_{12} quantifies the interaction between surfactant molecules of different types. Rewriting this energy density in terms of the order parameter ϕ, we obtain

$$f(\phi) = \frac{k_B T}{\ell^2} \left[\frac{1+\phi}{2} \ln \frac{1+\phi}{2} + \frac{1-\phi}{2} \ln \frac{1-\phi}{2} \right.$$
$$\left. - \frac{1}{4}(b_1 + b_2 + b_{12}) - \frac{\phi}{2}(b_1 - b_2) - \frac{\phi^2}{4}(b_1 + b_2 - b_{12}) \right]. \qquad (3)$$

The diffusive current is now driven by a gradient in the chemical potential derived from the total free energy functional $F[\phi] = \iint f(\phi)\, dA$,

$$\mathbf{j}_D = -\lambda \nabla_s \frac{\delta F}{\delta \phi} = -\lambda f''(\phi) \nabla_s \phi$$
$$= -D \left[\frac{1}{1-\phi^2} - \frac{1}{2}(b_1 + b_2 - b_{12}) \right] \nabla_s \phi, \qquad (4)$$

where $f''(\phi)$ means second derivative with respect to ϕ and the Einstein relation $D = \lambda k_B T/\ell^2$ relates the interfacial diffusion constant D to the mobility λ. To rule out a double well form of $f(\phi)$, which would generate phase separation of the surfactant mixture already in thermal equilibrium, we only consider $b_1 + b_2 - b_{12} < 2$ or with $b_{12} = (b_1 + b_2)/2$ require $b_1 + b_2 < 4$. The convex free energy $(f''(\phi) > 0)$ also means that the diffusive current $\mathbf{j}_D \propto -\nabla_s \phi$ is for all ϕ directed against $\nabla_s \phi$.

From the free energy density f of Eq. (2) we derive the surface tension, the gradient of which drives the Marangoni flow. The surface tension σ is the thermodynamic force conjugate to the surface area. This gives

$$\sigma = f - \frac{\partial f}{\partial c_1} c_1 - \frac{\partial f}{\partial c_2} c_2, \tag{5}$$

which we identify as the Legendre transform of the free energy (2) to the chemical potentials $\mu_i = \frac{\partial f}{\partial c_i}$. Hence, $\sigma = \frac{k_B T}{\ell^2}(b_1 c_1^2 + b_2 c_2^2 + b_{12} c_1 c_2)$, and in terms of ϕ using again $b_{12} = (b_1 + b_2)/2$, we arrive at the equation of state

$$\sigma(\phi) = \frac{k_B T}{\ell^2}(b_1 - b_2)\left(\frac{3}{8}\frac{b_1 + b_2}{b_1 - b_2} + \frac{1}{2}\phi + \frac{1}{8}\frac{b_1 + b_2}{b_1 - b_2}\phi^2\right). \tag{6}$$

We formulate the thermal noise term in Eq. (1) as Gaussian white noise with zero mean following Ref. 26,

$$\langle \zeta \rangle = 0, \tag{7a}$$

$$\langle \zeta(\mathbf{r}, t)\zeta(\mathbf{r}', t') \rangle = -2k_B T \lambda \nabla_s^2 \delta(\mathbf{r} - \mathbf{r}')\delta(t - t'). \tag{7b}$$

Here, the strength of the noise correlations is connected to the mobility λ of the diffusive current via the fluctuation–dissipation theorem.

2.3. Marangoni flow

The advective current for the order parameter ϕ is given by

$$\mathbf{j}_A = \phi \mathbf{u}|_R, \tag{8}$$

where $\mathbf{u}|_R$ is the flow field at the droplet interface. It is driven by a non-uniform surface tension σ and therefore called Marangoni flow.[25,27] Since surfactants constantly enter or leave the interface, the surface flow has a non-zero surface divergence $\nabla_s \cdot \mathbf{u}|_R \neq 0$. In fact, it can be shown that an incompressible surface flow is not compatible with self-propulsion of the droplet.[28]

In order to evaluate $\mathbf{u}|_R$, one has to solve the Stokes equation for the flow field $\mathbf{u}(\mathbf{r})$ surrounding the spherical droplet $(r > R)$ as well as for the flow field $\hat{\mathbf{u}}(\mathbf{r})$ inside the droplet $(r < R)$. Both solutions are matched at the droplet interface $r = R$ by the condition[25]

$$\nabla_s \sigma = \mathbf{P}_s \left. (\mathbf{T} - \hat{\mathbf{T}})\mathbf{n}\right|_{r=R}, \tag{9}$$

where $\mathbf{P}_s$ is the surface projector introduced after Eq. (1) and $\mathbf{n} = \mathbf{e}_r$ the radial unit vector. Equation (9) means that a gradient in surface tension σ is compensated by a jump in viscous shear stress. Here, $\mathbf{T} = \eta[\nabla \otimes \mathbf{u} + (\nabla \otimes \mathbf{u})^T]$ is the viscous shear stress tensor of a Newtonian fluid with viscosity η outside of the droplet and the same relation holds for $\hat{\mathbf{T}}$ of the fluid with viscosity $\hat{\eta}$ inside the droplet. We have performed the tedious evaluation of the complete flow field in Ref. 14 for an arbitrary but given surface tension field and only summarize the relevant results here. Alternative derivations are found in Ref. 29–32.

The Marangoni flow field $\mathbf{u}|_R$ at the interface can be expanded into spherical harmonics $Y_l^m(\theta, \varphi)$, where θ, φ are the spherical angular coordinates

$$\mathbf{u}|_R = -\frac{\eta}{2(\eta + \hat{\eta})}\mathbf{v}_D + \frac{1}{\eta + \hat{\eta}}\sum_{l=1}^{\infty}\sum_{m=-l}^{l}\frac{Rs_l^m}{2l+1}\nabla_s Y_l^m. \tag{10}$$

The surface tension comes in through the expansion coefficients

$$s_l^m = \iint \sigma(\theta, \varphi)\overline{Y}_l^m(\theta, \varphi)\,\mathrm{d}\Omega, \tag{11}$$

where bar means complex conjugate. Equation (10) shows that Marangoni flow is directed along gradients in surface tension, but higher expansion modes contribute with a decreasing strength $\propto$

$1/(2l + 1)$. Equation (10) also contains the droplet velocity vector $\mathbf{v}_D$. Using the Lorentz reciprocal theorem, one can calculate it by averaging the surface flow field $\mathbf{u}|_R$ or ultimately the surface tension gradient $\nabla_s \sigma$ over the whole interface,[14,33]

$$\mathbf{v}^D = -\frac{1}{4\pi R^2} \frac{3\eta + 3\hat{\eta}}{2\eta + 3\hat{\eta}} \iint \mathbf{u}|_R dA = -\frac{1}{4\pi R(2\eta + 3\hat{\eta})} \iint \nabla_s \sigma dA.$$

$$(12)$$

The first relation reduces to the expression given in Ref. 28 for rigid active spherical swimmers ($\hat{\eta} \to \infty$). Finally, one realizes that the droplet velocity vector with its propulsion speed $v_D \geq 0$ and the swimming direction $\mathbf{e}$ (with $|\mathbf{e}| = 1$) is solely determined by the dipolar coefficients ($l = 1$) of the surface tension,[14,30,32]

$$\mathbf{v}_D = v_D \mathbf{e} = \frac{1}{\sqrt{6\pi}} \frac{1}{2\eta + 3\hat{\eta}} \begin{pmatrix} s_1^1 - s_1^{-1} \\ i\left(s_1^1 + s_1^{-1}\right) \\ -\sqrt{2} s_1^0 \end{pmatrix}. \qquad (13)$$

By setting $m = 0$, Eqs. (10) and (13) reduce to the case of an axisymmetric droplet swimming along the z-direction, which we studied in Ref. 2.

Together with the equation of state $\sigma(\phi)$ from Eq. (6), the advective current $\mathbf{j}_A$ in Eq. (8) is now fully specified. Using the diffusion current $\mathbf{j}_D$ from Eq. (4), the diffusion–advection-reaction equation (1) becomes a closed equation in ϕ. We solve it numerically by a finite-volume method as detailed in Ref. 5.

2.4. Relevant system parameters

One can formulate Eqs. (1) and (7(b)) in dimensionless quantities by rescaling time by the characteristic diffusion time $\tau_D = R^2/D$ and lengths by droplet radius R. Thereby, the relevant system parameters become visible. The most important one is the Marangoni number

$$M = \frac{(b_1 - b_2)R}{\lambda(\eta + \hat{\eta})}, \qquad (14)$$

which compares the strength of the advective current or Marangoni flow to the diffusive current, which scale, respectively, as $k_B T(b_1 - b_2)/[(\eta + \hat{\eta})\ell^2]$ and $D/R = \lambda k_B T/(\ell^2 R)$. Furthermore, one has the ratio of shear viscosities, $\nu = \hat{\eta}/\eta$, for the fluids inside and outside of the droplet, respectively, and $\kappa = \tau_D/\tau_R$ tunes the ratio between diffusion and relaxation time. Finally, the reduced noise strength $\xi = \ell/R \propto 1/\sqrt{N}$, where N is the total number of surfactants at the droplet interface, connects the droplet size R to the molecular length scale ℓ.

3. Swimming Active Emulsion Droplets

In the introduction we explained that a bromination reaction inside the emulsion droplet creates a mixture of two types of surfactants. Initially, the mixture will be uniform and then an instability arises, where it starts to demix. This creates gradients in surface tension and ultimately gives rise to the directed droplet motion. Suppose demixing spontaneously arises locally. The diffusive current $\mathbf{j}_D \propto -\nabla\phi$ will smooth out this demixing. For the advective current, we roughly found $\mathbf{j}_A \propto \phi\nabla\phi$ in Eq. (10), strictly valid for each lm mode. This means that for $\phi > 0$, the current $\mathbf{j}_A$ amplifies gradients in ϕ by moving more $+1$ surfactants along the gradient than -1, which enhances demixing. Thus, $\mathbf{j}_D$ and $\mathbf{j}_A$ are competing and as soon as $\mathbf{j}_A$ dominates over $\mathbf{j}_D$ as quantified by the Marangoni number M, the surfactant mixture phase-separates. The resting state becomes unstable and the droplet starts to swim.

In the following, we summarize the possible motional states and study the coarsening dynamics of the surfactant mixture as a result of compositional fluctuations.

3.1. *Motional states and state diagram*

All the results in this subsection are obtained for $\kappa = \tau_D/\tau_R = 0.1$ and $b_1 = 2$, $b_2 = 1$ so that $b_1 + b_2 = 3$. In Ref. 2, the dynamic equation without noise was solved for the uniaxial case. Starting from

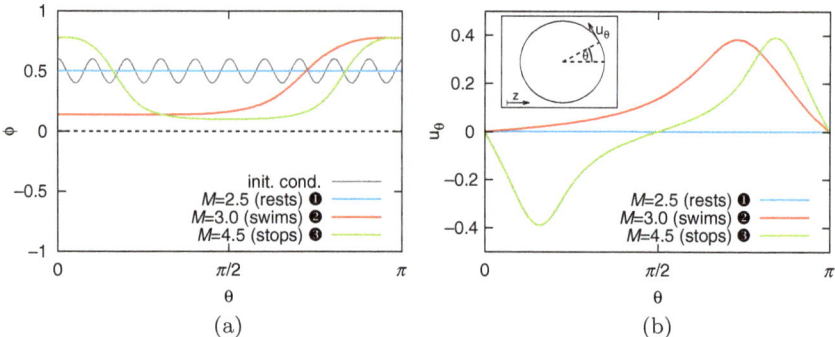

Fig. 1. (a) Stationary order-parameter profiles for $\phi_{eq} = 0.5$ and several Marangoni numbers M. Modulated line: Initial condition. (b) Corresponding longitudinal interface velocity profiles. Inset: Droplet geometry. (Reproduced from Ref. 2 with kind permission of *EPL*; copyright (2013).)

a modulated order-parameter field (see Fig. 1(a)), ϕ becomes uniform and the droplet does not move at low Marangoni numbers, it rests. At medium values it swims due to a smooth step in ϕ. Then, at high M the order-parameter field becomes symmetric about $\theta = \pi/2$ and the droplet stops. The corresponding longitudinal surfactant velocity profiles are plotted in Fig. 1(b) and the resulting droplet speeds in Fig. 2(a). In addition to the motional states as already discussed, we observe an oscillation state. The corresponding order-parameter field color-coded in Fig. 2(b) is no longer stationary but also varies periodically in time.

Finally, Fig. 3 shows a state diagram in the relevant parameter space ϕ_{eq} versus M with the four motional states: resting, swimming, stopping, and oscillating occuring for increasing M. There is no swimming motion possible for negative ϕ_{eq} since both the advective and diffusive current are directed against the gradient in ϕ. In particular, the transition from resting to swimming shows a subcritical bifurcation, i.e., there is a region where both states are possible and the observed state depends on the initial order-parameter profile.[5] Similar state diagrams occur for smaller values of κ. For $\kappa = 0.01$, the swimming region increases in size and then shrinks again for $\kappa = 0.001$ until for $\kappa = 0$ swimming solutions are no longer

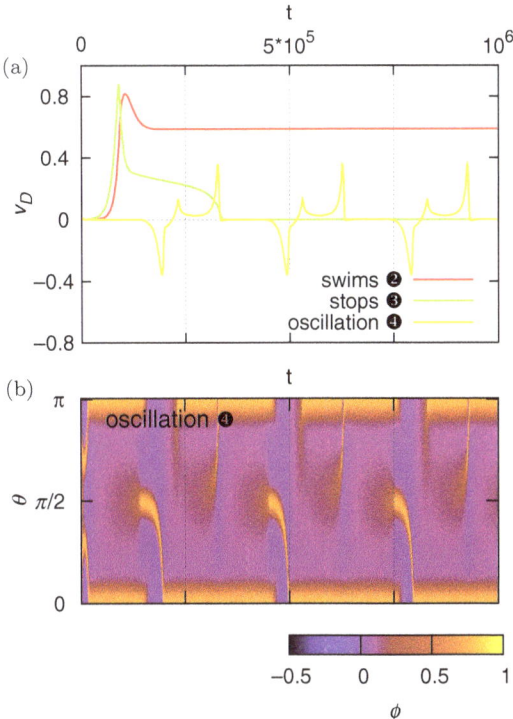

Fig. 2. (a) Droplet swimming speed v_D versus time for swimming, stopping, and oscillating droplets. Parameters are the same as in Fig. 1 and case 4 belongs to $M = 10.5$. (b) Illustration of the spatiotemporal variation of $\phi(\theta, t)$ for case 4. Same time scale as in (a). (Reproduced from Ref. 2 with kind permission of *EPL*; copyright (2013).)

possible. On the other hand, for $\kappa = 1$ and 10, i.e., in the limit of fast bromination reaction and exchange of surfactants, only resting, stopping, and oscillating solutions but no stable swimming solutions occur.

3.2. *Fluctuating order-parameter field and coarsening dynamics*

For the following, we solved the full dynamic equation (1) including thermal noise. We studied the coarsening dynamics of the surfactant mixture starting from a uniform mixture at $\phi_{eq} = 0.5$ with

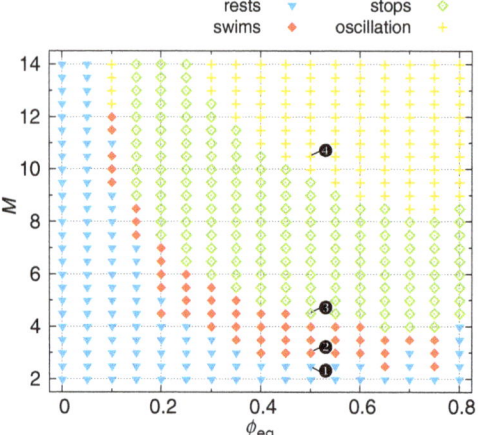

Fig. 3. State diagram of the active droplet for M versus ϕ_{eq}. Examples for the order parameter profiles at the positions marked with numbers are given in Fig. 1(a) (states 1–3) and Fig. 2(b) (state 4). (Reproduced from Ref. 2 with kind permission of *EPL*; copyright (2013).)

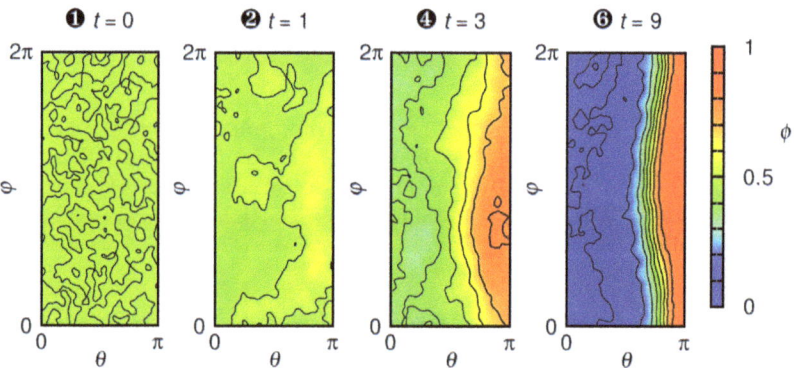

Fig. 4. Color-coded order-parameter profile $\phi(\theta, \varphi)$ at various time steps in the coordinate frame of the droplet. The front of the droplet is located at $\theta = 0$. Lines of equal ϕ are drawn. The relevant parameters are: $M = 3$, $\kappa = 0.1$, $\phi_{eq} = 0.5$, and $\xi = 10^{-3}$. (Reproduced from Ref. 5 with kind permission of *EPJ*; copyright (2016).)

very small random fluctuations. Figure 4 shows the order parameter profile $\phi(\theta, \varphi)$ at various time steps. Shortly after the simulation starts, small islands or domains with $\phi > \phi_{eq}$ and $\phi < \phi_{eq}$ emerge,

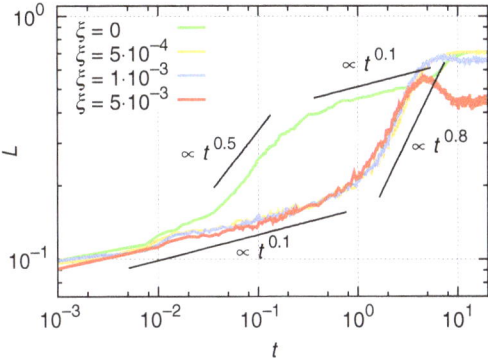

Fig. 5.   Mean domain size L averaged over 200 simulation runs plotted versus reduced time in units of τ_D for different noise strengths ξ. A domain is defined by a compact region with $\phi > \phi_{\mathrm{eq}}$. Same parameters as in Fig. 4 are used. (Reproduced from Ref. 5 with kind permission of *EPJ*; copyright (2016).)

which rapidly grow until $t \approx 1$. In this period, the droplet hardly moves. Then the coarsening or demixing process is slowed down. The domains coalesce on larger scales and the droplet speeds up significantly. Since the droplet interface area is finite, the domains ultimately coalesce to two large regions with $\phi < \phi_{\mathrm{eq}}$ and $\phi > \phi_{\mathrm{eq}}$ with each of them covering about half of the interface. This happens at about $t = 5$ (not shown), where the droplet speed is maximum. Then, the domain wall between the two regions moves to its final position closer to the south pole at $\theta = \pi$ and reaches it at $t \approx 9$. Simultaneously, the droplet speed decreases to its final value since closer to the south pole the area of the domain wall and therefore the region of non-uniform surface tension is smaller. The surface or Marangoni flow is concentrated on the region of the domain wall and points toward the south pole, where surface tension is largest. Thus, the droplet moves with the north pole in front and, in the language of microswimmers, it therefore is a pusher.

We also quantified the temporal evolution of the coarsening dynamics by introducing an average domain size L, which measures the mean lateral extension of all compact regions with $\phi > \phi_{\mathrm{eq}}$.[5] Figure 5 shows $L(t)$ versus time for different noise strengths ξ.

A separation of time scales in the coarsening dynamics is clearly visible. At early times, we find a power-law behavior $L(t) \propto t^{0.1}$. Without noise, coarsening then quickly speeds up at a rate $L(t) \propto t^{1/2}$ and then slows down again to $L(t) \propto t^{0.1}$. In contrast, thermal fluctuations in the order-parameter-profile hinder early coarsening and the mean domain size continues to grow slowly with $L(t) \propto t^{0.1}$ over several decades and then crosses over to a fast final coarsening with rate $L(t) \propto t^{0.8}$. The crossover time at $t \approx 1$ is only determined by the diffusion time τ_D and does not depend on noise strength ξ.

Interestingly, a similar scenario occurs for coarsening in the dynamic model H, where the Cahn–Hilliard equation couples to fluid flow at low-Reynolds number via an advection term. A slow coarsening rate $L(t) \propto t^{1/3}$ in a diffusive regime at short times is followed by an advection-driven regime with $L(t) \propto t$ at later times.[34,35] Although we cannot simply reformulate our model as an advective Cahn–Hilliard equation, since the phase separation in our case is driven by the interfacial flow $\mathbf{u}|_R$ itself, we observe similar coarsening regimes when we include some noise.

3.3. *Rotational diffusion*

Our solution of the dynamic equation (1) including thermal noise shows that in the stationary state of the order-parameter profile, the droplet performs Brownian motion in space when the droplet velocity $v_D\mathbf{e}(t)$ is integrated in time.[5] The reason is that the profile as illustrated in Fig. 4 at time $t = 9$ or ultimately the domain wall exhibit thermal fluctuations. As a result, the swimming direction $\mathbf{e}(t)$ performs a random walk on the unit sphere. In Ref. 5, we confirmed this scenario by directly measuring the orientational correlation function, which behaves as $\langle \mathbf{e}(0) \cdot \mathbf{e}(t) \rangle = e^{-t/\tau_r}$. Here, the decorrelation time $\tau_r = 1/(2D_r)$ is directly related to the rotational diffusion coefficient D_r. For small thermal noise strengths ξ, we observed a clear scaling $\tau_r \propto 1/\xi^2$. To rationalize the scaling, we succeeded to extract a stochastic equation for the Brownian motion on the unit sphere from Eq. (1). For times beyond the duration of coarsening, one projects the fluctuations about the stationary solution on the spherical harmonics $Y_{l=1}^{\pm 1}$, since they are responsible for fluctuations in $\mathbf{e}(t)$. In particular, we found $D_r \propto \xi^2$ and thereby confirmed the observed scaling.

4. Light-Induced Motion of Emulsion Droplets

In this section, we consider the case where the mixture of surfactants is initiated locally by laser light. Certain surfactants are known to be photosensitive.[36–39] For example, surfactants based on azobenzene can undergo photoisomerization, where UV light (365 nm) transforms a *trans* to a *cis* isomer and blue light (450 nm) causes a transformation from *cis* to *trans*. During the *trans-cis* isomerization, subunits within the molecule change their relative orientation. Naturally, a different molecular structure also affects the surface tension of a surfactant-covered interface. Experiments showed that surfactants in the *cis* state cause a higher surface tension compared to the ones in the *trans* state.[37] This effect has recently been used to generate Marangoni flow.[22]

In Ref. 14, we took up this idea and applied it to droplets in order to study their motion initiated by a laser beam. Figure 6 summarizes the four situations we have looked at: the droplets are either adsorbing so that only one light spot at the interface transforms the surfactants or they are transparent, where two light spots occur and where refraction has to be taken into account. In (a), the surrounding fluid and thereby the droplet interface is laden with *trans* surfactants. Then, UV light causes isomerization at the interface to the *cis* isomere, which results in a higher surface tension, and Marangoni flow toward the light spots at the interface is initiated. The droplet is pushed out of the beam. In (b), we consider that fluid and droplet interface are laden with *cis* isomers assuming that they are sufficiently stable against spontaneous conversion to *trans* isomers. Here, Marangoni flow goes away from the light spots and the droplet is pulled into and thereby kept in the laser beam.

We describe the dynamics of the surfactant mixture again by Eq. (1), neglecting any thermal noise. The order parameter is

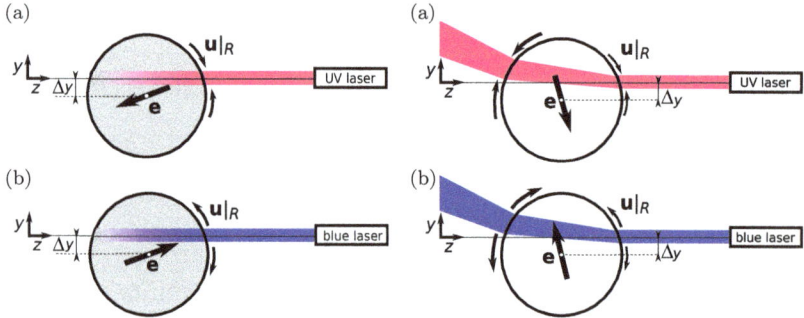

Fig. 6. Emulsion droplets with photosensitive surfactants in four situations with an offset Δy of the droplet center with respect to the laser beam. The Marangoni flow $\mathbf{u}|_R$ initiated by laser light and the resulting swimming direction $\mathbf{e}$ are indicated. Left: Strongly adsorbing droplets, (a) Initial *trans* surfactant and UV light: Droplet is pushed out of the laser beam. (b) Initial *cis* surfactant and blue light: Droplet is pulled into the laser beam. Right: Transparent droplet is pushed out (a) or pulled into (b) the laser beam. (Reproduced from Ref. 14 with kind permission of AIP Publishing; copyright (2016).)

chosen $\phi = +1$ or -1 in regions where all surfactants are either in the *cis* or *trans* state, respectively, while in mixtures of both surfactants ϕ is in the range $-1 < \phi < 1$. Choosing again $b_1 = 2 > b_2 = 1$, we assure that the surface tension of Eq. (6) is larger for pure *cis* isomeres. The source term in Eq. (1) couples the interface to the outer fluid and thus $\phi_{eq} = -1$, when the fluid is laden with *trans* molecules, and $\phi_{eq} = 1$ for *cis* isomers. The cross-section of the light beam typically has a radius $\rho = 0.2R$ and all the presented results are determined with $M = 1$ and $\kappa = \tau_D/\tau_R = 1$ if not stated otherwise.

4.1. *Adsorbing emulsion droplets*

Figure 7 shows the order parameter and flow fields for adsorbing droplets either pushed by UV light (a) or pulled by blue light (b), when the laser beam is directed toward the droplet center ($\Delta y = 0$). At $\theta = 0$, one recognizes the action of the laser beam and then ϕ relaxes toward its bulk value. The Marangoni flow is largest where ϕ and surface tension have a large gradient. In (a), the droplet is pushed away from the light source and acts as a pusher, while in (b), the droplet is a puller moving toward the source.

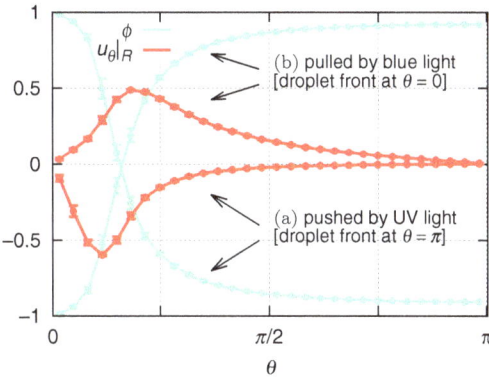

Fig. 7.   Stationary solutions of the order-parameter field ϕ and the flow field $u_\theta|_R$ for (a) A droplet pushed by UV light. (b) A droplet pulled by blue light. In both cases, the laser light hits the droplet interface at $\theta = 0$ with offset $\Delta y = 0$. (Reproduced from Ref. 14 with kind permission of AIP Publishing; copyright (2016).)

Pushing the droplet is an unstable motion since for a non-zero off-set Δy between beam axis and droplet center, the droplet moves out of the laser beam (Fig. 8, left (a)), while the pulling motion is stable against $\Delta y \neq 0$ (Fig. 8, left (b)). Here, the droplet relaxes back to the on-axis motion moving straight toward the light source. However, the pulled droplet develops an instability for decreasing $\kappa = \tau_D/\tau_R$ and an oscillatory motion about the laser beam occurs (see $\kappa = 0.1$, in Fig. 8, right). At $\kappa = 0.1$, the relaxation of the photoinduced *trans* surfactants toward the *cis* isomers is much slower than diffusive spreading. Thus, the swimming direction does not follow instantaneously the location of the light spot on the droplet interface but shows some persistence. This allows the droplet to cross the beam axis and to ultimately develop an oscillatory trajectory. In Ref. 14, we showed that the transition to the oscillatory motion occurs via a subcritical Hopf bifurcation.

4.2.   *Transparent emulsion droplets*

We now discuss the case of an emulsion droplet with negligible light absorbance, where the laser beam traverses the droplet and also

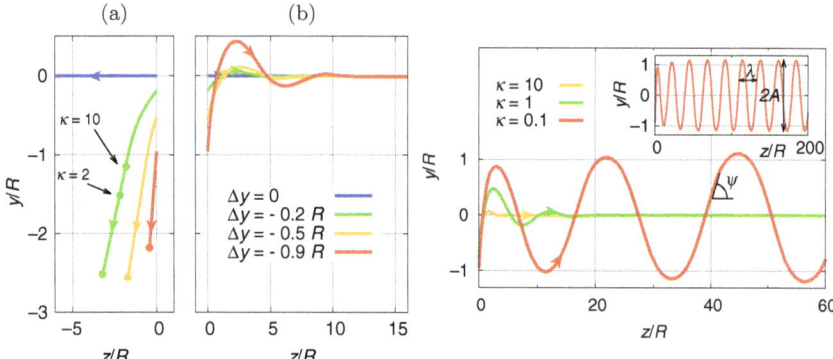

Fig. 8. Left: Trajectories of a droplet either pushed by UV light (a) or pulled by blue light (b) for different offsets Δy. The droplet initially starts at $z = 0$ and $y = \Delta y$ and stops at the positions marked by dots. The laser is positioned at $y = 0$ and shines from right to left (compare Fig. 6, left). Right: Trajectories of a droplet pulled by blue light for different relaxation rates κ. Initially, the droplet is placed at $z = 0$ and $y = -0.9R$. The laser is positioned at $y = 0$ and shines from right to left (compare Fig. 6). Inset: For $\kappa = 0.1$, a stable oscillation develops. (Reproduced from Ref. 14 with kind permission of AIP Publishing; copyright (2016).)

actuates it at a second light spot (see Fig. 6, right). We focus on a water droplet immersed in a transparent oil phase. Due to the different refractive indices of oil and water, the transmitted beam is refracted at each interface according to Snell's refraction law. We apply the refraction law to partial beams of the incident light so that it widens while crossing and leaving the droplet. In what follows, we use the refraction indices for oil ($n = 1.45$) and water ($\hat{n} = 1.35$), respectively. In our treatment, we neglect any reflection except for total reflection above the critical angle $\alpha_{\max} = \arcsin(\hat{n}/n)$. Since the two light spots on the droplet interface (see Fig. 6, right) are well separated from each other, the droplet velocity vector is a superposition of the vectors induced by each spot. Note, for $\Delta y \neq 0$, the light spot for the laser beam leaving the droplet is noticeably larger and its induced Marangoni flow dominates the droplet velocity.

In the case (a), where UV light transforms *trans* to *cis* isomers, the droplet is ultimately pushed out of the laser beam as indicated in Fig. 6, right. In case (b), where the oil phase is laden with *cis*

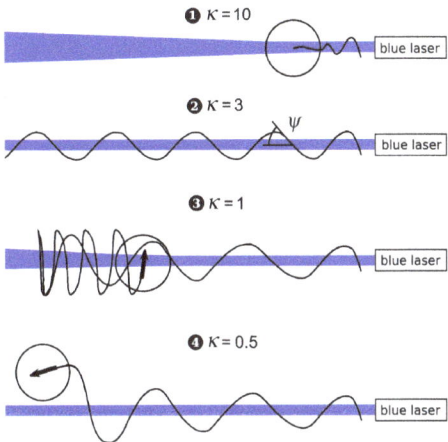

Fig. 9. Trajectories of a transparent water droplet suspended in oil, which is actuated by blue light, for decreasing κ. All droplets start at the right with an initial offset $\Delta y = -0.5$ and in all cases $M = 1$. (Reproduced from Ref. 14 with kind permission of AIP Publishing; copyright (2016).)

surfactants and blue light is applied, the droplet is pulled toward the beam axis. As for the adsorbing droplet, there is a pronounced dependence of the observed droplet trajectories on $\kappa = \tau_D/\tau_R$ as illustrated in Fig. 9. In the case of strong coupling to the bulk phase, $\kappa = 10$, the droplet performs a damped oscillation about $y = 0$. Upon decreasing the relaxation rate to values below $\kappa = 4.5$, the droplet undergoes a subcritical Hopf bifurcation (for a detailed discussion see Ref. 14) and the droplet starts to oscillate about the laser beam while moving away from the light source. This is nicely illustrated for $\kappa = 3$. Below $\kappa = 2.2$, the droplet changes its dynamics completely (see $\kappa = 1$ in Fig. 6). After moving away from the light source for a few droplet radii R with increasing oscillation amplitude, the droplet reverses its swimming direction and reaches a stationary oscillating state. The reversal occurs because the amplitude of the oscillation is so large that the size of the second spot decreases due to total reflection and the first spot pulls more strongly. Finally, at relaxation rates below $\kappa = 0.54$, the droplet eventually leaves the beam and stops.

5.　Conclusions

We have reviewed the theory to treat active emulsions with the droplet interface fully covered by a surfactant mixture. In addition to diffusive and advective currents, a phenomenological term describes the exchange of surfactants with the surrounding fluid through adsorption and desorption, while it also includes a chemical reaction, which changes the surfactant type at the interface.

We have applied the theory to bromine water droplets in a surfactant-laden oil phase. The bromination reaction locally changes the surfactant type and the resulting gradient in surface tension causes Marangoni flow which in turn enhances the gradient. We have demonstrated that this gives rise to a symmetry-breaking transition and the droplet starts to swim when the advective current dominates the smoothing diffusive current as quantified by the Marangoni number M. Increasing M further, the droplet stops again and for even larger M performs oscillations due to a spatio-temporal variation of the mixture order-parameter field. Taking into account thermal fluctuations, we can describe the coarsening dynamics from a uniform mixture to a fully phase-separated coverage of the droplet interface. The mean domain size grows slowly over several decades and then crosses over to a fast final coarsening. Interestingly, this two-stage coarsening process is reminiscent of the coarsening scenario of model H with a diffusive regime followed by an advective regime. In steady state, the domain wall, separating the regions of the two surfactant types from each other, fluctuates. Thereby the swimming direction performs a random walk on the unit sphere and gives rise to a persistent random walk of the swimming droplet.

In the second example, the droplet interface is fully covered with a photosensitive surfactant. It can be switched reversibly between a *trans* and *cis* configuration under illumination of light with a specific wavelength. This also changes surface tension. In our setup, a laser beam creates a patch of switched surfactant. The surface tension becomes non-uniform and Marangoni flow occurs. We have looked at strongly adsorbing and transparent droplets. In both cases, droplets with initial *trans* surfactants, meaning lower surface tension, are pushed out of the beam, whereas droplets are pulled into the beam when they are initially covered by *cis* surfactants. Both for adsorbing and transparent droplets, a subcritical Hopf bifurcation toward oscillating trajectories occurs.

Active emulsions provide a playground for inducing self-propulsion with an interesting dynamics, as demonstrated by the two examples, which we reviewed here. Inspired by this work, we recently looked at the simpler geometry of a plane fluid interface and generalized the theory to the case where the interface is not fully covered by surfactants.[21] Combining the idea of a photoresponsive interface with the method of feedback control, we are able to induce regular and irregular oscillatory dynamics in an array of activating light spots. This demonstrates the potential of photoresponsive interfaces for basic research in pattern formation and nonlinear dynamics but also for light-driven microfluidic applications.

Acknowledgments

We thank S. Herminghaus, U. Thiele, S. Thutupalli, and A. Zöttl for helpful discussions and acknowledge financial support by the Deutsche Forschungsgemeinschaft in the framework of the collaborative research center SFB 910 and the research training group GRK 1558.

References

1. S. Thutupalli, R. Seemann, and S. Herminghaus, Swarming behavior of simple model squirmers, *New J. Phys.* **13**, 073021 (2011).
2. M. Schmitt and H. Stark, Swimming active droplet: A theoretical analysis, *Europhys. Lett.* **101**, 44008–1–6 (2013).
3. S. Herminghaus, C. C. Maass, C. Krüger, S. Thutupalli, L. Goehring, and C. Bahr, Interfacial mechanisms in active emulsions, *Soft Matter.* **10**, 7008–7022 (2014).
4. C. C. Maass, C. Krüger, S. Herminghaus, and C. Bahr, Swimming droplets, *Annu. Rev. Condens. Matter.* **7** (2016).

5. M. Schmitt and H. Stark, Active Brownian motion of emulsion droplets: Coarsening dynamics at the interface and rotational diffusion, *Eur. Phys. J. E.* **39**, 80–1–15 (2016).

6. A. Najafi and R. Golestanian, Simple swimmer at low reynolds number: Three linked spheres, *Phys. Rev. E.* **69**, 062901–1–4 (2004).

7. R. Dreyfus, J. Baudry, M. L. Roper, M. Fermigier, H. A. Stone, and J. Bibette, Microscopic artificial swimmers, *Nature.* **437**, 862–865 (2005).

8. E. Gauger and H. Stark, Numerical study of a microscopic artificial swimmer, *Phys. Rev. E.* **74**, 021907–1–10 (2006).

9. E. Lauga and T. R. Powers, The hydrodynamics of swimming microorganisms, *Rep. Prog. Phys.* **72**, 096601–1–36 (2009).

10. R. Vogel and H. Stark, Rotation-induced polymorphic transitions in bacterial flagella, *Phys. Rev. Lett.* **13**, 158104–1–5 (2013).

11. J. Elgeti, R. G. Winkler, and G. Gompper, Physics of microswimmers — single particle motion and collective behavior: A review, *Rep. Prog. Phys.* **78**, 056601–1–50 (2015).

12. D. Alizadehrad, T. Krüger, M. Engstler, and H. Stark, Simulating the complex cell design of trypanosoma brucei and its motility, *PLoS Comput. Biol.* **11**, e1003967–1–13 (2015).

13. A. Zöttl and H. Stark, Emergent behavior in active colloids, *J. Phys. Condens. Matter.* **28**, 253001–1–28 (2016).

14. M. Schmitt and H. Stark, Marangoni flow at droplet interfaces: Three-dimensional solution and applications, *Phys. Fluids.* **28**, 012106–1–29 (2016).

15. S. Shinkai, K. Matsuo, A. Harada, and O. Manabe, Photocontrol of micellar catalyses, *J. Chem. Soc. Perkin Trans.* **2**, 1261–1265 (1982).

16. H. Sakai, H. Ebana, K. Sakai, K. Tsuchiya, T. Ohkubo, and M. Abe, Photo-isomerization of spiropyran-modified cationic surfactants, *J. Colloid Interf. Sci.* **316**, 1027–1030 (2007).

17. J. Eastoe, M. S. Dominguez, P. Wyatt, A. Beeby, and R. K. Heenan, Properties of a stilbene-containing gemini photosurfactant: Light-triggered changes in surface tension and aggregation, *Langmuir.* **18**, 7837–7844 (2002).

18. T. Shang, K. A. Smith, and T. A. Hatton, Photoresponsive surfactants exhibiting unusually large, reversible surface tension changes under varying illumination conditions, *Langmuir.* **19**, 10764–10773 (2003).

19. E. Chevallier, A. Mamane, H. A. Stone, C. Tribet, F. Lequeux, and C. Monteux, Pumping-out photo-surfactants from an air-water interface using light, *Soft Matter.* **7**, 7866–7874 (2011).

20. D. Baigl, Photo-actuation of liquids for light-driven microfluidics: State of the art and perspectives, *Lab Chip.* **12**, 3637–3653 (2012).

21. J. Grawitter and H. Stark, Feedback-control of photoresponsive fluid interfaces, *Soft Matter* **14**, 1856–1869 (2018).

22. A. Diguet, R.-M. Guillermic, N. Magome, A. Saint-Jalmes, Y. Chen, K. Yoshikawa, and D. Baigl, Photomanipulation of a droplet by the chromocapillary effect, *Angew. Chem. Int. Ed.* **48**, 9281–9284 (2009).

23. N. Kavokine, M. Anyfantakis, M. Morel, S. Rudiuk, T. Bickel, and D. Baigl, Light-driven transport of a liquid marble with and against surface flows, *Angew. Chem. Int. Ed.* **55**, 11301–11301 (2016).

24. M. Anyfantakis and D. Baigl, Dynamic photocontrol of the coffee-ring effect with optically tunable particle stickiness, *Angew. Chem. Int. Ed.* **53**, 14077–14081 (2014).

25. V. A. Nepomniashchii, M. G. Velarde, and P. Colinet, *Interfacial Phenomena and Convection*, Chapman & Hall. 1st edn. (2002).

26. R. C. Desai and R. Kapral, *Dynamics of Self-Organized and Self-Assembled Structures.* Cambridge University Press (2009).

27. S. Chandrasekhar, *Hydrodynamic and Hydromagnetic Stability.* Oxford University Press (1961).

28. H. A. Stone and A. D. Samuel, Propulsion of microorganisms by surface distortions, *Phys. Rev. Lett.* **77**, 4102 (1996).

29. J. Bławzdziewicz, P. Vlahovska, and M. Loewenberg, Rheology of a dilute emulsion of surfactant-covered spherical drops, *Physica A.* **276**, 50–85 (2000).

30. J. A. Hanna and P. M. Vlahovska, Surfactant-induced migration of a spherical drop in stokes flow, *Phys. Fluids.* **22**, 013102 (2010).

31. J. T. Schwalbe, F. R. Phelan Jr, P. M. Vlahovska, and S. D. Hudson, Interfacial effects on droplet dynamics in Poiseuille flow, *Soft Matter.* **7**, 7797–7804 (2011).

32. O. S. Pak, J. Feng, and H. A. Stone, Viscous Marangoni migration of a drop in a Poiseuille flow at low surface Péclet numbers, *J. Fluid. Mech.* **753**, 535–552 (2014).

33. R. S. Subramanian, The Stokes force on a droplet in an unbounded fluid medium due to capillary effects, *J. Fluid Mech.* **153**, 389–400 (1985).

34. A. Bray, Coarsening dynamics of phase-separating systems, *Phil. Trans. R. Soc. Lond. A.* **361**, 781–792 (2003).

35. A. J. Bray, Theory of phase-ordering kinetics, *Adv. Phys.* **51**, 481–587 (2002).

36. O. Karthaus, M. Shimomura, M. Hioki, R. Tahara, and H. Nakamura, Reversible photomorphism in surface monolayers, *J. Am. Chem. Soc.* **118**, 9174–9175 (1996).

37. J. Y. Shin and N. L. Abbott, Using light to control dynamic surface tensions of aqueous solutions of water soluble surfactants, *Langmuir.* **15**, 4404–4410 (1999).

38. K. Ichimura, S.-K. Oh, and M. Nakagawa, Light-driven motion of liquids on a photoresponsive surface, *Science.* **288**, 1624–1626 (2000).

39. J. Eastoe and A. Vesperinas, Self-assembly of light-sensitive surfactants, *Soft Matter.* **1**, 338–347 (2005).

Chapter 23

Hydrodynamic Theory of Phoretic Propulsion: Clarifications, New Concepts, and Reassessment

Sergey D. Traytak

Semenov Institute of Chemical Physics
of Russian Academy of Sciences,
4 Kosygina St., 119991 Moscow, Russia

sergtray@mail.ru

We describe theoretically the thermophoretic propulsion in the system of two spherical particles due to bulk chemical reaction occurring in one of them. To solve the arising heat and hydrodynamic problems, the generalized method of separation of variables is used. In essence, the original boundary value problems we reduce to solution of the much easier infinite set of linear algebraic equations. Moreover, the classical hydrodynamic theory of thermophoresis is subjected to criticism. In particular, we demonstrate that the existing classical theory requires a fundamental reassessment and development to include high-order multipoles corrections to the phoretic velocity due to thermal and hydrodynamics interactions.

1. Introduction, Motivation, and Goals

Nowadays, the term "phoresis" signifies the drift motion of small objects suspended in gas (or liquid) medium[a] due to an applied

---

[a]We do not distinguish between gas and liquid medium assuming they are only non-ionic.

uniform gradient of some scalar field of physical quantities like electric potential, concentration, temperature, etc.[1] Thus, one deals with the object propulsion in gradients: Of electric potential (*electrophoresis*), concentration of diffusing particles (*diffusiophoresis*), temperature (*thermophoresis*), etc. It is common knowledge that different regimes of the particle phoretic motion are characterized by the *Knudsen number* $Kn = \lambda/l$, defined as the ratio of the mean free path of a gas molecule λ to a characteristic size of the problem l.[2]

We shall investigate here so-called *continuum limit*[3] (or *near-continuum regime*[1]), i.e., considering all appropriate physical parameters as $Kn \to 0+$. Theoretical descriptions of the phoretic propulsions in continuum limit are rather similar, so for the definiteness, we restrict ourselves to the case of thermophoresis only.

The theory of thermophoresis in the continuum limit stems from the seminal work by Epstein.[4] It is generally accepted that he "was the first to rigorously derive an expression for the thermophoretic velocity, ... in gases for the large particle regime", i.e., in continuum limit.[5] Therefore, his theory is still of considerable current use (see, e.g., Refs. 6 and 7 and references therein). Note in passing that Epstein's theory and its ensuing refinements[8–10] are termed in various ways: *macroscopic approach,*[11] the *thermophoresis theory in the near-continuum regime,*[1] *hydrodynamic theory of thermophoresis,*[6] or most commonly, *thermophoresis theory of large particles* (or just *at small Knudsen numbers*).[9,12] Throughout this chapter, however, the *classical hydrodynamic theory of thermophoresis* (CHTT) will be understood to mean theory of suspended particles motion in a prescribed external steady temperature gradient for Stoke's slip-flow regime.

On the other hand, the literature on the *self-propulsion* (*propelling*) *of small objects* over the last decades has become more and more voluminous, studying various facets of this motion and its applications.[13–18] The term "self-propulsion" is used in reference to a large amount of natural and artificial phenomena, where the autonomous propulsion of small objects occurs by means of some of their own ability for motion. Phoretic effects play an important role among these various mechanisms of self-propulsion.[18–22]

For example, the directed motion of a Janus particle is induced due to a temperature gradient generated by itself. So the latter motion is naturally called *self-thermophoresis*.[19,23,24] Note that presently corresponding *theory of self-thermophoresis* (TST) seems to be well elaborated. However, this is quite strange, but a simple inspection of references shows that these two branches of science almost do not interact with each other, as if CHTT and TST are completely different theories. If the papers on the CHTT do not cite any works on the TST, the works on the TST are mainly limited to the relevant references (see, for example, the works of A. Würger and of R. Golestanian[25–28]).

In 1982, Fuchs noted that the theory of thermophoresis of aerosol particles at large Kn agrees well with experimental data, whereas the CHTT for small Kn and high thermal conductivity of the particles reveals serious difficulties.[29] Many years later, this question still remains to be unclarified and, e.g., in 2006 Prodi *et al.* pointed out: "While in the free-molecule regime (Kn $\gg$ 1) ... theory seems to be in good agreement with the experimental results, in the slip-flow regime (Kn $<$ 0.1) ... previously published measurements are contradictory".[30] Recently, Sagot once again called attention to some discrepancy between estimated and observed values of the thermophoretic velocity.[31] In this connection, we should note that very often, to overcome experimental difficulties, indirect measures of phoretic velocity are used.[32,33] However, this kind of experiment cannot identify the proper phoretic effects among other existing competing effects (see also Section 6).

In 2011, Young presented an interesting investigation concerning the Grad 13-moment method of solution to the Boltzmann equation with applications to the theory of thermophoresis for all Knudsen numbers.[34] Nevertheless, the theoretical reassessments proposed by Young have almost not referred to the principal ideas of the CHTT, if at all.[34] Moreover, according to Zheng and Davis: "Thermophoresis theories for spherical particles are fairly well developed".[35]

An additional point to emphasize is that in the present study we do not discuss Brenner's unified theory of phoretic phenomena in gases and liquids[36] and various aspects of phoresis in liquids.[37–39]

Thus, the main motivation for writing this chapter is threefold: (a) to highlight close relations between the CHTT and the TST; (b) to put forward possible ways to revise the existing theory to make its predictions be in good agreement with the experimental data; (c) to draw the attention to intrinsic difficulties of the CHTT, including the infinite temperature paradox, and resolve it.

Last but not least is that this chapter in some sense may be treated as a "journey to the past". The point is that many years ago the author had been studying some problems on applications of the CHTT, namely, capture of fine aerosol particles by drops and by coarse rigid particles in laminar and turbulent flows.[40–46] However, the results obtained show that, particularly, all the applications of the CHTT should be thoroughly revised.

Thus, we intend to take here a fresh look at the CHTT and its applications to correct their drawbacks.

2. Theoretical Model

Suppose that the global Cartesian coordinate system $(O; \mathbf{x})$ is defined in 3D Euclidean space $\mathbb{R}^3$, where point $\mathbf{x} = (x_1, x_2, x_3) \in \mathbb{R}^3$ is relative to the standard basis $(\mathbf{e}_1, \mathbf{e}_2, \mathbf{e}_3) \subset \mathbb{R}^3$. For the sake of definiteness, we shall treat here the two-sphere system $\Omega_1 := \{\mathbf{x} : \|\mathbf{x}\| < R_1\}$ and $\Omega_2 := \{\mathbf{x} : \|\mathbf{x} - \mathbf{L}_{21}\| < R_2\}$ with the origin $\{O\}$ at the centre of sphere 1 and $R_1 := R \neq R_2 := a$. Here, $\| \cdot \|$ denotes the Euclidean norm and

$$\mathbf{L}_{ij} = \mathbf{r}_j - \mathbf{r}_i, \quad \mathbf{L}_{ij} = -\mathbf{L}_{ji}, \tag{1}$$

where $\mathbf{L}_{ij}$ is the vector, connecting the centers of spheres $\{O_j\}$ and $\{O_i\}$ (hereafter $i \neq j = 1, 2$) as depicted in Fig. 1, whereas $\|\mathbf{L}_{ij}\| = L$; P is the observation point. Assume that the spheres 1 and 2 can be either freely suspended in the liquid (case A) or connected by an infinitely thin rigid rod, forming an asymmetric rigid dumbbell called *Janus complex*[20–23] (case B). Moreover, let the position of particle 1 be fixed in case A. Note that the statement of the problem bears some resemblance to that studied in Ref. 47. It is clear that case

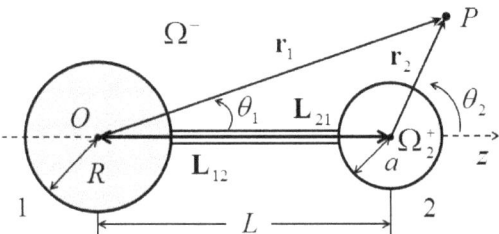

Fig. 1. Sketch of the rigid dumbbell model of the active self-propelled particle.[19]

B corresponds to so-called *rigid dumbbell model of the active self-propelled particle* which, in fact, is a widely used model in theoretical works[19-21] (see Fig. 1). The domain Λ_{ij} defined by the constraint $\Lambda_{ij} := \{\mathbf{x} : L > a + R_1\} \subset \mathbb{R}^3$ is said to be the *configuration manifold* of the system at issue.

As it is common, $\Omega^{\pm} \subset \mathbb{R}^3$ will denote 3D domains: either bounded (Ω^+) or exterior to a bounded region (Ω^-) with a smooth boundary $\partial\Omega$ of class C^2. For the model at issue: $\Omega_i^+ \equiv \Omega_i$ are interiors; and $\Omega^- := \mathbb{R}^3 \backslash \overline{\Omega}_1 \cup \overline{\Omega}_2$ exterior of particles 1 and 2 with smooth boundaries. Sometimes it is expedient to use also the *partial domains* $D_i := \mathbb{R}^3 \backslash \overline{\Omega}_i$, whereas $\Omega^- = D_1 \cap D_2$. It is natural to term domain Ω^- as 3D *manifold for the temperature field* or *temperature manifold* for short.[48]

We need to determine the thermophoretic force acting on this system and its subsequent motion. Note that for both the cases A and B, sphere 2 moves because of temperature gradient, so it is often addressed as *phoretic (thermophoretic) particle* (PP).[18,36]

Provided

$$(T_s - T_0)/T_0 \ll 1, \quad \text{where} \quad \lim_{r_i \to \infty} T = T_0, \quad \lim_{r_1 \to R+0} T = T_s \quad (2)$$

and $T(P)$ is the absolute temperature of the fluid at $P \in \Omega^-$; all physical properties are assumed to be constant.[34] Further, it is expedient to consider the problem with respect to the *disturbance of temperature fields*

$$T^-(P)\{T^+(P)\} := T(P) - T_0 \quad \text{in} \quad \Omega^-\{\Omega_i^+\}. \quad (3)$$

Besides, for the sake of clarity and simplicity we shall adopt original Epstein's statement of the problem.[4] So we can suppose that the heat equation is uncoupled with the velocity field and, therefore, the problem consists of two separate parts: (a) The calculation of the temperature field and (b) the incorporation of these results into the relevant hydrodynamic boundary conditions to solve the viscous flow equations.

Furthermore, assume that temperature inhomogeneity $T^-(P)$ in the manifold Ω^- is caused by the bulk quasi-steady chemical reaction occurring in particle 1.[49] The chemical reaction taking place within Ω_1^+ can be either endothermic or exothermic with the volumetric thermal energy generation rate $\Phi(P; T^+)$ which, using condition (2), can be simplified as follows:[49]

$$\Phi(P; T^+) := gR^{-2}(T^+ + R_g T_0^2 E_a^{-1}), \qquad (4)$$

where R_g and E_a are the gas constant and activation energy; g is a complex reaction parameter given explicitly in Ref. 49.

3. The Heat Problem

If temperature disturbance caused by thin rod is negligibly small, the calculation of the quasi-steady temperature field for above two cases A and B may be performed entirely in the same manner. Therefore, we shall not distinguish heat parts for these two cases.

3.1. *Statement of the heat problem*

Taking representation (4) into account, the quasi-steady temperature distribution in the whole space is governed by

$$-\kappa_1^+ \nabla^2 T^+ = \Phi(P; T^+) \quad \text{in} \quad \Omega_1^+, \quad -\nabla^2 T^+ = 0 \text{ in } \Omega_2^+, \qquad (5)$$

$$-\nabla^2 T^- = 0 \text{ in } \Omega^-, \qquad (6)$$

with *conjugate (transmission) boundary conditions* on the particle surfaces

$$T^-\big|_{\partial\Omega_i^-} - T^+\big|_{\partial\Omega_i^+} = 0, \tag{7}$$

$$-\partial_{r_i}T^-\big|_{\partial\Omega_i^-} + \chi_i\partial_{r_i}T^+\big|_{\partial\Omega_i^+} = 0. \tag{8}$$

Here, $(\cdot)\big|_{\partial\Omega_i^\pm} := \lim_{r_i \to R_i \mp 0}(\cdot)$ are the limits from $\Omega_i^\pm$; $\chi_i = \kappa_i^+/\kappa^-$, where κ^- and κ_i^+ are the respective fluid and the ith particle thermal conductivities. We complete the statement of the above heat problem with conditions at the origin of Ω_i^+, i.e.,

$$T^+\big|_{r_i \to 0} < +\infty \tag{9}$$

and regularity condition at infinity

$$T^-\big|_{r_i \to \infty} \rightrightarrows 0. \tag{10}$$

Rigorously speaking, condition on the uniform vanishing (10) is equivalent to the regularity condition.[50] It may be proved that boundary value problem (5)–(10) is well-posed,[50] therefore, we can seek for its *classical (regular) solution* $T^\pm(\mathbf{x}) \in C^0(\overline{\Omega}^\pm) \cap C^2(\Omega^\pm)$.

3.2. *Semi-analytical solution of the heat problem*

The *method of bispherical coordinates* is the commonly used one to obtain the exact solution to the posed heat problem (5)–(10) in the two-sphere exterior.[22,52] However, for a number of reasons, the *generalized method of separation of variables* (GMSV) offers the greatest promise. This method allows one to find the semi-analytical solutions of various problems for the Laplace or Stokes equations in domains with complex boundaries consisting of many connected components to any degree of accuracy.[51,52]

Often, for real reactions one can find the solution in a very simple form,[49]

$$T^+(P) \equiv T_s - T_0 \quad \text{in} \quad \Omega_1^+, \tag{11}$$

where $T_s = T_0 - R_g T_0^2 E_a^{-1}$ is the constant temperature of particle 1.

3.2.1. *Derivation of the resolved ISLAE*

By means of superposition principles it may be shown that general solution to (5) in the temperature manifold Ω^- and domain Ω^+ may be uniquely represented as follows[48]:

$$T^-(P) = T_1^- + T_2^- \quad (P \in \Omega^-), \quad T^+(P) = T_2^+ \quad (P \in \Omega^+), \quad (12)$$

where T_i^- are the *partial solutions* corresponding to the *i*th particle.

Scalar axially symmetric *regular* and *irregular solid spherical harmonics* of the degree n and $-n-1$ with respect to a bound domain Ω^+ read

$$\psi_n^+(r,\theta) := r^n P_n(\cos\theta), \qquad \psi_n^-(r,\theta) := r^{-n-1} P_n(\cos\theta), \quad (13)$$

where $P_n(\mu)$ is Legendre polynomial of n degree from a canonical basis solution $\{\psi_n^\pm\}_{n=0}^\infty$ to Laplace's equations (5), (6) in domains $\Omega^\pm$, respectively.

"*Irregular to regular*" *translation addition theorem* for solid harmonics holds true[51]

$$\psi_n^-(\xi_j,\theta_j) = \sum_{k=0}^\infty W_{nk}^{(j,i)} \psi_k^+(\xi_i,\theta_i) \quad \text{for} \quad \epsilon_i\xi_i < 1, \quad (14)$$

where $\xi_i := r_i/R_i$, $(R_1 = R, R_2 = a)$ are dimensionless radial variables and so-called *mixed-basis matrices elements*

$$W_{nk}^{(j,i)} \equiv W_{nk}^{(j,i)}(\mathbf{L}_{ji}) = (-1)^k \binom{k+n}{k} \epsilon_i^k \epsilon_j^{n+1} P_{k+n}(\cos\Theta_{ji}), \quad (15)$$

whereas $\Theta_{ji} = 0$ if $j < i$ and $\Theta_{ji} = \pi$ if $j > i$ (see Fig. 1) and $\epsilon_i := R_i/L_{ji}$.

Addition theorem (14) allows us to reexpand any partial solution written in local coordinates connected with the *j*th particle, i.e., $T_j^-(r_j,\theta_j)$ solely in terms of variables of another local coordinate system $(O_i; r_i, \theta_i)$.

Partial solutions we are looking for in the dimensionless form include

$$T_i^-(\xi_i, \theta_i) = \sum_{n=0}^{\infty} A_n^i \psi_n^-(\xi_i, \theta_i), \quad T_2^+(\xi_2, \theta_2) = \sum_{n=0}^{\infty} \widetilde{A}_n^2 \psi_n^+(\xi_2, \theta_2),$$

(16)

where A_n^1, A_n^2, $\widetilde{A}_n^2$ are unknown coefficients to be determined from the boundary conditions on the particles' surfaces. Substituting these solutions into the boundary conditions with the aid of addition theorem (14), one gets the so-called *resolved infinite set of linear algebraic equations* (ISLAE) of the II kind

$$A_n^1 + \sum_{m=0}^{\infty} A_m^2 W_{mn}^{(2,1)} = (T_s - T_0)\delta_{0n},$$

(17)

$$A_n^2 + \sum_{m=0}^{\infty} A_m^1 W_{mn}^{(1,2)} = \widetilde{A}_n^2,$$

(18)

$$-(n+1)A_n^2 + n \sum_{m=0}^{\infty} A_m^1 W_{mn}^{(1,2)} = n\chi \widetilde{A}_n^2.$$

(19)

Hereafter, $\chi = \chi_2$. For the slip condition (38) posed upon the PP surface, we need only external field $T^-(P)$. So we can eliminate sequence $\{\widetilde{A}_n^2\}_{n=0}^{\infty}$ in (18), (19) and then recast the result in a more convenient matrix notation

$$(\mathbf{I} + \mathbf{W}) \, \mathbf{A} = \mathbf{B}.$$

(20)

Here, we denoted $\mathbf{A} := \left(\mathbf{A}^{(1)}, \mathbf{A}^{(2)}\right)^T$, $\mathbf{B} := \left(\mathbf{B}^{(1)}, \mathbf{B}^{(2)}\right)^T$,

$$\mathbf{I} := \begin{pmatrix} \mathbf{E} & \mathbf{0} \\ \mathbf{0} & \mathbf{E} \end{pmatrix}, \mathbf{W} := \begin{pmatrix} \mathbf{0} & \mathbf{W}^{(1,2)} \\ \widetilde{\mathbf{W}}^{(2,1)} & \mathbf{0} \end{pmatrix},$$

$$\widetilde{W}_{mn}^{(2,1)} := \frac{n(\chi - 1)}{[1 + (1+\chi)n]} W_{mn}^{(1,2)},$$

where $\mathbf{E}$ is the infinite unit matrix.

3.2.2.   Solvability of the resolved ISLAE

In derived resolved ISLAE (20), we deal with the direct sum of
Hilbert spaces of infinite sequences $\ell_2^2 := \ell_2 \oplus \ell_2$. It is evident that
$\mathbf{B} = (T_s - T_0, 0, ...)^T \in \ell_2^2$, so, assuming that $\mathbf{A} \in \ell_2^2$, we have the
infinite matrix operator in (20): $W : \ell_2^2 \to \ell_2^2$. Provided $\overline{\Omega}_1 \cap \overline{\Omega}_2 = \varnothing$,
one can straightforwardly prove the absolute convergence of the dou-
ble series

$$\sum_{n,m=0}^{\infty} |W_{nm}| < +\infty. \tag{21}$$

Hence, denoting $\mu := \sup_{n,m} |W_{nm}|$, the known *Koch condition*
implies[53,54]

$$\sum_{n,m=0}^{\infty} W_{nm}^2 \le \sum_{n,m=0}^{\infty} \mu |W_{nm}| < +\infty, \tag{22}$$

which proves that $\mathbf{W}$ in (20) is a *compact infinite matrix operator* in the Hilbert space ℓ_2^2.[54] Therefore, the ISLAE (20) is of
normal Poincare–Koch type and the *Fredholm–Hilbert alternative*
holds true.[53] Hence, a unique solution of class ℓ_2^2 exists and this
sequence may be found by the *method of reduction* to any degree of
accuracy.[53,54]

Other important cases of the infinite matrix operators: $W : \ell_p \to$
ℓ_p $(1 \le p < +\infty)$ and $W : \ell_\infty \to \ell_\infty$ may be considered similarly.

3.3.   The infinite temperature paradox and its resolution

In the rest of this section, we shall study the case A. This system
allows us to elucidate a paradox concerning the boundary condition
at the infinite point $\{\infty\}$ within the scope of the CHTT.

One of the most important postulates of the CHTT states: a
*steady-state gradient of constant value has been created and main-
tained in unbound spaces filled with the quiescent liquid.* Thus, since
Epstein's pioneering work, the commonly accepted condition at $\{\infty\}$

in local coordinates of the PP reads[4,11,34]

$$T^-(r_2, \theta_2) \sim T_\infty + |\nabla T_\infty| r_2 \cos \theta_2 \quad \text{as} \quad r_2 \to \infty. \quad (23)$$

However, one can see that this condition cannot be adequate from both physical and mathematical points of view. Really, *it follows from (23) that, e.g., along the z axis, i.e., at $\theta_2 = 0$ (see Fig. 1) one has the linear rate temperature divergence as $r_2 \to \infty$ and this divergence does not make any physical sense.* Hence, the classical statement of the heat problem (5)–(9) and (23) is the ill-posed one. It is expedient to term this fact the *infinite temperature paradox*.

Alternately, sometimes it is assumed that the temperature gradient is subject to the following boundary condition at infinity[12,13,55,56]

$$\nabla T|_{r_1 \to \infty} \to \nabla T_\infty, \quad (24)$$

where ∇T_∞ is a prescribed constant gradient at $\{\infty\}$.

In statement (24), we do not deal with the infinite temperature paradox explicitly, but its failure immediately follows from the non-uniqueness of the solution. Indeed, let $\Omega^- = \{r > R\}$ be the exterior of the ball of radius R, then both functions $u_1 = R/r$ and $u_2 \equiv 1$ obey Laplace's equation under the same boundary conditions at $r = R$ and condition (24) for $\nabla T_\infty = 0$.

Concerning this matter, we call attention to page 96 of Ref. 57, where we find: "If the posed boundary value problem possesses several solutions, then expression 'solution of the problem' makes no sense. Therefore, it is first necessary to prove uniqueness of the solution before speaking about its solution." Besides, we try to be strictly faithful to the basic principle: If a mathematical problem is well posed and tractable by means of rigorous mathematical methods, there is no use to invent *ad hoc* hypothesis based on some additional physical assumptions.

Nevertheless, despite the ill-posed heat problem (5)–(9) and (23) (or (24)), CHTT sometimes arrived at a reasonable result (40) giving rather good support by experiments (see discussion in Section 6).

3.4.   *Solution by the reflection method*

To clarify the mathematical and physical sense of the CHTT and to resolve the above paradox, we apply the *method of reflections* (*Schwarz's method*)[58,59] to solve posed heat boundary value problem (5)–(10). Note in passing that iterative solution of the resolved ISLAE (20) leads to the same series as the method of reflections.[48]

Zero approximation consists of the sum of the non-interacting temperature fields

$$T^{-(0)}(P) = T_1^{-(0)} + T_2^{-(0)}, \qquad T^{+(0)}(P) = T_2^{+(0)}, \qquad (25)$$

where partial zero approximations satisfy conditions on $\partial\Omega_1$ and $\partial\Omega_2$

$$T_1^{-(0)}(\xi_1, \theta_1) = (T_s - T_0)\,\xi_1^{-1}, \quad T_2^{-(0)}(\xi_2, \theta_2) \equiv 0 \equiv T_2^{+(0)}(\xi_2, \theta_2).$$

Introduce the *first reflections* $(T_2^{-(1)}, T_2^{+(1)})$ to compensate the changes in the boundary conditions on $\partial\Omega_2$ due to the term $T_1^{-(0)}$

$$T_2^{-(1)}(\xi_2, \theta_2) = \sum_{n=0}^{\infty} A_n^{(1)}\psi_n^-(\xi_2, \theta_2), \qquad (26)$$

$$T_2^{+(1)}(\xi_2, \theta_2) = \sum_{n=0}^{\infty} \tilde{A}_n^{(1)}\psi_n^+(\xi_2, \theta_2). \qquad (27)$$

Thus, the first approximation near the PP reads

$$T^{-(1)}(P) = T_1^{-(0)} + T_2^{-(1)}, \quad T^{+(1)}(P) = T_2^{+(1)}. \qquad (28)$$

Recasting $T_1^{-(0)}(\xi_1, \theta_1)$ in terms of local coordinates $(O_2; \xi_2, \theta_2)$, we have

$$T_1^{-(0)}(\xi_1, \theta_1) = \sum_{n=0}^{\infty} Q_n\psi_n^+(\xi_2, \theta_2), \quad \text{for} \quad \epsilon_2\xi_2 < 1, \qquad (29)$$

where $Q_n = (-1)^n (T_s - T_0)\,\epsilon_1\epsilon_2^n$. Unknown coefficients in (26) and (27) may be found by substitution expressions (28) and (29) to the conjugate boundary conditions (7) and (8)

$$A_0^{(1)} = 0, \quad \widetilde{A}_0^{(1)} = Q_0, \quad \text{and for } n \geq 1$$

$$A_n^{(1)} = \frac{(1-\chi)\,n}{[1+(1+\chi)\,n]}Q_n, \quad \widetilde{A}_n^{(1)} = \frac{(1+2n)}{[1+(1+\chi)\,n]}Q_n.$$

For particular case of monopole ($n = 0$) and dipole approximation ($n = 1$) in a neighborhood of the PP surface and its inside Ω_2^+, one has[55]

$$T^{-(1)}(r_2, \theta_2) \approx T_\infty + \left[1 + \frac{(1-\chi)}{(2+\chi)}\left(\frac{a}{r_2}\right)^3\right]\nabla T_\infty r_2 \cos\theta_2, \quad (30)$$

$$T_2^{+(1)}(r_2, \theta_2) \approx \frac{3}{2+\chi}\nabla T_\infty r_2 \cos\theta_2, \quad (31)$$

where $T_\infty = (T_s - T_0)\,\epsilon_1$ and $\nabla T_\infty = -T_\infty L^{-1}$.

It is noteworthy that Yang and Ripoll considered self-propelled motion of a nanodimer (case B at $R = a$) assuming "the spherical symmetry of the temperature distribution around the sphere 1".[20] One can see, however, that their solution amounts to the unperturbed zero approximation (25).

Thus, within the scope of the CHTT, the external temperature field represents an approximation (30), comprising two terms: (a) classical expression for the temperature "at infinity" (23) and (b) its first reflection on the PP surface. Nevertheless, in the CHTT these fields are considered to be exact and it is universally accepted to use them for calculating the "exact" value of thermophoretic velocity (see also Section 6).

4. Stokes' Flows for Phoretic Motion

4.1. *Statement of the stokes problem*

For definiteness, consider now the case A, i.e., a propulsion of a rigid spherical PP of radius a due to reaction in the static particle 1. Moreover, here we shall study only the axially symmetric translational motion without rotation. We assume that at large distances from the PP, the original motion is unaffected by its presence and motion of the PP is slow with some constant velocity $\mathbf{U}$ along the z-direction $\mathbf{U} = U\mathbf{e}_z$, $U = \text{const}$. Note that $\mathbf{U}$ is defined as the translational velocity of the PP center of mass. In spherical coordinates of PP, one has: $\mathbf{U} := U\cos\theta_2\mathbf{e}_r^2 - U\sin\theta_2\mathbf{e}_\theta^2$, where $\mathbf{e}_r^j$ and $\mathbf{e}_\theta^j$ are unit vectors of spherical coordinates $(O_j; r_j, \theta_j)$ of the jth sphere.

As is customary, consider the local fluid velocity vector $\mathbf{v}(P)$, and scalar pressure $p(P)$ fields in the manifold Ω^-, forming pair $(\mathbf{v}, p)$ called the *Stokes flow*. Provided stabilization of the time-dependent Stokes flow holds, the stationary slow motion of liquid is governed by the coupled system of Stokes equations[61–63]

$$-\eta\nabla^2\mathbf{v} + \nabla p = \mathbf{f} \quad \text{in} \quad \Omega^-, \tag{32}$$

$$\nabla \cdot \mathbf{v} = 0 \quad \text{in} \quad \Omega^-. \tag{33}$$

In the preceding, η is the fluid dynamic viscosity, and $\mathbf{f}$ the external force acting on the fluid. The incompressibility constraint (33) is used to determine the pressure from the velocity field. System (32)–(33) provides a good approximation to the slow flows around particles 1 and 2, i.e., when the Reynolds number $\text{Re} := aU/\nu$ (where $\nu = \eta/\rho$ is the kinematic viscosity and ρ the gas mass density) is small enough $(\text{Re} \ll 1)$.[62]

Append the appropriate Dirichlet boundary conditions for velocity

$$\mathbf{v}|_{\partial\Omega_i} = \mathbf{U}_{is}(\theta_i; T^-), \tag{34}$$

where $\mathbf{U}_{is} \in C^1(\partial\Omega_i)$ are given velocity boundary values.

In addition, the fluid is prescribed to be at rest far from the particles and also pressure vanishes at infinity

$$\mathbf{v}|_{r_i\to\infty} \rightrightarrows \mathbf{0}, \quad p|_{r_i\to\infty} \rightrightarrows 0. \tag{35}$$

Thus, so-called *external Stokes boundary value problem* (32)–(35) is imposed and we look for its classical solution $(\mathbf{v}, p)$ in unbound manifold Ω^-, i.e., $\mathbf{v} \in C^0(\overline{\Omega}^-) \cap C^2(\Omega^-)$ and $p \in C^0(\overline{\Omega}^-) \cap C^1(\Omega^-)$. It is clear that the equations of motion for spheres positions $\mathbf{x}_i$ in the quasi-steady regime are

$$\dot{\mathbf{x}}_i = \mathbf{U}_i, \quad \mathbf{x}_i|_{t=0} = \mathbf{x}_i^0 \text{ in } \Lambda_{ij}, \tag{36}$$

$$\mathbf{v} = \mathbf{U}_i \text{ in } \Omega^+. \tag{37}$$

4.2. *Phoretic effects due to the thermal slip*

It is well known that thermophoresis is essentially related to an asymmetry in the fluid temperature field $T^-(P)$ adjacent to the particles or some asymmetry in physical or chemical properties on their surfaces.[13]

Thus, mathematically the main mechanisms of the thermophoresis are caused by inhomogeneity due to temperature $T^-(P)$: in equation (32), $\mathbf{f} = \mathbf{f}_F(\mathbf{x}; T^-) \neq 0$,[37,38] and in boundary conditions on the particles surfaces (34), $\mathbf{U}_{is}(\theta_i; T^-) \neq 0$.[4]

Consider now both cases of the model setting the slip conditions on the second sphere.

(1) Case A *corresponds to the CHTT in gases.* Assuming the stick boundary condition on the first sphere surface $\partial\Omega_1$ and impose *Maxwell's first-order thermal slip velocity condition* upon the PP surface, we have for the right-hand side of (34)

$$\mathbf{U}_{1s}(\theta_1) = \mathbf{0}, \tag{38}$$

$$\mathbf{U}_{2s}(\theta_2; T^-) = \mathbf{U}_T + C_s \frac{\nu}{T_0} \nabla_s T_s^-(\theta_2), \tag{39}$$

where $\mathbf{I}_s := \mathbf{I} - \mathbf{e}_r^2 \mathbf{e}_r^2$ is the unit surface dyadic, with $\mathbf{I}$ the idemfactor, $\nabla_s := \mathbf{I}_s \cdot \nabla$ is the surface-gradient operator[62]; C_s is the thermal slip coefficient,[12] $\mathbf{U}_T$ is the thermophoresis velocity and the temperature along the surface

$$T_s^-(\theta_2) := \lim_{r_2 \to a+0} T^-(r_2, \theta_2).$$

Use of the thermal slip condition (38) yields the famous Epstein's formula

$$\mathbf{U}_T = -C_s \frac{\nu}{T_0} \left(\frac{2}{2+\chi} \right) \nabla T_\infty^-. \tag{40}$$

(2) Case B *corresponds to the TST in liquids*. The propulsive velocity $\mathbf{U}_T$ for a self-thermophoretic particle is given by a surface integral of effective surface slip velocity $\mathbf{U}_s(\theta_2; T^-)$ driven by local temperature gradient

$$\mathbf{U}_{1s}(\theta_1) = \mathbf{U}_T, \tag{41}$$

$$\mathbf{U}_{2s}(\theta_2; T^-) = \mathbf{U}_T + b(\theta_2)\nabla_s T_s^-(\theta_2), \tag{42}$$

where $b(\theta_2)$ is the mobility characterized by interactions between the particle and fluids.[13]

5. Use of the GMSV to the Stokes Problem

It is common knowledge that the general solution to the exterior Stokes problem (32)–(35) may be expressed in a linear combination of solid spherical harmonics and their derivatives in any partial domain D_i by means of known Lamb representation. Therefore, the desired solution in the whole manifold Ω^- may be found semi-analytically with the aid of the GMSV.

Thus, for unique expansions of Stokes flow $(\mathbf{v}, p)$ regular at infinity, we have[61–63]

$$\mathbf{v}(P) = \sum_{j=1}^{2}\sum_{n=1}^{\infty}\{\nabla\varphi_{-n-1}^{j} - \frac{(n-2)}{2\eta n(2n-1)}r_j^2\nabla p_{-n-1}^{j}$$

$$+ \frac{(n+1)}{\eta n(2n-1)}\mathbf{r}_j p_{-n-1}^{j}\} \text{ in } \Omega^-, \tag{43}$$

$$p(P) = \sum_{j=1}^{2}\sum_{n=1}^{\infty}p_{-n-1}^{j} \text{ in } \Omega^-. \tag{44}$$

Here, $\{\varphi_{-n-1}^{j}\}_{n=0}^{\infty}$ and $\{p_{-n-1}^{j}\}_{n=0}^{\infty}$ are complete sets of exterior harmonic functions $\varphi_{-n-1}^{j} : D_j \to \mathbb{R}$ and $p_{-n-1}^{j} : D_j \to \mathbb{R}$ written in local coordinates as follows:

$$\varphi_{-n-1}^{j} = M_n^j \psi_n^-(r_j, \theta_j), \quad p_{-n-1}^{j} = N_n^j \psi_n^-(r_j, \theta_j), \tag{45}$$

where M_n^j, N_n^j are unknown coefficients to be determined from the boundary conditions. It is evident that (see Fig. 1)

$$\mathbf{r}_j = \mathbf{r}_i + \mathbf{L}_{ij}, \quad r_j^2 = r_i^2 + L^2 - 2r_i L \cos\theta_i. \tag{46}$$

In case A, e.g., the appropriate resolved ISLAE can be now obtained by substituting expansions (43) and (44) along with temperature field $T^-(P)$ (12) into the slip boundary conditions (39) transformed with the help of Brenner's procedure[64] and the addition theorem (14), taking into consideration equality (46) and well-known recurrence relations for the Legendre polynomials. Finally, it may be shown that for the configuration manifold Λ_{ij}, the corresponding resolved ISLAE possesses solution by the method of reduction. Here, we shall not go into rather cumbersome and lengthy technical details concerning features of the GMSV in application to the Stokes problem (32)–(35), especially since this aspect became quite clear, taking into account solution to the heat problem given in Section 3. Moreover, there exist excellent works where this procedure is performed in full details for similar problems (see, e.g., Ref. 65 on axially symmetric two-sphere system phoresis and Ref. 66 on phoretic motion of arbitrary clusters of N spheres).

6. Discussion

It has been known that the CHTT describes the thermophoretic propulsion owing to an abstract external temperature gradient applied in a quiescent unbounded medium without any reasons for creation of this very gradient. For many decades, enormous efforts were mounted to calculate the thermophoretic velocities of different particles (rigid, volatile, compounded, etc.) under various boundary conditions containing refined appropriate physical and chemical parameters, but still the main and unaltered requirement was existence of a fixed external temperature gradient. As a rule, one reads something like: "Let a steady-state gradient be created and maintained in unbounded space filled with the quiescent binary gas mixture".[67] Performing theoretical studies, most researchers usually made the inference that numerous practical applications of obtained theoretical results completely justified these serious efforts. Nevertheless, are these expectations true?

Taking into account results obtained here, we can conclude that for the external temperature field $T^-(P)$ concerning the case A of the model at issue, the CHTT yields an approximate result (30). We showed that this result corresponds to the dipole approximation for the first reflection of the zero approximation to the temperature field in a neighborhood of the PP surface (see Section 3.4 for details).

It is clear that in real physical systems one usually deals with a geometrically complex confined space and only real physical objects (bodies, walls, channels, heat sources, etc.) can create and maintain the temperature gradient. This circumstance affects the results of experiments, too. In his review[1] concerning existing theories and experiments on thermophoresis in page 267, Zheng notes: "... the theories ... are based on the assumption that the particle is surrounded by an infinite expanse of gas. However, the gas is always enclosed by some boundaries separated by finite distance in experiments." Below in page 269 he also added: "Quantitative measurements of the effect of finite gas volume are scarce".

Inhomogeneities of the temperature field may be caused by various roots: (1) inert obstacles and boundaries (e.g., between phases); (2) heat sources and sinks; (3) edges of liquid. It is natural to call these physical objects the *phoretic surrounding particles (bodies)*(PSP). The full collection of the PSP we shall call the *phoretic surroundings* of a given problem. Consider 3D phoretic surroundings $\Sigma_P \subseteq \mathbb{R}^3$ consisting of N_s static non-overlapping PSP Σ_i: $\Sigma_P := \bigcup_{i=1}^{N_s} \overline{\Sigma}_i$ such that $\overline{\Sigma}_i \cap \overline{\Sigma}_j = \varnothing$, if $i \neq j$, $(i = \overline{1, N_s})$. For example, in case A of the problem under consideration, particles 1 and 2 constitute the PSP and the PP, respectively (see Fig. 2).

Further, it is clear that the CHTT approximate results are of little use unless we estimate the range of their validity, including,

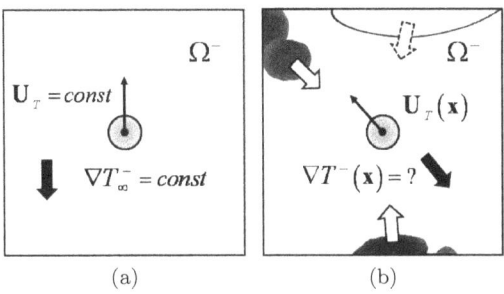

(a)   (b)

Fig. 2. Schemes, illustrating physical models for the thermophoretic propulsion:
(a) Classical with a prescribed constant temperature gradient. (b) Temperature
gradient field due to phoretic surrounding particles (PSP), where solid arrows —
action from the active PSP, dashed arrows — action from the passive PSP (obstacles). Temperature gradient should be calculated for a given configuration.

of course, accuracy of this approximation. To carry out required estimations, one has to solve relevant many particles problem, taking into consideration *hydrodynamic* and *thermal (diffusion) interactions* in the whole system "phoretic surrounding particles–phoretic particles".[58,62,68]

It is interesting that PPS acting on themselves through other PP or PSP play the role of self "phoretic surrounding particles" being, essentially self-propelling particles. Moreover, in the quasi-steady-state regime there is no principal difference in calculations of the temperature field $T^-(P)$ for both case A and case B of the model. In this sense, there should be no significant difference between the CHTT and TST. On the other hand, in case A we have propulsion of sphere 2 and in case B both spheres drift as a whole dumbbell particle.

Thus, the CHTT persuasively requires a fundamental revision and development to include first of all the high-order multipoles corrections on temperature and hydrodynamics interactions.

The new approach is essential to the development and use of the phoretic propulsion theory for different applications, e.g., evaluation of capture efficiency for aerosols by various particles and obstacles.[40–45] More general problems within the scope of linear models (investigated, e.g., in Refs. 7, 8, 47, 66 and 69) may be straightfor-wardly treated in the similar manner.

6.1.   The electrostatic analogy

It seems that the CHTT was developed in the image and likeness of classical electrostatics, where problems with a given constant electric field regardless of its origin are very common.

Nonetheless, in his classical book[60] concerning the boundary-value problem on the conducting sphere in a uniform electric field, Jackson specified the physical system, which creates the uniform electric field. In page 62, we read: "A uniform field can be thought of as being produced by appropriate positive and negative charges at infinity". As an example, he considered potential of two-point charges $\pm q$, located at positions $z = \mp z_0$ "then in a region near the origin whose dimensions are very small compared to z_0, there is an approximately constant electric field $E_0 \approx 2q/z_0{}^2$ parallel to the z axis. In the limit as $z_0, q \to \infty$, with $q/z_0{}^2$ constant, this approximation becomes exact". (Note that for convenience sake the half distance between charges $\pm q$ we denote here as z_0). Really, in spherical coordinates $(O; r, \theta)$ connected with the origin $\{O\}$ the potential function reads

$$\Phi(r, \theta; z_0, q) = \frac{q}{\sqrt{r^2 + z_0{}^2 + 2rz_0 \cos\theta}} - \frac{q}{\sqrt{r^2 + z_0{}^2 - 2rz_0 \cos\theta}}, \tag{47}$$

where $r < z_0$. Consider the asymptotics of this expression

$$\Phi(r, \theta; z_0, q) \sim -2qz_0{}^{-2} r \cos\theta, \quad \text{as} \quad z_0 \to \infty. \tag{48}$$

Double limit $\lim_{q, z_0 \to \infty} 2qz_0{}^{-2} := E_0 = const$ leads to the exact relation

$$\Phi(r, \theta) := \lim_{q, z_0 \to \infty} \Phi(r, \theta; z_0, q) = -E_0 r \cos\theta \equiv -E_0 z. \tag{49}$$

Meanwhile, for all $|q| < \infty$, one gets the regularity condition

$$\Phi(r, \theta; z_0, q) \rightrightarrows 0 \quad \text{as} \quad z_0 \to \infty. \tag{50}$$

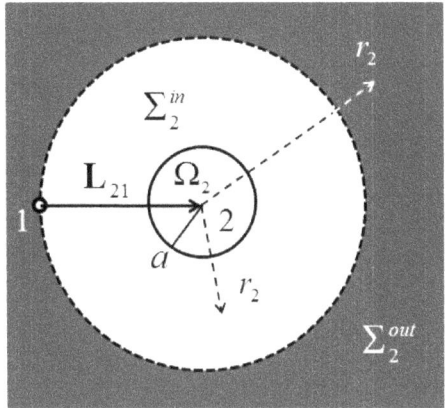

Fig. 3. Interaction and far-field sets for the problem with a point sink of heat.

6.2. *Exact analytical solution for the point sink of heat*

The *constant gradient approximation* (CGA) works only inside the *interaction set* (IS) Σ_i^{in} (e.g., in case of a point sink $\Sigma_2^{\mathrm{in}} := \{a < r_2 < L\}$). The complement of Σ_i^{in} to the whole domain Ω^- one can call the *far-field set* (FFS) Σ_i^{out} ($\Sigma_2^{\mathrm{out}} := \{L < r_2\}\backslash\overline{\Omega}_1$) (see Fig. 3).

To clarify the main ideas under consideration, simplify the case A of the model, assuming that PSP 1 is a static point sink of strength $Q_0 > 0$ located at the origin, i.e., consider Q/r_1, where $Q := -Q_0/4\pi\kappa^-$. It turns out that analytical solutions of this problem are feasible. It is significant that fundamental solution to the 3D Laplace equation does not possess uniformly valid expansion in $\mathbb{R}^3\backslash\{0\}$. Indeed, in local coordinates $(O_2; r_2, \theta_2)$, the textbook expansions can be written as

$$\frac{L}{r_1} = \frac{1}{\sqrt{1 + \tilde{\xi}_2^2 + 2\tilde{\xi}_2 \cos\theta_2}}$$

$$= \begin{cases} \sum_{k=0}^{\infty} (-1)^k \, \tilde{\xi}_2^k P_k\left(\cos\theta_2\right) \text{ in } \Sigma_2^{\mathrm{in}}, \\ \sum_{k=0}^{\infty} (-1)^k \, \tilde{\xi}_2^{-1-k} P_k\left(\cos\theta_2\right) \text{ in } \Sigma_2^{\mathrm{out}}. \end{cases} \quad (51)$$

For simplicity, hereafter $\epsilon \equiv \epsilon_2 = a/L < 1$ and $\widetilde{\xi}_2 := \epsilon\xi_2$ is the *compressed radial variable* when $\epsilon \ll 1$. As $\epsilon \to 1$, the interaction set $\Sigma_2^{in} := \{\epsilon < \widetilde{\xi}_2 < 1\}$, where CGA works, shrinks to zero. Then, it is clear that we cannot use the CGA in the FFS $\Sigma_2^{out} := \{\widetilde{\xi}_2 > 1\}$.

This problem is similar to that treated in Subsection 3.4 and it may be easily solved analytically by means of the GMSV. So, the temperature fields in the manifold $\Omega^- = \mathbb{R}^3\backslash\{0\} \cup \overline{\Omega}_2$ and domain Ω_2^+ can be given in the form

$$T^- (\xi_2, \theta_2) = \frac{Q}{r_1} + \sum_{n=0}^{\infty} A_n^2 \psi_n^- (\xi_2, \theta_2),$$

$$T^+ (\xi_2, \theta_2) = \sum_{n=0}^{\infty} \widetilde{A}_n^2 \psi_n^+ (\xi_2, \theta_2).$$

For the external field $T^-(P)$, we obtain the desired exact solution

$$T^- (\xi_2, \theta_2) = \frac{Q}{L} \left\{ 1 + \sum_{n=1}^{\infty} (-1)^n \, \epsilon^n \left[\xi_2^n + \frac{(1-\chi) \, n\xi_2^{-1-n}}{1 + (1+\chi) \, n} \right] P_n (\mu_2) \right\},$$

$$(52)$$

with $\mu_2 := \cos\theta_2$. Hence, the CGA is

$$T^-(\xi_2, \theta_2) \approx \frac{Q}{L} - \left[1 + \frac{(1-\chi)}{(2+\chi)} \left(\frac{a}{r_2} \right)^3 \right] \frac{Q}{L^2} r_2 \cos\theta_2. \qquad (53)$$

The structure of this approximation is the same as in the more general case (31) if we set $T_\infty = QL^{-1}$ and $\nabla T_\infty = -T_\infty L^{-1}$. It is evident that the CGA for the slip velocity reads

$$U_s^a (\theta_2; \chi) = \frac{3}{2+\chi} C_s \frac{\nu}{T_0} \frac{Q}{L} \epsilon \sin\theta_2. \qquad (54)$$

Hence, with the aid of expression (52) and definition (38), one obtains the exact expression for the slip velocity

$$U_s (\theta_2; \chi) = C_s \frac{\nu}{T_0} \frac{Q}{L} \epsilon \omega_s (\theta_2; \chi) \sin\theta_2. \qquad (55)$$

In (55), function $w_s(\theta_2; \chi)$ is a correction factor for the multipole effects

$$w_s(\theta_2; \chi) = -\frac{2+\chi}{3\epsilon} \frac{\partial}{\partial \mu_2} \left\{ \sum_{n=1}^{\infty} (-1)^n \epsilon^n \left[\frac{2n+1}{1+(1+\chi)n} \right] P_n(\mu_2) \right\}. \tag{56}$$

One can see that in a specific case at $\chi = 1$, series (56) yields the explicit formula

$$w_s(\theta_2; 1) = (1 + \epsilon^2 + 2\epsilon \cos \theta_2)^{-3/2}. \tag{57}$$

Plots in Fig. 4 show that multipole character of slip velocity becomes more profound for short distances L (as $\epsilon \to 1$) and close to the front point $\theta_2 = \pi$, i.e., the CGA behaves rather badly unless $\epsilon \to 0$. In other words, the CGA is valid when L is much larger than a. Furthermore, one can obviously expect that the corresponding thermophoretic force increases as the distance L decreases.

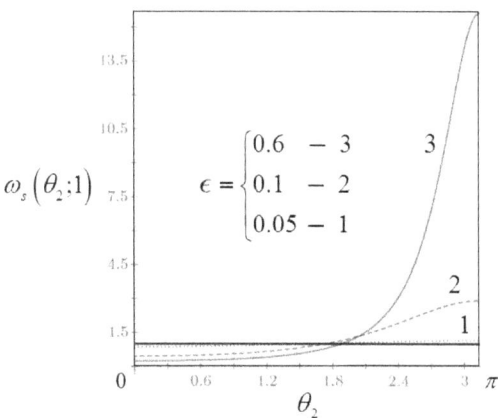

Fig. 4. The correction factor w_s as a function on θ_2 at $\chi = 1$ ((57)) for several values of ϵ. Doted curve 1 corresponds to $\epsilon = 0.05$, dashed curve 2 and solid curve 3 to $\epsilon = 0.1$ and $\epsilon = 0.6$, respectively. A line trace $w_s \equiv 1$ is depicted by the solid horizontal line.

7. Conclusions

We present here new physical and mathematical insight into describing the phenomenon of thermophoretic propulsion. Using the mathematically simplest two-sphere axially symmetric rigid dumbbell model, we studied thermophoretic motion problem under the classical Epstein's assumptions. As the main result, the many-particle nature of the problem at issue was highlighted. We also revealed the close relation between the CHTT and the TST.

Our study showed that the beforehand derived formulae for thermophoresis velocity, under prescribed constant temperature gradients do not provide reliable results for the real physical systems. To obtain correct values of the thermophoretic velocity, one should perform complete procedure for solution of the heat and hydrodynamic problems, posed for a real physical system at issue.

Thus, we demonstrate that the existing CHTT requires a fundamental reassessment and development to include high-order multipole corrections on temperature and hydrodynamics interactions.

Finally, taking these grave circumstances in mind, we can conclude that all applications of the classical theory should be revised as well.

Acknowledgments

This work was performed within the framework of the state task program of the FASO Russia (Theme 0082-2014-008, No AAAA-A17-117040310008-5). The author is also very grateful to the reviewer for valuable comments and appropriate and constructive suggestions.

References

1. F. Zheng, Thermophoresis of spherical and non-spherical particles: A review of theories and experiments, *Adv. Colloid Interf. Sci.* **97**, 255–278 (2002).
2. C. Y. Cha and B. J. McCoy, Thermal force on aerosol particles, *Phys. Fluids* **17**, 1376–1380 (1974).

3. Y. Sone, Flows induced by temperature fields in a rarefied gas and their ghost effect on the behavior of a gas in the continuum limit, *Annu. Rev. Fluid Mech.* **32**, 779–811 (2000).
4. P. S. Epstein, Zur theorie des radiometers, *Z. Phys.* **54**, 537–563 (1929).
5. G. S. McNab and A. Meisen, Thermophoresis in liquids, *J. Colloid Interf. Sci.* **44**, 339–346 (1973).
6. E. R. Shchukin, N. V. Malay, and Z. L. Shulimanova, On the velocity of thermophoresis of solid two-layer large and moderately large aerosol particles, *Vestnik AGU.* **2**(161), 23–30 (2015).
7. C. Y. Li and H. J. Keh, Thermophoresis of a particle in a concentric cavity with thermal stress slip, *Aerosol Sci. Technol.* **52**, 269–276 (2018).
8. J. R. Brock, On the theory of thermal forces acting on aerosol particles, *J. Colloid Interf. Sci.* **17**, 768–780 (1962).
9. W. F. Phillips, Motion of aerosol particles in a temperature gradient, *Phys. Fluids* **18**, 144–147 (1975).
10. L. Talbot, R. K. Cheng, R. W. Schefer, and D. R. Willis, Thermophoresis of particles in a heated boundary layer, *J. Fluid Mech.* **101**, 737–758 (1980).
11. V. S. Galoyan and Y. I. Yalamov, *Dynamics of Droplets in Inhomogeneous Viscous media.* Luys, Yerevan (1985).
12. S. P. Bakanov, Thermophoresis in gases at small Knudsen numbers, *Aerosol Sci. Technol.* **15**, 77–92 (1991).
13. J. L. Anderson, Colloid transport by interfacial forces, *Ann. Rev. Fluid Mech.* **21**, 61–99 (1989).
14. S. E. Spagnolie and E. Lauga, Hydrodynamics of self-propulsion near a boundary: predictions and accuracy of far-field approximations, *J. Fluid Mech.* **700**, 105–147 (2012).
15. W. E. Uspal, M. N. Popescu, S. Dietrich, and M. Tasinkevych, Self-propulsion of a catalytically active particle near a planar wall: from reflection to sliding and hovering, *Soft Matter.* **11**, 434–438 (2015).
16. J. L. Moran and J. D. Posner, Phoretic self-propulsion, *Annu. Rev. Fluid Mech.* **49**, 511–540 (2017).
17. G. Oshanin, M. N. Popescu, and S. Dietrich, Active colloids in the context of chemical kinetics, *J. Phys. A. Math. Theor.* **50**, 134001(56) (2017).
18. N. Yoshinaga, Simple models of self-propelled colloids and liquid drops: From individual motion to collective behaviors, *J. Phys. Soc. Japan* **86**, 101009–101010 (2017).
19. A. Baskaran and M. C. Marchetti, Statistical mechanics and hydrodynamics of bacterial suspensions, *Proc. Natl. Acad. Sci. USA* **106**, 15567–15572 (2009).
20. M. Yang and M. Ripoll, Simulations of thermophoretic nanoswimmers, *Phys. Rev. E.* **84**, 061401 (2011).
21. P. de Buyl and R. Kapral, Phoretic self-propulsion: A mesoscopic description of reaction dynamics that powers motion, *Nanoscale* **5**, 1337–1344 (2013).
22. S. Michelin, E. Lauga, Autophoretic locomotion from geometric asymmetry, *Eur. Phys. J. E.* **38**, 7 (2015).

23. H.-R. Jiang, N. Yoshinaga, and M. Sano, Active motion of a Janus particle by self-thermophoresis in a defocused laser beam, *Phys. Rev. Lett.* **105**, 268302(4) (2010).

24. X. Lin, T. Si, Z. Wu, and Q. He, Self-thermophoretic motion of controlled assembled micro-/nanomotors, *Phys. Chem. Chem. Phys.*, DOI: 10.1039/c7cp02561k (2017).

25. S. N. Rasuli and R. Golestanian, Soret motion of a charged spherical colloid, *Phys. Rev. Lett.* **101**, 108301(4) (2008).

26. A. Wurger, Thermal non-equilibrium transport in colloids, *Rep. Prog. Phys.* **73**, 126601(35) (2010).

27. R. Golestanian, Collective behavior of thermally active colloids, *Phys. Rev. Lett.* **108**, 038303(5) (2012).

28. T. Bickel, A. Majee, and A. Wurger, Flow pattern in the vicinity of self-propelling hot Janus particles, *Phys. Rev. E.* **88**, 012301(6) (2013).

29. N. A. Fuchs, Thermophoresis of aerosol particles at small Knudsen numbers: theory and experiment, *J. Aerosol Sci.* **13**, 327–330 (1982).

30. F. Prodi, G. Santachiara, S. Travaini, A. Vedernikov, F. Dubois, C. Minetti, and J. C. Legros, Measurements of phoretic velocities of aerosol particles in microgravity conditions, *Atmos. Res.* **82**, 183–189 (2006).

31. B. Sagot, Thermophoresis for spherical particles, *J. Aerosol Sci.* **65**, 10–20 (2013).

32. G. N. Lipatov and E. A. Chernova, Thermophoresis of highly dispersed aerosols, *J. Aerosol Sci.* **20**, 931–933 (1989).

33. D. Gonzalez, A. G. Nasibulin, A. M. Baklanov, S. D. Shandakov, D. P. Brown, P. Queipo, and E. I. Kauppinen, A new thermophoretic precipitator for collection of nanometer-sized aerosol particles, *Aerosol Sci. Technol.* **39**, 1064–1071 (2005).

34. J. B. Young, Thermophoresis of a spherical particle: Reassessment, clarification, and new analysis, *Aerosol Sci. Technol.* **45**, 927–948 (2011).

35. F. Zheng and E. J. Davis, Thermophoretic force measurements of aggregates of micro-spheres, *J. Aerosol Sci.* **32**, 1421–1435 (2001).

36. H. Brenner, Phoresis in fluids, *Phys. Rev. E.* **84**, 066317(8) (2011).

37. A. F. Andreev, Thermophoresis in liquids, *Sov. Phys. JETP.* **67**, 117–120 (1988).

38. R. Piazza, Thermophoresis: Moving particles with thermal gradients, *Soft Matter* **4**, 1740–1744 (2008).

39. S. D. Traytak. Hydrodynamic effects for propulsion of a molecular motor by asymmetric reaction in fluids in: *Books of Abstracts (RKCM-10)*, Moscow, Russia (2010), pp. 146–147.

40. Y. I. Yalamov, E. R. Shchukin, and S. D. Traytak, Role of the thermodiffusiophoresis and Brownian motion in the capture of aerosol particles by drops, *Izv., Atmos. Ocean. Phys.* **15**, 122–125 (1979).

41. S. D. Traytak, Problem of interaction of clouds with surrounding aerosol. *Izv. Atmos. Ocean. Phys.* **24**(12), 1298–1306 (1988).

42. S. D. Traytak, Capture of hydrosol particles by a moving drop, *High Temp.* **23**(6), 885–888 (1985).

43. S. D. Traytak, Capture of aerosol particles by a drop for large values of the particle Peclet number and small values of the heat and mass transfer Peclet number, *High Temp.* **26**(6), 923–927 (1988).

44. S. D. Traytak, Nonsteady capture of aerosol particles by thermodynamically nonequilibrium drops, *High Temp.* **28**(4), 587–596 (1990).

45. S. D. Traytak, Deposition of aerosol particles on a coarse spherical particle in uniform fields of temperature and concentration gradients, *Colloid J.* **52**, 1142–1149 (1990).

46. A. G. Kudryavtsev and S. D. Traytak, Diffusion of aerosol particles in a stochastic nonisothermal medium, *Sov. Phys. JETP.* **69**, 987–988 (1989).

47. S. H. Chen and H. J. Keh, Axisymmetric thermophoresis of multiple aerosol spheres, *Aerosol Sci. Technol.* **24**, 21–35 (1996).

48. S. D. Traytak, Methods for solution of the boundary value problems in domains with disconnected boundary, *J. Compos. Mech. Des.* **9**, 495–521 (2003).

49. T. H. Hsieh and H. J. Keh, Thermophoresis of an aerosol sphere with chemical reactions, *Aerosol Sci. Technol.* **46**, 361–368 (2012).

50. S. Axler, P. Bourdon and W. Ramey, *Harmonic function theory.* Springer (2001).

51. M. Galanti, D. Fanelli, S. D. Traytak, and F. Piazza, Theory of diffusion-influenced reactions in complex geometries, *Phys. Chem. Chem. Phys.* **18**, 15950–15954 (2016).

52. S. D. Traytak and D. S. Grebenkov, Diffusion-influenced reaction rates for active "sphere-prolate spheroid" pairs and Janus dimers, *J. Chem. Phys.* **148**, 024107(11) (2018).

53. L. V. Kantorovich and V. I. Krylov, *Approximate methods of higher analysis.* Interscience Publishers, New York (1964).

54. L. V. Kantorovich and G. P. Akilov, *Functional analysis.* Pergamon Press, Oxford (1982).

55. W.-K. Li, C.-Y. Soong, C.-H. Liu, and P.-Y. Tzeng, Thermophoresis of a micro-particle in gaseous media with effect of thermal stress slip, *Aerosol Sci. Technol.* **44**, 1077–1082 (2010).

56. J. L. Anderson and D. C. Prieve, Diffusiophoresis: Migration of colloidal particles in gradients of solute concentration, *Separ. Purif. Meth.* **13**, 67–103 (1984).

57. E. M. Kartashov, *Analytical Methods in Theory of Heat Conduction in Solids.* Vysshaya Shkola Press, Moscow, 3rd edn. (2001).

58. S. D. Traytak and M. Tachiya, Diffusion-controlled reactions in an electric field: Effects of an external boundary and competition between sinks, *J. Chem. Phys.* **107**, 9907–9920 (1997).

59. S. D. Traytak, Convergence of a reflection method for diffusion-controlled reactions on static sinks, *Physica A* **362**, 240–248 (2006).

60. J. D. Jackson, *Classical Electrodynamics.* John Wiley and Sons, New York, 3rd edn. (1962).

61. H. Lamb, *Hydrodynamics*. Cambridge University Press, Cambridge, 6th edn. (1932).

62. J. Happel and H. Brenner, *Low Reynolds Number Hydrodynamics with Special Applications to Particulate Media*. Prentice-Hall, Englewood Cliffs, NJ, (1965d).

63. S. Kim and S. J. Karrila, *Microhydrodynamics: Principles and Selected Applications*. Butterworth-Heinemann, Boston, MA (1991).

64. H. Brenner, The Stokes resistance of a slightly deformed sphere, *Chem. Eng. Sci.* **19**, 519–539 (1964).

65. D. W. Mackowski, Phoretic behavior of asymmetric particles in thermal nonequilibrium with the gas: Two-sphere aggregates, *J. Colloid Interf. Sci.* **140**, 138–157 (1990).

66. A. V. Filippov, Phoretic motion of arbitrary clusters of N spheres, *J. Colloid Interf. Sci.* **241**, 479–491 (2001).

67. S. N. D'yakonov and A. N. Nikolsky, The thermophoresis of the volatile sphere in the binary gas mixture taking into account thermodiffusion and Stefan effects, *Diff. Eqs. Control Processes* **1**, 17–26, (2001) (in Russian).

68. A. Z. Zinchenko, An efficient algorithm for calculating multiparticle thermal interaction in a concentrated dispersion of spheres, *J. Comp. Phys.* **111**, 120–135 (1994).

69. S. P. Bakanov, The thermophoresis of solids in gases, *J. Appl. Math. Mech.* **69**, 767–772 (2005).

Index